Problem-Based Learning Approach in Microbiology

Developments in Microbiology

Problem-Based Learning Approach in Microbiology

Khalid Mubarak Bindayna
Microbiology, College of Medicine, Arabian Gulf University, Manama, Kingdom of Bahrain

Jameela Al-Salman
Ministry of Health, Manama, Kingdom of Bahrain

ELSEVIER

ACADEMIC PRESS
An imprint of Elsevier

Academic Press is an imprint of Elsevier
125 London Wall, London EC2Y 5AS, United Kingdom
525 B Street, Suite 1650, San Diego, CA 92101, United States
50 Hampshire Street, 5th Floor, Cambridge, MA 02139, United States
The Boulevard, Langford Lane, Kidlington, Oxford OX5 1GB, United Kingdom

ISBN: 978-0-323-95092-3

For Information on all Academic Press publications
visit our website at https://www.elsevier.com/books-and-journals

Publisher: Stacy Masucci
Acquisitions Editor: Linda Versteeg-Buschman
Editorial Project Manager: Michaela Realiza
Production Project Manager: Fahmida Sultana
Cover Designer: Christian J. Bilbow

Typeset by MPS Limited, Chennai, India

Dedication

To my father who taught me manhood, to my mother who taught me love, to my wife and my soulmate, to my children and grandchildren who are my hope, to the soul of my brother, Khalifa, who taught me the meaning of devotion, to all my brothers and sisters to whom I will always sacrifice. To the soul of my friend, Hisham; he is the man whom I will never forget.

—Khalid Mubarak Bindayna

I dedicate this work to the aspiring researchers, medical professionals, and healthcare workers in the hopes that they will find the contents of this book insightful and illuminating. I hope that after exploring this book, you begin to challenge yourself to learn things through a problem-solving lense and approach situations in microbiology more critically.

—Jameela Al-Salman

Contents

Acknowledgment

We want to acknowledge the contribution of Dr. Mazoun Said Rashid Al Alawi, Dr. Rawan Alagha, and Dr. Marwa Khalifa Bindayna and thank them for their efforts.

How to use this book

The book was designed as a material to teach students how to be active participants in their learning and use critical thinking through problem-solving. Therefore the cases (Problems) should be delivered as a series of separate but interrelated problems or the whole cases should be presented in one session, with a reading list or other learning resources to support the student's inquiry.

The problems are selected based on their importance and commonality; they are not selected to cover all aspects (subjects) of microbiology but rather how to learn.

The problems should be presented to the students in a small group tutorial. The group should not exceed eight students to maximize fruitful discussion. In this setup, the tutor is a facilitator of the interaction and avoids direct teaching. Tutors need to facilitate students to develop their learning strategies focused on self-directed learning, motivation, and empowerment. Tours are encouraged to use the Q&A section with each problem. The exact learning needs of a problem are held by the tutors and not provided to students directly but are explored through personal inquiry. The listed steps should be followed.

1. The group analyzes the problem, often by brainstorming ideas on a whiteboard or a digital board.
2. The group generates their learning needs.
3. These are transformed into hypothesis-driven reasoning or questions.
4. The group then collaborates to set their agreed learning needs with help from a tutor.
5. This should be followed by prioritizing learning needs to be completed during a specific self-study time.
6. During the following session, parts (3) and (4) are revisited as students' effort to make last understandable explanations of the problem.

Since the issue is crucial and takes a few days between sessions, students should be given enough time to conduct necessary study. This will allow them to work either independently or jointly.

The student should have access to a variety of reference materials, including text books, websites, online journals, and other works of literature. These materials should allow students to apply new information to the issue along with resource sessions or seminars.

CHAPTER 1

Basic concepts in microbiology and immunology

Problem number 1.1: antibiotics

A 70-year-old man presented with fever, altered mental status, and hiccups. He had a past medical history of seizures, kidney stones, recurrent urinary tract infections (UTIs), removal of a brain cyst 8 years ago, and placement of a brain shunt one year after the removal of the brain cyst.

The patient was admitted to the hospital and the following empiric treatment was started immediately:

1. Vancomycin (VA)
2. Metronidazole
3. Cefepime

The patient continued to deteriorate on the empiric treatment. The possibility of pneumonia was considered and cefepime was changed to ceftriaxone while continuing VA.

The following are the reports of the investigations carried out:

1. Vitals
   a. Temperature: 39.4°C
   b. Blood pressure: 90/60 mm Hg
   c. Pulse: 110/min
   d. Respiratory rate: 25/min
2. Physical examination
   a. Appearance: Lethargic and pale
   b. Extremities: Cold
   c. Abdomen: Normal
   d. Chest: Normal
3. Urine culture: Negative (Day 1)
4. Respiratory culture: Negative (Day 1)
5. Blood culture and sensitivity

CONTENTS

Problem-Based Learning Approach in Microbiology. DOI: https://doi.org/10.1016/B978-0-323-95092-3.00001-9

a. (Day 1) 2 sets: Positive—(1) Clusters of Gram-positive cocci (2) Methicillin-resistant *Staphylococcus aureus* (MRSA) susceptible to VA (MIC—2 μg/mL)
b. (Day 3) 2 sets: Positive—MRSA susceptible to VA (MIC—2 μg/mL)
c. (Day 6) 1 set: Positive—MRSA susceptible to VA (MIC—4 μg/mL)

The cultured organism was thus identified as VA-intermediate *S. aureus* (VISA).

Diagnosis

The final diagnosis was that of bacteremia caused by a VISA blood infection.

Treatment

VA was stopped and the patient was treated with linezolid instead.

Follow-up and outcome

1. Three subsequent blood cultures: Negative.
2. Patient became afebrile.
3. Mental status improved and became normal.
4. The patient was discharged after 1 week of hospitalization.

Question and answer

1. What is the significance of performing blood cultures in this case?
   Blood cultures can reveal the presence of bacteria in the bloodstream, which are not present in sterile blood. The presence of bacteria in the blood is indicative of bacteremia, which may advance into potentially life-threatening conditions, such as septicemia and septic shock if left untreated.
2. How is antimicrobial susceptibility determined?
   Antimicrobial susceptibility is determined through various tests, namely disk diffusion (DD), broth microdilution (BMD), and minimal inhibitory concentration (MIC).
   BMD involves the use of a 96-well microtiter plate inoculated with varying concentrations of different antimicrobial agents. The MIC is

the lowest concentration of an antimicrobial agent that prevents the growth of bacteria, indicated by no turbidity or growth.
DD involves inoculating and spreading bacteria on an agar plate followed by the addition of different antimicrobial disks. The diameter of the zone of inhibition (no growth) around each antimicrobial disk is measured after appropriate incubation.
MIC or zone of inhibition measurements are divided into susceptible, intermediate, and resistant categories depending on certain predetermined criteria.

3. What is VA?
   VA is a glycopeptide antibiotic and is the drug of choice for the treatment of Gram-positive bacteria, including MRSA. Development of resistance to VA is very rare among MRSA strains, although a handful of VA-resistant *S. aureus* (VRSA) (MIC $\geq$ 16 μg/mL) cases have been reported.
4. What are VSSA, VISA, and VRSA?
   VSSA is VA-susceptible *S. aureus* (MIC $\leq$ 2 μg/mL), VISA is VA-intermediate *S. aureus* (MIC: 4–8 μg/mL), and VRSA is VA-resistant *S. aureus* (MIC $\geq$ 16 μg/mL).
5. What is the recommended method for the detection of VISA?
   Samples can be tested using either the broth dilution (BD) method or the DD method. A VA agar screen plate is also used in the test in case the DD method is used or if the testing laboratory is not validated for the detection of VRSA. MIC $\leq$ 2 μg/mL (zone of inhibition $\geq$ 15 mm) and no growth on the VA screen plate is reported as VSSA and treated accordingly. However, if MIC $\geq$ 4 μg/mL (zone of inhibition $</\geq$ 15 mm) and/or there is growth on the VA screen plate, the possibility of VISA/VRSA is to be considered. In that case, the sample is retested using the BD method and then processed for storage. Infection control authorities, local health departments, and Centers for Disease Control and Prevention (CDC) should be notified about the possibility of VISA/VRSA. In case of MIC $\geq$ 8 μg/mL, the sample should be sent to CDC for MIC confirmation and *van* gene detection.
6. Why did the culture report MRSA initially, but VISA in the later cultures?
   The selective pressure due to the extended VA therapy may have caused the emergence of VISA. Initially, the isolated strain was susceptible to VA (VSSA) (MIC—2 μg/mL), but the strain isolated in the later culture had MIC—4 μg/mL (VISA). Extended VA therapy is often known to result in the isolation of VISA.
7. What is the difference in the colony characteristics of VISA and MRSA?
   VISA colonies have a pinpoint appearance, which differs from the spherical shape of the MRSA colonies.

8. What are various treatment options available for VISA infections?
   Oxazolidinones, such as linezolid, are used in the case of VISA infections due to the rare occurrence of resistance. Tigecycline is considered only for select cases due to its side effects. Combination therapy, using rifampin and fusidic acid, is another option being explored, although it has not been approved by FDA (U.S. Food and Drug Administration) for use in the United States.

Learning objectives

1. Define antibiotics.
2. Describe the classification of antibiotics.
3. Differentiate between bactericidal and bacteriostatic antibiotics.
4. Summarize the various mechanisms of action of antibiotics.
5. Explain the pharmacodynamics and pharmacokinetics of antibiotics.
6. Describe the clinical significance of antibiotics.
7. Enumerate the various adverse reactions attributable to antibiotic therapies.
8. Define antibiotic resistance.
9. Enumerate various antibiotic-resistant bacterial strains.
10. Summarize the mechanisms of antibiotic resistance.

Discussion

Differential diagnosis

Pneumonia	Pneumonia presents with general malaise, fever, sweating, chills, rapid and shallow breaths, loss of appetite, fatigue, productive cough, nausea, and vomiting.
Lung abscess	Lung abscesses present themselves with chest pain that worsens breathing, fever, night sweats, fatigue, loss of appetite, weight loss, and a productive cough with sputum that may or may not be blood-streaked.
Empyema	Empyema presents with a fever, night sweats, fatigue, dyspnea, chest pain, weight loss, and a productive or nonproductive cough.
Renal insufficiency	Renal insufficiency presents with fatigue, weakness, nausea, dyspnea, an irregular heartbeat, fluid retention causing peripheral edema, altered mental status, and decreased urine output.
Perinephric abscess	Perinephric abscesses cause a fever, flank pain extending to the groin regions or the legs, fatigue, and abdominal pain.
Pyelonephritis	Pyelonephritis presents itself with fever, chills, flank/back/groin/abdominal pain, nausea, vomiting, turbid urine, a foul smell in urine, dark-colored urine, frequent urination, and pain during urination.

Background

Antibiotics are metabolic products generated by some microorganisms that inhibit certain pathogenic bacteria. The discovery of antibiotics has ushered in an era of effective therapy and cure against several potentially life-threatening infections. Antibiotics may either be cytotoxic, meaning that they kill the microorganism, or cytostatic, indicating that they inhibit the growth or replicate the microorganism. Antibiotics exert their therapeutic effect through several means, such as inhibition of bacterial cell wall synthesis, inhibition of protein or nucleic acid synthesis, and membrane disorganization. The widespread use of antibiotics has also promoted the emergence of multidrug resistant (MDR) pathogens. Pathogens have evolved to evade the antimicrobial effects of antibiotics by various means, such as antibiotic inactivation, target modification, altered permeability, and eluding the metabolic pathways. The nonjudicial use of antibiotics has also resulted in complications, such as antibiotic resistance, which has increased the challenges involved in the effective treatment of potentially life-threatening infections. Antibiotic resistance has been responsible for the emergence and reemergence of several pathogens globally.[1]

Classification

Antibiotics may destroy bacterial cells either by preventing cell reproduction or by inhibiting an indispensable cellular process. Based on the means employed by antibiotics to exert their antimicrobial effects, antibiotics are classified into bactericidal and bacteriostatic antibiotics. MIC and minimum bactericidal concentration (MBC) are important parameters that define the bactericidal and bacteriostatic effects of antibiotics. MIC is the lowest concentration of an antibiotic that inhibits visible bacterial growth at 24 hours, while MBC is the concentration of an antibiotic that reduces bacterial density by 1000-fold at 24 hours.[2] The MBC to MIC ratio is instrumental in defining both bacteriostatic and bactericidal activities of antibiotics. An MBC to MIC ratio greater than 4 is indicative of bacteriostatic activity, while an MBC to MIC ratio less than or equal to 4 defines bactericidal activity.[2] Tetracyclines, lincosamides, macrolides, oxazolidinones, sulfonamides, and glycylcyclines are classified as bacteriostatic, while bactericidal antibiotics include aminoglycosides, ß-lactams, glycopeptides, nitroimidazoles, fluroquinolones, and cyclic lipopeptides. Although classified based on their mode of action (bactericidal or bacteriostatic), some bacteriostatic antibiotics exhibit bactericidal activities against certain bacteria, and some bactericidal antibiotics exhibit bacteriostatic activity against certain particular bacteria, thereby highlighting the indefinite nature of the classification system. Therefore antibiotic efficacy is a function of several variable factors, such as the bacterial target, the host environment, the site of infection, and the infecting dose.[3]

The mechanism of action by which antibiotics exert their antibacterial effects also influences the classification of antibiotics. Antibiotics such as the ß-lactams and glycopeptides target the bacterial cell wall by inhibiting its synthesis or by weakening the peptide cross-links, thereby destroying the integrity of the cell wall. Inhibition of protein synthesis processes that are integral to the growth and replication of bacteria targeting the 50 S and 30 S ribosomal subunits is another common mechanism of action among antibiotics such as aminoglycosides, tetracyclines, chloramphenicol, macrolides, and oxazolidinones. Certain classes of antibiotics, such as quinolones, inhibit the processes involved in bacterial DNA replication, while some others, such as sulfonamides and trimethoprim, inhibit indispensable metabolic processes like folic acid metabolism (Fig. 1.1).

Pharmacokinetics and pharmacodynamics

Although MBC and MIC describe antibiotic efficacy, they are in vitro parameters that paint an incomplete picture of the actual clinical dynamics of antibiotics

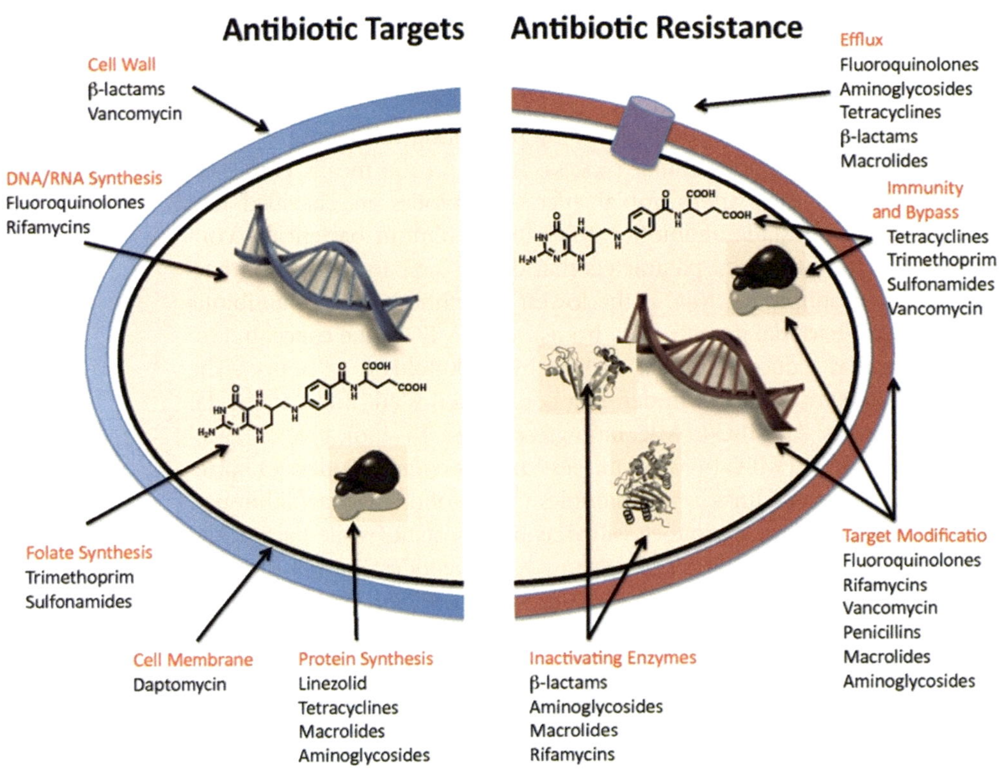

FIGURE 1.1

Antibiotic targets and mechanisms of resistance. *Credit: Gerard D Wright/Own work/https://creativecommons.org/licenses/by/2.0/deed.en. https://commons.wikimedia.org/wiki/File:Antibiotic_resistance_mechanisms.jpg.*

in vivo. Pharmacokinetic and pharmacodynamic parameters, on the other hand, provide a more rational basis for dosage optimization, considering both the dose concentration and the dosing interval. Pharmacokinetic and pharmacodynamic parameters are thus crucial for maximizing the efficacy of antimicrobial therapy.

The pharmacokinetic parameters, such as absorption, distribution, metabolism, and elimination, describe the time course and concentration dynamics of the passage of an antibiotic through the host body from administration to excretion. These parameters affect the effective antibiotic concentration over time, and are, therefore crucial to the clinical efficacy of the administered antibiotic.[4,5] Pharmacokinetics of an antibiotic facilitates the determination of the most effective and safe mode of administration, the association of the drug efficacy with food, and the effective drug concentration, depending on the volume of body fluid into which the drug dissolves. Perfusion rates of various organs; the ability to diffuse across biological membranes; solubility; protein-binding percentage and concentration of a free unbound drug; the effect of local factors such as pH or the presence of a foreign body at the site of infection; the ability of antibiotics targeting intracellular pathogens to enter and be active in the intracellular space, and the ionization state of the antibiotic are important determinants of the clinical efficacy of the drug. The elimination of administered drugs is another crucial pharmacokinetic parameter. The administration of a drug is followed by a peak plasma level, which declines over time due to drug distribution and elimination. The administered drugs are eliminated either in the form of metabolites or the unchanged form by excretory organs, such as the kidney, liver, and intestines. Enterohepatic circulation allows for drugs and their metabolites excreted in the bile to be reabsorbed into the bloodstream and recycled, which positively affects the pharmacokinetics of certain drugs. Compromised hepatic or renal function results in increased plasma and tissue drug concentrations that mediate several toxic and potentially life-threatening side effects. Therefore dosage adjustment proportionate to the decline in hepatic or renal functioning is crucial for ensuring the safety of the administered antibiotic. Monitoring plasma concentrations of the administered antibiotics at regular time intervals and adjusting the dosage based on a combination of calculated estimates is indispensable in patients with a history of hepatic or renal insufficiency.[6]

Pharmacodynamics of an antibiotic describes the in vivo pharmacological effect of the drug on the infection etiology after administration. Bactericidal and bacteriostatic antibiotics exhibit different pharmacodynamics. Bactericidal drugs may either be concentration-dependent or time-dependent. While the efficacy of concentration-dependent bactericidal drugs is directly proportional to the concentration of the administered drug, the efficacy of time-dependent bactericidal drugs is determined by the duration for which the effective drug concentration is maintained.[4] The high levels of the unbound and active form of the antibiotic exceeding the MBC immediately after administration causes a drastic decrease in the bacterial count in the initial phase. This phase is followed by a gradual decrease in the levels of unbound drug until it falls below the MBC, although the levels continue to exceed the MIC. During this phase, the bacterial count may either remain stable due to the pharmacodynamics of the administered drug or the host's immune defenses may cause the bacterial count to decline. Bacteriostatic drugs cause the bacterial count to decline in the initial phase immediately after administration when the unbound drug concentrations exceed the MIC. The unbound drug levels fall below the MIC over time after which the sustained antibacterial effect can be mediated by multiple factors, including the host's defense mechanisms. Therefore the postantibiotic effect (PAE) prevalent in the initial phase immediately after drug administration facilitates the primary suppression of bacterial growth while the host's defense mechanisms are being elicited. PAE also promotes post-antibiotic leukocyte enhancement that increases the susceptibility of the pathogen to the phagocytic elements of the host's immune system. In addition, even at concentrations below MIC, antibiotics can alter bacterial morphology and slow the rate of bacterial growth, thus prolonging the PAE period. Minimal antibacterial concentration (MAC) describes the minimal drug concentration that alters bacterial cell morphology. The efficacy of the administered drug, however, reduces to ineffective levels gradually, which allows the surviving pathogens to resume growth and replication. Thus the dosing interval of antibiotics is determined by the inherent replication rate of the pathogen, the availability of growth-supporting nutrients in the infected host's tissues, and the immunocompetency of the host.[6]

The clinical efficacy of an antibiotic as a bactericidal or a bacteriostatic agent is a summative effect of several bacterial and host factors, such as the size of the initial infectious inoculum; the MIC and MBC of the antibiotic; the pharmacokinetic characteristics of the antibiotic; the duration of effective PAE; the replication rate and requirements of the pathogen; and the adequacy of the host's immune system.

Clinical significance

Antibiotic therapy may be empiric, prophylactic, or targeted, depending on the clinical presentation and the availability of microbiological reports.

Empiric antibiotic therapy is instrumental during the initial phase of clinical intervention when the exact etiology or source of the infection may not be known. Since the rapid initiation of antibiotic therapy is crucial for thwarting the growth and replication of pathogens, commencement of broad-spectrum empiric therapy even before the microbiological reports are available plays a significant role in ensuring better prognosis. Empiric therapy should be modified appropriately to directed antibiotic therapy once microbiological culture and sensitivity reports indicate the exact etiology and susceptibilities of the infection.[7]

Prophylactic antibiotic therapy is generally prescribed for immunocompromised patients to prevent the pathologies caused by opportunistic infections, even in the absence of an active infection. Certain congenital anomalies and abnormalities of organ systems also increases the predisposition to various opportunistic infections, for example, recurrent UTIs in children with vesicoureteral reflux. Such individuals may also require prophylactic antibiotic therapy to prevent organ damage due to increased vulnerability to infections. Prophylactic therapy is also recommended prior to invasive medical and surgical procedures, in cases of trauma injuries such as accidents, burn injuries, and animal bites.[7]

The route of administration of the antibiotic therapy, the aggressiveness of the therapy, and the activity spectrum of the administered antibiotics are influenced by the severity of the suspected diagnosis. Physiological parameters that influence the pharmacodynamic and pharmacokinetic parameters of antibiotics, such as age, drug allergies, medical history, adequacy of renal and hepatic functions, immunocompetency, and past history of antibiotic usage affect the process of antibiotic selection and dosage determination.[4,8]

Adverse reactions

Adverse reactions to antibiotic therapies compromise the efficacy of the antibiotic, thereby allowing pathogens to proliferate and exert potentially life-threatening pathological effects. In addition, adverse reactions are also responsible for serious medical emergencies such as hypersensitivity reactions and damage to organ systems. Uncontrolled inflammatory reactions elicited by antibiotics result in potentially fatal IgE-mediated anaphylactic shock.[9] Defective metabolism and elimination of antibiotics from the host's system, due to hepatic or renal insufficiencies, cause accumulation of the drugs and their metabolites, which exerts toxic effects on the host system. Certain reactions known as side-effects of an antibiotic may not be elicited by either the host's immune system or the pharmacokinetic and pharmacodynamic parameters of the antibiotic.[10]

Adverse reactions to antibiotics are enhanced further in immunocompromised individuals, those with complicated comorbidities, and in individuals belonging to age extremes. Regular monitoring of plasma concentrations is instrumental in guiding effective antibiotic therapy and in the prevention of adverse reactions. Emphasis should be placed on regular monitoring of functionalities of organs, such as the kidney and liver, involved in the metabolism and elimination of drugs.

The clinical manifestations of adverse reactions to antibiotic therapy vary depending on the affected organ system and can provide clinically significant warning signs. Dermatologic manifestations of adverse reactions include rash, erythema multiforme, Stevens – Johnson syndrome, and toxic epidermal necrolysis, while differential blood counts indicative of thrombocytopenia, leukopenia, agranulocytosis, abnormal platelet aggregation, and increased international normalized ratio are some of the hematological manifestations. Renal failure, interstitial nephritis, crystallization in renal tubules, and acute tubular necrosis indicate renal damage. Cardiac involvement is often presented as QT prolongation. Potential involvement of the central nervous system is indicated by seizures, ototoxicity, peripheral neuropathy, and vestibular dysfunction. Hepatotoxicity, myopathy, electrolyte abnormalities, drug-induced fever, and drug-induced diarrhea are some of the systemic manifestations of adverse reactions to antibiotics.[4]

Antibiotic resistance

Antibiotic resistance is the phenomenon by which antibiotics become ineffective and allow the proliferation of pathogens, even at therapeutic levels. Pathogens may either develop complete resistance to antibiotics or may become less sensitive, thereby warranting the need for increased dosages. Increased dosages are often responsible for toxic side-effects and adverse drug reactions. Basic low levels of antibiotic resistance, such as the absence of a cell wall in bacteria belonging to the *Mycoplasma* genus, are conferred upon bacteria as part of the natural selection process. Genetic mutations and the ability of bacteria to transfer genetic material through plasmid transfer promotes the evolution of antibiotic resistance further. Antibiotic abuse, nonadherence to antibiotic regimes, and prolonged broad-spectrum empiric or prophylactic therapies are largely responsible for the emergence of antibiotic-resistant bacterial strains. The presence of even a single resistant bacterial colony can render the administered antibiotic ineffective, and subsequently lead to increased resistance of the bacteria to the antibiotic.[11] Antibiotic resistance acquired against a particular antibiotic often results in resistance against all structurally similar antibiotics belonging to the same class. Frequent or irrational use of antibiotics promote the emergence of resistant bacterial strains through the exertion of selective pressure. Bacterial strains that have accumulated multiple resistant traits can develop resistance against multiple classes of antibiotics. MDR, extensively drug resistant (XDR), and pan-drug resistant (PDR) bacterial strains, such as MRSA, VRSA, certain tuberculosis (TB) bacteria, and many more, pose a formidable threat to public health globally. Bacteria mediates antibiotic resistance through several mechanisms, such as hydrolysis, efflux pumps, target modification, phosphorylation, acetylation, nucleotidylation, altered biosynthetic pathways, monooxygenation, carbon-oxygen lyase, and ADP-ribosylation[12] (Fig. 1.1). The emergence of resistant bacterial populations depends on several factors, such as the bacterial propensity toward acquiring resistance; the rate of spontaneous genetic mutations that confer antibiotic resistance; the immunocompetency of the host, and the pharmacokinetic and pharmacodynamic parameters of the antibiotic at the site of infection. Better understanding of pharmacological attributes of antibiotics such as MIC, MBC, and MAC play an important role in the development of effective and safe strategies for the clinical use of antibiotics.

Student notes

Key take-aways

1. Antibiotics are the "new era miracle drugs" that provide effective therapy and cure several potentially life-threatening infections.
2. Antibiotics are classified as bacteriostatic and bactericidal depending on whether they kill the pathogen or prevent its growth and replication.
3. Mechanisms of action is when antibiotics exert their therapeutic effect, including inhibition of the bacterial cell wall synthesis, inhibition of protein or nucleic acid synthesis, membrane disorganization, and inhibition of essential metabolic pathways.
4. Pharmacokinetic and pharmacodynamic parameters of antibiotics play an important role in the development of effective and safe treatment strategies and prevent the emergence of antibiotic resistant bacterial strains.
5. Bacteria mediates antibiotic resistance through several mechanisms, such as hydrolysis, efflux pumps, target modification, phosphorylation, acetylation, nucleotidylation, altered biosynthetic pathways, monooxygenation, carbon-oxygen lyase, and ADP-ribosylation.
6. The development of antibiotic resistance due to antibiotic abuse, nonadherence to antibiotic treatment regimes, and prolonged broad-spectrum empiric or prophylactic therapies has increased incidence of treatment failures, resulting in higher infection-associated mortality rates.
7. Amalgamation of pharmacological attributes of antibiotics, bacterial properties, and physiological and immunological factors of the host is important in ensuring effective and safe clinical use of antibiotics.

Problem number 1.2: sterilization

An incident of failed sterilization procedures was investigated in a dental clinic. The incident occurred due to a rare lapse of monitoring during the autoclaving cycle. Two hundred and forty patients had been exposed to the unsterilized equipment due to the sterilization lapse.

The incident was immediately followed by proper sterilization of all equipment. The cause of the lapse and the risks associated with it were identified to ensure appropriate postexposure management of affected patients and to develop strategies and recommendations for the prevention of similar occurrences in future. Within 36 hours of the incident, 120 sources and the 240 exposed patients were traced and contacted for risk assessment. The exposed patients were tested for blood-borne infections, including hepatitis B virus (HBV), hepatitis C virus (HCV), and human immunodeficiency virus (HIV). Immunization and hyperimmune globulin for hepatitis B and tetanus toxoids were administered to susceptible exposed patients, especially those undergoing invasive dental procedures, such as extraction, implant, scaling, and oral surgery. All exposed patients were followed up for 6 months, and none of the exposed patients seroconverted to HBV, HCV, or HIV.

Question and answer

1. What is sterilization and how is a sterilization cycle measured?
   Sterilization is defined as the process of using various chemical and physical methods by which vegetative and spore forms of microorganisms are destroyed and eliminated. A ≥ 106 log reduction of colony-forming units (CFU) of the most resistant microbial spores is achieved at the half-time of a regular sterilization cycle.
2. How does disinfection differ from sterilization?
   Sterilization involves the use of various methods that ensure the elimination of all vegetative and spore forms of microorganisms. On the other hand, disinfection does not destroy spores existing on inanimate surfaces, although it results in the complete elimination of vegetative forms of microorganisms. Disinfection processes facilitate a ≥ 103 log reduction of CFU of the nonspore-forming microorganisms.
3. How are disinfectants classified?
   Disinfectants can be classified as high-level disinfectant (HLD), intermediate-level disinfectant (ILD), and low-level disinfectant (LLD) based on their ability to eliminate microorganisms from inanimate surfaces.
4. What is the difference between decontamination and cleaning?
   Decontamination is the process by which pathogenic microorganisms are eliminated from inanimate surfaces, while cleaning gets rid of visible dirt and grime from various inanimate surfaces and objects. A minimum reduction of ≥ 1 log CFU is usually achieved by cleaning processes.
5. How are medical devices classified?
   Medical devices are classified into critical, semicritical, and noncritical based on their role in patient care and the potential risk posed by the devices in the transmission of infections. This classification is indicative of the appropriate level of sterilization or disinfection required for these devices.
6. How are patient care devices reprocessed?
   Patient care devices are reprocessed as per their classification into critical, semicritical, and noncritical devices. Critical devices come in direct contact with sterile body parts and sterilization is, therefore, the method of choice for reprocessing these devices. Semicritical devices encounter mucous membranes and nonintact skin. These devices are, therefore, reprocessed by heat sterilization or cleaning followed by disinfection using HLDs. Noncritical devices are restricted to contact only with intact skin. However, these devices can contribute to indirect means of transmission of infection. Although these devices do not need sterilization due to their low risk of transmission of infection, regular cleaning and disinfection with LLDs is important.

7. Which are the pathogens that are the most susceptible and most resistant to the antimicrobial effects of sterilization methods and disinfectants?
   Enveloped viruses are the most susceptible, while prions are the most resistant to the antimicrobial effects of sterilization methods and disinfectants.
8. How are critical and semicritical patient care devices treated for the elimination of prions?
   Critical and semicritical devices used for the care of patients with prion diseases are sterilized using a combination of steam sterilization and sodium hydroxide (1 N NaOH) at 121°C for 30 minutes. Devices sensitive to these sterilization techniques are cleaned twice, treated with peracetic acid (PAA), iodophors, 3% sodium dodecyl sulphate, or 6 M urea, and autoclaved at 121°C for 30 minutes to ensure the neutralization of prions. Noncritical devices are used on a disposable basis as far as possible. Liquid chemical sterilants such as PAA (0.2%), glutaraldehyde (≥2.4%), ortho-phthalaldehyde (OPA) (0.55%), and hydrogen peroxide (7.5%) are used for the sterilization of heat-sensitive critical and semicritical patient care devices that can be immersed in liquids to ensure the elimination of prions.
9. What is the method of choice for the sterilization of pharmacological products?
   Pharmacological products such as serum, vaccines, and antibiotics are sterilized by filtration.
10. How are disinfection methods determined based on the types of microorganisms to be targeted?
    Disinfection of heat-tolerant semicritical patient care devices is often based on the types of microorganisms that are to be targeted. Pasteurization of heat-tolerant devices for approximately 50 minutes ensures riddance of all vegetative forms of microbes, although this method does not affect the spore forms of the microbes. Heat-sensitive semicritical patient care devices are disinfected using HLDs that target Gram-positive bacteria, Gram-negative bacteria, mycobacteria, lipid-enveloped viruses, large nonenveloped viruses, microbial spores, cysts, trophozoites, and coccidia. ILDs such as chlorine-based products and phenolics are used to disinfect noncritical soiled devices, and surfaces and devices contaminated with bodily fluids such as blood, feces, sputum, respiratory droplets, and mucus. These disinfectants target Gram-positive bacteria, Gram-negative bacteria, enveloped and nonenveloped viruses, and mycobacteria. Noncritical patient care devices are disinfected using LLDs such as 70% alcohol, quaternary ammonium compounds, phenolics, and chlorine-based products that do not have tuberculocidal activity, but ensure the elimination of vegetative bacterial forms, lipophilic viruses, and are also effective against some fungi.

Learning objectives

1. Define sterilization.
2. Differentiate between sterilization and disinfection.
3. Explain the clinical significance of sterilization.
4. Classify medical devices and equipment as per their sterilization needs.
5. Describe the methods involved in reprocessing patient care devices.
6. Summarize the different sterilization methods.
7. Justify the preferences of sterilization methods for different medical devices and equipment.
8. Explain the processes required for cleaning and decontamination of inanimate surfaces.
9. Summarize the management of resistant pathogens.
10. Describe the methods involved in the prevention of nosocomial infections.

Discussion

Background

Hospital-acquired infections or nosocomial infections are defined as localized or systemic infections acquired during medical procedures that may be conducted on an in-patient or an out-patient basis or during hospital stays from the infectious agent(s) or toxins present in the healthcare environment. It is important to ensure that the presenting infection was not acquired during a procedure or during the hospital stay period before a definitive diagnosis of a nosocomial infection can be made. Contaminated devices and environments in healthcare establishments are usually responsible for such healthcare-associated nosocomial infections. Although such nosocomial infections are usually mild, incidences involving potentially severe and life-threatening infections are also known. Critical patient care areas such as intensive care units (ICUs), nurseries, surgical units, dialysis departments, chemotherapy areas, isolation wards, and burn wards are especially prone to the spread of nosocomial infections. The burden of nosocomial infections is reflected in the increased duration of hospital stay and the high financial implications of the management and treatment associated with incidences of such infections.

Sterilization is defined as the process of using various chemical and physical methods by which vegetative and spore forms of microorganisms are destroyed and eliminated. A $\geq 10^6$ log reduction of CFU of the most resistant microbial spores is achieved at the half-time of a regular sterilization cycle. Sterilant chemicals such as PAA (0.2%), glutaraldehyde (≥2.4%), OPA (0.55%), and hydrogen peroxide (7.5%) can destroy all vegetative and spore forms of microorganisms when used for 3–12 hours.

Sterilization is indispensable in healthcare facilities to prevent the spread of hospital-acquired infections among patients during invasive medical and surgical procedures. Invasive medical and surgical procedures involve the use of various instruments, devices, and equipment that come in direct contact with various tissue and mucosal surfaces within patient bodies, and can, therefore, become potential carriers of infectious agents. In addition, infectious agents can also be transmitted among patients and the interacting healthcare personnel either through accidental exposures such as needle-stick injury (HBV and HIV) or through direct exposure to respiratory exudates (influenza, TB). Exposure to infectious environments such as contaminated surfaces, inadequately or inappropriately sterilized instruments and devices can also result in the spread of nosocomial infections caused by opportunistic pathogens such as *Pseudomonas aeruginosa, Acinetobacter baumannii,* especially among immunocompromised and high-risk patients.

In addition to sterilization, disinfection, decontamination, cleaning, and antisepsis are also important contributors to an infection-free safe healthcare environment. Disinfection does not destroy spores existing on inanimate surfaces, although it results in the complete elimination of vegetative forms of microorganisms. A $\geq 10^3$ log reduction of CFU of the nonspore-forming microorganisms can be achieved through disinfection. Disinfectants can be classified as HLD, ILD, and LLD based on their ability to eliminate microorganisms from inanimate surfaces. Decontamination is the process by which pathogenic microorganisms are eliminated from inanimate surfaces, while cleaning gets rid of visible dirt and grime from various inanimate surfaces and objects. A minimum reduction of ≥ 1 log CFU is usually achieved by cleaning processes. Reduction of the microbial load existing on epithelial surfaces of patients and healthcare personnel achieved through the process of antisepsis is mandatory before invasive procedures to prevent the entry of potentially pathogenic microorganisms inside a patient's body.

The alarmingly large number of incidences of hospital-acquired infections and outbreaks associated with inappropriate and inadequate sterilization and disinfection processes warrants the need for better strategies and recommendations for sterilization and disinfection of equipment, devices, instruments, surfaces, and the environment in healthcare facilities. In addition, precise and strict protocols are required to ensure the role of healthcare personnel in the prevention of transmission of nosocomial infections in the healthcare environment.

While there are no definite standards and rules for sterilization, disinfection, and cleaning processes, the Clinical and Laboratory Standards Institute provides some recommendations for MICs of sterilizing agents standardized by the International Organization of Standardization.[13]

The emergence of novel pathogens and MDR strains of existing pathogens globally has led to an increase in the incidence of nosocomial infections associated with such resistant pathogens. The management and treatment of such resistant nosocomial infections pose a formidable challenge for clinicians and other healthcare personnel due to the pan-resistance to existing treatment strategies and enhanced transmissibility. The reduction or elimination of these pathogens, thus, requires appropriately modified sterilization and disinfection strategies.

Classification of medical devices and equipment

Medical devices and equipment are crucial components of patient care and are often used by multiple patients, thereby making them sources of potential infection transmission. Appropriate cleaning, disinfection, and sterilization of these devices and equipment aimed at reducing or eliminating the microbial load are, therefore, crucial for preventing transmission of infection.

Medical devices are classified into critical, semicritical, and noncritical based on their role in patient care and the potential risk posed by the devices in the transmission of infections. This classification indicates the required level of sterilization or disinfection for these devices.[14] Critical devices are defined as those that are intended for use in invasive techniques, thereby coming in direct contact with a normally sterile environment such as vasculature and tissues. Surgical instruments, cardiac catheters, implants, needles, ultrasound probes used in sterile body cavities are categorized as critical devices. Semicritical devices are those that come in contact with mucous membranes or nonintact skin such as a flexible endoscope, respiratory therapy equipment, manometry probes, diaphragm-fitting rings, and laryngoscope blades. On the other hand, the use of noncritical devices, such as blood pressure cuffs and stethoscopes are restricted to intact skin.

In addition, environmental surfaces that do not come directly in contact with the patient have been classified into clinical contact surfaces and housekeeping surfaces.[15] Clinical contact surfaces include the hands of healthcare personnel that have a high microbial load and come in direct contact with the patient, and high-touch surfaces, such as bed rails, telephones, computers, door handles, light and fan switches, and medical equipment like ventilator, X-ray machines, hemodialysis machines that result in indirect patient contact. Housekeeping surfaces, such as sinks and floors precipitate at a very low risk of infection transmission.

Reprocessing of patient care devices

Patient care devices are reprocessed as per their classification into critical, semicritical, and noncritical devices.

Critical devices come in direct contact with sterile body parts, and therefore, they pose the highest threat of transmission of infection. Sterilization is, therefore, the method of choice for reprocessing these devices. Ethylene oxide, plasma sterilization, liquid sterilization with glutaraldehyde or PAA for heat-sensitive devices is recommended.[16,17] Catheters, needles, implantable devices, and intravascular devices are used as sterile, disposable, single-use items. Needles used for neurological tests such as lumbar puncture may either be disposable or may be sterilized using heat, steam, or ethylene oxide and reused. Surgical instruments are heat sterilized in autoclaves and reused. Reprocessing of arterial pressure transducers, heart-lung oxygenator surfaces, hemodialysis and plasmapheresis equipment involves heat or low-temperature sterilization. Arterial pressure transducers may also be sterilized using ethylene oxide. Ultrasound probes are disinfected using HLDs as per the manufacturer's instructions. In addition, adequate precautions are taken to prevent environmental contamination of packed sterile devices.

Semicritical devices come in contact with mucous membranes and nonintact skin. The recommended methods for reprocessing these devices are heat sterilization or cleaning followed by disinfection using HLDs. Rinsing with sterile water or alcohol followed by forced air drying reduces the contamination level. Respiratory support and therapy equipment are disinfected using HLDs, while the disinfection of laryngoscopes and the blades may need liquid germicides to be combined with HLDs. Sterilization is recommended for anesthetic equipment such as airways and endotracheal tubes; however, disinfection using HLDs may be used alternately. Heat sterilization, low-temperature sterilization, or HLDs such as OPA, glutaraldehyde, and PAA may be used for endoscopes. Nebulizers, on the other hand, are reprocessed by cleaning and disinfecting. Resuscitation accessories are washed with detergent and hot water or subjected to heat disinfection.

Noncritical devices are restricted to contact only with intact skin. These devices are responsible for indirect means of transmission of infection. Clinically challenging infections such as MRSA and VA-resistant *Enterococci* (VRE) are often spread through such noncritical devices. Due to their low risk of transmission of infection, these devices do not need sterilization, although regular cleaning and disinfection with LLDs is important. Blood pressure apparatus and cuff, doppler, stethoscope bell, thermometer, trolleys, and high-touch surfaces are cleaned with 70% alcohol after every use and on a daily basis. Ambu bags and masks are cleaned with detergent, dried, and thermally disinfected, while intravenous (IV) stands, IV monitoring pumps, walls, and washbasins are cleaned with detergent and water and dried. Surgical masks and gowns are usually disposable and if they are to be reused, they are washed in 0.5% bleaching solution and dried in the sun or the clothes drier. Reusable cloth appliances are cleaned with detergent and water followed by drying and disinfection with 70% alcohol.

Sterilization and disinfection methods based on types of microorganisms

Effective sterilization and disinfection are dependent on the susceptibilities and resistance of the various microorganisms to the different methods of sterilization and different levels of disinfectants. Adequacy of the decontamination processes is subject to the compatibility, toxicity, odor, and irritability associated with various agents and methods. While enveloped viruses are the most susceptible, prions are the most resistant to the antimicrobial effects of sterilization methods and disinfectants.

Heat-tolerant surgical instruments and semicritical patient care devices are sterilized at high temperatures using steam or dry heat. High-temperature sterilization targets and eliminates both vegetative and spore forms of microbes. Heat-sensitive critical and semicritical patient care devices, on the other hand, are sterilized at low temperatures using ethylene oxide gas, hydrogen peroxide, ozone, gas plasma, gaseous chlorine dioxide, ionizing radiation, or pulsed light. Gram-positive bacteria, Gram-negative bacteria, mycobacteria, lipid-enveloped viruses, large nonenveloped viruses, microbial spores, cysts, trophozoites, and coccidia are susceptible to this method of sterilization. However, some pathogens such as prions resist all forms of sterilization and require the use of liquid chemical sterilants. Heat-sensitive critical and semicritical patient care devices that can be immersed in liquids are treated with chemical sterilants to eliminate prions. Serum, vaccines, and antibiotics also need to be sterilized before being used for pharmacological purposes. Filtration is the recommended means of sterilization for these pharmacological products.

Some semicritical and all noncritical patient care devices are disinfected instead of being sterilized. Disinfection involves the use of HLDs, ILDs, and LLDs. Heat-tolerant semicritical patient care devices, such as respiratory therapy equipment, is pasteurized for approximately 50 minutes to get rid of all vegetative forms of microbes. This method, however, does not affect the spore forms of the microbes. Heat-sensitive semicritical patient care devices such as gastrointestinal (GI) endoscopes and bronchoscopes are disinfected using HLDs that target Gram-positive bacteria, Gram-negative bacteria, mycobacteria, lipid-enveloped viruses, large nonenveloped viruses, microbial spores, cysts, trophozoites, and coccidia. ILDs such as chlorine-based products and phenolics with proven tuberculocidal activity are used to disinfect noncritical soiled devices, and surfaces and devices contaminated with bodily fluids such as blood, feces, sputum, respiratory droplets, and mucus. These disinfectants target Gram-positive bacteria, Gram-negative bacteria, enveloped and nonenveloped viruses, and mycobacteria. Noncritical patient care devices are disinfected using LLDs such as 70% alcohol, quaternary ammonium compounds, phenolics, and chlorine-based products that do not have tuberculocidal activity. Such LLDs eliminate vegetative bacterial forms, lipophilic viruses, and are also effective against some fungi.

Antisepsis for skin disinfection of patients and healthcare personnel requires the use of chloroxylenols, polymeric biguanides, diamidines, anilides, alexidine, hexachloraphene, and triclosan.[18]

Cleaning and decontamination of inanimate surfaces

Cleaning and disinfection of inanimate and housekeeping surfaces in addition to appropriate cleaning and reprocessing of medical devices is crucial for drastically reducing the transmission of nosocomial infections. The detergents and disinfectants used for routine cleaning and disinfection in healthcare facilities should be EPA (The Environmental Protection Agency)-registered. Adequate knowledge among the housekeeping staff about the cleaning and decontamination protocols and appropriate methods of preparation of disinfectants and detergents is key to ensuring adherence to the recommended decontamination measures. Mandated use of recommended personal protective equipment (PPE) such as gloves, masks, gowns, boots, and face shields is important for ensuring the safety of the healthcare personnel and the patients. Demarcation and the separation of the PPE removal area are also recommended. Accurate dilutions, pH, exposure time, storage conditions, and fresh preparation of detergents and disinfectants are important determinants of the effectiveness of the decontamination processes. Preference for wet mopping over dry mopping minimizes the generation of dust aerosols and is therefore encouraged. Detergents are recommended for disinfection of tabletops and counters in administrative offices, cafeterias, and cabins of healthcare personnel, while HLDs are used for disinfection of high-exposure areas such as operation theaters, ICUs, isolation wards, transplant units, burn wards, dialysis departments, and chemotherapy areas. Such high-risk areas should also be equipped with vacuum pumps and exhausts with high-efficiency particulate air filters. Agents responsible for highly contagious infections, such as a respiratory syncytial virus, severe acute respiratory syndrome viruses, norovirus, *Clostridium difficile*, or resistant *Staphylococci* such as MRSA, VRE, VISA, VRSA should be treated with a high level of caution.

Management of resistant pathogens

The emergence of novel pathogens and MDR pathogens poses a formidable challenge for healthcare facilities. The difficulties associated with the management of such resistant pathogens promote their spread and transmission among patients and healthcare personnel in healthcare facilities. Nosocomial infections caused by such novel or resistant pathogens are often responsible for increased rates of mortality. MRSA, VRE, VISA, VRSA, metallo-ß-lactamase producers, agents producing extended-spectrum ß-lactamases, MDR- TB, XDR-TB, and PDR-TB are some of the high-risk pathogens that thrive in healthcare environments and require stringent cleaning and disinfection protocols. Surfaces and devices contaminated with HIV, HBV, HCV, and *Mycobacterium tuberculosis* also require rigorous cleaning and disinfection protocols. Glutaraldehyde with/without phenol/phenate/isopropyl alcohol, hydrogen peroxide, PAA, hypochlorite and hypochlorous acid, and OPA at the recommended dilutions are the commonly used disinfectants that target such resistant pathogens. Adherence to the recommended methods of preparation and contact time is crucial for ensuring the effectiveness of these disinfectants.

Prions responsible for transmissible spongiform encephalopathies such as the Creutzfeldt – Jacob Disease (CJD), variant CJD (vCJD), kuru, Gerstmann – Straussler – Scheinker syndrome, and Fatal Familial Insomnia are resistant to all generally employed forms of sterilization and disinfection. The causative agent responsible for prion diseases is altered protein fragments known as prions that have an affinity to neurological tissues. Therefore the cerebrospinal fluid and components of the neurological system such as the brain tissue, the spinal cord, retina, and optic nerve are considered to be at high risk for the transmission of prion diseases. Patient risk assessment based on the criteria of family history of prion diseases, transplant history involving neurological tissues, pituitary hormone injections derived from cadavers, cases of unusually progressive dementia with ataxia and myoclonus, and cases suspected or diagnosed of transmissible spongiform encephalopathy is the key to determining the appropriate protocols for the management of such patients and their environments. All critical and semicritical devices used for the care of patients with prion diseases must be sterilized using a combination of steam sterilization and sodium hydroxide (1 N NaOH) at 121°C for 30 minutes. Devices sensitive to these sterilization techniques must be cleaned twice, treated with PAA, iodophors, 3% sodium dodecyl sulphate, or 6 M urea, and autoclaved at 121°C for 30 minutes to ensure the neutralization of prions.[18] Noncritical devices should be used on a disposable basis as far as possible.

Sterilization, disinfection, and cleaning are therefore crucial for the prevention of transmission and spread of nosocomial infections among patients and healthcare personnel. Adherence to recommendations of sterilization and disinfection protocols by trained and knowledgeable healthcare personnel and housekeeping staff ensures effective infection control in healthcare establishments. In addition to sterilization, disinfection, cleaning of environmental surfaces, patient care devices, and housekeeping surfaces, rigorous implementation of hand hygiene by healthcare personnel continues to be the cornerstone of all infection control measures in healthcare establishments.

Student notes

Key take-aways

1. Infections caused due to infectious agents present in healthcare environments are known as nosocomial infections.
2. Nosocomial infections are usually mild and can be managed well. However, incidences of resistant and potentially life-threatening nosocomial infections pose a formidable challenge for healthcare personnel and are often responsible for an extended hospital stay, high financial implications, and increased rates of mortality.
3. Critical patient care areas such as ICUs, burn wards, isolation wards, surgical units, dialysis departments, and chemotherapy units are at high risk of transmission of nosocomial infections.
4. Stringent implementation of recommended sterilization, disinfection, and cleaning protocols is crucial for hindering the spread of nosocomial infections.
5. Usage of EPA-registered disinfectants and FDA-approved sterilization processes ensures effective infection control.
6. Training of healthcare personnel and housekeeping staff in sterilization and disinfection protocols is important for the effective implementation of the recommended protocols.
7. The concentrations of the disinfectants, their preparation protocols, storage conditions, pH, and contact time are critical determinants of the effectiveness of the decontamination processes.
8. Special precautions are required for the management of emerging and resistant pathogens such as MRSA, VISA, VRSA, VRE, MDR-, XDR-, PDR-TB, norovirus, HIV, HBV, HCV, and spore-producing microbes such as *C. difficile*.
9. Prions are resistant to the general sterilization and disinfection protocols, and therefore require the implementation of accurate risk assessment and specified protocols for risk mitigation.
10. Emphasis on stringent hand hygiene practices is the cornerstone for the prevention of the spread of nosocomial infections in healthcare establishments.

Problem number 1.3: fever

A 24-year-old man complained of malaise and anorexia that was existent for about a week. He mentioned an isolated vomiting episode without blood in the vomit. Although he reported feeling feverish through this duration, he had not clinically checked his temperature. The patient also complained of aches in several body joints, although there was no evidence of swelling. He did not report any change in urine or bowel habits.

The patient was unemployed and shared living quarters with four other people. He mentioned a serologically confirmed diagnosis of infectious mononucleosis almost 6 years ago and disclosed a strong addiction to smoking (~20 cigarettes/day) and alcohol consumption (an average of 30 units of alcohol per week). Although his history revealed occasional consumption of marijuana and various forms of ecstasy without adequate knowledge of the ingredients, he denied any occasion of IV drug use. The patient also revealed indulgence in occasional sexual practices with multiple homosexual partners. He mentioned having taken a negative HIV test about 6 months ago, although he did not produce a report of the test. He did not have any international travel history in the last 3 years, and his family history did not reveal any relevant medical history.

The following are the reports of the investigations carried out:

1. Vitals
   a. Temperature: 38.8°C
   b. Blood pressure: 90/60 mm Hg
   c. Pulse: 110/min
   d. Respiratory rate: 25/min
2. Physical examination
   a. Appearance: Jaundiced
   b. Abdomen: Tenderness in the right upper quadrant
   c. All other joints and organ systems were found to be normal on physical examination.
3. Complete blood count (CBC)

Hemoglobin (g/dL)	15.4	13–17
Mean corpuscular volume (MCV) (fL)	90	80–100
WBC (/L)	12×10^9	$4.0–11.0 \times 10^9$
Platelets (/μL)	3,00,000	150,000–400,000

4. Prothrombin time: 16 s (10–14 s)
5. Biochemistry

Urea (mmol/L)	4.3	3–6.3
Sodium (mmol/L)	139	135–145
Potassium (mmol/L)	4.0	3.5–5.3
Creatinine (μmol/L)	70	60–110
Fasting glucose (mmol/L)	4.5	4.0–6.0
Bilirubin (mmol/L)	60	3–17
Alkaline phosphatase (IU/L)	385	30–300
Alanine aminotransferase (IU/L)	550	5–35

6. Hepatitis antibody tests

Anti-hepatitis A	Negative
Anti-hepatitis B	Positive
Anti-hepatitis C	Negative

Diagnosis

The final diagnosis is that of acute viral hepatitis B infection.

Treatment and support

1. Entecavir.
2. Recommendation to avoid smoking, drinking alcohol, or using drugs until the infection is treated effectively.
3. The patient was counseled about the long-term ill-effects of smoking, alcohol, and drug abuse.
4. Counseling regarding the importance of safe sexual practices was also given.

Follow-up and outcome

The liver function tests and prothrombin time gradually started to normalize over the monthly follow-ups, and the patient reported back after 6 months with normal liver function tests and prothrombin time.

Question and answer

1. What are pyrogens and how do they play a role in the mechanism of fever?
   Pyrogens are chemical mediators that are released in response to various pathological processes in the body and are responsible for increasing the core body temperature by more than 0.5°C. This increase in the core body temperature above the "set-point" is known as a "fever."
2. What is the difference between a fever and hyperthermia?
   Fever and hyperthermia are responsible for an elevation in the core body temperature. Although the result of both fever and hyperthermia is the same, the mechanisms differ considerably in their pathophysiology. Fever is an increase in the core body temperature beyond 0.5°C elicited by the hypothalamus and is a controlled process wherein the body temperature does not exceed 106°C. Hyperthermia, on the other hand, is an uncontrolled increase in the core body temperature beyond the body's ability to lose heat,

which results when the hypothalamus is overwhelmed and loses its ability to regulate the core temperature.

3. What are exogenous pyrogens and endogenous pyrogens?
   Exogenous pyrogens are substances like lipopolysaccharides (LPS) on the cell wall of several pathogens that trigger the production of lipids, known as prostaglandins (PGs), at the site of the tissue injury or infection, and are responsible for the regulation of processes, such as inflammation, recruitment of leukocytes and lymphocytes, and the formation of blood clots. Endogenous pyrogens, on the other hand, are certain chemical mediators known as cytokines, like interleukin-1, that are released due to the interaction of the leukocytes, tissue macrophages, and granular lymphocytes, recruited because of the pathophysiological processes involved in the elicitation of fever. The invading pathogens are phagocytosed and digested by the recruited cells, and in the process, the hypothalamus is triggered to elicit an increase in the core body temperature.
4. Which are the physiological reactions to the pyrogenic processes?
   Pyrogenic processes result in some specific physiological reactions, such as systemic vasoconstriction, which results from the stimulation of the posterior hypothalamic sympathetic center, piloerection, for example, the phenomenon of hair standing on end due to the contraction of the arrector pili muscles mediated by the excited sympathetic system, and thermogenesis, for instance, the phenomenon of enhanced heat production, resulting from sympathetic stimulation, shivering, and thyroxine discharge.
5. How do the hypothalamic temperature control processes maintain a constant core body temperature?
   Pathological processes may affect the pace of heat loss and heat production, causing a fluctuation in the "set-point" of the core body temperature. The pace of heat loss is more prominent than that of heat production when the core body temperature rises considerably above the "set point," ensuring that the core body temperature reduces until it approaches 37°C. On the other hand, the pace of heat production is more prominent than that of heat loss when the core body temperature falls considerably below the "set point," guaranteeing that the core body temperature rises to reach 37°C. This constant control of the core body temperature is very important for optimal cellular and tissue level functioning.
6. How does a fever affect the metabolic processes and the immune responses of an individual?
   A fever causes the oxygen demand to increase, resulting in an increased heart rate, an amplified respiratory rate, and also causes the metabolic processes to use fats and proteins as primary sources

of energy rather than carbohydrates. The switch-over from carbohydrates to proteins and fats compromises the growth of pathogens due to the reduced amount of glucose available for use as a growth medium for the pathogens.

The functionality of the immune system is enhanced by a fever due to the increased motility and activity of leukocytes, activation of T-cells, production of antibodies, and stimulation of antiviral interferon production. In addition, the increased body temperature inhibits the growth of certain microbial agents that proliferate at the normal body temperature.

7. What is hyperpyrexia and its clinical relevance?

   Hyperpyrexia is an extremely high fever ($> 41.5°C$) elicited by conditions such as heatstroke. This condition can cause local hemorrhages and a systemic parenchymatous degeneration of cells, especially targeted at the central nervous system. Neurons do not regenerate and cannot be replaced in the event of severe damage and loss of normal function. Extensive damage to certain other vital organs, such as the liver, kidneys, or the cardiac muscle, may also prove to be potentially fatal.

8. Why is the pattern of a fever clinically important?

   The pattern of a fever is clinically important as it enables differentiation and diagnosis of various pathologies. A waxing and waning pattern of a fever that occurs every $48 - 72$ hours is usually indicative of certain types of malaria, while increased body temperature in the evening is suggestive of a TB infection. Temperature-pulse dissociation is a characteristic in typhoid fever, brucellosis, leptospirosis, factitious fever, and some drug-induced fevers.

9. What is the mode of action of the commonly prescribed antipyretic medicines, such as paracetamol?

   Paracetamols, such as aspirin and acetaminophen, are cyclooxygenase (COX) inhibitors. Cyclooxygenase facilitates the conversion of the PGE2 that interacts with the hypothalamus to elicit the increase in the core body temperature to PGH2, which isomerizes further to PGE2. Cyclooxygenase inhibitors prevent the continued production of PGE2, thereby preventing further increase in the core body temperature.

Learning objectives

1. Define pyrogen.
2. Summarize the etiology of fever.
3. Describe the immune responses in fever.

4. Describe the pathophysiology of a pyrogenic reaction.
5. Summarize the clinical significance of a fever.
6. Describe the diagnostic approach to a fever.
7. Enumerate the management options for a fever.

Discussion

Background

The hypothalamus acts as the thermoregulatory center, which maintains the body's core temperature at a "set-point." Although the "set-point" at which the body temperature is regulated is 37°C, minor variations (~0.5°C) in this "set-point" through the course of the day occurs as a result of several metabolic and physiological processes of the human body. Certain substances, known as "pyrogens," can increase the core body temperature by more than 0.5°C, a condition known as "fever." A fever can therefore be defined as an elevation of the core body temperature above the "set-point." This elevation in the core body temperature is caused by the release of chemical mediators that are triggered by various pathological processes in the body. Although both are an elevation in the core body temperature, a fever and hyperthermia differ considerably in their pathophysiology. A fever is an increase in the core body temperature beyond 0.5°C provoked by the hypothalamus and is a controlled process wherein the body temperature does not exceed 106°C. Hyperthermia, on the other hand, is an uncontrolled increase in the core body temperature beyond the body's ability to lose heat, which results when the hypothalamus is overwhelmed and loses its ability to regulate the core temperature.

Pathophysiology

A fever is stimulated by the host immune response as a result of a pathogenic invasion in various tissues or the bloodstream but may also be a response to systemic inflammation. LPS on the cell wall of several pathogens are exogenous pyrogens that trigger the production of lipids, known as PGs, at the site of tissue injury or infection, and are responsible for regulation of processes, such as inflammation, recruitment of leukocytes and lymphocytes, and formation of blood clots.[19] PGE2, a PG, demonstrates pyrogenic activity through the PGE2-EP3 receptor, which interacts with the hypothalamus, causing an increase in the core body temperature.[20,21,22] The process of PGE2 synthesis comprises of the conversion of the membrane phospholipids to arachidonic acid, which is further converted to PGH2 by COX. PGH2 then undergoes isomerization to form PGE2 by a terminal PGE synthase.[22,23] In addition, the recruited leukocytes, tissue macrophages, and granular lymphocytes phagocytose digest the invading pathogens, which release certain cytokines, such as interleukin-1, known as endogenous pyrogens, also known to trigger the hypothalamus to elicit an increase in the core body temperature.[24,25] An increase in the core body temperature also requires an increased heat preservation and production.

Physiologic reactions to the pyrogenic process

The sudden elevation in the core body temperature triggers several physiological reactions.

1. Stimulation of the posterior hypothalamic sympathetic center results in systemic vasoconstriction.
2. The excited sympathetic system causes the arrector pili muscles to contract, resulting in a condition known as piloerection, wherein hair "stands on end."
3. Sympathetic stimulation, shivering, and thyroxine discharge also enhance heat production or thermogenesis.[25]

Clinical significance

The "set point" maintained by the hypothalamus is crucial for the regulation of the core body temperature to 37°C. Pathological processes that cause the "set point" to change by more than 0.5°C affect the pace of both heat loss and heat production. The pace of heat loss is more prominent than that of heat production when the core body temperature rises considerably above the "set point," ensuring that the core body temperature reduces until it approaches 37°C. On the other hand, the pace of heat production is more prominent than that of heat loss when the core body temperature falls considerably below the "set point," guaranteeing that the core body temperature rises to reach 37°C. Therefore the hypothalamic temperature control processes maintain optimal cellular and tissue level functioning by a constant endeavor to keep the core body temperature as close to the "set-point" level as possible.

Hyperthermia, however, is a condition wherein the core body temperature is raised uncontrollably without any alteration in the "set-point," either due to exogenous heat exposure or endogenous heat production. Clinical differentiation between hyperthermia and a fever is important to ensure rapid and appropriate treatment as the treatment options for a fever and hyperthermia vary tremendously.

The ability to develop a fever becomes impaired with increasing age. Geriatric individuals usually have a low baseline body temperature compared to younger adults, leading to the development of only a modest fever despite a serious underlying pathology. This can be misleading and generally results in a diagnostic delay, often precipitating life-threatening complications. Similarly, infants, chronic kidney disease patients, and individuals who take corticosteroids on a regular basis may not develop a fever in case of infections, contradictorily these patients may get hypothermia.

Clinical manifestations and complications

A fever reflects in the form of several clinical manifestations, varying based on the organ system involved. Effects on metabolism are reflected by the increased need for oxygen, which results in the heart rate, rate of respiration, and the use of body proteins as an energy source to increase. The switch-over of the body metabolism from carbohydrates to fats and proteins is a characteristic metabolic effect of fever, reducing the amount of glucose available for pathogens to use as a growth medium. A fever enhances the functionality of the immune system by increasing the motility and activity of the leukocytes, triggering the activation of T-cells, eliciting the production of antibodies, and stimulating the antiviral interferon production. The increased body temperature inhibits the growth of certain microbial agents that proliferate at the normal body temperature.

Although a fever is the body's defense mechanism and has several beneficial effects in reducing the potential damages caused by various pathologies, the increase in the core body temperature can also have certain adverse effects. Hyperpyrexia, an extremely high fever (>41.5°C) provoked by conditions such as heatstroke, is known to cause local hemorrhages and a systemic parenchymatous degeneration of cells, especially targeted at the central nervous system. Neurons do not regenerate and cannot be replaced in the event of severe damage and loss of normal function. Extensive damage to certain other vital organs, such as the liver, kidneys, or the cardiac muscle, may also prove to be potentially fatal.

The pattern of a fever is also clinically important for the diagnosis of certain pathologies. A waxing and waning pattern of a fever that occurs every 48–72 hours is usually indicative of certain types of malaria, while increased body temperature in the evening is suggestive of a TB infection. Temperature-pulse dissociation is characteristic in typhoid fever, brucellosis, leptospirosis, factitious fever, and some drug-induced fevers. Healthy individuals show a direct proportionality between body temperature and pulse, with every 1°C rise in temperature causing the pulse rate to increase by 4.4 beats/minute.

Diagnostic approach

Patient history and physical examination are crucial primary steps in the diagnostic approach toward a fever or hyperthermia. Laboratory investigations include an array of tests, such as a CBC with erythrocyte sedimentation rate (ESR) and C-reactive protein (CRP). Elevated leukocyte count, increased ESR, and high CRP levels are significative to an infection or an inflammatory condition. Elevated procalcitonin is usually suggestive of bacterial infections, while tuberculin skin test is a specific test for TB infections. HIV tests are critical in diagnosing HIV infections and facilitates the detection of seroconversion in an HIV positive individual. Serum lactate dehydrogenase uncovers the lactic acid levels in the blood and elevated levels are indicative of systemic infections, such as septicemia, which can be potentially life-threatening. Blood cultures enable the detection of pathogens in the bloodstream, while specific antibody tests reveal the presence of autoimmune antibodies, such as the rheumatoid factor (RF), antinuclear antibody (ANA), and heterophile antibody, pinpointing autoimmune conditions. Creatinine phosphokinase is a test that indicates cardiac pathologies. Serum protein electrophoresis (SPEP) measures certain protein levels in the blood and can be instrumental in diagnosing several diseases. Imaging studies like Computed Tomography (CT) scans, X-rays, magnetic resonance imaging, ultrasonography, and positron emission tomography scans may be required in certain cases as warranted by the patient history and physical examination details. Lumbar puncture is often recommended in cases related to the central nervous system, and tests for parasites are also recommended in cases that have a history of likely exposure due to travel to parasite-endemic regions.

Certain cases, such as patients with cancer, HIV, a history of drug abuse, a history of organ transplantation, a history of prolonged hospital stay, and other anomalies or abnormalities that predispose them to infections or inflammatory conditions, should be considered for common opportunistic infections, nosocomial infections, thrombophlebitis, and infective carditis. The diagnostic approaches mentioned here are not exhaustive, and several more specific tests may be recommended based on the patient history and findings of physical examination.

Treatment and management

The management and treatment of a fever is generally focused on reducing the core body temperature and bringing it as close to the "set point" as possible. This is important for ensuring optimal functioning of the body's metabolism and all the organs at the cellular and tissue level. Inhibitors of COXs, such as aspirin and acetaminophen, commonly known as paracetamol, are generally used for temperature reduction in fevers.[26] High intake of fluids and adequate rest are also recommended for the management of a fever. Investigations to reveal the underlying cause of fevers are emphasized upon to guarantee effective treatment.

Student notes

Key take-aways

1. The hypothalamus is the thermoregulatory center of the body that maintains the core body temperature at a particular "set-point," conducive for optimal cellular and tissue level functioning.
2. Metabolic functions cause minor variations in the "set-point" through the course of the day and are physiologically normal.
3. The increase of the core body temperature beyond the "set-point" is known as a "fever."
4. Several pathological processes trigger the release of chemical mediators known as "pyrogens," which elicit an increase in the core body temperature, thereby causing a fever.
5. Pyrogens may be endogenous, such as interleukin-1, or exogenous, such as LPS.
6. Pyrogens trigger the production of PGs, such as PGE2, which on interaction with the hypothalamus, stimulates a rise in the normal core body temperature.
7. Unlike a fever, hyperthermia is an uncontrolled increase in the core body temperature beyond the body's ability to lose heat, which occurs when the hypothalamus is overwhelmed and loses its ability to regulate the core temperature.
8. Clinical differentiation between a fever and hyperthermia is important as treatment for the two pathological manifestations contradict each other.
9. A fever enhances the functionality of the immune system by increasing the motility and activity of leukocytes, triggering the activation of T-cells, eliciting the production of antibodies, and stimulating the antiviral interferon production.
10. The increased body temperature inhibits the growth of certain microbial agents that proliferate at the normal body temperature.
11. Hyperpyrexia, an extremely high fever ($> 41.5°C$) caused by conditions like heatstroke can cause extensive and often irreversible tissue damage.
12. Patient history and physical examination are crucial primary steps in the diagnosis of a fever or hyperthermia.
13. The management and treatment of a fever is generally focused on reducing the core body temperature and bringing it as close to the "set point" as possible.
14. Inhibitors of COXs, such as aspirin and acetaminophen, commonly known as paracetamol, are generally used for temperature reduction in a fever.
15. Investigations to reveal the underlying cause of a fever are emphasized upon to ensure effective treatment.

Problem number 1.4: food poisoning caused by *Shigella flexneri*

A 10-year-old girl presented with a high-grade fever, abdominal pain, and complaints of frequent foul-smelling watery motions with traces of blood and mucus. The patient's history revealed the consumption of food eaten out 2 days ago, although no other family member had complaints of diarrhea. The patient's father had complained of abdominal pain, but did not have diarrhea. The attending physician suspected acute GI infection related to the intake of contaminated food.

The following are the reports of the investigations carried out:

1. Physical examination
   a. Flushed and mildly dehydrated appearance
   b. Abdomen: Tender to touch
2. Vitals
   a. Temperature: 39.4°C
   b. Pulse: 95/min
   c. Blood pressure: 95/65 mmHg
3. CBC

Test	Value	Reference range
Hemoglobin (g/dl)	12.2	12–16
Total WBC (/L)	17.5×10^9	$4.5–11 \times 10^9$
RBC (/L)	4.8×10^{12}	$4–5.2 \times 10^{12}$
Hematocrit (%)	34	36–46
PMN (%)	82	40–70
Lymphocyte (%)	11	22–44
Monocyte (%)	2	4–11
Eosinophil (%)	5	0–8
Basophil (%)	2	0–3

4. Serum electrolytes

Test	Value	Reference range
$Na+$ (mmol/L)	138	136–145
$K+$ (mmol/L)	4.2	3.5–5.0
Cl^- (mmol/L)	102	98–106

5. Antibodies to *Salmonella*: a titer of 1/40 (not significant)

6. Stool routine
   a. Macroscopic:
   b. Liquid consistency; blood-stained
   c. Microscopic:
   d. Mucus + +; WBC + + +; RBC + + +
   e. No visible ova or yeast cells
7. Stool culture and sensitivity:
   a. Presence of Gram-negative *S. flexneri* rods
   b. Ciprofloxacin: Sensitive
   c. Cotrimoxazole: Sensitive
   d. Chloramphenicol: Sensitive
   e. Ampicillin: Resistant
8. Urine culture: Negative
9. Blood culture: Negative

Diagnosis

Based on the clinical presentation of the disease, a viral etiology was ruled out and the final diagnosis was acute GI tract infection caused by *S. flexneri*.

Treatment

1. Fluid replacement therapy: oral rehydration solution
2. Cotrimoxazole (sulfamethoxazole + trimethoprim in a fixed dose combination)
3. The patient was asked to attend for a follow-up as an outpatient 5 days later.

Follow-up and outcome

The patient presented as afebrile and started having normal bowel movements after 5 days of the follow-up. A continuation of cotrimoxazole was prescribed for 1 week.

Question and answer

1. What is the most likely cause of acute (bloody) diarrhea associated with a fever?

   GI infection is the most likely cause of acute (bloody) diarrhea associated with a fever.
2. What types of microorganisms cause diarrhea? Name examples.

Bacteria, parasites, and viruses can cause diarrhea. *Escherichia coli, Salmonella, Shigella, campylobacter, rotavirus, Entamoeba histolytica, and Giardia* are some examples of microorganisms that can cause diarrhea.

3. What is the function of the mucous membranes?

   The mucus membranes provide a natural barrier against invading microorganisms.

4. Why did the clinical manifestations of the infection vary between the patient and the father, although both consumed the same food?

   Clinical manifestations of an infection may vary between individuals depending on the infectious dose and the individual's susceptibility to the infection, which is attributed to the natural defenses and immunity from previous infections.

5. Why was there a 24-hour time lag between the consumption of the contaminated food and the development of symptoms?

   The time lag occurs because it takes time for an infectious agent to cause the first symptom of the disease, known as the incubation period of the organism.

6. Why was a urine culture requested?

   Although the patient shows GI symptoms, a UTI may be responsible for the fever. Since urine is usually sterile, the presence of organisms in urine could be considered as a sign of infection and a probable cause, resulting in the symptoms manifesting by the patient. However, although urine is sterile, the collection process by urination may contaminate the infection. Therefore organisms must be found in sufficiently large numbers to suspect an infection (few organisms suggest contamination while collection).

7. Why was a blood culture requested?

   Bacteria can be invasive, thereby passing through the mucosa and circulating in the blood stream. The fever manifested by the patient suggests the possibility of bacteremia. Blood culture can detect and identify the bacteria circulating in the blood stream.

8. Do all microorganisms have an invasive capacity?

   No, not all microorganisms have an invasive capacity. The invasive organisms usually have virulence factors that allow them to destroy the mucous membranes and invade deeper tissues. Certain other organisms can produce potent exotoxins that can harm the host without invading the tissues (diphtheria, tetanus). Food poisoning is an example of a disease caused by such potent exotoxins produced by certain microorganisms.

9. Which defenses come into action if microorganisms penetrate mucous membranes and invade into deeper tissues?

   Natural or innate defenses are divided into cellular, including phagocytes and natural killer cells, and humoral (soluble), including antibodies, complement proteins, and interferons. These defenses together with our anatomic and physiologic barriers constitute our primary (or innate) defenses.

10. What allowed the possibility of a viral infection to be ruled out?
    The bloody diarrhea and fever are not typical of a viral GI infection, thereby ruling out the possibility of a viral infection.
11. What is titer?
    Antibodies are quantitated functionally, for example, by their reaction with antigens. Antibodies are serially diluted to determine their effective concentration, then their activity is determined. The titer is the reciprocal of the highest dilution of the patient's serum at which the reaction can be detected.
12. What is *Salmonella* and how is the serological test for *Salmonella* interpreted?
    Salmonella is a bacterium that causes an invasive GI infection. A serology test result with titers below 1/80 indicates that the patient does not have an active *Salmonella* infection. (Significant titer is $> 1/80$.)
13. What does the presence of RBC and WBC in stool indicate?
    The presence of RBC and WBC in stool implies damage to surface layers of the GI tract—that is, invasive infection, and the presence of WBCs also suggest an infection.
14. What does the presence of mucus in the stool indicate?
    Mucus production is one of the natural defenses in reaction to irritation of mucosal membranes, and is useful in the clearance of organisms or material causing irritation. The mucus in the stool therefore specifies irritation to the GI mucosal membranes caused by the infection.
15. What are ova and cysts, and what is their source?
    Ova are eggs of worms residing mainly in the intestine, but they can also be found in the stools. Cyst is a stage of protozoan life cycle highly resistant to the environment.
16. Name examples of GI protozoan and helminthic infections and explain how these parasitic infections are diagnosed.
    E. histolytica, *Giardia lamblia*, *Ascaris*, and *Ankylostoma* are some examples of GI protozoan and helminthic infections, which are diagnosed by microscopic examination of stools and through other approaches, including serological, molecular, and special stains.
17. What is an acute infection? What are the other patterns of infection?
    An infection with sudden onset and severe symptoms is known as an acute infection. Acute infections result either in prompt recovery or death, although some untreated cases may also cause chronic disease. Infection patterns may also be chronic, persistent, and latent.

18. How are GI infections acquired?

    GI infections are acquired through the ingestion of contaminated food or water, and this route of transmission is known as the fecal-oral route of transmission. This route is facilitated from a carrier of the disease, such as insects infecting food or drink or through sewage contamination in a water supply.

19. How is *S. flexneri* transmitted? What are other modes of transmission of infectious diseases?

    S. flexneri is transmitted from person-to-person through direct contact, contaminated food, sexual contact, and drinking water. Infectious diseases may be blood-borne, sexually transmitted, vector (insect)-borne, or may be transmitted through droplets and aerosols. Vertical transmission (mother-to-child) is another mode of transmission, which may occur in the case of certain infectious diseases.

20. What is cotrimoxazole and why was the need for an antimicrobial therapy indicated for this patient?

    Cotrimoxazole is an antimicrobial drug, and considering the invasive nature of the infection indicated by the fever and the presence of blood in the stool, antimicrobial therapy is suggested.

21. What is the principle of antimicrobial therapy?

    Antimicrobial therapy relies on the selective targeting of pathogens without affecting the host by identifying a structure or step in the metabolism, not shared between microbe and host.

22. Is it always necessary to have information regarding culture and sensitivity before starting therapy?

    Although culture and sensitivity reports are important for ensuring guided and effective therapy, empirical therapy can be started based on the most likely suspected causative agent, determined in accordance with the clinical manifestations. This prevents loss of important treatment initiation time, and the therapy can be modified later for compliance with the culture and sensitivity reports.

23. Why was *S. flexneri* sensitive to ciprofloxacin, cotrimoxazole, and chloramphenicol, but resistant to ampicillin?

    The sensitivity report revealed that the in vitro growth of the organism (*Shigella*) was arrested by ciprofloxacin, cotrimoxazole, and chloramphenicol, but was not affected by ampicillin.

24. How are antimicrobial sensitivities determined?

    Antimicrobial sensitivities are determined in the microbiology lab by culturing organisms in the presence of the various antimicrobials and observing their effects, eventually resulting in the determination of the MIC, which facilitates the determination of the most effective therapeutic dose.

25. Why was the patient advised to continue therapy despite being symptom-free?

The continuation of therapy is important as clinical improvement does not necessarily lead to the eradication of the organisms, and may result in a recurrence of the infection, which also increases the possibility of the microorganism becoming resistant to the antimicrobial therapy.

Learning objectives

1. Describe the role of mucous membranes as a natural barrier against infection.
2. Define incubation period.
3. Describe host susceptibility to infection and the host – pathogen relationship in acute/chronic infections.
4. Explain microbial pathogenicity and virulence factors.
5. Define titer.
6. Define and characterize the patterns and sources of infection.
7. Describe the modes of transmission of infectious diseases.
8. Explain the mechanism of action of antimicrobials on microbes.
9. Explain the principles and clinical importance of antibiotic sensitivity testing, resistance, and MIC.
10. Describe the clinical manifestations that lead to the diagnosis of acute GI infection before laboratory diagnostic reports are obtained.
11. Describe the characteristics of *S. flexneri*, the organism responsible for the manifestation of GI infection.
12. Interpret the correlation between the clinical manifestations and the pathogenicity of *S. flexneri*.
13. Determine the appropriate course of treatment and the most effective antibiotic regimen to ensure prompt and effective resolution of the infection.

Discussion

Differential diagnosis

Nontyphoidal *Salmonella* infection	Nontyphoidal *Salmonella* infection can manifest as gastroenteritis, focal infections, or bacteremia. The infection presents itself with a fever and GI symptoms, such as abdominal pain and diarrhea, or symptoms of focal infection.
Typhoid fever	Typhoid fever presents itself with a fever, abdominal pain, nausea, vomiting, headache, diarrhea, constipation, weakness, and skin rash.
Escherichia coli infection	*E. coli* infection develops with abdominal cramps, sudden and severe watery diarrhea with traces of blood, fever, fatigue, loss of appetite, nausea, gas, and vomiting.

Continued

Continued

Campylobacter infection	*Campylobacter* infection manifests with bloody diarrhea, fever, stomach cramps, nausea, and vomiting.
Aeromonas infection	*Aeromonas* infection presents itself with watery diarrhea, fever, and abdominal cramps.
Clostridium difficile infection	*C. difficile* infection presents with frequent diarrhea, high-grade fever, general malaise, loss of appetite, and stomach pain.
Entamoeba histolytica infection	*E. histolytica* infection progresses with stomach cramps, loose stools, stomach pain, fever, and bloody diarrhea.

Background

S. flexneri is one of the bacterial species responsible for Shigellosis, a form of bacterial diarrhea common in developing countries. The infection spreads through contaminated food and water, direct contact with infected individuals, and poor sanitation conditions. While GI infections caused by the *Shigella* bacterial species can affect all age groups—due to the high resistance of the bacteria in the stomach acid and the lower infectious dose required[27] to cause infection—the extremes of the age spectrum, individuals with underlying GI conditions, and immunocompromised individuals are at higher risk of developing the infection.

Etiological agent

Shigella flexneri

Classification and morphology

S. flexneri are Gram-negative, nonmotile, nonsporing, noncapsulated, and short rod-shaped bacteria (Fig. 1.2) belonging to the genus *Shigella* in the *Enterobacteriaceae* family. The fimbriae present in some of the bacterial strains enables the bacteria to adhere to the GI mucosa, thereby facilitating the establishment of the bacterial infection. The bacteria are facultative anaerobes, which supports their survival in the hostile GI conditions. *S. flexneri* is classified as serotype B of the four *Shigella* serotypes, and is further classified into six serotypes. The serotypic classification is based on the nature of the O-antigen present on the cell envelope.

Epidemiology

S. flexneri is almost endemic to developing countries and is responsible for the highest levels of mortality among all GI infections. The prevalence of poor hygiene and sanitation, malnutrition, compromised quality of drinking water, and overcrowding contribute to the endemic status of *S. flexneri* infections. The human GI tract is the only known reservoir for the *S. flexneri*

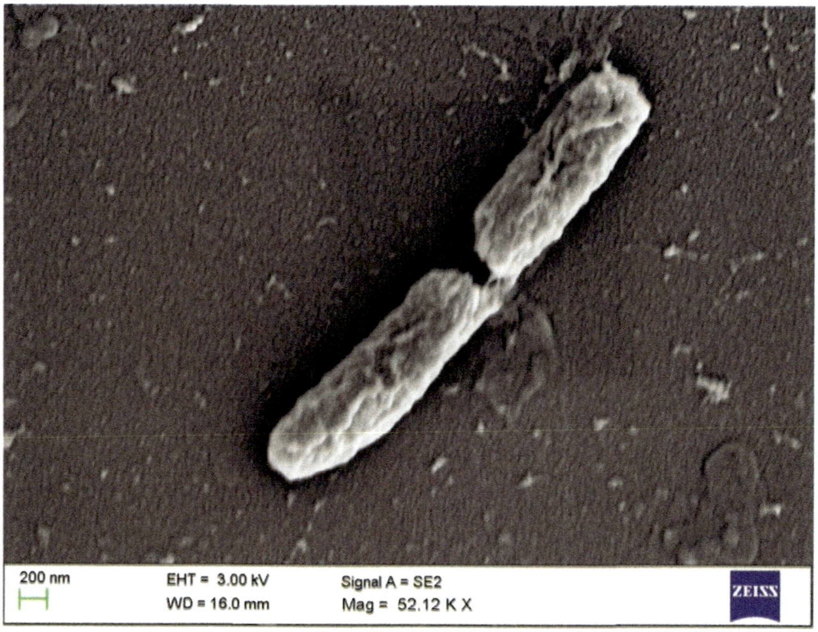

FIGURE 1.2
Shigella flexneri visualized using scanning electron microscopy. *Credit: Vader 1941/Own work /https://creativecommons.org/licenses/by-sa/4.0/. https://en.wikipedia.org/wiki/File:Shigella_flexneri.tif.*

bacteria, one of the most invasive GI pathogens. While most cases are self-limiting, severe complications, resulting even in fatal outcomes, occur in high-risk individuals. The burden of the disease in endemic areas is therefore high.

S. flexneri is generally responsible for acute infection and a chronic carrier state is rare, with an exception in malnourished children and immunocompromised individuals. While contaminated food and water are the main sources of infection, fomites, such as contaminated door handles, water taps, flush handles, and lavatory seats are also an important means of hand-to-mouth infection. Vector-borne transmission from flies and sexual transmission among homosexual men as part of the gay bowel syndrome, are also known.[28,29]

Virulence and pathogenicity

S. flexneri is notoriously resistant to the stomach acid due to an ability to upregulate the acid resistance genes.[30] This ability to survive the acidic gastric conditions facilitates infection via low infective dose,[27] thereby augmenting the highly infectious nature of *S. flexneri*. In addition to upregulating the acid-resistant genes, *S. flexneri* also downregulates the expression of the host antibacterial proteins, which further assists the establishment of an infection.

Ingestion of the bacteria is followed by the invasion of the small intestine where it multiplies before entering the large intestine. This process takes an average of 3 days, which is usually the incubation period for *S. flexneri* infections. Direct invasion of the colonic mucosa results in cell injury and cytopathic effects attributed to the production of bacterial enterotoxins. The colon provides the optimal conditions for the bacteria to replicate and spread among the mucosal epithelial cells after the penetration of the mucosal barrier. The host inflammatory response elicited by the bacterial invasion of the colon contributes to the destruction of the colonic epithelial layer, and is responsible for the clinical manifestations of *Shigella* infections.

S. flexneri invades the basolateral epithelium of the colon by transcytosis, thereby gaining access to the mucosa-associated lymph nodes containing the highly endocytic M-cells. M-cells are membranous epithelial cells that play an important role in antigen recognition, resulting in immune activation of intestinal lymphoid tissue, causing the induction of macrophages and cellular apoptosis.[31,32] An untreated *Shigella* infection allows the bacteria to exploit the host inflammatory response to further increase bacterial penetration of the colonic epithelium.[31] Proinflammatory cytokines, such as IL-1 and IL-8, are released as a consequence of the inflammatory responses and result in the recruitment of polymorphonuclear (PMN) cells to the infected area. The influx of the PMN cells increases the level of inflammation, thereby compromising the integrity of the epithelium (Fig. 1.3). Diarrhea, a common clinical manifestation of *Shigella* infections, is attributable to the impaired ability of the damaged epithelium to absorb adequate and appropriate nutrition. While the PMN influx promotes the spread of *S. flexneri*, the PMN cells have been found to be responsible for localized restriction of the infection, thereby preventing systemic dissemination of the infection.[33]

The role of the enterotoxins produced by *S. flexneri* in impaired nutrient absorption and manifestation of diarrhea has been proven as an alternate pathogenic mechanism for the bacteria.[34]

Immune response

Shigella infections result in severe inflammation of the gut mucosa, which persists for more than 1 month even after the resolution of the infection. Upregulation of pro-inflammatory cytokines, such as IL-1, IL-8, TNF-α, IL-10, IL-4, IFN-γ, IL-6, and TNF-ß, while being responsible for the extensive inflammation levels, also plays an important role in containing the infection to the localized tissue. Although macrophages and monocytes are recruited for killing the bacteria, they themselves succumb to apoptosis. This induces the production of IL-18, which targets NK cells and T-lymphocytes, causing the induction of IFN-γ. IFN-γ, facilitating bacterial clearance and replication inhibition within the epithelial cells through the activation of fibroblasts and macrophages.[35,36] The NF-κß activation in the *Shigella*-infected epithelial cells is responsible for the production of IL-8 and IL-1, which mediate the recruitment of the PMN cells.

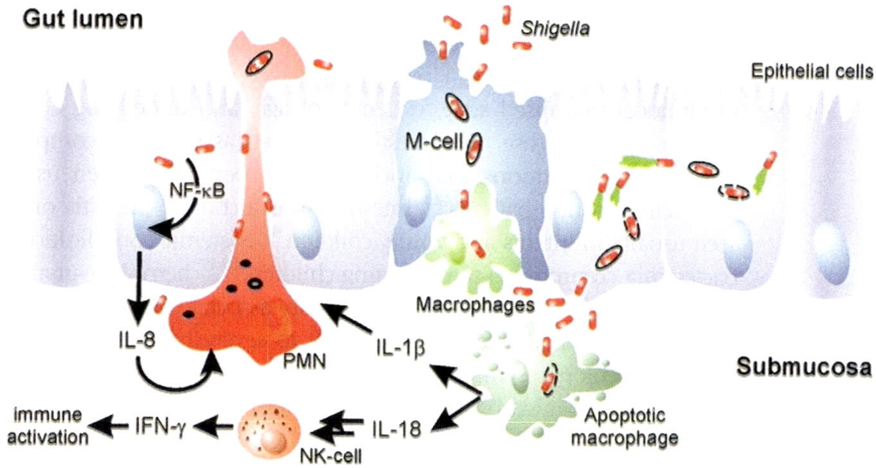

FIGURE 1.3

Shigella shown translocating across the intestinal epithelium and inducing host immune response. *Credit: Jake. P. Morgan/Contributed to microbewiki/ From Clin. Microbiol. Rev. January 2008 vol. 21 no. 1 134–156. https://microbewiki.kenyon.edu/index.php/File: Shigella_pathogenesis.jpeg.*

The human glycoprotein, lactoferrin present in mucosal secretions, breast milk, and phagocytic cells can disrupt the bacterial surface in addition to intestinal defensins that show antibiotic properties against enteric bacteria.[37,38]

Serotype-specific immunity against *S. flexneri* has been attributed to the upregulation of serum IgG, IgM, and IgA, thereby implying an evident role of humoral immunity in *S. flexneri* infections.[31]

Clinical manifestations and complications

S. flexneri infections manifest with mild symptoms, such as mild abdominal discomfort in a vast majority of healthy adults, while severe diffuse colicky abdominal pain may develop in children, immunocompromised individuals, and those with underlying GI conditions. Scanty mucoid diarrhea preceding the appearance of blood in stools is a characteristic feature of invasive *S. flexneri* infections. Patients may also get a fever, nausea, vomiting, loss of appetite, anorexia, lethargy, fatigue, and tenesmus.[39,40] *S. flexneri* infections rarely cause severe systemic symptoms, such as seizures, encephalopathy, delirium, meningismus, anuria, and coma.[41,42]

The indication of toxicity upon physical examination, and the vital signs revealing a fever, tachycardia, tachypnea, and hypotension, plus the patient history of probable exposure to *S. flexneri*, should be considered indicative of an invasive infection. The presentation of a distended bowel with hyperactive bowel sounds and tenderness in the lower abdomen due to the involvement of sigmoid colon and rectum are highly suggestive of *S. flexneri* infections.

A timely diagnosis and treatment of *S. flexneri* infections usually results in a good prognosis and a favorable outcome with the patients recovering without any debilitating complications. However, a delayed diagnosis, inappropriate treatment, immunocompromised state, extremes of age, and underlying GI conditions can lead to severe disease and intestinal, as well as systemic, complications. While intestinal complications may include colon perforation seen usually in infants and malnourished patients;[43] intestinal obstruction;[44] proctitis or rectal prolapse seen usually in infants and young children.[45] Systemic complications may include bacteremia commonly seen in young children,[46,47] hemolytic-uremic syndrome—an uncommon form of acute kidney injury associated with *S. flexneri* infections in young children and infants.[48] Moderate to severe hypovolemia, leukemoid reaction,[49] neurologic symptoms,[50,51] Reiter syndrome—a form of inflammatory arthritis accompanied by conjunctivitis and urethritis,[52] vulvovaginitis—presenting as painless vaginal discharge seen in young girls with or without diarrhea,[53] and keratitis.[54]

Diagnosis

Histopathological analysis of the *S. flexneri* infected colon, rectal, or distal ileal tissues generally reveal evidence of infiltration of PMN cells into the epithelium, resulting in the formation of inflammatory patch pseudomembranes. Besides, serological investigations, such as a CBC, may show leukocytosis with left shift, leukopenia, anemia, and thrombocytopenia.[55,56] Stool analysis is routinely performed for diagnosing *S. flexneri* infections and usually shows the presence of fecal leukocytes and blood (Fig. 1.4). While liver function tests may display mild elevation in bilirubin levels, renal function tests indicate elevated blood urea

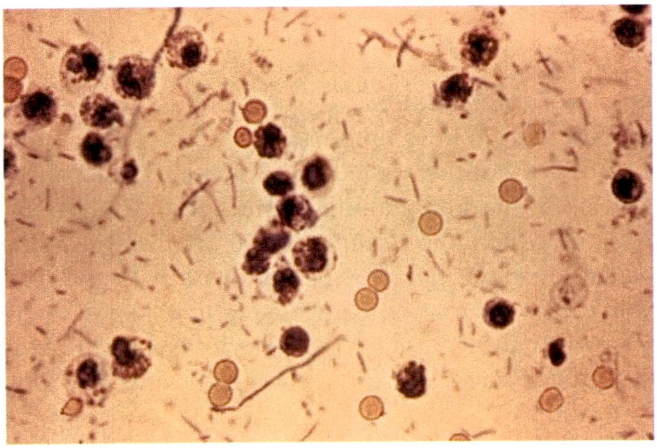

FIGURE 1.4

Photomicrograph of a stool sample from a *Shigella* infected patient, showing the presence of exudates. *Credit: USCDCP/ https://creativecommons.org/licenses/publicdomain/. https://pixnio.com/science/microscopy-images/shigellosis-shigella/photomicrograph-revealed-stool-exudates-in-a-patient-with-shigellosis-which-is-also-known-as-shigella.*

nitrogen and creatinine in some cases. The inflammatory effects of the infection result in an elevation of the levels of certain inflammatory markers, such as ESR and CRP. Blood culture reports are usually negative, bar in rare, complicated cases where bacteremia may be indicated. The presence of alpha-1 antitrypsin in stool is indicative of acute phase of the infection, and continued high levels are suggestive of ineffective or absent therapeutic intervention.[57] Enzyme-linked immunoassay (ELISA) and polymerase chain reaction are not usually recommended as diagnostic tests for *S. flexneri* infections, but may be indicated in a minority of patients.

Treatment and prevention

A rapid diagnosis and timely medical intervention are the mainstays of treatment for *S. flexneri* infections. Hydration and electrolyte management through oral rehydration or IV interventions are the most prescribed means of management that prevent further deterioration of the patient's condition. Antimotility drugs, such as loperamide, paregoric, or diphenoxylate, are usually contraindicated in *S. flexneri* infections as they prolong the infection, thereby further increasing the damages caused by the inflammatory processes and also facilitate the transmission of the pathogen by prolonged shedding of the bacteria.[58]

Antibiotic susceptibility tests are recommended over empirical antibiotic treatment due to the prevalence of regional resistance to antibiotics. The preferred antibiotic regimen also differs depending on the age of the patient, with fluoroquinolone being recommended for patients with no risk factors for resistance while a third-generation cephalosporin being recommended for high-risk patients. Second-generation cephalosporin, ampicillin, and trimethoprim-sulfamethoxazole are used only if susceptibility is indicated. On the other hand, the first-line drug for empirical therapy recommended for use in children is azithromycin. Cefixime and ceftibuten are generally prescribed as first-line antibiotics in regions reporting wide-spread antibiotic resistance.[59] Pivmecillinam, an extended-spectrum penicillin, can be used to decrease the duration of diarrhea, and is also known to eradicate the bacteria from the stool.[60]

Ceftriaxone in the parenteral form is indicated only in children with severe infection showing signs of bacteremia, such as lethargy and temperature >39°C (102.2°F), immunocompromised patients, and those unable to take oral medications.[47]

As there is no vaccine for *S. flexneri*, the following activities can help to prevent the infection from spreading: good hand hygiene after bathroom use and before food preparation; avoiding the handling of food until stool cultures test negative for *S. flexneri*; supervising hand hygiene in young children; quickly disposing soiled diapers; consuming boiled or treated water and hygienically cooked food; practicing safe sex; and avoiding public pools.

Student notes

Key take-aways

1. *S. flexneri* is one of the etiological agents responsible for a diarrheal disease known as Shigellosis.
2. *S. flexneri* infections usually spread through the fecal-oral route, and are usually food and water borne.
3. Poor hygiene and sanitation conditions, in addition to overcrowding in developing countries, results in the endemicity of *S. flexneri* infections.
4. While *S. flexneri* infections are self-limiting and resolve without causing much damage to the host, extremes of age, an immunocompromised state, and underlying GI conditions increases the risk of severe disease.
5. *S. flexneri* is especially notorious for its invasive properties and its ability to penetrate the mucosal epithelium after replication in the small intestine, thereby invading the colon.
6. The cascade of immune responses from the recognition of *S. flexneri* antigens causes an inflammatory state, which is responsible for the clinical manifestations of the infection as well as containing its spread.
7. A rapid diagnosis and appropriate treatment often result in a good prognosis and favorable outcome of the disease.
8. Hydration and maintenance of electrolyte levels are often sufficient for the management of the disease; however, antibiotic therapy may be indicated in some cases.
9. While empirical antibiotic therapy may be used initially, susceptibility testing should be referred to in order to ensure effective therapy.
10. Antibiotic resistance in *S. flexneri* shows regional variation.
11. The prevention of *S. flexneri* infections relies solely on good hand and washroom hygienic practices.
12. The consumption of appropriately cooked food and boiled water is known to reduce the chances of contracting *S. flexneri* infections.

Problem number 1.5: immunodeficiency

A 4-year-old girl presented with pneumonia and indicated a failure at consistent growth. The patient had a history of recurrent episodes of otitis media, sinusitis, and pneumonia since the age of 1.5 years. She had been hospitalized previously for a week for pneumonia. Investigations during the recurrent infections indicated low serum immunoglobulin levels; however, there had been no medical intervention at that time. The child had been conceived through in vitro fertilization and had a full-term gestation. The child's birth weight was normal, and she had been breastfed until 3 years of age. Her immunizations were up to date. The patient is the youngest of three healthy half siblings and is the only joint child of her healthy parents. The patient did not have primary contact with any other sick children.

The following are the reports of the investigations carried out:

1. General
   a. Weight: 15 kg
   b. Height: 100 cm
   c. Nasal speech and slurring of words (onset reported after adenoidectomy and tonsillectomy at 2 years of age; receiving speech therapy)
2. Vitals
   a. Temperature: 39°C
   b. Blood pressure: 90/60 mm Hg
   c. Pulse: 110/min
   d. Respiratory rate: 25/min
3. Physical examination
   a. Moderate generalized lymphadenopathy
   b. Chest: Crepitations in left lung base
   c. Ear: Grommets in situ (2nd set)
4. CBC

Test	Values	Reference values
Hemoglobin (g/dL)	10.9	11–15
Mean corpuscular volume (MCV) (fL)	80	80–100
WBC (/L)	20.5×10^9	$4.0–11.0 \times 10^9$
Platelets (/μL)	3,00,000	150,000–400,000

5. Sweat test: Normal
6. Immunological studies

Test	Values	Reference range
IgG (g/L)	0.23	6–16
IgA (g/L)	0.09	0.6–3
IgM (g/L)	1.52	0.5–2.6
CD3	1287	1800–3000
CD4	497	1000–1800
CD4:CD3	0.38	—
CD8	810	800–1500
CD19	155	700–1300
NK	168	200–600
Tetanus antibodies (IU/ml)	3.27	≥ 0.1
Streptococcus pneumoniae IgG antibodies (mg/L)	48.19	≥ 0.2

Diagnosis

Agammaglobulinemia and transient hypogammaglobulinemia were ruled out due to the presence of CD19, IgM, and IgA and age above 2 years, respectively. Therefore the final diagnosis was concluded as hypogammaglobulinemia. In addition, balance problems and elevated alpha fetoprotein levels reported 2 years later resulted in an added diagnosis of ataxia – telangiectasia (A – T) immunodeficiency.

Treatment and therapy

1. Regular IV immunoglobulin (IVIg) transfusion (12 g IVIg every 3 weeks with trough levels of 7–8 g IgG).
2. Regular monitoring of serum IgG levels.
3. Occupational therapy for partial hearing loss and head control.

Follow-up and outcome

The patient remained free of any significant infections, and chest CT scans demonstrated clear lungs. Auscultation examinations indicated a clear chest; however, there were clean perforations noted in both ears with partial loss of hearing on the left side. After 2 years of continued treatment and therapy, the patient reported problems related to balance, and alpha fetoprotein levels were found to be elevated. The case was then referred to a neurologist and A – T immunodeficiency was also diagnosed.

Question and answer

1. How do low antibody levels make an individual vulnerable?
   Antibodies are instrumental in the effective functioning of the adaptive immune system. These are proteins that are specifically targeted to recognize and bind pathogenic antigens, thus aiding the elimination process. The elimination of extracellular pathogens, such as certain encapsulated bacteria, primarily depends on such antigen – antibody binding mechanisms. Phagocytic cells often fail to recognize and ingest encapsulated bacteria due to the presence of the external capsule. Therefore antibodies and other attributes of the immune system, such as the complement system, are required to enable the recognition and ingestion of such bacteria by phagocytic cells. In addition to bacteria, the control and elimination of other pathogens, such as viruses, also require antibodies.

2. What is hypogammaglobulinemia and what are the various causes?
   Hypogammaglobulinemia is a deficiency of the immune system characterized by low levels of the gamma immunoglobulin (IgG). This condition is caused due to a decreased production of IgG precipitated by various malignancies, such as lymphoma, thymoma, leukemia, and multiple myeloma; certain medications, such as carbamazepine, oxcarbazepine, and some immunosuppressive agents; acquired infections, such as pediatric HIV; extreme starvation and malnutrition; enteropathies resulting in protein-loss; chylothorax; and hypercatabolic syndrome.
3. What is A − T immunodeficiency?
   A − T, also known as the Louis − Bar syndrome, is an autosomal recessive genetic disorder characterized by immune dysfunction, progressive neurodegeneration (especially cerebellar degeneration), cutaneous abnormalities other than telangiectasia, cancer predisposition, radiosensitivity, and sterility. Immune dysfunction in A − T is precipitated by reduced levels of IgA, IgE, IgG2, absolute lymphopenia, and a decreased ratio of CD4 + helper cells to CD8 + suppressor T-cells, affecting both cellular and humoral immunity. A − T patients are highly susceptible to recurrent upper and lower respiratory tract infections, especially to the infections caused by encapsulated bacteria.
4. How does the genetic defect in A − T affect DNA repair mechanisms?
   A − T is caused by mutations in the *ATM* gene located on the long arm of chromosome 11 at 11q22−23. The *ATM* gene regulates the phosphatidylinositol 3 kinase-like enzyme, which coordinates cellular response to DNA double-stranded breaks. The formation of DNA double-stranded breaks due to various extrinsic and intrinsic factors precipitates a cell cycle arrest mediated by the ATM protein, thereby preventing the production of new DNA. Proteins are then recruited and activated to repair the damage, thus ensuring DNA repair before the completion of cell division. High severity of DNA damage elicits the *ATM* gene to mediate cell apoptosis and eliminates the defective cell. However, the A − T mutation in the *ATM* gene prevents the formation and normal functioning of the ATM protein, thereby facilitating genomic instability. Diseases and conditions like cancers that occur in response to DNA damage and cellular stress cannot be prevented due to the absence of the ATM protein. Therefore A − T is also known as a genome instability syndrome, a DNA repair disorder, or a DNA damage response (DDR).
5. How is A − T inherited?
   A − T is inherited as an autosomal recessive trait. Since A − T is a recessive trait, both parents of an affected individual are required to

be carriers of the mutated *ATM* gene. This means that each parent has one normal and one mutated abnormal copy of the *ATM* gene. A child born to such carrier parents will have the disorder only if the child inherits the same abnormal *ATM* gene from both parents.

6. Why does A – T selectively affect only the IgG, IgA, and IgE immunoglobulin isotypes?
   IgM and IgD immunoglobulin isotypes are generated by RNA splicing-mediated isotype switching occurring at the RNA level, while IgG, IgA, and IgE immunoglobulin isotypes are generated by an ATM, protein-dependent DNA recombination of the switch regions. Since mutations in the *ATM* gene affect the formation of the ATM protein and compromise the cellular DNA repair mechanisms, only the DNA-dependent immunoglobulin isotypes are affected.
7. What is IVIg therapy?
   IVIg is a blood product prepared by pooling IgG immunoglobulins extracted from the serum of over a thousand blood donors. IVIg is given as a plasma protein replacement therapy to individuals affected by various forms of immunodeficiencies, which compromise the antibody generation capabilities of the individual. The administration of IVIg at regular intervals of 3–4 weeks ensures the maintenance of protective antibody levels required for the prevention of infections and conference of passive immunity.
8. What is immunoglobulin class switching?
   Immunoglobulin class switching is the mechanism by which an antibody changes from one immunoglobulin class to another. The process of immunoglobulin class switching affects the constant region portion of the antibody heavy chain, although the variable region of the heavy chain remains unaffected. This ensures that antigen specificity of the immunoglobulins remains unaffected.

Learning objectives

1. Enumerate the various immunodeficiency disorders.
2. Describe the clinical manifestations of an immunodeficiency disorder.
3. Define A – T.
4. Summarize the etiology of immunodeficiency disorders.
5. Describe the epidemiology of immunodeficiency disorders.
6. Justify the predisposition of immunodeficiency affected individuals to infections.
7. Describe the role of immunodeficiencies in autoimmune diseases.
8. Justify the failure to thrive in individuals with immunodeficiencies.
9. Discuss the prognosis of immunodeficiency disorders.

10. Identify the characteristic clinical manifestations of immunodeficiency disorders.
11. Enumerate the methods involved in the diagnosis of immunodeficiency disorders.
12. Summarize the treatment and management options for immunodeficiency disorders.

Discussion

Background

The normal and effective functioning of the immune system is important for eliminating life-threatening pathogenic infections and preventing hyperinflammatory autoimmune conditions. Gene or protein defects resulting in deficiencies of various components of the immune system affect the effectiveness of immune responses. Immunodeficiencies may be primary or secondary depending on the factors responsible for the condition. Primary immunodeficiencies comprise of T-cell deficiency, B-cell deficiency, combined T-cell and B-cell deficiency, complement deficiency, phagocyte deficiency, and immunoglobulin A (IgA) deficiency. Secondary immunodeficiencies, on the other hand, are those acquired during the lifetime an individual's life, either due to the use of certain drugs that suppress the immune system or because of certain infections and conditions that compromise the normal functioning of the immune system.

Etiology

Primary immunodeficiencies resulting from T-cell deficiency include DiGeorge syndrome, also known as congenital thymic aplasia, chronic mucocutaneous candidiasis, hyper-immunoglobulin M syndrome, and interleukin-12 receptor deficiency. On the other hand, primary immunodeficiency from B-cell deficiency causes X-linked agammaglobulinemia (XLA), also known as Bruton agammaglobulinemia. Combined T-cell and B-cell deficiency can cause severe combined immunodeficiency disease (SCID), Wiskott – Aldrich syndrome, immunodeficiency with A – T, and major histocompatibility complex (MHC) deficiency known as bare leukocyte syndrome. Complement deficiency causes hereditary angioedema, recurrent bacterial and viral infections due to C3 and C5 deficiencies, respectively, bacteremia attributable to deficiency of membrane attack complex (MAC), and autoimmune conditions caused by C2 and C4 deficiencies. Combined T-cell and B-cell deficiency and complement deficiency are both primary immunodeficiencies. Chronic granulomatous disease (CGD) and leukocyte adhesion deficiency syndrome caused by phagocyte deficiencies are also primary immune deficiencies. Deficiency of the IgA is more common compared to deficiencies of the other immunoglobulins, and is known as selective IgA deficiency, another form of primary immunodeficiency.

Secondary immunodeficiencies can be drug-induced, such as those precipitated by steroids, chemotherapeutic agents, and immunosuppressive agents administered to transplant recipients. Several other factors, such as malnutrition, obesity, malignancies, and acquired infections like acquired immune deficiency syndrome (AIDS) also result in secondary immunodeficiencies.

Pathophysiology

Intrinsic defects in components of the immune system, such as T-cells, components of the complement system, and phagocytes, present as primary immunodeficiencies. Although all immunodeficiencies increase predisposition to infections, the type of infection is suggestive of the specific deficiency. While recurrent infections caused by extracellular bacteria suggest antibody deficiency, recurrent fungal infections suggest a lack of T-lymphocytes.

A congenital anomaly of the thymus, known as the DiGeorge anomaly, results in T-cell deficits, which vary depending on the severity of the anomaly. This anomaly is caused by a microdeletion on chromosome 22 that compromises the appropriate development of the pharyngeal pouches. The pharyngeal pouches are instrumental in the embryonic development of many important body parts, such as the tonsils, jaws, ears, thyroid, parathyroid, thymus, and some parts of the heart, like the aortic arch and cardiac outflow tract. This condition presents itself with a broad range of phenotypic expressions, including cardiac anomalies, recurrent infections, hematological abnormalities, abnormal facies, thymic hypoplasia or aplasia, palatal abnormalities, developmental delay, autoimmune cytopenias, and predisposition to malignancies.[61]

The hyper-IgM syndrome is a primary immunodeficiency that causes a deficiency of IgG, IgE, and IgA, while causing an elevation in the IgM levels. Several X-linked and autosomal recessive or dominant mutations have been associated with the etiology of the disease. The clinical manifestations vary from pulmonary disorders, GI disorders, autoimmune conditions, hematologic abnormalities, lymphoproliferation, and malignancies depending on the specific mutation responsible for the condition.[62]

XLA, also known as Bruton disease, is caused by a mutation in the Bruton's tyrosine kinase (*Btk*) gene. This mutation prevents the normal development and maturation of B-cells. B-cells manufacture immunoglobulins or antibodies, which ensure an effective humoral immune response. Therefore XLA predisposes affected individuals to serious and even fatal infections.[63]

Certain presentations of combined T-cell and B-cell deficiencies, such as the various forms of SCIDs, are fatal within the first few years of life due to the extensive inability of the immune system to elicit effective immune responses. SCID is a result of mutations in genes that facilitate the differentiation and proliferation of T-cells, B-cells, and NK cells. The immunoglobulin recombination-activating gene (RAG)-1 and RAG-2 endonuclease proteins mediate the process of recombination of the variable (V), diversity (D), and joining (J) gene segments of immunoglobulins by cleaving DNA at recombination signal sequence sites. Mutations in RAG-1 or RAG-2 genes compromise the process of repair of the cleaved DNA, causing the immunoglobulin recombination to be functionally impaired. Various forms of SCID are caused by the inability of the functionally impaired immunoglobulins to recognize a wide variety of antigenic epitopes. Although SCID typically results in complete deficiency of an effective immune system, certain mutations that permit partial enzymatic activity result in "leaky" SCID or Omenn syndrome. The V(D)J recombination of immunoglobulins in "leaky" SCID is restricted, thereby generating a limited repertoire of T-cells and B-cells, which often fail to distinguish between self and foreign antigens precipitating autoimmune conditions.[64]

A – T, also known as the Louis – Bar syndrome, is an autosomal recessive genetic disorder caused by mutations in the *ATM* gene located on the long arm of chromosome 11 at 11q22–23. The *ATM* gene regulates the phosphatidylinositol 3 kinase-like enzyme, which coordinates a cellular response to DNA double-stranded breaks. The A – T mutation in the *ATM* gene prevents the formation and normal functioning of the ATM protein, thereby facilitating genomic instability. Diseases and conditions such as cancers that occur in response to DNA damage and cellular stress cannot be prevented in absence of the ATM protein. Therefore A – T is also known as a genome instability syndrome, a DNA repair disorder, or a DDR. This condition is characterized by immune dysfunction, progressive neurodegeneration (especially cerebellar degeneration), cutaneous abnormalities other than telangiectasia, cancer predisposition, radiosensitivity, and sterility. Immune dysfunction in A – T is precipitated by reduced levels of IgA, IgE, IgG2, absolute lymphopenia, and a decreased ratio of CD4 + helper cells to CD8 + suppressor T-cells, affecting both cellular and humoral immunity. A – T patients are highly susceptible to recurrent upper and lower respiratory tract infections, especially to the infections caused by encapsulated bacteria.[65]

MHC class I and II deficiency associated with primary immunodeficiencies are caused by mutations in the genes that encode for the transport-associated protein and MHC class II transactivator, respectively. MHC class I deficiency results in the lack of MHC class I molecule expression, and manifests as a deficiency of CD8 + T-lymphocytes. On the other hand, MHC class II deficiency results in the absence of MHC class II molecule on antigen-presenting cells, including macrophages and dendritic cells, and manifests as bare lymphocyte syndrome.[66]

CGD is a primary immunodeficiency inherited in a X-linked recessive manner, resulting in a deficiency of phagocytes. Nicotinamide adenine dinucleotide phosphate (NADPH) oxidase enzyme complex in phagocytes, also known as phagocyte oxidase (phox), mediates the elimination of the ingested microorganisms. Defects in this enzyme complex results in the impaired killing and elimination of microbes. Phagocytosis normally causes neutrophils to increase their oxygen consumption through a process termed as the respiratory or oxidative burst. Phagocytosis acts as a physiological stimulus that results in the generation of superoxide anion by the transfer of electrons from reduced NADPH to molecular O_2. This superoxide anion is then converted to bactericidal hydrogen peroxide and other reactive oxidants, causing the oxidative burst. Defects in the enzyme complex results in the dysfunctional process of phagocytosis, thereby causing the individual to be susceptible to a broad range of bacterial and fungal infections, which may even be life-threatening.[67]

Complement deficiencies are primary immunodeficiencies that affect the complement system, which is a crucial part of the innate humoral immune system. The clinical manifestation varies depending on the deficient complement protein. The complement system plays a crucial role in the processes of opsonization, cytotoxic destruction, formation of MACs, and release of peptides that promote the inflammatory response. Deficits in the complement pathways can increase susceptibility to a wide range of infections as well as local or systemic inflammatory and thrombotic conditions.[68]

Secondary immunodeficiency is acquired during an individual's life, rather than being inherited or caused by a congenital defect. Such immunodeficiencies may be induced by the prolonged use of certain drugs, such as steroids; chemotherapeutic agents like cyclophosphamide; antirheumatic drugs like mycophenolate; immunomodulatory drugs such as leflunomide; and immunosuppressants such as rapamycin, which affect the immune responses elicited by both T and B-lymphocytes. Viral infections, such as HIV infection, that affects CD4 + T cells downregulates cellular immune responses, resulting in life-threatening opportunistic infections and cancers. Malnutrition-associated conditions, such as protein-energy malnutrition, can result in secondary immunodeficiency by affecting cell-mediated immunity and phagocytosis. Nutritional deficiencies resulting from cancer, autoimmune conditions, chronic renal disease, trauma, burns, chronic infections, and deficiencies of zinc, iron, selenium, copper, and vitamins B6 and B12 can also hamper the normal effective functioning of the immune system.[69]

Clinical manifestations and complications

Immunodeficiencies can present themselves as a broad spectrum of clinical manifestations, depending on the specific condition. The patient history is often indicative of an investigation for immunodeficiencies. Since most immunodeficiencies are caused by X-linked recessive or dominant mutations, the male sex is an important factor for the consideration of immunodeficiencies. A history of recurrent and refractory infections, especially those caused by rare and opportunistic microorganisms, is a crucial indicator of immunodeficiencies. Hematologic abnormalities and chronic inflammatory conditions also result in a high index of suspicion for immunodeficiencies. The presence of circumstances or medical conditions that could precipitate malnutrition also warrant investigations for immunodeficiencies.[70,71]

Respiratory tract infections, usually caused by encapsulated bacteria, are the most commonly occurring infections in patients with immunodeficiencies. Infections like otitis media, meningitis, tonsillitis, TB, recurrent and refractory candidiasis, lymphocytic interstitial pneumonitis, septicemia, and bacteremia, are also very common. Patients often develop a fever, hypothermia, chronic fatigue, malaise, extreme unexplained weight loss, arthritic symptoms, chronic cough, chronic headaches, intestinal malabsorption, dyspnea, multiple episodes of convulsions, purulent conjunctivitis, bronchiectasis, recurrent abscesses, severe atopy, cachexia, chronic diarrhea, hematological abnormalities such as petechiae and anemia, laryngeal edema, abnormal bleeding, tetany, ataxia, telangiectasia, angioedema, asthenia, anorexia, anaphylaxis, hypoparathyroidism, aphthous stomatitis, pruritus, vasculitis, macroglossia, poor wound healing, nail dystrophy, and venous telangiectasias of the trunk and limbs.

Certain congenital abnormalities often increase the index of clinical suspicion for immunodeficiencies. Such congenital abnormalities include failure to thrive, stunted growth, global development delay, gross congenital defects like cleft palate, facial abnormalities like dysmorphisms, denture abnormalities, hearing impairment, multiple organ dysfunction, absence of tonsils, cardiac abnormalities such as cardiac murmur and conotruncal malformation, absence of lymph nodes, thymic aplasia or hypoplasia, dermatologic abnormalities such as pyodermitis and eczema, congenital ichthyosis, microcephaly, exocrine pancreatic insufficiency, albinism, spondiloepiphyseal dysplasia, cafe-au-lait spots, bamboo hair, scoliosis, hyperextensible joints, mental retardation, urogenital abnormalities, chondrodysplasia, and delayed cord separation.

The existence of some chronic conditions, such as rheumatic diseases, systemic granulomas, opportunistic malignancies, lymphoproliferative disorders, bone marrow failure, lupus-like syndrome, hydrops fetalis, costochondral junction flaring, hypohidrosis, hypotrichosis, alopecia, glomerulonephritis, hemolytic-uremic syndrome, serositis, osteoporosis, frequent fractures, adrenal disease, late-onset primary encephalopathy, early-onset diabetes, thyroiditis, palmoplantar hyperkeratosis, amyloidosis, and periodontitis, further warrant investigations for the diagnosis of immunodeficiencies.[72]

Immunodeficiencies predispose affected individuals to several life-threatening complications. Recurrent and refractory infections caused by bacteria, viruses, fungi, and parasites, opportunistic malignancies, septic shock, anaphylactic shock, bleeding disorders, renal insufficiency, respiratory insufficiency, liver diseases, multiorgan failure, intrauterine growth retardation, hydrops fetalis, systemic rheumatic disorders, endocrinopathy, congenital abnormalities, metabolic disturbances, neurological disorders, acidosis, and alkalosis, are some of the most common complications seen in immunodeficient individuals.[73]

Diagnostic approach

Immunodeficiencies are diagnosed primarily based on immunological investigations, including quantitative serum immunoglobulin, isohemagglutinins, and postimmunization antibody activity assessment, total lymphocyte counts and stimulation assays, assessment of the complement system and phagocytic activity.

Levels of serum immunoglobulins (IgG, IgM, IgA, IgE, IgG1, IgG2, IgG3, IgG4) can demonstrate any abnormalities in normal immunoglobulin levels. Isohemagglutinins, antiA and antiB that are attributes of the A, B, AB, and O blood groups are also assessed to identify hematologic disorders and deficiencies. Coagulation studies are important for the diagnosis of hematologic abnormalities. Factor V assay, fibrinogen level, prothrombin time, thrombin time, and bleeding time are the most recommended coagulation studies.

Immunizations and exposure result in the generation of IgG and IgM antibodies toward the specific antigens. However, abnormalities or disorders in the purview of immunodeficiencies compromise the ability of the immune system to generate effective antibody levels in response to immunizations or exposures. Therefore evaluation of the antibody levels generated against routine immunizations, such as tetanus toxoid, diphtheria toxoid, pneumococcal polysaccharide, polio, rubella, measles, varicella zoster, heterophile antibodies, and antistreptolysin O titer, indicate compromised elicitation of immune responses, in case of abnormalities related to immune system deficiencies.

Viral infections like HIV, hepatitis B, cytomegalovirus, dengue, and Epstein – Barr virus are known to compromise the normal functioning of the immune system. Therefore diagnosis of such viral infections is important in cases of suspected acquired immunodeficiencies. Total lymphocyte counts comprising of T-lymphocytes (CD3, CD4, and CD8), B-lymphocytes (CD19 and CD20), and CD4/CD8 ratio, as well as lymphocyte stimulation assays, are important indicators of immunodeficiencies. SPEP, phorbol ester and ionophore test, hemolytic assays, and phytohemagglutinin are some more clinical diagnostic tests that are indicated in the diagnosis of hematologic and immunologic abnormalities.

Phagocytes, such as macrophages and neutrophils, are important in primary immune responders in eliminating some bacteria. The nitroblue tetrazolium test is the gold standard of diagnosis for phagocytic abnormalities.

The complement system, comprising of several components, is an important element of the innate immune system. Measurement of the serum levels of individual components, such as C3, C4, Factor B, C1 inhibitor, assays for complement-binding, and complement-activating agents by immunoprecipitation tests, ELISA, or western blotting is useful in identifying deficits of the immune system. Assays that evaluate the various complement system pathways, such as the classical pathway assay, the alternative pathway assay, specific properdin test, and the mannose pathway assay, facilitate the identification of deficits or abnormalities in effective functioning of the complement pathways.

Autoimmunity studies for the detection of circulating immune complexes, cold agglutinins, C1q autoantibody, C1 inhibitor autoantibody, ANA, anti-ds DNA, RF, antihistones, anti-Smith, anti-(SS-A), anti-(SS-B), anti-RBC, antiplatelet, antineutrophil, and organ-specific auto-immune antibodies are recommended for the diagnosis of systemic autoimmune disorders.

In addition to the specialized tests for evaluating immunological disorders, some general diagnostic tests are required for detecting pathologies generally associated with immunodeficiencies. Complete blood cell count, tuberculin test, bone marrow biopsy, histopathological studies, liver function test, blood chemistry, tumoral markers, cytokine levels, chest X-ray, ultrasound, CT scan, fluorescent in situ hybridization, and genetic testing are some commonly recommended diagnostic tests.[74]

Treatment and management

Primary and secondary immunodeficiencies are very complicated and affect multiple organ systems. However, several treatment and management strategies have evolved that facilitate better prognosis and improved quality of life.

Immunoglobulin therapy, also known as IVIg therapy, is the most prescribed management strategy in various immunodeficiencies. A mixture of antibodies, generally comprising of gamma globulins, is administered intravenously to facilitate effective elicitation of immune responses in individuals affected by various forms of immunodeficiencies.[75]

Transfer factors, also known as dialysable leukocyte extract, are protein molecules obtained from individuals who have developed immunity to a specific infection and are capable of transferring antigen-specific immunity to immunodeficient individuals. Several infections and chronic conditions, such as interstitial pneumonia, recurrent viral infections, chronic mucocutaneous candidiasis, primary TB, Wiskott – Aldrich syndrome, SCID, chronic active hepatitis, coccidioidomycosis, Behcet disease, aphthous stomatitis, familial keratoacanthoma, and certain malignancies, benefit from treatment with transfer factor.[76]

Antibiotics, antiviral agents, and antifungal agents are commonly used for treating bacterial, viral, and fungal infections in individuals affected by immunodeficiencies. Systemic inflammatory and autoimmune conditions, such as systemic lupus erythematosus, autoimmune polyendocrinopathy candidiasis ectodermal dystrophy, and autoimmune lymphoproliferative syndrome, require the use of immunosuppressants to ensure the elicitation of controlled immune responses that can provide effective protection while preventing inflammatory damage. Extensive forms of immunodeficiencies, such as RAG-1/RAG-2 SCID, adenosine deaminsase-SCID, Artemis-deficient SCID (ART-SCID), Wiskott – Aldrich syndrome, XLA, and acute leukemia, often benefit from bone marrow transplant.[77] Thymus transplant is recommended as an effective treatment strategy in DiGeorge syndrome.

Malignancies often require the use of cytokines, including interleukin-2, interleukin-7, interleukin-12, interleukin-18, and interleukin-21 for immunotherapy. Interferon gamma therapy is recommended in specific conditions, such as chronic granulomatous disease, bladder carcinoma, melanoma, Chagas disease, HIV/AIDS, and cryptococcal meningitis. Nutritional supplements comprising of vitamins A, C, E, and B6 and minerals like iron, zinc, selenium, and copper are also used as supportive therapy in immunodeficiency-associated conditions.

Clinical trials are ongoing for the use of ibrutinib, a Bruton tyrosine kinase inhibitor in the management of B-cell malignancies, such as relapsed or refractory chronic lymphocytic leukemia, small lymphocytic lymphoma, relapsed or refractory mantle cell lymphoma, and nongerminal center B-cell subtype of diffuse large B-cell lymphoma.[78]

Student notes

Key take-aways

1. Normal and effective immune responses are instrumental in eliminating life-threatening pathogenic infections and preventing hyperinflammatory autoimmune conditions.
2. Gene or protein defects and congenital abnormalities may result in deficiencies of various components of the immune system, thus affecting the effectiveness of immune responses.
3. Immunodeficiencies may be primary or secondary depending on the factors responsible for the condition.
4. Primary immunodeficiencies comprise of T-cell deficiency, B-cell deficiency, combined T-cell and B-cell deficiencies, complement deficiency, phagocyte deficiency, and IgA deficiency.
5. Secondary immunodeficiencies are those acquired during the lifetime of an individual, either due to the use of certain drugs that suppress the immune system or because of certain infections and conditions that compromise the normal functioning of the immune system.
6. Immunodeficiencies can present as a broad spectrum of clinical manifestations, depending on the specific condition.
7. The patient history is often indicative of reasons for a high index of clinical suspicion for immunodeficiencies.
8. Recurrence of rare and opportunistic infections, chronic inflammatory conditions, congenital defects, and hematologic abnormalities are hallmark signs of immunodeficiencies.
9. Immunodeficiencies are diagnosed primarily based on immunological investigations, although specific forms of immunodeficiencies require targeted diagnostic methods.
10. Although primary and secondary immunodeficiencies can result in life-threatening complications that affect multiple organ systems, the judicial use of treatment and management strategies can improve the prognosis in affected individuals.

References

1. Davies J, Davies D. Origins and evolution of antibiotic resistance. *Microbiol Mol Biol Rev.* 2010;74(3):417–433. Available from: https://doi.org/10.1128/MMBR.00016-10.

2. Pankey GA, Sabath LD. Clinical relevance of bacteriostatic vs bactericidal mechanisms of action in the treatment of Gram-positive bacterial infections. *Clin Infect Dis.* 2004;38 (6):864–870. Available from: https://doi.org/10.1086/381972.

3. Nemeth J, Oesch G, Kuster SP. Bacteriostatic vs bactericidal antibiotics for patients with serious bacterial infections: systematic review and *meta*-analysis. *J Antimicrob Chemother.* 2015;70 (2):382–395. Available from: https://doi.org/10.1093/jac/dku379.

4. Calhoun C, Wermuth HR, Hall GA. *Antibiotics. StatPearls*. Treasure Island (FL): StatPearls Publishing; 2021, January. Available from: https://www.ncbi.nlm.nih.gov/books/NBK535443/.

5. Sy SK, Zhuang L, Derendorf H. Pharmacokinetics and pharmacodynamics in antibiotic dose optimization. *Expert Opin Drug Metab Toxicol*. 2016;12(1):93–114. Available from: https://doi.org/10.1517/17425255.2016.1123250.

6. Levison ME, Levison JH. Pharmacokinetics and pharmacodynamics of antibacterial agents. *Infect Dis Clin North Am*. 2009;23(4). Available from: https://doi.org/10.1016/j.idc.2009.06.008. 791-vii.

7. Leekha S, Terrell CL, Edson RS. General principles of antimicrobial therapy. *Mayo Clin Proc*. 2011;86(2):156–167. Available from: https://doi.org/10.4065/mcp.2010.0639.

8. Lynch TJ. Choosing optimal antimicrobial therapies. *Med Clin North Am*. 2012;96(6):1079–1094. Available from: https://doi.org/10.1016/j.mcna.2012.08.006.

9. Gruchalla RS, Pirmohamed M. Clinical practice. Antibiotic allergy. *N Engl J Med*. 2006;354(6):601–609. Available from: https://doi.org/10.1056/NEJMcp043986.

10. Granowitz EV, Brown RB. Antibiotic adverse reactions and drug interactions. *Crit Care Clin*. 2008;24(2). Available from: https://doi.org/10.1016/j.ccc.2007.12.011. 421-xi.

11. Alanis AJ. Resistance to antibiotics: are we in the post-antibiotic era? *Arch Med Res*. 2005;36(6):697–705. Available from: https://doi.org/10.1016/j.arcmed.2005.06.009.

12. Zaman SB, Hussain MA, Nye R, Mehta V, Mamun KT, Hossain N. A review on antibiotic resistance: alarm bells are ringing. *Cureus*. 2017;9(6):e1403. Available from: https://doi.org/10.7759/cureus.1403. Published June 28, 2017.

13. Rutala WA, Weber DJ, Society for Healthcare Epidemiology of America. Guideline for disinfection and sterilization of prion-contaminated medical instruments. *Infect Control Hosp Epidemiol*. 2010;31(2):107–117. Available from: https://doi.org/10.1086/650197.

14. Spaulding EH. Chemical disinfection of medical and surgical materials. In: Lawrence C, Block SS, eds. *Disinfection, Sterilization, and Preservation*. Philadelphia (PA): Lea & Febiger; 1968:517–531.

15. Favero MS, Bond WW. Chemical disinfection of medical and surgical materials. In: Block SS, ed. *Disinfection, Sterilization, and Preservation*. 5th ed. Philadelphia (PA): Lippincott Williams & Wilkins; 2001:881–917.

16. Laboratory centres for the Disease Control Bureau of Infectious Diseases. Laboratory centre for Disease Control Bureau of Infectious Diseases; Ottawa (Canada): 1998. Infection control guidelines. Communicable disease report.

17. WHO. *Practical Guidelines for Infection Control in Health Care Facilities*. Geneva (Switzerland): WHO; 2003.

18. Mohapatra S. Sterilization and disinfection. *Essent Neuroanesth*. 2017;929–944. Available from: https://doi.org/10.1016/B978-0-12-805299-0.00059-2.

19. Steiner AA, Ivanov AI, Serrats J, et al. Cellular and molecular bases of the initiation of fever. *PLoS Biol*. 2006;4(9):e284. Available from: https://doi.org/10.1371/journal.pbio.0040284.

20. Elmquist JK, Scammell TE, Saper CB. Mechanisms of CNS response to systemic immune challenge: the febrile response. *Trends Neurosci*. 1997;20(12):565–570. Available from: https://doi.org/10.1016/s0166-2236(97)01138-7.

21. Coceani F, Akarsu ES. Prostaglandin E2 in the pathogenesis of fever. An update. *Ann N Y Acad Sci*. 1998;856:76–82. Available from: https://doi.org/10.1111/j.1749-6632.1998.tb08315.x.

22. Ivanov AI, Romanovsky AA. Prostaglandin E2 as a mediator of fever: synthesis and catabolism. *Front Biosci*. 2004;9:1977–1993. Available from: https://doi.org/10.2741/1383. Published 2004 May 1.

23. Murakami M, Kudo I. Recent advances in molecular biology and physiology of the prostaglandin E2-biosynthetic pathway. *Prog Lipid Res*. 2004;43(1):3–35. Available from: https://doi.org/10.1016/s0163-7827(03)00037-7.

24. Justiz Vaillant AA, Qurie A. *Interleukin* August 30 *StatPearls*. Treasure Island (FL): StatPearls Publishing; 2020.

25. Balli S, Sharan S. *Physiology, fever*. *StatPearls*. Treasure Island (FL): StatPearls Publishing; 2021, Jan-. Available from. Available from: https://www.ncbi.nlm.nih.gov/books/NBK562334.

26. Przybyła GW, Szychowski KA, Gmiński J. Paracetamol—an old drug with new mechanisms of action [Published online ahead of print, August ·7, 2020] *Clin Exp Pharmacol Physiol*. 2020;. Available from: https://doi.org/10.1111/1440-1681.13392.

27. DuPont HL, Levine MM, Hornick RB, Formal SB. Inoculum size in Shigellosis and implications for expected mode of transmission. *J Infect Dis*. 1989;159(6):1126–1128. Available from: https://doi.org/10.1093/infdis/159.6.1126.

28. Hines JZ, Pinsent T, Rees K, et al. Notes from the field: Shigellosis outbreak among men who have sex with men and homeless persons—Oregon, 2015 – 2016. *MMWR Morb Mortal Wkly Rep*. 2016;65(31):812–813. Available from: https://doi.org/10.15585/mmwr.mm6531a5. Published August 12, 2016.

29. Centers for Disease Control and Prevention (CDC). *Shigella flexneri* serotype 3 infections among men who have sex with men–Chicago, Illinois, 2003 – 2004. *MMWR Morb Mortal Wkly Rep*. 2005;54(33):820–822.

30. Small P, Blankenhorn D, Welty D, Zinser E, Slonczewski JL. Acid and base resistance in *Escherichia coli* and *Shigella flexneri*: role of rpoS and growth pH. *J Bacteriol*. 1994;176 (6):1729–1737. Available from: https://doi.org/10.1128/jb.176.6.1729-1737.1994.

31. Jennison AV, Verma NK. *Shigella flexneri* infection: pathogenesis and vaccine development. *FEMS Microbiol Rev*. 2004;28(1):43–58. Available from: https://doi.org/10.1016/j.femsre.2003.07.002.

32. Zychlinsky A, Thirumalai K, Arondel J, Cantey JR, Aliprantis AO, Sansonetti PJ. In vivo apoptosis in *Shigella flexneri* infections. *Infect Immun*. 1996;64(12):5357–5365. Available from: https://doi.org/10.1128/IAI.64.12.5357-5365.1996.

33. Sansonetti PJ, Arondel J, Huerre M, Harada A, Matsushima K. Interleukin-8 controls bacterial transepithelial translocation at the cost of epithelial destruction in experimental shigellosis. *Infect Immun*. 1999;67(3):1471–1480. Available from: https://doi.org/10.1128/IAI.67.3.1471-1480.1999.

34. Aslam A, Okafor CN. *Shigella*. *StatPearls*. Treasure Island (FL): StatPearls Publishing; 2020, January. Available from. Available from: https://www.ncbi.nlm.nih.gov/books/NBK482337/.

35. Biet F, Locht C, Kremer L. Immunoregulatory functions of interleukin 18 and its role in defense against bacterial pathogens. *J Mol Med (Berl)*. 2002;80(3):147–162. Available from: https://doi.org/10.1007/s00109-001-0307-1.

36. Way SS, Borczuk AC, Dominitz R, Goldberg MB. An essential role for gamma interferon in innate resistance to *Shigella flexneri* infection. *Infect Immun*. 1998;66(4):1342–1348. Available from: https://doi.org/10.1128/IAI.66.4.1342-1348.1998.

37. Gomez HF, Ochoa TJ, Carlin LG, Cleary TG. Human lactoferrin impairs virulence of *Shigella flexneri*. *J Infect Dis*. 2003;187(1):87–95. Available from: https://doi.org/10.1086/345875.

38. Salzman NH, Ghosh D, Huttner KM, Paterson Y, Bevins CL. Protection against enteric salmonellosis in transgenic mice expressing a human intestinal defensin. *Nature*. 2003;422 (6931):522–526. Available from: https://doi.org/10.1038/nature01520.

39. Mathan VI, Mathan MM. Intestinal manifestations of invasive diarrheas and their diagnosis. *Rev Infect Dis*. 1991;13(Suppl 4):S311–S313. Available from: https://doi.org/10.1093/clinids/13.supplement_4.s311.

40. Niyogi SK. Shigellosis. *J Microbiol.* 2005;43(2):133–143.

41. Erqou S, Teferra E, Mulu A, Kassu A. A case of shigellosis with intractable septic shock and convulsions. *Jpn J Infect Dis.* 2007;60(5):314–316.

42. Ferrera PC, Jeanjaquet MS, Mayer DM. Shigella-induced encephalopathy in an adult. *Am J Emerg Med.* 1996;14(2):173–175. Available from: https://doi.org/10.1016/S0735-6757(96)90127-8.

43. Azad MA, Islam M, Butler T. Colonic perforation in *Shigella dysenteriae* 1 infection. *Pediatr Infect Dis.* 1986;5(1):103–104. Available from: https://doi.org/10.1097/00006454-198601000-00019.

44. Bennish ML, Azad AK, Yousefzadeh D. Intestinal obstruction during shigellosis: incidence, clinical features, risk factors, and outcome. *Gastroenterology.* 1991;101(3):626–634. Available from: https://doi.org/10.1016/0016-5085(91)90518-p.

45. Khan WA, Griffiths JK, Bennish ML. Gastrointestinal and extra-intestinal manifestations of childhood shigellosis in a region where all four species of Shigella are endemic. *PLoS One.* 2013;8(5):e64097. Available from: https://doi.org/10.1371/journal.pone.0064097. Published May 17, 2013.

46. Martin T, Habbick BF, Nyssen J. Shigellosis with bacteremia: a report of two cases and a review of the literature. *Pediatr Infect Dis.* 1983;2(1):21–26.

47. Struelens MJ, Patte D, Kabir I, Salam A, Nath SK, Butler T. Shigella septicemia: prevalence, presentation, risk factors, and outcome. *J Infect Dis.* 1985;152(4):784–790. Available from: https://doi.org/10.1093/infdis/152.4.784.

48. Siegler RL. The hemolytic uremic syndrome. *Pediatr Clin North Am.* 1995;42(6):1505–1529. Available from: https://doi.org/10.1016/s0031-3955(16)40096-9.

49. Butler T, Islam MR, Bardhan PK. The leukemoid reaction in shigellosis. *Am J Dis Child.* 1984;138(2):162–165. Available from: https://doi.org/10.1001/archpedi.1984.02140400044010.

50. Ashkenazi S, Dinari G, Zevulunov A, Nitzan M. Convulsions in childhood shigellosis. Clinical and laboratory features in 153 children. *Am J Dis Child.* 1987;141(2):208–210. Available from: https://doi.org/10.1001/archpedi.1987.04460020098036.

51. Khan WA, Dhar U, Salam MA, Griffiths JK, Rand W, Bennish ML. Central nervous system manifestations of childhood shigellosis: prevalence, risk factors, and outcome. *Pediatrics.* 1999;103(2):E18. Available from: https://doi.org/10.1542/peds.103.2.e18.

52. Porter CK, Choi D, Riddle MS. Pathogen-specific risk of reactive arthritis from bacterial causes of foodborne illness. *J Rheumatol.* 2013;40(5):712–714. Available from: https://doi.org/10.3899/jrheum.121254.

53. Murphy TV, Nelson JD. *Shigella vaginitis*: report of 38 patients and review of the literature. *Pediatrics.* 1979;63(4):511–516.

54. Tobias JD, Starke JR, Tosi MF. *Shigella keratitis*: a report of two cases and a review of the literature. *Pediatr Infect Dis J.* 1987;6(1):79–81.

55. Fried D, Maytal J, Hanukoglu A. The differential leukocyte count in shigellosis. *Infection.* 1982;10(1):13–14. Available from: https://doi.org/10.1007/BF01640829.

56. Halpern Z, Averbuch M, Dan M, Giladi M, Levo Y. The differential leukocyte count in adults with acute gastroenteritis. *Scand J Infect Dis.* 1992;24(2):205–207. Available from: https://doi.org/10.3109/00365549209052613.

57. Bennish ML, Salam MA, Wahed MA. Enteric protein loss during shigellosis. *Am J Gastroenterol.* 1993;88(1):53–57.

58. DuPont HL, Hornick RB. Adverse effect of lomotil therapy in shigellosis. *JAMA.* 1973;226(13):1525–1528.

59. Rahman M, Shoma S, Rashid H, et al. Increasing spectrum in antimicrobial resistance of Shigella isolates in Bangladesh: resistance to azithromycin and ceftriaxone and decreased susceptibility to ciprofloxacin. *J Health Popul Nutr.* 2007;25(2):158–167.

60. Salam MA, Dhar U, Khan WA, Bennish ML. Randomised comparison of ciprofloxacin suspension and pivmecillinam for childhood shigellosis. *Lancet.* 1998;352(9127):522–527. Available from: https://doi.org/10.1016/S0140-6736(97)11457-X.

61. Lambert MP, Arulselvan A, Schott A, et al. The 22q11.2 deletion syndrome: cancer predisposition, platelet abnormalities and cytopenias. *Am J Med Genet A.* 2018;176(10):2121–2127. Available from: https://doi.org/10.1002/ajmg.a.38474.

62. Yazdani R, Fekrvand S, Shahkarami S, et al. The hyper IgM syndromes: epidemiology, pathogenesis, clinical manifestations, diagnosis and management. *Clin Immunol.* 2019;198:19–30. Available from: https://doi.org/10.1016/j.clim.2018.11.007.

63. Hernandez-Trujillo VP, Scalchunes C, Cunningham-Rundles C, et al. Autoimmunity and inflammation in X-linked agammaglobulinemia. *J Clin Immunol.* 2014;34(6):627–632. Available from: https://doi.org/10.1007/s10875-014-0056-x.

64. Chinn IK, Shearer WT. Severe combined immunodeficiency disorders. *Immunol Allergy Clin North Am.* 2015;35(4):671–694. Available from: https://doi.org/10.1016/j.iac.2015.07.002.

65. Rothblum-Oviatt C, Wright J, Lefton-Greif MA, McGrath-Morrow SA, Crawford TO, Lederman HM. Ataxia telangiectasia: a review. *Orphanet J Rare Dis.* 2016;11(1):159. Available from: https://doi.org/10.1186/s13023-016-0543-7. Published November 25, 2016.

66. Reith W, Mach B. The bare lymphocyte syndrome and the regulation of MHC expression. *Annu Rev Immunol.* 2001;19:331–373. Available from: https://doi.org/10.1146/annurev.immunol.19.1.331.

67. Arnold DE, Heimall JR. A review of chronic granulomatous disease. *Adv Ther.* 2017;34(12):2543–2557. Available from: https://doi.org/10.1007/s12325-017-0636-2.

68. Schröder-Braunstein J, Kirschfink M. Complement deficiencies and dysregulation: pathophysiological consequences, modern analysis, and clinical management. *Mol Immunol.* 2019;114:299–311. Available from: https://doi.org/10.1016/j.molimm.2019.08.002.

69. Bourke CD, Berkley JA, Prendergast AJ. Immune dysfunction as a cause and consequence of malnutrition. *Trends Immunol.* 2016;37(6):386–398. Available from: https://doi.org/10.1016/j.it.2016.04.003.

70. Murata C, Ramírez AB, Ramírez G, Cruz A, Morales JL, Lugo-Reyes SO. Análisis discriminante para predecir el diagnóstico clínico de inmunodeficiencias primarias: reporte preliminar [Discriminant analysis to predict the clinical diagnosis of primary immunodeficiencies: a preliminary report]. *Rev Alerg Mex.* 2015;62(2):125–133.

71. Routes J, Abinun M, Al-Herz W, et al. ICON: the early diagnosis of congenital immunodeficiencies. *J Clin Immunol.* 2014;34(4):398–424. Available from: https://doi.org/10.1007/s10875-014-0003-x.

72. Barandun S. Clinical manifestations of immunodeficiency disorders. *Monogr Hum Genet.* 1972;6:160–171. Available from: https://doi.org/10.1159/000392686.

73. Justiz Vaillant AA, Qurie A. *Immunodeficiency. StatPearls.* Treasure Island (FL): StatPearls Publishing; 2021, January. Available from: https://www.ncbi.nlm.nih.gov/books/NBK500027/.

74. Locke BA, Dasu T, Verbsky JW. Laboratory diagnosis of primary immunodeficiencies. *Clin Rev Allergy Immunol.* 2014;46(2):154–168. Available from: https://doi.org/10.1007/s12016-014-8412-4.

75. Jolles S, Sewell WA, Misbah SA. Clinical uses of intravenous immunoglobulin. *Clin Exp Immunol.* 2005;142(1):1–11. Available from: https://doi.org/10.1111/j.1365-2249.2005.02834.x.

76. Rocklin RE. Use of transfer factor in patients with depressed cellular immunity and chronic infection. *Birth Defects Orig Artic Ser.* 1975;11(1):431–435.

77. Shamriz O, Chandrakasan S. Update on advances in hematopoietic cell transplantation for primary immunodeficiency disorders. *Immunol Allergy Clin North Am.* 2019;39(1):113–128. Available from: https://doi.org/10.1016/j.iac.2018.08.003.

78. Ponader S, Burger JA. Bruton's tyrosine kinase: from X-linked agammaglobulinemia toward targeted therapy for B-cell malignancies. *J Clin Oncol.* 2014;32(17):1830–1839. Available from: https://doi.org/10.1200/JCO.2013.53.1046.

CHAPTER 2

Infections of the central and sensory nervous system

Problem number 2.1: Tetanus caused by *Clostridium tetani* infection

A 28-year-old male presented with a complaint of general body ache, stiffness, excessive sweating, palpitation, and difficulty in swallowing solids and liquids for 1 week. He also mentioned that a needle had penetrated his right foot about 10 days ago. He had visited a healthcare facility before this presentation; however, he had not been provided proper wound management.

He did not report having any known chronic medical illnesses and was not taking any medications. His history of previous vaccination schedules was, however, unknown.

The following are the reports of the investigations conducted:

1. Vitals
   Blood pressure: Normal
   Heart rate: 105/min (tachycardic)
   Breathing pattern: Normal.
2. Physical examination
   Respiratory system: Maintaining an open airway with a normal breathing pattern
   Gastrointestinal system: Normal
   Cardiovascular system: Normal
   Neurological examination
   Conscious, alert, oriented, and following commands, albeit with some communication problems probably attributable to some language barrier, lower limb stiffness without any signs of upper limb stiffness
   Power examination of the upper limbs: 5/5 power
   Power examination of the lower limbs: Stiffness prevented power examination

CONTENTS

Problem-Based Learning Approach in Microbiology. DOI: https://doi.org/10.1016/B978-0-323-95092-3.00007-X

Reflexes: Reduced in the lower limbs, normal in the upper limbs
No meningeal signs were observed
Normal pupils and eye movement
Spatula test: Positive
A small clean scar was seen on the sole of his right foot.

3. Creatine kinase level: High.
4. Radiological examination
   X-ray: Normal.

Diagnosis

The final diagnosis was that of tetanus caused by *C. tetani* infection.

Treatment

The patient was admitted to the intensive care unit (ICU).

The following courses of treatment were given based on the reports of the investigations:

1. Elective intubation to avoid the risk of spasming of respiratory muscles and upper airway with subsequent depression.
2. Sedation using Dormicum (Midazolam) and Nimbex (Cisatracurium Besyalate) infusions.
3. Nasogastric (NGT) feed calculated as per the dietary requirements of the patient.
4. Penicillin G (0.5 million Units, IV Q4 hourly) × 14 days.
5. Flagyl (Metronidazole 500 mg, IV Q8 hourly) × 14 days.
6. Magnesium sulfate.
7. DVT prophylaxis.
8. Immunoglobulin (IVIG, intravenous immunoglobulin 400 mg/kg/day for 5 days).
9. Propofol (Diprivan) and Fentanyl infusions to control patient's spasms.
10. Several septic workups were conducted during the ICU stay and antibiotic regimes were modified accordingly.
    Antibiotics: Meropenem, Azithromycin, and Colistin was given as per the patient's culture sensitivity reports.
11. Regular monitoring of heart rate and blood pressure.
12. Tetanus vaccination.

Question and answer

1. What is tetanus?
   Tetanus, also known as "lockjaw" is a bacterial infection caused by *C. tetani*. The infection is mediated by a poison (toxin) produced by the bacteria that compromise the normal functioning of muscles, often primarily resulting in a stiffening of the neck and jaw muscles.
2. What are the risk factors instrumental that facilitate the development of a tetanus infection?
   Inadequate or a complete lack of immunization, lack of timely booster doses (every 10 years), presence of a puncture wound, presence of foreign bodies such as a splinter or a nail, coinfection with other pathogens, lack of appropriate treatment for necrotic or gangrenous tissue, injuries caused due to crushing trauma, road accidents, burns, exposure to bats, dental and surgical procedures lacking important sterilization procedures, animal bites, inappropriately performed abortions and circumcisions, intravenous drug use and tattooing, and untreated foot ulcers and open wounds are some of the risk factors that facilitate the development of a tetanus infection.
3. How has the inclusion of the tetanus vaccine in routine immunization affected the incidence of tetanus?
   Tetanus is a vaccine-preventable disease. The inclusion of the tetanus vaccine in routine immunization has led to a dramatic decline in the incidence of the disease. Most cases of tetanus are seen to occur in unvaccinated individuals.
4. Which are the toxins produced by the *C. tetani* bacteria?
   The *C. tetani* bacteria produces two types of toxins that are antigenically and pharmacologically distinct. Tetanolysin, mediates local tissue destruction, while tetanospasmin is a powerful neurotoxin responsible for clinical tetanus.
5. Why do the clinical manifestations of tetanus differ from individual to individual?
   Tetanolysin and tetanospasmin are the two antigenically and pharmacologically distinct toxins that are produced by *C. tetani* in a mutually independent manner. The differences in the physiological and pathological effects mediated by these toxins cause the clinical manifestations of the disease to vary from individual to individual.
6. What are the factors that determine the incubation period of tetanus?
   The incubation period of tetanus varies from 2 days to several weeks, generally being 6–12 days. The site and nature of the wound, the dose, virulence of the contaminating organism, and the immunity of the patient, are all factors that contribute to the determination of the incubation period.

7. How is the prognosis affected by the incubation period of the disease in tetanus?
   A long incubation period is usually associated with a milder form of the disease leading to a better prognosis, while a shorter incubation period is usually associated with a grave prognosis.
8. Which is the most commonly seen first symptom of tetanus?
   Trismus, also known as "lockjaw" resulting from the stiffening of the neck and jaw muscles is the most commonly seen first symptom of tetanus.
9. What is the classification of tetanus based on the clinical manifestations of the disease?
   Tetanus can be classified into generalized tetanus, local tetanus, and cephalic tetanus. Generalized tetanus manifests autonomic overactivity symptoms such as irritability, restlessness, sweating, and tachycardia in the initial stages of the disease, followed by manifestations such as trismus, opisthotonus (backward arching of the head, neck, and spine), dysphagia, risus sardonicus (sustained spasm of facial muscles that resembles a grin), nuchal rigidity, systemic rigidity, and tonic contractions in the clinical stage of the disease. Dyspnea, airway obstruction, and apnea caused by a compromised airway is also a common presentation of the disease. Excruciating pain and tendon ruptures and fractures are often caused by extreme muscle contractions. Local tetanus, on the other hand, presents with muscle contractions localized to the affected region of the body. Cephalic tetanus is a specific form of local tetanus that primarily affects the nervous system, presenting with trismus, dysphagia, and focal cranial neuropathies.
10. Why is elective intubation considered in tetanus?
    Elective intubation is often recommended in progressive tetanus as a preventive measure against the risk of spasming of respiratory muscles and upper airway with subsequent depression. Elective intubation ensures effective respiratory functions, even in the event of respiratory depression.
11. What is the role of TIG in the treatment of tetanus?
    TIG is an antitetanus immunoglobulin that is administered as an infusion to inactivate the unbound toxin and prevent any further toxin production. However, TIG may not be effective in neutralizing the toxin that is already bound to the nervous tissue.
12. Does a tetanus infection confer life-long immunity from further infections?
    A tetanus infection does not confer life-long immunity from further tetanus infections, thereby emphasizing the importance of a full course of active immunization despite getting a tetanus infection.

Learning objectives

1. Describe the clinical manifestations that lead to the diagnosis of tetanus before laboratory diagnostic reports are obtained.
2. Determine the logical flow of differential diagnosis.
3. Describe the characteristics of *C. tetani,* the etiological agent of tetanus.
4. Predict the prognosis of the disease based on the clinical manifestations of the case.
5. Interpret the correlation between the clinical manifestations and the pathogenicity of *C. tetani.*
6. Justify the importance of active immunization and regular administration of booster doses for the prevention of tetanus.

Discussion

Differential diagnosis

Etiology	Rationale
Drug-induced dystonia (such as those due to phenothiazones)	Drug-induced dystonia presents with a deviation of eyes, writhing movements of the head and neck, and an absence of tonic muscular contraction between spasms. Tetanus, however, does not cause eye deviations and the muscles show characteristic tonic contraction between spasms. The administration of anticholinergic medications can reverse the spasms due to drug-induced dystonia.
Trismus caused by a dental infection	Trismus caused by a dental infection presents with an obvious dental abscess, and the lack of progression or superimposed spasms usually distinguish it from tetanus.
Strychnine poisoning due to ingestion of rat poison	Accidental or intentional strychnine poisoning often presents with a clinical syndrome similar to tetanus. Blood, urine, and tissue investigations, however, can differentiate between the two conditions.
Malignant neuroleptic syndrome	The malignant neuroleptic syndrome presents with fever, muscle rigidity, autonomic dysfunction, and an altered mental status. Patient history of administration of antipsychotics that are known to induce the syndrome can help in the differential.
Stiff-person syndrome	Stiff-person syndrome presents with clinical manifestations similar to tetanus; however, the rapid response to benzodiazepines characteristic of this condition distinguishes the spasms seen in stiff-person syndrome from those seen in tetanus.

Background

Tetanus is a leading cause of morbidity and mortality in developing countries.[1] The introduction of the tetanus toxoid in routine childhood vaccination in the 1940s resulted in a steady decline in incidence rates, and the death-to-case ratio has also shown a drastic decline in recent years.[2]

Tetanus toxoid was first produced in 1924 and consists of a formaldehyde-treated toxin. There are two types of toxoids available—adsorbed (aluminum salt precipitated) toxoid and fluid toxoid. Although the rates of seroconversion are about equal, the adsorbed toxoid is preferred because the antitoxin response reaches higher titers and is longer lasting than that elicited by the fluid toxoid,[3] although the antitoxin levels decrease with time. While some individuals may be protected for life, most individuals show minimally protective antitoxin levels by 10 years after the last dose. Hence routine boosters are recommended every 10 years.[4]

Field injuries are common among outdoor workers, and lack of awareness of the injury consequences and the importance of seeking medical attention is a major reason behind the rise in morbidity/mortality rates. In addition, diagnosis of tetanus depends heavily on clinical manifestations, and therefore early detection is important for the effective treatment of the disease. Hence, awareness about the disease and its clinical manifestations among physicians is very important to ensure accurate diagnosis and effective treatment of the disease. In developed countries where the incidence is low due to vaccine programs, any new case should be reported immediately. This helps those unexposed to familiarize themselves with the disease, facilitate further management, and develop approaches for a superior primary prevention protocol. The importance of vaccinating people at risk of infection as it is one of the diseases that does not confer immunity postinfection is also emphasized.

Etiological agent

Clostridium tetani

Classification and morphology

C. tetani is an anaerobic Gram-positive bacillus belonging to the *Clostridiaceae* family. The bacterium is characterized by straight parallel sides and rounded ends, which usually occurs singly and occasionally in chains. The bacillus has a characteristic "drumstick" appearance due to the terminal spherically shaped bulging spores (Figs. 2.1 and 2.2). Peritrichous flagella aid in the motility of the noncapsulated bacillus. The Gram staining characteristics of this organism are very peculiar and depend on the age of the culture. While young cultures are strongly Gram-positive, older cells show variable staining and may even be Gram-negative.

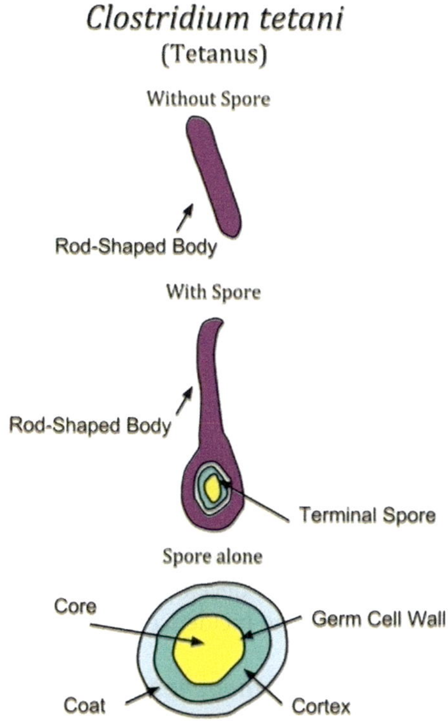

FIGURE 2.1

Representative image of *Clostridium tetani* showing morphology with and without spores. Source*: From Kholmes16/Own work/; https://creativecommons.org/licenses/by-sa/4.0/deed.en; https://commons.wikimedia.org/wiki/File:Clostridium_Tetani.svg.*

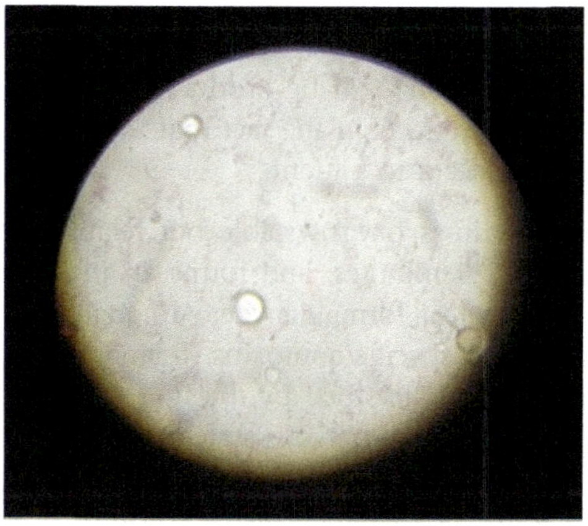

FIGURE 2.2

Clostridium tetani in culture medium. Source*: From Netha Hussain/Own work/; https://creativecommons.org/licenses/by-sa/4.0/deed.en; https://commons.wikimedia.org/wiki/File:Clostridium_tetani_in_culture_medium.jpg.*

Ten serological types of *C. tetani* have been identified (types I – X), of which type VI contains nonflagellated strains while all other types possess type-specific flagellar antigens. The toxin produced by all the serological types is the same and can be neutralized by the antitoxin produced against any one type.[5]

Epidemiology

Tetanus is a disease affecting the nervous system and is characterized by muscle spasms caused by a toxin-producing bacteria, *C. tetani*. The causative organism is widely distributed in soil and the intestines of human beings and animals. This organism is ubiquitous and has been recovered from street and hospital dust, plaster of Paris, unclean bandages and cotton wool, wall plaster, rusted equipment/surfaces, and catgut. This organism usually occurs as a contaminant in apparently harmless wounds.

The disease usually presents four different clinical manifestations: neonatal, cephalic, localized, and generalized. Although tetanus is now rare in the developed world, the disease remains a threat for all unvaccinated individuals, particularly in developing countries. Universal vaccination of children with tetanus toxoid in developed countries has resulted in a dramatic and steady reduction in the incidence of tetanus in these regions since 1940.

Pathogenicity and pathogenesis

The spores of *C. tetani* germinate and produce toxins under favorable conditions, such as anaerobic or low oxygen conditions, which develop in injured tissues consisting of large numbers of dead cells. The bacteria colonize small, apparently harmless wounds and produce two antigenically and pharmacologically distinct toxins, namely tetanolysin, responsible for local tissue destruction and a powerful neurotoxin (tetanospasmin), which causes clinical tetanus. The production of these two toxins is mutually independent and therefore, one can be produced with or without the production of the other, which reflects in the fact that the clinical manifestations of the disease vary from patient to patient.

Tetanospasmin is oxygen stable but relatively heat-labile, being inactivated at 65°C within 5 minutes, and forms an inactive toxoid in the presence of low concentrations of formaldehyde. Although it is an effective antigen, it can be neutralized by specific antitoxins. Tetanospasmin is produced locally at the site of injury and is absorbed by the motor nerve endings, eventually being transported to the central nervous system (CNS) where it targets the spinal cord. The toxin resembles strychnine in its effects, differing in its presynaptic site of action as against the postsynaptic site of action of strychnine. The tetanus toxin targets and blocks synaptic inhibition in the spinal cord by interfering with the neurotransmitters, resulting in the uncontrolled spread of impulses generated in the CNS. The clinical manifestations characteristic of tetanus such as muscle rigidity, uncontrolled muscle contractions, tonic muscle spasms, and severe seizures seen in some cases are a consequence of the lack of reciprocal inhibition.

Trivial puncture wounds or lapses in aseptic protocols during surgeries, result in tissue vulnerability leading to colonization by *C. tetani* followed by production and release of the toxin under favorable conditions. The incubation period varies from 2 days to several weeks, generally being 6–12 days. The site and nature of the wound, the dose, virulence of the contaminating organism, and the immunity of the patient, are all factors important for the determination of the incubation period. The prognosis of the disease varies depending on the incubation period, the prognosis being grave when the incubation period is short. Neonatal tetanus, often attributable to unsterile transection of the umbilical cord following delivery in unvaccinated mothers, has a short incubation period. Infants present with irritability and develop severe rigidity within the first week of infection. Tetanus generally culminates in a poor outcome in infants. The period of disease onset, that is the interval between the first symptom of the disease, usually, trismus, commonly known as "lockjaw" (Fig. 2.3) and the onset of the spasms is also of prognostic significance.

Resistance

C. tetani spores show varied resistance to heat subject to strain differences. While most are killed by boiling for 10–15 minutes, some resist boiling for up to 3 hours. Therefore autoclaving at 121°C for 20 minutes is recommended for ensuring destruction of spores. The spores are very resilient and can survive in soil for years, and are resistant to most antiseptics. Exposure to iodine (1% aqueous solution) and hydrogen peroxide (10 volumes) kill the spores within a few hours and are therefore considered more effective than 5% phenol or 0.1% mercuric chloride solutions.

Clinical manifestations

Generalized tetanus is characterized by peculiar clinical manifestations, such as trismus, opisthotonus (backward arching of the head, neck, and spine), dysphagia, risus sardonicus (sustained spasm of facial muscles that resembles a grin), nuchal rigidity, systemic rigidity, and tonic contractions. Compromised airway manifesting as dyspnea, eventually resulting in airway obstruction and apnea is also a common presentation of the disease. These extreme muscle contractions cause excruciating pain, often precipitating tendon rupture and fractures. The clinical manifestations are preceded by an early phase manifesting autonomic overactivity symptoms, such as irritability, restlessness, sweating, and tachycardia.

Local tetanus is relatively rare and presents with muscle contractions in the localized affected region of the body. Cephalic tetanus is an extremely rare and specific form of local tetanus affecting the nervous system and is

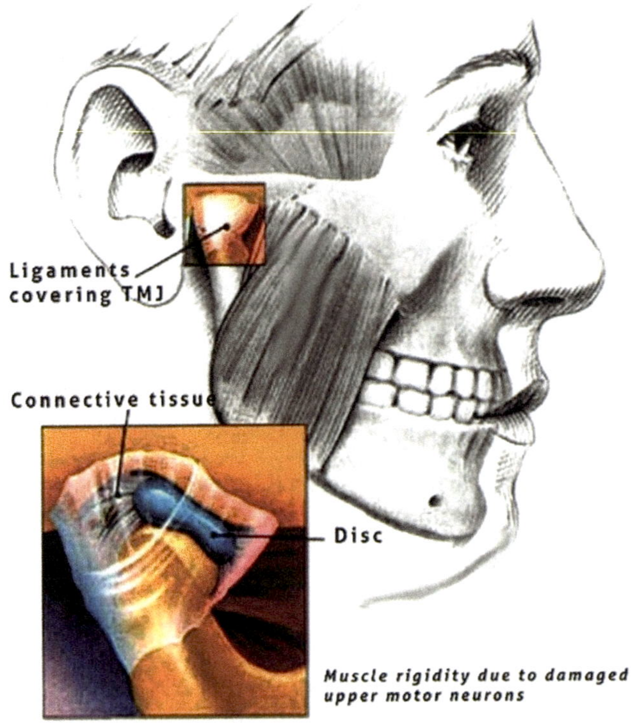

FIGURE 2.3
Representative image showing damage to the upper motor neurons responsible for muscular rigidity leading to lockjaw. Source*: From Paul/; http://www.lockjawsymptoms.net/; https://creativecommons.org/licenses/by/3.0/deed.en; https://commons.wikimedia.org/wiki/File: Lockjawsymptoms.jpg.*

generally preceded by an ear infection or a head trauma. This form of tetanus presents with trismus, dysphagia, and focal cranial neuropathies.[6]

Diagnosis

The diagnosis of tetanus largely depends on clinical grounds and laboratory tests can only be used for confirmation, as failure to establish a laboratory diagnosis for tetanus is not uncommon. While microscopy may demonstrate the presence of the typical "drumstick" bacilli in wounds, this is not a reliable diagnosis of the disease itself as the bacilli may be present in some wounds without tetanus developing. The microscopic distinction between *C. tetani* and morphologically similar bacilli can also be a diagnostic challenge. Diagnosis by culture, however, is more dependable as the isolation of the organism from excised bits of tissue from necrotic wounds is more likely than from wound swabs.

Treatment

Convulsions in tetanus patients are known to be invoked by exposure to noise and light, thus requiring patients to be treated preferably in special units in hospitals. Although the patients may be isolated for clinical reasons, person to person transmission of tetanus is not a matter of concern.

Treatment consists of clinical practices to ensure controlled spasms and a maintained airway. Elective intubation may be considered to ensure effective respiratory functions, in anticipation of subsequent respiratory depression characteristic of the disease progression. Appropriate NGT feed is recommended to maintain a good nutritional status.

TIG infusion may be administered to inactivate the unbound toxin and prevent any further toxin production, even though TIG may not neutralize the toxin already bound to the nervous tissue. Rapid initiation of antibiotics and continued regimes as required is crucial for effective treatment.

Patients recovering from tetanus infection require a full course of active immunization as one infection with *C. tetani* does not confer immunity to further potential infections.

Prevention

Despite the ubiquitous nature of the causative organism, tetanus is a preventable disease. The clinical manifestations of the disease are attributable to the action of the toxin. Therefore prevention aims at the development of antitoxic immunity through active immunization and administration of appropriate booster doses consisting of intramuscular injections of tetanus toxoid.

Surgical prophylaxis aimed at the removal of the foreign bodies, necrotic tissue, and blood clots to prevent an anaerobic environment favorable for the tetanus bacillus is recommended in cases of severe injury or when the patient is immunocompromised.

Antibiotic prophylaxis to prevent the production of the toxin by destruction or inhibition of the tetanus bacilli and pyogenic bacteria in wounds may also be considered depending on the severity of the injury and the immunity of the patient.

Passive immunization using tetanus antitoxin such as TIG can prove useful not only in reducing the incidence of tetanus but also in prolonging the incubation period and reducing mortality. However, passive immunization should be considered only in rare and emergency cases owing to complications related to its frequent use.

Student Notes

Key take-aways

- The differential diagnosis of tetanus is very broad and may be confusing. The disease prognosis depends largely on the quick initiation of effective treatment using the appropriate antitoxin and antibiotics.
- High doses of sedatives and muscle relaxants, as well as prolonged mechanical ventilation, are usually necessary and expected in most cases.
- Intensivists have an important role to play in the management of these patients.
- This is a disease that can be largely prevented by adequate active immunization and the administration of regular booster doses.
- Targeted active immunization of high-risk groups can facilitate the process of disease prevention.

Problem number 2.2: Shingles caused by herpes zoster virus infection

A 12-year-old boy with a history of relapsed acute lymphoblastic leukemia (ALL) undergoing chemotherapy presented with a lesion on his right cheek, which appeared about 1 week ago as a small, red, itchy, but painless pimple akin to a mosquito bite. Gradually over the week, the lesion increased in size up to 2×2 cm, and at the time of the presentation, it appeared necrotic and black in color, was nonitchy and painless, and was surrounded by a minimal red rim.

There were no other lesions elsewhere on the patient's body, and the patient was not febrile, although the possibility of a fever being masked due to the regular intake of dexamethasone (steroid) existed. The patient complained of mild throat pain with an occasional dry cough. There was no sign of gastrointestinal distress, such as vomiting or loose motions, and the patient had not encountered any sick individuals. None of the other family members had similar complaints. There were no known food/drug allergies or history of any other known medical illnesses. All immunizations were up-to-date.

The patient had been diagnosed with ALL at the age of 6 years, after which complete treatment had resulted in remission for 3 years. However, the ALL had relapsed in the bones, and the patient was currently undergoing chemotherapy. There was no history of reduced G6PD activity.

The following are the reports of the investigations conducted:

1. Vitals
   Temperature: 37.2°C
   Blood pressure: 115/75 mmHg

2. Physical examination
   The right cheek showed a lesion that grew in size with a necrotic center that had sloughing surrounded by vesicles, which were drying up.
   The central part of chest showed an active lesion with vesicles and surrounding erythema.
3. Radiological investigations
   Ultrasound:
   No underlying collection of fluid/abscess.
   No lymph nodes.
4. A complete sepsis workup was done.
5. Consults were taken from:
   a. Plastic surgeon
      Possible debridement of lesion negated.
      Conservative management with local flumazenil cream recommended.
   b. Dermatologist
      Examined the vesicles around the lesions in dermatomal distribution.
      Review since lesion spread to the chest.

Diagnosis

The final diagnosis concluded a herpes zoster virus (HZV) infection based on the clinical manifestations.

Treatment

1. Meropenem 1500 mg q8
2. Voriconazole 200 mg BD
3. Linezolid 400 MG q8
4. Acyclovir 400 mg q8.

Question and answer

1. What is the difference between shingles caused by the HZV infection and chickenpox caused by the varicella zoster virus (VZV)?
   The HZV is the reactivated form of the VZV. VZV establishes a persistent and dormant state in the sensory ganglia of the cranial nerve or the dorsal root ganglia following a childhood chickenpox infection. A competent immune system controls the latent replication of the virus, ensuring that the virus is not reactivated in an

immunocompetent individual. However, an immunocompromised status of the host facilitates reactivation, resulting in the HZV infection manifesting as shingles.

2. How does the sensory ganglia of the human nervous system become a reservoir for latent VZV despite the primary mode of transmission of the virus being through the respiratory route?

   The upper respiratory tract is the route of entry for the virus where the epithelial cells are infected. Infection is followed by an incubation period of about 2 weeks, after which skin lesions appear due to the infection of the capillary endothelial cells. While the cutaneous vesicles appear as blisters, vesicles on mucosal surfaces develop into shallow ulcers. After the primary infection, the virus establishes a latent infection in the sensory ganglia of the human nervous system.

3. What causes the severe pain associated with HZV infections?

   A primary chickenpox infection is followed by the establishment of a latent VZV infection in the sensory ganglia of the human nervous system. The host immune system ensures control on viral replication during latency, resulting in dormancy of the virus within the host. Reactivation of this endogenous virus triggered by a weakened immune system results in viral replication in neuronal cell bodies. Virions are released by lysis of the infected neuronal cells and travel to the area of skin innervated by the infected sensory ganglion through the nerve. A vesicular rash and blistering of the skin manifests and inflammation of the affected nerves results in the severe pain associated with HZV infections.

4. What is known to be a typical pattern of the appearance of HZV-associated vesicles?

   A centripetal distribution of the vesicles resulting from the involvement of a particular dermatome is typical of HZV infections. The lumbar and cervical roots are most commonly involved, while the involvement of the motor roots is rare.

5. Why is exogenous reinfection of HZV infections not known to occur?

   Exogenous reinfection of HZV does not occur after a primary VZV infection due to the life-long immunity bestowed on the host by the primary infection, following which the virus becomes latent in the host nervous system. The humoral, as well as the cell-mediated immune systems, play an important role in the establishment of the life-long immunity to the virus. Antibodies to the virus, in addition to proliferative and cytotoxic T-cells that are detectable in serological investigations immediately after the appearance of symptoms, continue to increase during convalescence and maintain life-long effective protection levels. Therefore HZV infections are a result of reactivation of the endogenous virus due to a weakened immune system, rather than an exogenous infection.

6. What is postherpetic neuralgia (PHN)?
   Clinical manifestations, such as paresthesia and dysesthesia caused due to the damage of the peripheral nerves, debilitating pain, and movement restriction, may persist for prolonged periods of time in some cases of HZV infection. PHN is a result of pain that persists for more than 3 months after the resolution of the rash. PHN is preventable by ensuring an early diagnosis and treatment of HZV infections.
7. How are HZV infections associated with damage to skeletal structure?
   HZV infections usually affect specific organ systems, and the clinical manifestations are related to the functions associated with the particular organ system. However, systemic manifestations, such as ischemic necrosis, may be seen owing to the close association between nerves and blood vessels. Ischemic necrosis can result in long-term effects, like osteonecrosis, resulting in the loss of good bone structure and may even result in developmental anomalies associated with the skeleton and teeth.
8. What is zoster sine herpete (ZSH)?
   ZSH is an atypical presentation of an HZV infection in the cranial nerve, spinal nerve, viscera, and autonomic nerve characterized by pain in the affected dermatome without the appearance of skin lesions.

Learning objectives

1. Determine the appropriate differential diagnosis based on the patient history, known underlying medical conditions, and clinical manifestations.
2. Differentiate lesions caused by HZV from other causes based on the morphology of the vesicles and distribution pattern.
3. Diagnose the cause for the clinical manifestations based on the presented symptoms and laboratory diagnosis.
4. Describe the characteristics of the HZV, the reactivated form of the VZV.
5. Predict the prognosis of the case based on the clinical manifestations and the extent of the pathology presented.
6. Correlate the clinical manifestations with the pathogenicity of the HZV.
7. Determine and administer the appropriate antiviral therapy based on the case presentation to avoid complications and long-term effects of the infection.
8. Assess the candidature of the patient for vaccine administration to prevent recurrence of infections, especially in at-risk patients.

Discussion

Differential diagnosis

Cellulitis	Cellulitis is a type of bacterial infection affecting the deep layers of the skin. The disease presents with swelling, redness, warmth, and pain in the infected area.
Chickenpox	Chickenpox is a contagious viral disease caused by a primary infection with the varicella zoster virus. The characteristic presentation of the disease includes a skin rash with small, itchy blisters, which eventually develop scabs.
Dermatologic manifestations of herpes simplex	Herpes simplex infections caused by the herpes simplex virus can manifest in many different forms depending on the organ system affected. The dermatologic manifestations of herpes simplex infections present as localized blisters, also known as cold sores or fever blisters, and may be accompanied with a fever and general malaise.
Ecthyma	Ecthyma is a skin infection caused by the bacteria responsible for impetigo. The disease is characterized by the formation of ulcers covered with crusted sores and erodes deeper layers of the skin into the dermis.
Erysipelas	Erysipelas is a common bacterial infection that affects the superficial layers and lymphatic vessels within the skin. The infection is identified by a tender, bright red rash with well-defined margins and swelling in the affected area.
Mucosal candidiasis	Candidiasis is an infection of the skin and mucosal membranes caused by fungi belonging to the *Candida* species. The clinical manifestations include white patches inside the mouth accompanied by a cotton-like feeling, loss of taste, soreness, pain while eating and swallowing, and cracked and red skin at the corners of the mouth.
Lichen striatus	Lichen striatus is a rash that can have idiopathic causes or can be caused by the administration of certain vaccines, viral infections, and some kinds of injuries or trauma. The rash presents itself as light-pink colored, inflamed, scaly, itchy, and flat bumps, which form a line or band on the skin over time.
Insect bites	Insect bites can present themselves as inflamed, itchy, swollen, sore, and painful areas of skin, sometimes with an eschar present at the bite site.
Erysipeloid	Erysipeloid is an acute bacterial infection that affects traumatized and injured skin and organs. It usually presents itself with inflamed red skin, swelling, throbbing or pulsating pain, burning or itching sensation, fluid-filled blisters, low-grade fever, and occasionally swollen lymph nodes.

Continued

Continued

Folliculitis	Folliculitis is an inflammation of the hair follicles caused by a bacterial or fungal infection. The infection usually develops small red bumps or white-headed pimples around the hair follicles, which turn into pus-filled blisters that break open and form crusts. A large swollen bump or mass with painful, tender, itchy, and burning skin may also manifest.
Irritant contact dermatitis	Irritant contact dermatitis is a nonallergic skin condition from prolonged exposure to common irritants. The condition usually causes a dry, cracked, scaly skin, red rash, severe itching, bumps and blisters that may ooze pus and later form crusts, swelling, and burning or tenderness.

Background

The HZV is the reactivated form of the VZV, which is the etiological agent for chickenpox. VZV establishes a persistent and dormant state in the sensory ganglia of the cranial nerve or the dorsal root ganglia following a childhood chickenpox infection. A competent immune system controls the latent replication of the virus, ensuring that it is not reactivated in an immunocompetent individual. Shingles, the disease caused by HZV, generally occurs in adulthood, and is thus strongly associated with the immunocompromised status of an individual.[7,8]

Etiological agent

Herpes zoster virus

Classification and morphology

HZV, the reactivated form of the VZV, is also known as the *human alphaherpesvirus 3*, and belongs to the genus *Varicellovirus* in the *Alphaherpesvirinae* subfamily of the *Herpesviridae* family. Dormancy is a characteristic of all herpes viruses, and VZV is well-known for its ability to remain dormant in the nerves after a childhood chickenpox infection. Reactivation of VZV triggered by an immunocompromised status of the host results in HZV infection, causing shingles in adults.

HZV is spherical-shaped and comprises of a lipid bilayer envelope studded with glycoproteins that facilitate the process of host cell invasion. The lipid envelope encloses a nucleocapsid of 162 hexametric and pentameric capsomeres arranged in an icosahedral form. The nucleocapsid is surrounded by

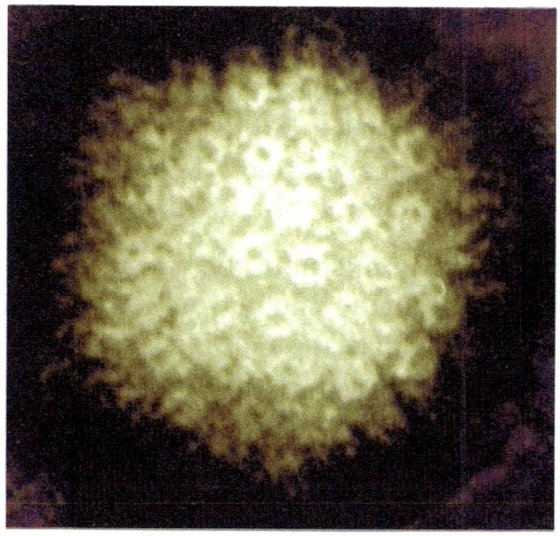

FIGURE 2.4
Fiery-looking electron micrograph of herpes simplex virus. Source*: From NIAID/; https://creativecommons.org/licenses/by/2.0/; https://www.flickr.com/photos/niaid/5614251360.*

the tegument made up of loosely associated proteins that are important for the process of virus reproduction in the host cell and contains a double-stranded linear DNA within its core (Fig. 2.4).

Epidemiology

HZV is a human virus and therefore infects only human hosts. As HZV infection is a reactivation of the VZV due to the immune system's failure to control the replication of the latent virus, the occurrence of shingles is not associated with any seasonal variation. The sensory ganglia of the human nervous system act as the reservoir for the latent virus.

The majority of the world's population gets infected with VZV by late middle age, thereby providing life-long immunity to the virus. Immunocompromised status triggered by emotional stress, certain medications, such as immunosuppressants and steroids, acute or chronic illnesses, malignancies, and old age, results in infection by HZV.

Pathogenicity

The transmission of VZV is primarily airborne through infected respiratory droplets, although infection through direct skin contact is also possible. The upper respiratory tract is the route of entry for the virus where the epithelial cells are infected. Infection is followed by an incubation period of about

2 weeks, after which, skin lesions appear due to an infection of the capillary endothelial cells. While the cutaneous vesicles appear as blisters, vesicles on mucosal surfaces develop into shallow ulcers. The large number of free virus particles in the vesicles, in the oral cavity, and upper respiratory tract, is the main source of transmission. The infectious period ranges from a few days before lesions appear until a few days after new lesions cease to appear.

After the primary infection, the virus establishes latent infection in the sensory ganglia of the human nervous system. The host immune system ensures control on viral replication during latency, resulting in dormancy of the virus within the host. Reactivation of this endogenous virus triggered by a weakened immune system results in viral replication in neuronal cell bodies. Virions are released by lysis of the infected neuronal cells and travel to the area of skin innervated by the infected sensory ganglion through the nerve. The virus causes a vesicular rash and skin blistering (Fig. 2.5), and inflammation of the affected nerves, resulting in the severe pain associated with HZV infections.[9]

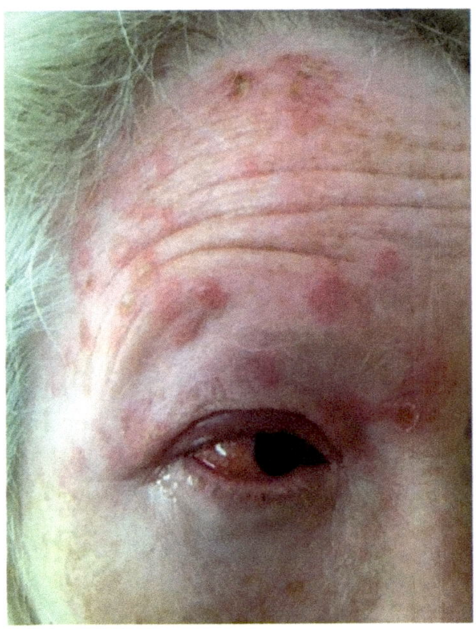

FIGURE 2.5
Shingles affecting the eye and the forehead region. Source*: From Burntfingers/Own Work/; https://creativecommons.org/licenses/by-sa/4.0/deed.en; https://commons.wikimedia.org/wiki/File:Day06_shingles_or_Herpes_Zoster_Virus_attacking_forehead_and_eye.jpg.*

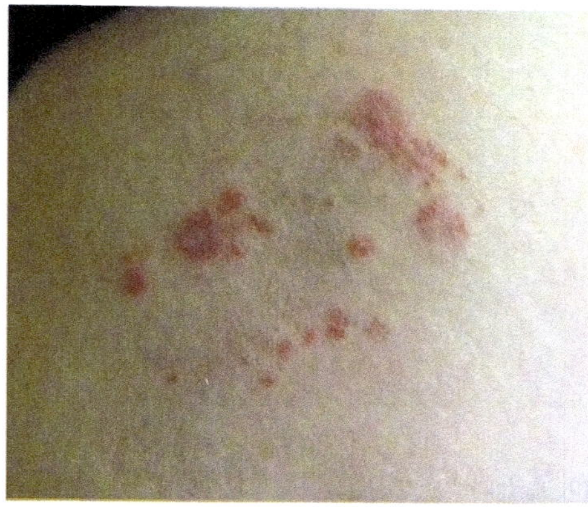

FIGURE 2.6
Characteristic centripetal appearance of a shingles rash. Source*: From Rosser1954/Own work/; https://creativecommons.org/licenses/by-sa/4.0/deed.en; https://commons.wikimedia.org/wiki/File:Shingles_rash,_thigh_Caused_by_the_herpes_zoster_virus.jpg.*

The involvement of a particular dermatome with a centripetal distribution of the vesicles is typical of HZV infections (Fig. 2.6). The lumbar and cervical roots are most commonly involved, while the involvement of the motor roots is rare.[10]

Immune responses

Exogenous reinfection of HZV does not occur after a primary VZV infection because of the life-long immunity bestowed on the host by the primary infection, so the virus becomes latent in the host nervous system. The humoral, as well as the cell-mediated immune systems, play an important role in the establishment of the life-long immunity to the virus. Antibodies of the virus, in addition to proliferative and cytotoxic T-cells that are detectable in serological investigations, immediately after the appearance of symptoms, continue to increase during convalescence and maintain life-long effective protection levels. Therefore HZV infections are a result of reactivation of the endogenous virus due to a weakened immune system, rather than an exogenous infection. Interferon responses, triggered by the infected cells, also play a significant role in strengthening the host immune responses. The increased frequency of HZV infections in geriatric and immunocompromised individuals is attributable to the reduced efficacy of cell-mediated responses against HZV in these groups.[11]

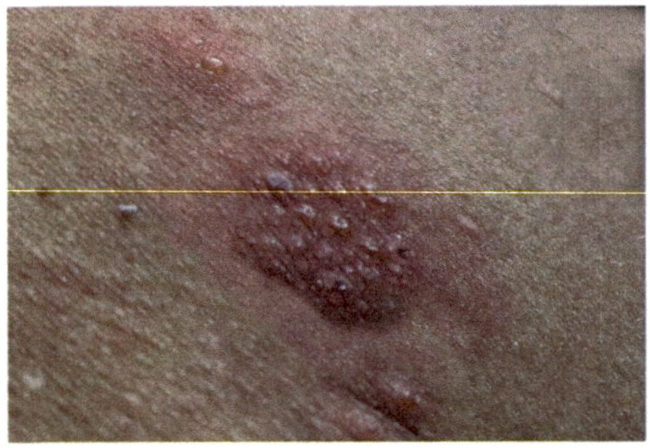

FIGURE 2.7
Shingles lesions clustered together to form erythematous papule. Source*: From NIAID/; https://creativecommons.org/licenses/by/2.0/; https://www.flickr.com/photos/niaid/46509144172.*

Clinical manifestations

The appearance of vesicles in HZV infections occurring during the acute eruptive phase is typically preceded by the preeruptive phase by about 2–3 days. The preeruptive phase causes a fever, malaise, headaches, photophobia, and a tingling sensation in the skin followed by excruciating burning pain. The symptoms seen in the preeruptive phase continue to present in the eruptive phase in addition to the actual appearance of obvious lesions. The vesicles are formed from lesions bunched close together as erythematous papules (Fig. 2.7). The region on which the vesicles are formed generally shows severe edema, and the distribution of the vesicles is unilateral, centripetal, and restricted to a single dermatome at the initial stages of the infection. HZV vesicles can appear on various dermatomes, with the thoracic, cervical, and trigeminal dermatomes being the most affected. As the infection progresses, the vesicles may rupture or ulcerate, allowing the infection to spread further due to the release of the free viral particles in the vesicles. The infectious stage lasts until the vesicles dry out and eventually form crusts or scabs. Although this phase typically lasts between 2 weeks and 4 weeks, the pain may continue and even become chronic in some cases. Clinical presentations, such as paresthesia and dysesthesia due to damage to the peripheral nerves, debilitating pain, and movement restriction, may be seen in some cases for prolonged periods of time. Pain that persists for more than 3 months after the resolution of the rash is known as PHN, which can be prevented with an early diagnosis and treatment of HZV infections.[12]

HZV infections that affect specific organ systems cause symptoms and aftereffects related to the functions associated with the particular organ system. Systemic manifestations, such as ischemic necrosis, may be seen owing to the close association between nerves and blood vessels. Ischemic necrosis can result in long-term effects, such as osteonecrosis, resulting in the loss of good bone structure and may even result in developmental anomalies associated with the skeleton and teeth.[10] Complications of systemic infections may manifest as secondary bacterial infections, PHN, scarring, and deformities.

HZV is a neurotropic virus, so the involvement of the CNS is common in HZV infections. Complications of CNS infections may manifest as cranial nerve palsy, polyneuritis, paralysis of the diaphragm, muscle weakness, neurogenic bladder, partial facial paralysis, Guillain Barre syndrome, aseptic meningitis, encephalitis, and myelitis.[10]

ZSH is an atypical presentation of HZV infection in the cranial nerve, spinal nerve, viscera, and autonomic nerve, characterized by pain in the affected dermatome without the appearance of skin lesions.[12] Disseminated HZV infections are defined as the cases with the appearance of more than 20 skin lesions on areas other than the primarily affected dermatome or those directly adjacent to it.[12] Although vertical transmission of VZV infections from mother to child are known to occur, reactivated HZV infections are not associated with fetal VZV infection. Coinfection with multiple herpes viruses, such as herpes simplex viruses, cytomegalovirus, and Epstein – Barr virus is also possible in individuals manifesting shingles.[10]

Laboratory diagnosis

The diagnosis of HZV infections relies mainly on clinical diagnosis based on disease presentation. A history of fever and malaise preceding the appearance of lesions, complaints about excruciating burning pain, abnormal skin sensations, characteristic morphology of lesions and vesicles, and a typical pattern of distribution of vesicles should be considered as indicators of HZV infection, especially in elderly and immunocompromised patients. The typical appearance of the cutaneous lesions and vesicles, along with the peculiar distribution pattern, is important in differentiating between chickenpox, shingles, herpes simplex infections, dermatitis herpetiformis, impetigo, contact dermatitis, candidiasis, allergic hives, and insect bite eschars.

As HZV is caused by the same etiological agent responsible for VZV, laboratory diagnosis techniques used for both the diseases are the same. Vesicular fluid can be used to perform the Tzanck smear test, which shows multinucleated giant cells. This test, however, has lower sensitivity and specificity compared to the direct fluorescent antibody and polymerase chain reaction (PCR) tests, and therefore has limited use. The VZV – IgM antibody test can detect the virus in the blood during active infection but is not effective during dormancy.

PCR tests can be conducted using small quantities of samples obtained from vesicular fluid, corneal lesions, and blood, and are the most reliable.

Real-time PCR tests are rapid, easy to perform, have high levels of sensitivity and specificity, and are therefore used frequently as the method of choice for laboratory diagnosis of VZV/HZV infections in correlation with clinical manifestations.[10]

Treatment

Although antiviral therapy for a vast majority of viral infections has broad-spectrum effects, specific and effective antiviral treatment is available against the herpes family of viruses. The morbidity associated with debilitating pain and scarring, resulting from HZV infections, can be reduced to a large extent by providing timely and effective antiviral therapy. The course of the disease can be expedited by rapid resolution of lesions, decreased pain, and prevention of complications and long-term effects, such as PHN, especially in elderly and immunocompromised patients.

Acyclovir, valacyclovir, and famciclovir are the antiviral drugs of choice for the treatment of HZV infections. Topical application of certain antibiotic creams, like mupirocin and soframycin, is also recommended for the prevention of secondary bacterial infections in the affected area. Analgesics and antipyretics may be prescribed for relief from pain, fever, and malaise. Severe cases with complaints of excruciating and unbearable pain may indicate the need for administration of opioids, topical lidocaine, and nerve blocks. Topical application of capsaicin and Emla is recommended after the formation of crusts or scabs at the site of the lesions.[10]

Prevention

There are currently two vaccines, Zostavax and Shingrix, available to prevent HZV infections, especially in elderly patients.

Zostavax is a live virus vaccine and is therefore contraindicated in pregnant women and immunocompromised patients. However, routine vaccination of geriatric individuals above 60 years of age with Zostavax is recommended. Due to the short-term efficacy of Zostavax, its use in individuals between 50 years and 59 years of age is limited because the efficacy of the vaccine may reduce by the time the vaccinated individual is at the highest risk of HZV infections.

Shingrix is a recombinant virus vaccine recommended for individuals above 50 years of age, to be administered 2–6 months apart. This vaccine has been known to provide strong protection against shingles and PHN, and is not contraindicated in immunocompromised individuals. As it is a recombinant vaccine, the risks and side-effects associated with it are lower, which makes it the safest and preferred vaccine out of the two.[13]

Student Notes

Key take-aways

- Shingles, caused by the HZV, is caused by the VZV, responsible for chickenpox, reactivating.
- The HZV is primarily neurotropic, and therefore infections caused by it frequently manifest the involvement of the CNS.
- While chickenpox infections generally occur in childhood and affect the majority of the world's population, shingles occurs in adulthood and is generally a result of a weakened immune system.
- Shingles is not usually caused by the exogenous HZV, but is rather a result of the immune system's failure to keep the replication of endogenous VZV from a primary chickenpox infection under control.
- The geriatric and immunocompromised individuals are at higher risk of reactivation and recurrence of infections and complications associated with the disease.
- The lesions and vesicles of the HZV infections have a typical morphology and follow a peculiar centripetal dermatome restricted distribution pattern, which facilitates the differentiation of the HZV infections from other infections.
- Severe pain and neuralgia associated with the herpes zoster infections result in high levels of morbidity associated with the disease.
- Diagnosis of herpes zoster infections is primarily dependent on clinical manifestations, although laboratory diagnosis by PCR may be opted for confirmation.
- Effective and timely antiviral treatment reduces the debilitating pain and scarring associated with herpes zoster infections and expedites the disease process, thereby decreasing the morbidity associated with the long-term effects and complications of the disease.
- Vaccination is recommended only for high-risk individuals as immunocompetent individuals are not at risk of developing the disease.

Problem number 2.3: Preseptal cellulitis caused by *Hemophilus influenzae* infection

A 17-year-old boy with a known history of sickle cell disease presented with pain in the lower limb and back. The pain started in the second half of the day in the lower back and progressed to the bilateral lower limbs. There was no history of trauma prior to the commencement of the pain, and no redness or swelling was noticed. The patient did not complain of any chest pain, cough, or fever.

The patient was a known case of sickle cell disease and was taking folic acid regularly and pain medication as needed. He did not have any known food/drug allergies. There was no history of surgeries or blood transfusions.

The following are the reports of the investigations carried out:

1. Vitals
   Temperature: 36.6°C
   Blood pressure: 130/85 mmHg
   Weight: 45 kg.
2. Physical examination
   Pain
   No respiratory distress
   Capillary refill: < 2 seconds
   No edema or redness on the back or the limbs.
3. Respiratory examination: Equal air entry bilaterally; no added sounds.
4. Cardiovascular examination: Normal S1, S2; no murmurs.
5. Abdominal examination: Soft; nontender; splenomegaly
   Vaso-occlusive crisis was suspected, so the patient was admitted to the hospital. On the fourth day of hospitalization, the patient spiked a fever of 39.4°C and complained of pain and swelling in the right eye, although there was no change in vision. Ophthalmologists were consulted on the case, and a complete eye assessment was done. A swab sample was collected from the right eye for culture and sensitivity tests. Meanwhile, Tobradex QID drops were prescribed for 1 week. Pediatrics ID were also consulted on the case.
6. C-reactive protein: 336 mg/L (reference value: < 10 mg/L).
7. Blood culture: Sterile.
8. Right eye swab (Culture + Sensitivity)
   Gram stain: No organism seen
   WBC: Occasional
   Culture: Sterile.

Diagnosis

The final diagnosis concluded preseptal cellulitis caused by *H. influenzae* infection.

Treatment

1. Rocephin 2 g OD

2. Clindamycin 450 mg Q8
3. Tobramycin—dexamethasone ophthalmic drops 1 drop Q6.

Question and answer

1. Why does sickle cell disease increase predisposition to complicated *H. influenzae* infections?
   H. influenzae can cause various types of infections depending on the organ system invaded by the bacteria. Preseptal cellulitis, also known as periorbital cellulitis, is an inflammatory condition resulting from a *H. influenzae* infection of the skin and soft tissues around the eye, and is commonly seen in children as a complication of acute sinusitis or trauma. The tissues affected due to preseptal cellulitis are anterior to the orbital septum, while an infection of the tissues' posterior to the orbital septum leads to a more serious condition known as orbital cellulitis. Vaso-occlusive crisis occurs commonly in sickle cell disease, and this increases predisposition to complicated *H. influenzae* infections.
2. What is the phenomenon of satellitism?
   H. influenzae are fastidious organisms requiring specific growth factors, such as the erythrocyte factors, factor X (hematin), and factor V (phosphopyridine nucleotide) released through the lysis of red blood cells. *H. influenzae* demonstrate high levels of growth and larger colonies in the vicinity of *Staphylococcus aureus* colonies because of the dependency of the bacteria on factor V. The *H. influenzae* colonies get progressively smaller farther away from the *S. aureus* growth. This phenomenon is known as satellitism.
3. What is the significance of the polyribosyl ribitol phosphate (PRP) antigen present on the surface of encapsulated *H. influenzae*?
   The presence of the PRP antigen present on the surface of encapsulated *H. influenzae* facilitates the invasive nature of the *H. influenzae B* (Hib) bacteria. This antigen is also instrumental in the elicitation of antibody responses through the induction of IgG, IgM, and IgA antibodies, and is therefore employed in the development of Hib vaccines.
4. Why is the capsule important for the virulence of *H. influenzae*?
   Encapsulated strains of the *H. influenzae* bacteria express proteins, such as protein H and surface fibrils (Hsf) on their capsular surface. These proteins facilitate adherence and virulence and are exclusive to the encapsulated strains. The capsule also enables immune evasion by conferring antiphagocytic properties on the bacteria. The capability of immune evasion is further enhanced by the lack of anticapsular antibody, resulting in increased bacterial proliferation.

5. Why have nontypeable *H. influenzae* strains (NTHi) emerged as causative agents for a large number of *H. influenzae* infections since the capsular *H. influenzae* vaccines have decreased the infections caused by the typeable strains?

   NTHi can invade the host cells by maintaining strong adhesion and mediating colonization, either by direct attachment to the surface epithelial cells or interaction with the underlying extracellular matrix layer. In addition to this, NTHi also have the ability to manipulate certain host serum factors. Evasion of host immune responses enables NTHi to form colonies that have characteristics of biofilms, which further enhances their immune evasive attributes.

6. How is preseptal cellulitis clinically differentiated from orbital cellulitis?

   The similarities in the clinical manifestations of preseptal cellulitis and orbital cellulitis can make the clinical differentiation between the diagnoses of the two conditions difficult to achieve. As the infection and inflammation are superficial and anterior to the periocular tissue in preseptal cellulitis, the clinical manifestations usually occur in either one of the eyes, and does not affect vision and intraocular pressures due to the extraocular muscles not being involved. Mild or early stages of preseptal cellulitis may also manifest without a fever. Rarely complicated preseptal cellulitis may affect vision and compromise the functions of the affected eye. Untreated preseptal cellulitis may lead to more severe conditions, such as orbital cellulitis, orbital abscesses, cavernous sinus thrombosis, and meningitis. Computed tomography (CT) scans of the orbits and sinuses is often indicated to enable differentiation between preseptal cellulitis and orbital cellulitis.

Learning objectives

1. Determine the appropriate differential diagnosis based on the patient history, known underlying medical conditions, clinical manifestations, and disease progression.
2. Differentiate the clinical manifestations of preseptal cellulitis from orbital cellulitis.
3. Determine the possible differential diagnoses for the preseptal cellulitis based on the clinical manifestations and the patient history.
4. Identify the etiological agent responsible for the particular case of preseptal cellulitis.
5. Describe the characteristics of *H. influenzae*.
6. Summarize the pathogenesis of preseptal cellulitis infection.
7. Predict the prognosis of the case based on the clinical manifestations and the extent of the pathology presented.

8. Determine the appropriate antibiotic regimen based on the clinical manifestations and the progression of the disease to ensure a rapid and effective resolution of the infection, and avoid any further debilitating complications.

Discussion

Differential diagnosis

Orbital cellulitis	Orbital cellulitis is an inflammatory condition caused by an infection of the soft tissues responsible for holding the eye in the socket. Clinical manifestations may include severe protrusion of the eye, pain and swelling in and around the infected eye, redness and soreness of the eye, tenderness of the nasal structures, the inability or trouble opening the infected eye, impaired or compromised vision, fever, headache, and discharge from the eye or nose.
Angioedema	Angioedema is a swelling of the deeper layers of skin and tissue caused by allergic reactions. The condition may manifest as swelling of the affected area, soreness, tenderness, itching, redness, and hives.
Internal and/or external hordeolum	A hordeolum results from bacterial infections of the glands associated with the eyelids. An infection of the oil glands within the eyelids leads an internal hordeolum, while an infection of the glands associated with the eyelashes causes an external hordeolum, commonly referred to as a stye. A hordeolum presents itself with a painful red bump at the base of eyelashes, either on the upper or the lower eyelids, swelling of and around the region surrounding the affected region, formation of crusts along the eyelids, light sensitivity, sore and itchy eye, discharge from the eye, and a constant feeling of discomfort in the eye.
Cavernous sinus thrombosis	Cavernous sinus thrombosis is a condition in which a blood clot forms in a vein in the cavernous sinus, a cavity at the base of the brain responsible for draining the deoxygenated blood from the brain back to the heart. The condition can manifest as swelling, redness, and irritation around the eye, drooping eyelids, severe headache, tearing of the eyes, difficulty or inability in moving the eye, high-grade fever, fatigue, pain or numbness in areas around the eyes, seizures, compromised or impaired vision, and altered mental status manifesting as confusion or even coma.
Insect bites	Insect bites around the eye often result in inflammation of the affected tissue, thereby presenting as swelling, redness, soreness, itching, and pain in and around the region affected by the insect bite.
Rhino-cerebral mucor mycosis	Rhino-cerebral mucor mycosis is an opportunistic fungal infection affecting the sinuses and parts of the brain. The infection presents itself with a fever, congestion of the nasal or sinus passages,

Continued

Continued

	headache, one-sided facial swelling, and appearance of black lesions on the nasal bridge or the upper palate of the mouth.
Chalazion	Chalazion is a cyst formed due to a blocked oil gland in either the upper or the lower eyelid. The cyst manifests as a small and usually painless lump on either the upper or the lower eyelid. Pain, swelling, or redness in a chalazion may indicate an infection.
Periorbital necrotizing fasciitis	Periorbital necrotizing fasciitis is an acute, invasive form of infection caused by multiple pathogens. The infection may present itself as pale red, swollen, and tense periorbital skin, severe often unbearable pain, cyanosis of the skin, rapid appearance of supralesional bullae, purpura, fever, chills, painless ulcers, drainage comprising of serosanguineous fluid, black necrotic eschars, cutaneous plaques, and tissue emphysema.
Nephrotic syndrome	Nephrotic syndrome is a constellation of symptoms resulting from impaired functioning of the kidneys. The condition can manifest with swelling of the limbs and/or the face, fatigue, observation of foamy or bubbly urine, loss of appetite, sudden weight gain, and fatigue.
Pott puffy tumor	Pott puffy tumor can occur as a complication of acute sinusitis or trauma. It is a nonneoplastic condition manifested by swelling and edema on the forehead and the scalp. Osteomyelitis of the frontal bone associated with subperiosteal abscess can result in Pott puffy tumor.

Background

Sickle cell disease predisposes children to exacerbated complications of infections caused by encapsulated bacteria, such as *H. influenzae*.[14] Vaso-occlusive crisis caused by sickle cell disease contributes further to the exacerbated manifestations of the complications. Preseptal cellulitis (Fig. 2.8), also known as periorbital cellulitis, is an inflammatory condition caused by an infection of

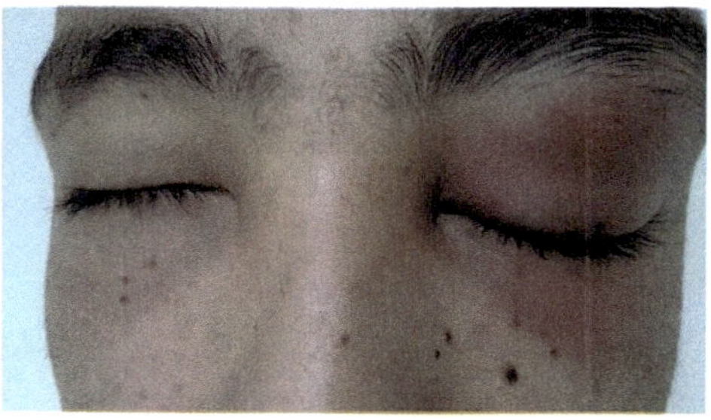

FIGURE 2.8

Representative image of preseptal cellulitis. Source*: From Afrodriguezg/Own work/; https://creativecommons.org/licenses/by-sa/3.0/deed.en; https://commons.wikimedia.org/wiki/File:Celulitis_Periorbitaria_%28Preseptal%29.JPG.*

the skin and soft tissues around the eye. The tissues affected due to preseptal cellulitis are anterior to the orbital septum, while an infection of the tissues posterior to the orbital septum results in a more serious condition, known as orbital cellulitis. Preseptal cellulitis is commonly seen in children as a complication of acute sinusitis or trauma. *H. influenzae* can cause various types of infections depending on the organ system invaded by the bacteria.

Etiological agent

Hemophilus influenzae

Classification and morphology

H. influenzae is a small, Gram-negative, nonmotile, coccobacillus (Fig. 2.9), belonging to the family *Pasteurellaceae* of the genus *Hemophilus*.

The bacteria are pleomorphic, facultatively anaerobic, and thrive well in environments with high carbon dioxide concentrations. *H. influenzae* are fastidious organisms requiring specific growth factors, such as the erythrocyte factors, factor X (hematin), and factor V (phosphopyridine nucleotide), released through the lysis of red blood cells. The dependency of the bacteria on factor V is especially well-exhibited by the high levels of growth and larger colonies in the vicinity of *S. aureus* colonies. The *H. influenzae* colonies get progressively smaller farther away from the *S. aureus* growth, a phenomenon known as satellitism (Fig. 2.10).

Both encapsulated as well as nonencapsulated forms of *H. influenzae* are human parasites and have pathogenic potential. Among the six encapsulated serotypes (a – f), *H. influenzae* type B (Hib) is of special clinical significance due to the presence of the PRP capsule, responsible for most of the invasive *H. influenzae* disease in children and more than half of the adult *H. influenzae* cases.

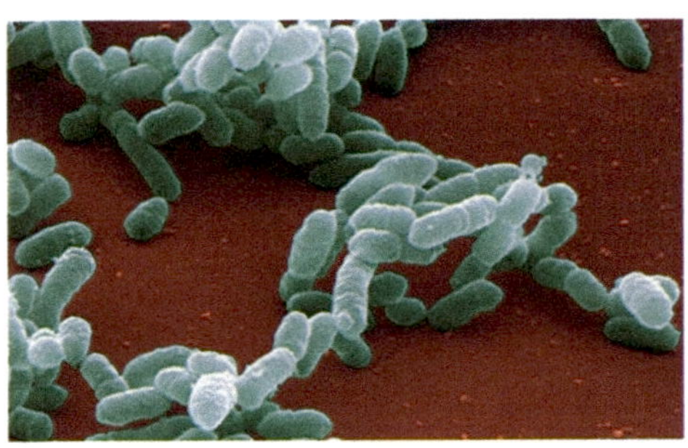

FIGURE 2.9
Hemophilus influenzae. Source: *From Juergen Berger/science photo library.*

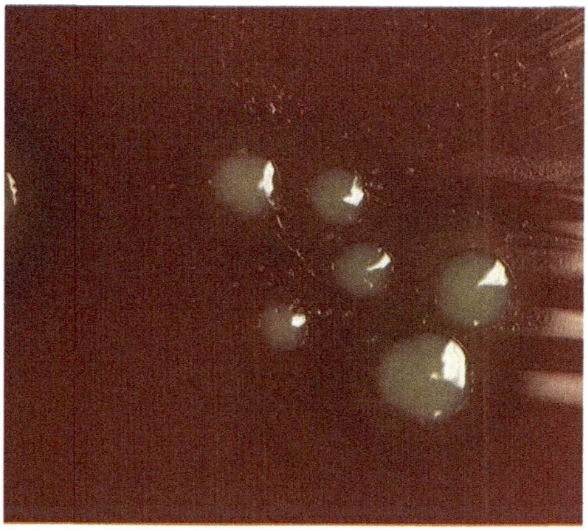

FIGURE 2.10

Hemophilus influenzae satelliting colonies (pinpoint) near *Staphylococcus aureus* (yellow) on blood agar plate. Source: *From Deminorwood/Own work/; https://creativecommons.org/licenses/by-sa/4.0/deed.en; https://commons.wikimedia.org/wiki/File: Hinfluenzae_satillite.jpg.*

Epidemiology

Humans are the natural hosts for the parasitic *H. influenzae* bacteria. Some non-NTHi are also present in the upper and lower respiratory tracts, the conjunctivae, and the genital tract of humans as part of the normal flora, which act as opportunistic pathogens in immunocompromised children.[15,16] The introduction of the Hib conjugate vaccine in the immunization schedule of children has resulted in a marked decrease in the incidence of invasive Hib disease.

Virulence and pathogenicity

There are three major surface antigens, the capsular polysaccharide, the outer membrane proteins, and lipooligosaccharide present on encapsulated strains of *H. influenzae*.[17] These surface antigens are responsible for virulence exhibited by *H. influenzae*. The classification of *H. influenzae* strains into six subtypes (a – f) is based on the different types of capsular polysaccharide, which is the major antigenic determinant in *H. influenzae*. The PRP antigen present on the surface of encapsulated *H. influenzae* is known to elicit antibody responses through the induction of IgG, IgM, and IgA antibodies, and is therefore employed in the development of Hib vaccines. Proteins, such as protein Hsf, facilitate adherence and virulence and are exclusive to the encapsulated strains. The capsule also facilitates immune evasion by conferring antiphagocytic properties on the bacteria. The capability of immune evasion is further enhanced by the lack of anticapsular antibody, resulting in increased bacterial proliferation.[18]

NTHi has been found to be responsible for many *H. influenzae* infections as the capsular *H. influenzae* vaccines have caused a decrease in infections caused by the typeable strains. An array of host proteins has been implicated in the facilitation of the pathogenic effects of NTHi. NTHi can invade the host cells by maintaining strong adhesion and mediating colonization, either by direct attachment to the surface epithelial cells or by interaction with the underlying extracellular matrix layer, in addition to manipulating certain serum factors. Evasion of host immune responses enables NTHi to form colonies that have characteristics of biofilms, which further enhances their immune evasive attributes.

Clinical manifestations

Preseptal cellulitis is generally preceded by a history of some upper respiratory tract infections, such as sinusitis or trauma, caused by an accident or an insect bite. The condition can present itself with complaints related to the ophthalmic region, discharge from the eye, tearing, fever, redness, swelling, and pain in the affected region. Physical examination is therefore essential for the clinical diagnosis of preseptal cellulitis.

The similarities in the clinical manifestations of preseptal cellulitis and orbital cellulitis can make the clinical differentiation between the diagnoses of the two conditions difficult to achieve. As the infection and inflammation are superficial and anterior to the periocular tissue in preseptal cellulitis, the clinical manifestations usually occur in either one of the eyes and does not affect vision and intraocular pressures due to the extraocular muscles not being involved. Mild or early stages of preseptal cellulitis may also manifest without a fever.

Rarely complicated preseptal cellulitis may affect vision and compromise the functions of the affected eye. Untreated preseptal cellulitis may lead to more severe conditions, such as orbital cellulitis, orbital abscesses, cavernous sinus thrombosis, and meningitis.[19]

Laboratory diagnosis

Preseptal cellulitis is usually diagnosed clinically based on the disease presentation and radiological findings. CT scans of the orbits and sinuses are often indicated to enable differentiation between preseptal cellulitis and orbital cellulitis, which may otherwise be difficult to differentiate. Occasionally, treating orbital cellulitis may be recommended as a safety measure in cases where the differentiation between preseptal cellulitis and orbital cellulitis is unclear. CT scans of the head may be suggested if intracranial involvement is suspected.

While in most cases of preseptal cellulitis, a CT scan will only show the swelling of the eyelid without any proptosis, fat stranding of orbital contents, or the involvement of the extraocular muscles, the scan may reveal evidence of sinusitis. Blood and swab cultures are generally not recommended for diagnosis as the collection of samples can be difficult in cases of preseptal cellulitis, and usually yield negative results.

Preseptal cellulitis caused by *H. influenzae* is usually caused by NTHi due to the advent of the Hib vaccines. Predisposing medical conditions, such as sickle cell disease, increase vulnerability to complications associated with preseptal cellulitis, and therefore need a quick diagnosis and effective therapeutic interventions.

Treatment and prevention

Appropriate and effective antibiotic treatment targeted against the etiological agent is the mainstay of treatment for preseptal cellulitis. Immunocompetent patients and those without any predisposing medical conditions can be treated on an outpatient basis with oral antibiotics. However, immunocompromised patients with underlying medical conditions, infants below 1 year of age, geriatric patients above 65 years of age, and those with severe manifestations or complications of the condition may need hospitalization and more aggressive antibiotic therapy.

Antibiotics, such as amoxicillin-clavulanic acid, cefpodoxime, and cefdinir, are not recommended routinely due to the possibility of antibiotic resistance. Trimethoprim-sulfamethoxazole (TMP-SMX), clindamycin, and doxycycline may be recommended, although doxycycline is not recommended for children below 8 years of age. Combination therapy, comprising of clindamycin or TMP-SMX along with amoxicillin-clavulanic acid or cefpodoxime or cefdinir, is considered effective and is widely recommended for preseptal cellulitis. Preseptal cellulitis caused by *H. influenzae* is usually treated using third-generation cephalosporin, subject to culture sensitivity reports. Antibiotic coverage using a beta-lactam may be indicated for patients who are not immunized against *H. influenzae*.[18,19]

Antibiotic therapy should usually result in the resolution of clinical symptoms; however, if the treatment does not show effectivity within 24 – 48 hours, intensive treatment with broad-spectrum antibiotics should be considered, in addition to a CT scan, and the possibility of surgical incision and drainage. Steroids are not routinely prescribed to avoid compromising the immune system.

Student Notes

Inclusion of the Hib vaccine in the immunization schedule in children has been shown to be effective in the reduction of *H. influenzae* cases, and is therefore considered effective in preventing preseptal cellulitis caused by *H. influenzae*.

Key take-aways

- Preseptal cellulitis is also known as periorbital cellulitis, and is an inflammatory condition resulting from an infection of the skin and soft tissues around the eye.
- The tissues affected due to preseptal cellulitis are anterior to the orbital septum, while an infection of the tissues' posterior to the orbital septum results in a more serious condition, known as orbital cellulitis.
- Preseptal cellulitis is commonly seen in children and individuals with underlying medical conditions, such as sickle cell disease as a complication of acute sinusitis or trauma.
- Although preseptal cellulitis is a mild disease and can usually be resolved by an appropriate and effective antibiotic course, untreated preseptal cellulitis can result in severe disease and life-threatening complications, such as cavernous sinus thrombosis and meningitis.
- *H. influenzae* is among the different pathogens that can cause preseptal cellulitis, especially among those unvaccinated against Hib, although NTHi has been found to be responsible for preseptal cellulitis caused by *H. influenzae* among those who are vaccinated.
- Diagnosis of preseptal cellulitis primarily depends on clinical manifestations and patient history.
- Quick, appropriate, and effective antibiotic therapy is considered the mainstay of treatment for preseptal cellulitis.
- Vaccination with the Hib vaccine is an effective means of reducing the number of cases of *H. influenzae* infections.

Problem number 2.4: Subdural empyema caused by *Streptococcus pyogenes* infection

A 13-year-old boy presented with a fever, cough, and persistent headache accompanied with numbness in the lower limbs, which radiated to the left upper limb associated with back pain, and an inability to walk or sit up. The fever, cough, and headache started about 3 weeks ago when the patient was in a normal state of health. The patient had taken paracetamol regularly for 2 weeks due to his febrile condition.

One week before the presentation, the patient had complained of ear pain associated with swelling in the right eye and a headache on the right side.

The swelling in the right eye was accompanied with redness and yellowish discharge, starting on the second day after the onset of the swelling. The patient was prescribed a course of Augmentin, which resolved the ear pain, the swelling, and the discharge in the eye, although the headache continued intermittently. The patient was then prescribed cefuroxime for 1 week following the diagnosis of sinusitis on a CT scan of the sinuses. However, 2 days later, the patient developed a fever, a mild cough, and chest pain in the emergency department. Based on primary physical examinations and clinical manifestations, the patient was diagnosed with atypical pneumonia, and discharged with a prescription of Augmentin, Pulmicort nebulizer, and Ventolin. The patient used three doses of the prescribed medications for 1 day. The next day, he developed numbness in his lower left limbs and weakness from his toes to his thigh, although he was still able to move independently.

On the day of presentation, the patient also showed urinary and stool incontinence and reported having observed a change in normal urine color to orange. There was a history of reduced oral intake and vomiting (three times). While the vomitus comprised of water twice, blood stains were visible the third time. The patient complained of photosensitivity, and was drowsy on presentation with difficulty of speech due to the onset of tongue numbness while in the emergency department.

The patient did not have any history of recent travel or exposure to sick individuals; there was no history of trauma or recent insect bite; the patient did not have any known medical illnesses or allergies, and had never been admitted to a hospital before. He had a normal birth history and his immunizations were up-to-date.

The patient's parents were in a nonconsanguineous marriage, and while his mother was healthy, his father had hypertension with renal failure requiring dialysis. He had three siblings; two sisters; and one brother.

The following are the reports of the investigations carried out:

1. Vitals
   Temperature: 39.4°C (at room temperature)
   Heart rate: 129
   SpO_2: 97 (room air)
   Respiratory rate: 41
   Blood pressure: 132/94 mmHg; 120/68 mmHg.
2. Physical examination
   The patient was conscious, alert, and well-oriented, but had difficulty with speech. He was lying on the bed, was not cyanosed, and did not complain of respiratory distress. His tongue was covered with whitish patches, but there were no signs of any deviation or fasciculations.
   Capillary refill: <2 seconds

Eye: Good bilateral eye reflex and normal fundoscopy
Respiratory: Equal air entry bilaterally; no added sounds
Cardiovascular system: Normal S1 + S2 + 0
Abdomen: Soft; lax; no organomegaly
Neurological: Normal cranial nerves; no sensation in left lower limbs; normal sensation in all other limbs; normal reflex in the upper limbs; absence of reflexes in both the lower limbs; normal tone and power in the left and right upper and lower limbs; power in the left lower limb was recorded at 4/5.
The patient was shifted to the pediatric intensive care unit (PICU) and kept on a cardiac monitor. An urgent brain CT scan was ordered and an MRI brain and spine were also recommended. High flow of oxygen was started.

3. Venous blood gas: pH—7.56; pCO_2−29.2; pO_2−152.
4. Chest X-ray: Resolving pneumonia.
5. Partial septic workup: Suspected presence of *S. pyogenes* (administration of antibiotics prior to sample collection may have resulted in inconclusive results).
6. A lumbar puncture was to be considered following imaging studies.
7. Brain CT: Frontal and para-falcine empyema with right maxillary and ethmoid sinusitis.
8. Brain MRI: Bifrontal and right-sided parasagittal interhemispheric subdural empyema; evidence of dural- and lepto-meningeal enhancement indicating the possibility of meningitis; early nonocclusive thrombosis at the right transverse sinus; bilateral asymmetrical right-sided paranasal sinusitis.

Diagnosis

The final diagnosis was that of meningitis with empyema secondary to acute sinusitis caused by *S. pyogenes* infection.

Treatment

1. IV antibiotics
   Rocephin
   Klacid
   Acyclovir
   Later changed to:
   Ceftriaxone
   Clindamycin
   Vancomycin

Metronidazole.
2. Dexamethasone was added to decompress the brain.
3. Surgical intervention for draining the empyema.
4. Right parasagittal interhemispheric subdural empyema evacuation.

Question and answer

1. What is a subdural empyema?
   A subdural empyema is a localized intracranial collection of pus and infected cell and tissue debris in the subdural space between the dura mater and the arachnoid.
2. How does meningitis facilitate the spread of infection in the brain?
   Meningitis is a condition that causes the meninges surrounding the brain and the spinal cord to be inflamed due to widespread inflammation of the fluid surrounding the meninges. This compromises the integrity of the meninges, and the lack of an anatomical barrier in the subdural space facilitates the spread of infection to different parts of the brain, resulting in widespread pathological effects.
3. Why are *S. pyogenes* characterized as group A ß-hemolytic streptococci?
   S. pyogenes are characterized as group A ß-hemolytic streptococci because they cause complete lysis of red blood cells.
4. How are *S. pyogenes* transmitted?
   S. pyogenes spread through direct or indirect contact with contaminated respiratory droplets. *S. pyogenes* can colonize the upper respiratory tract of humans without causing any symptoms and result in a carrier-state, which further facilitates the transmission of the organism. Trauma, abrasions, or lesions can compromise physical barriers and facilitate the entry of the bacteria into the host. Myositis, cellulitis, erysipelas, and necrotizing fasciitis are some of the manifestations of invasion of the muscle and the fascia by *S. pyogenes*.
5. How does the capsular layer of *S. pyogenes* protect the bacteria against phagocytosis?
   The capsular layer of the *S. pyogenes* bacteria is comprised of hyaluronic acid, which is not antigenic in humans. Therefore it facilitates immune evasion by providing protection against phagocytosis.
6. What is the significance of the M protein in the virulence of *S. pyogenes*?
   The M protein in the cell wall is known to be the most important virulence factor for *S. pyogenes*, and it enables the bacteria to evade the immune system by inhibition of opsonization. Adherence of the

bacteria to host cells is very important for infection to occur, and the M protein, lipoteichoic acid, and protein F in *S. pyogenes* facilitate this process.

7. How can a past *S. pyogenes* infection be detected?
   Streptokinase, streptodornase, and streptolysins are bacterial virulence factors that facilitate the invasion of tissues by *S. pyogenes*. Antibodies to either one of the antigenic streptolysin O, streptokinase, or streptodornase is indicative of a retrospective *S. pyogenes* infection.
8. How do streptokinase and streptodornase contribute to the virulence of *S. pyogenes*?
   Streptokinase has the ability to break down the fibrin barrier around lesions, thereby facilitating an unhindered spread of infection to the surrounding tissues. On the other hand, the ability of streptodornase to liquefy thick pus is responsible for the thin serous character of the exudate characteristic of streptococcal suppurative lesions caused by *S. pyogenes* infection.
9. How is the diagnosis of subdural empyema confirmed?
   The confirmation of the diagnosis of subdural empyema relies on cranial imaging. A hypodense area on the cerebral hemisphere, or along the falx with well-delineated collection margins seen in a computer tomography scan of the brain performed with an intravenous infusion of a contrast dye, is indicative of a subdural empyema. Such scans also facilitate the analysis of the extent and size of the subdural empyema, indicated by the shifting of midline structures. Magnetic resonance imaging with intravenous gadolinium enhancement is the diagnostic method of choice for intracranial subdural empyema, as it provides better visibility of interhemispheric small collections. Hypointense areas that appear crescent-shaped or elliptical underneath the skull or adjacent to the falx cerebri indicate a subdural empyema. MRI diffusion-weighted imaging sequences are now preferred over routine MRI sequences due to higher sensitivity profiles.

Learning objectives

1. Determine the appropriate differential diagnosis based on patient history, clinical manifestations, disease progression, and imaging studies.
2. Assess the primary cause of the subdural empyema.
3. Identify the etiological agent responsible for sinusitis and meningitis in the particular case.
4. Describe the characteristics of *S. pyogenes*, the bacteria responsible for the manifestation of subdural empyema secondary to sinusitis and meningitis.

5. Predict the prognosis of the case based on the clinical manifestations, disease progression, and the extent of pathology presented.
6. Assess the need and extent of surgical intervention required for the prevention of further pathological effects elicited by the subdural empyema.
7. Determine the appropriate antibiotic regimen based on the clinical manifestations, progression of the disease, and the septic workup reports to ensure rapid and effective resolution of the pathological manifestations, and avoid any further debilitating complications.

Discussion

Differential diagnosis

Cerebral abscess	Inflammation and accumulation of infected cell and tissue debris and pus from a primary infection of the ear, teeth, paranasal sinuses, epidural, lungs, heart, or kidneys can result in the formation of an abscess in the cerebrum. The condition manifests with a fever, headaches, nausea, vomiting, localized pain, alteration of mental status, muscle weakness, poor coordination, speech problems, vision-related problems, and seizures.
Epidural abscess	The accumulation of pus between the outer covering and the bones of the brain and the spinal cord results in the formation of an epidural abscess. The condition manifests with a fever, headaches, disturbed consciousness, inability to walk or paralysis, bladder and bowel incontinence, back pain, fatigue, weakness in the limbs, poor coordination, nausea, vomiting, and tingling or other unusual sensations.
Subdural hygroma	A subdural hygroma is a condition resulting from the accumulation of cerebrospinal fluid under the dural membrane of the brain due to trauma or infection. The condition may present with nausea, vomiting, neurological deficits, and seizures.
Meningitis	Meningitis is a condition resulting from an inflammation of the meninges surrounding the brain and the spinal cord. This inflammation is secondary to an infection of the fluid surrounding the meninges. Meningitis can manifest as a fever, headache, photosensitivity, neck stiffness, digestion problems, nausea, vomiting, numbness in different parts of the face, lack of thirst and appetite, and drowsiness.
Subdural hematoma	Tears in the bridging veins that cross the subdural space, following a trauma, result in bleeding and the accumulation of blood between the inner layer of the dura mater and the arachnoid mater of the meninges surrounding the brain, forming a subdural hematoma. The condition presents with loss of consciousness, altered mental status, seizures, numbness, headaches, pain, erratic breathing patterns, nausea, vomiting, personality changes, loss of muscle control, speech-related problems, vision-related problems, tinnitus, and compromised hearing ability.

Continued

Continued	
Leptomeningeal disease	Leptomeningeal disease results from multifocal metastases of cancerous cells to the leptomeninges, comprising of the arachnoid and pia mater. This complicated condition manifests with headaches, double vision, nausea, vomiting, weakness, fatigue, bladder and bowel incontinence, ataxia, neurological deficits, and several other neurological manifestations depending on the location of the metastasized cancer cells.
Cerebritis	Cerebritis is a condition characterized by the inflammation of the cerebrum, which may be secondary to an autoimmune inflammation of the brain. The condition may present with mild or severe symptoms such as headaches, vision-related problems, behavioral changes, anxiety, depression, dizziness, seizures, dementia, and stroke.
Brain tumor	A brain tumor results from an abnormal growth of cells, which may be benign or malignant in the brain. The clinical manifestations depend on the location and size of the tumor, headaches being the most common manifestation. The headaches caused by brain tumors occur even while sleeping, worsen when waking up, coughing, sneezing, or exercising. Some of the general manifestations include vomiting, confusion, seizures, double or blurred vision, altered mental status, numbness or partial paralysis, clumsiness, loss of taste, loss of smell, auditory problems, memory loss, difficulty in writing or reading, decreased alertness, loss of balance, difficulty swallowing, bladder and bowel incontinence, ataxia, speech-related problems, difficulty in communicating and understanding communications, and personality changes.

Background

Subdural empyema results from the collection of pus and infected cell and tissue debris in the subdural space between the dura mater and the arachnoid.[20] Bacterial infections are especially notorious for causing subdural empyemas as secondary complications of primary infections elsewhere in the body, such as sinusitis, otitis media, or even localized trauma due to accidents or surgeries. Immunocompromised individuals may develop subdural empyemas resulting from fungal infections as well, although this is rare in immunocompetent individuals. Subdural empyemas are usually preceded by meningitis, wherein the meninges surrounding the brain and the spinal cord are inflamed due to the fluid surrounding the meninges being swollen. The lack of an anatomical barrier in the subdural space facilitates the spread of the infection throughout the different parts of the brain, resulting in widespread pathological effects.[21] The rigid and confined space within the skull increases the risk of severe debilitating manifestations or even death by direct compression injury of the brain. Therefore prompt diagnosis and appropriate treatment is very important for ensuring favorable outcomes.

S. pyogenes is a bacterial human pathogen implicated in several invasive infections. Complications of infections resulting from untreated or ineffectively treated *S. pyogenes* infection are associated with high levels of morbidity and mortality. Subdural empyemas secondary to sinusitis caused by *S. pyogenes* need to be diagnosed and treated effectively to prevent the development of debilitating and life-threatening conditions.

Etiological agent

Streptococcus pyogenes

Classification and morphology

S. pyogenes are spherical or oval shaped cocci arranged in pairs within chains (Fig. 2.11). These are Gram-positive, catalase-negative, coagulase-negative pathogenic bacteria, and are facultative anaerobes. *S. pyogenes* are characterized as group A ß-hemolytic streptococci because they cause complete lysis of red blood cells. These bacteria are nonmotile and nonsporing organisms.

Epidemiology

S. pyogenes is a human-specific pathogen, which invades and colonizes the pharynx, anus, and genital mucosa. *S. pyogenes* can colonize the upper respiratory tract of humans without causing any symptoms, thus resulting in a carrier-state, which facilitate the transmission of the organism. Infections caused by *S. pyogenes* are extremely contagious, and can spread through

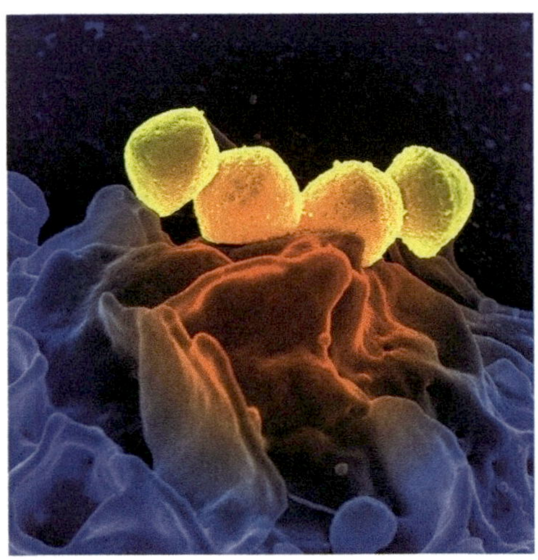

FIGURE 2.11
S. pyogenes bacteria on a neutrophil. Source: *From US Government department: Public Health Image Library, Centers for Disease Control and Prevention.*

transmission of infected droplets. Trauma, abrasions, or lesions can facilitate the entry of the bacteria into the host. Myositis, cellulitis, erysipelas, and necrotizing fasciitis are some of the manifestations of invasion of the muscle and the fascia by *S. pyogenes*. Septicemia and toxic shock syndrome are common complications arising from *S. pyogenes* infection of various internal organs.[22]

Virulence and pathogenicity

S. pyogenes is implicated in pyogenic infections that tend to spread locally, along lymphatics and through the bloodstream.[23] The clinical manifestations of *S. pyogenes* infections depend on the invaded organ system, and are elicited by multiple virulence factors. The *S. pyogenes* capsule is comprised of hyaluronic acid, which is not antigenic in humans. Therefore it facilitates immune evasion by providing protection against phagocytosis. The bacterial cell wall comprises of the outer layer of protein and lipoteichoic acid and an inner layer of peptidoglycan with a middle layer of a group-specific carbohydrate. The M protein in the cell wall is known to be the most important virulence factor for *S. pyogenes*, and it enables the bacteria to evade the immune system by inhibition of opsonization.[24] Adherence of the bacteria to host cells is very important for infection to occur, and the M protein, lipoteichoic acid, and protein F in *S. pyogenes* facilitate this process. The peptidoglycan layer is responsible for the pyrogenic and thrombolytic activities exhibited by *S. pyogenes*. Exotoxins, such as the erythrogenic toxin, have been implicated in the rashes seen in scarlet fever and in severe manifestations like toxic shock syndrome associated with *S. pyogenes* infection.[25]

Streptokinase, streptodornase, and streptolysins are some of the other virulence factors, which facilitate the invasion of tissues by *S. pyogenes*. Antibodies to either one of the antigenic streptolysin O, streptokinase, or streptodornase can indicate a retrospective *S. pyogenes* infection.[23] The ability to break down the fibrin barrier around lesions conferred on *S. pyogenes* by streptokinase facilitates the unhindered spread of infection to the surrounding tissues. *S. pyogenes* is known to cause suppurative lesions. The thin serous character of the exudate in characteristic of streptococcal lesions is attributable to the ability of streptodornase to liquefy thick pus. Localized collections of thick exudates, such as that seen in cases of empyema, are liquefied using a therapeutic preparation containing streptokinase and streptodornase.[26]

Antigenic cross-reactions between various virulence factors of *S. pyogenes* and various tissues of the human body have been implicated in the wide-spectrum of clinical manifestations of streptococcal infections, especially the late-onset complications, such as acute rheumatic fever and long-term injury to the cardiac valves, resulting in rheumatic heart disease[27] (Fig. 2.12).

Clinical manifestations and complications

A clinical diagnosis of *S. pyogenes* infection usually relies on patient history and findings upon physical examination. The clinical manifestations of *S. pyogenes* infections vary depending on the organ system that has been affected by the infection.

The most common presentation of streptococcal infections is streptococcal pharyngitis (Fig. 2.13), commonly known as a sore throat, which presents with a sudden onset of fever, chills, malaise, pharyngeal exudate, tender cervical lymphadenopathy, whitish patches at the back of the throat, and enlarged tonsils.[28] The spread of the bacteria from the throat to the surrounding tissues may occur in some cases, and such cases may present with otitis media, mastoiditis,

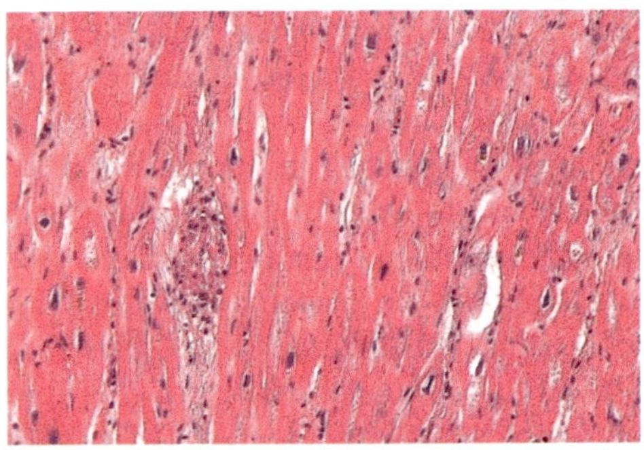

FIGURE 2.12

H & E stained micrograph of cardiac pathology seen in rheumatic heart disease, caused by *S. pyogenes*. The micrograph shows Anitschkow cells and Aschoff bodies characteristic of rheumatic heart disease. Source: *From Nephron/Own work/; https://creativecommons.org/licenses/by-sa/3.0/deed.en; https://commons.wikimedia.org/wiki/File:Rheumatic_heart_disease_-_3_-_high_mag.jpg.*

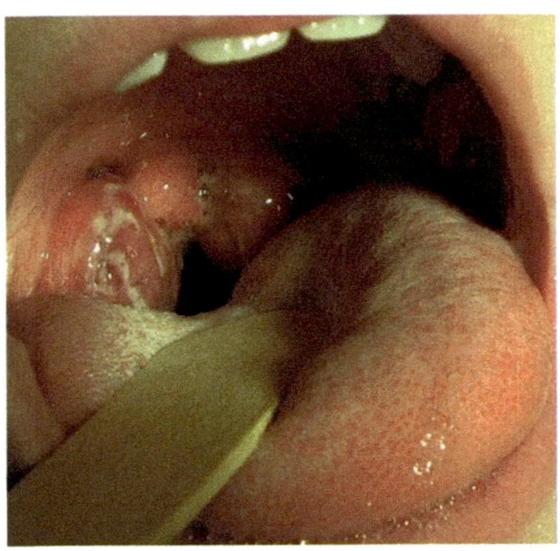

FIGURE 2.13

Tonsillar exudate in strep throat. Source: *From James Heilman/Own work/; https://creativecommons.org/licenses/by-sa/3.0/deed.en; https://commons.wikimedia.org/wiki/File:Strep_throat2010.JPG.*

suppurative adenitis, quinsy, Ludwig's angina, meningitis, glomerulonephritis, and streptococcal pneumonia.[23]

A type of skin infection caused by *S. pyogenes* in young children, known as nonbullous impetigo, typically presents with an itchy reddish rash that appears around the mouth or nose and goes on to form fluid-filled blisters later on, which rupture easily and are covered with a honey-colored crust. Impetigo lesions generally appear on the face and the lower extremities and are usually localized with no systemic manifestations. Rare complications, such as glomerulonephritis, can be associated with impetigo. Erysipelas is another skin infection caused by *S. pyogenes*, but unlike impetigo, it is usually seen in older patients. The superficial lymphatics are affected, resulting in a diffuse infection, presenting with red, swollen, indurated, and a sharply demarcated area of skin (Fig. 2.14).

Scarlet fever (Fig. 2.15), on the other hand, exhibits widespread systemic manifestations, such as a high-grade fever, sore throat, strawberry-like tongue (Fig. 2.16), and a blanchable, papular, nonconfluent rash that lasts for 7 – 10 weeks, followed by desquamation (Fig. 2.17) on the palms and soles.[29]

The *S. pyogenes* infections that result in invasion of the soft tissues exhibit severe symptoms and are associated with unfavorable outcomes, such as

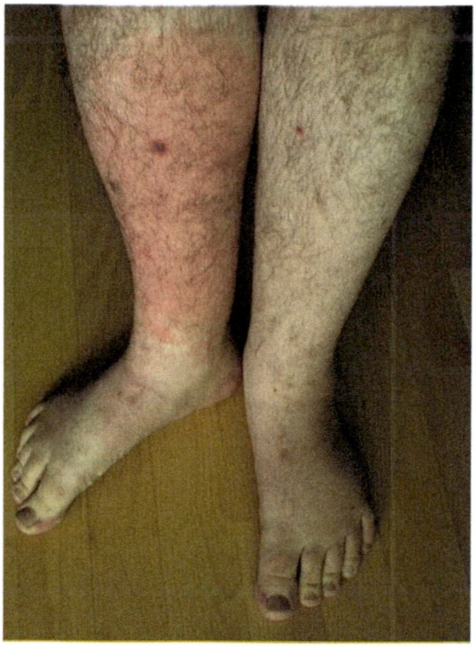

FIGURE 2.14

Erysipelas of the leg. Source*: From Tommi Nummelin/Own work/; https://creativecommons.org/licenses/by-sa/3.0/deed.en; https://commons.wikimedia.org/wiki/File:Erysipelas_in_a_foot.jpg.*

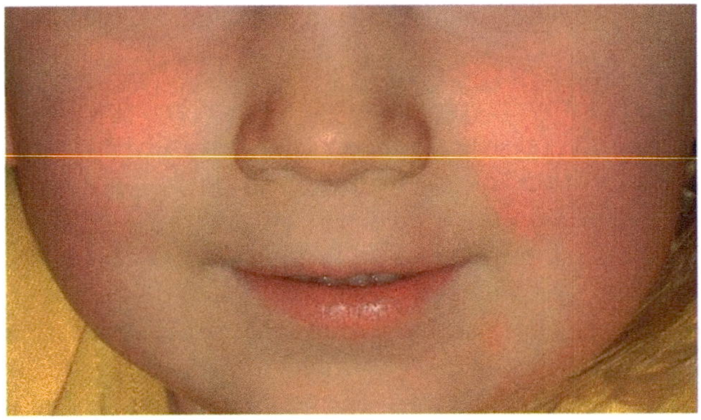

FIGURE 2.15
Scarlet fever rash. Source: *From Estreya at English Wikipedia|Permission = CC-BY-2.5; CC-BY-SA-2.5. Modified by Grook Da Oger/ Originally from en.wikipedia/; https://creativecommons.org/licenses/by-sa/4.0/; https://commons.wikimedia.org/wiki/File:Scarlet_fever_1.1.JPG.*

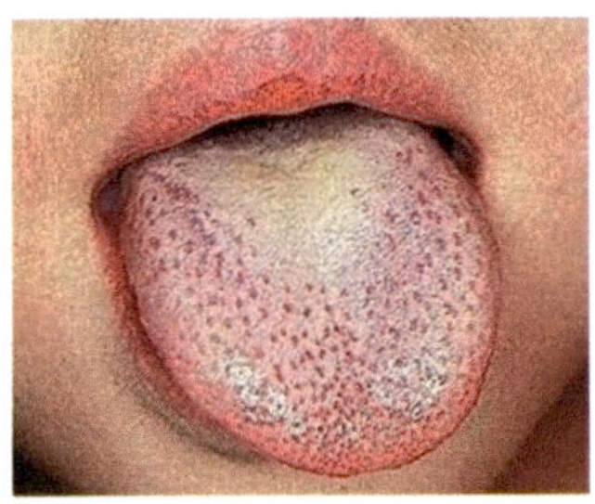

FIGURE 2.16
Strawberry-tongue typical of scarlet fever. Source: *From Afag Azizova/Own Work/; https://creativecommons.org/licenses/by-sa/3.0/deed.en; https://commons.wikimedia.org/wiki/File:Skarlatina.jpg.*

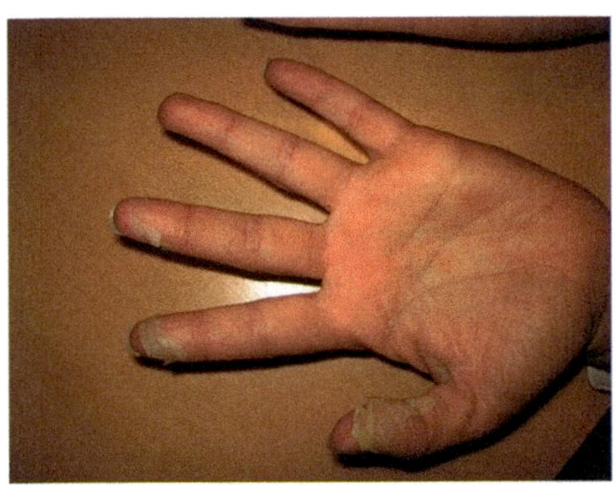

FIGURE 2.17
Desquamation of skin from finger-tips 8 days after scarlet fever. Source: *From Martin Kronawitter, Kellberg/Self-photographed/; https://creativecommons.org/licenses/by-sa/3.0/deed.en; https://commons.wikimedia.org/wiki/File:Scharlach2.jpg.*

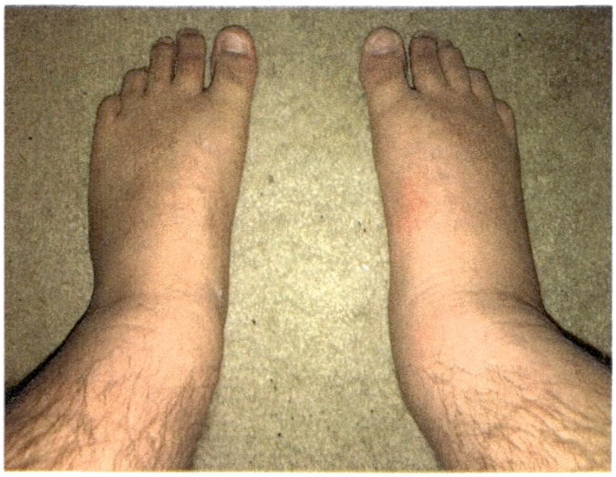

FIGURE 2.18
Cellulitis of the lower leg and foot with evident swelling. Source*: From cbinrva/Own work/; https://creativecommons.org/licenses/by-sa/4.0/deed.en; https://commons.wikimedia.org/wiki/File:Cellulitis_of_lower_leg_and_foot.jpg.*

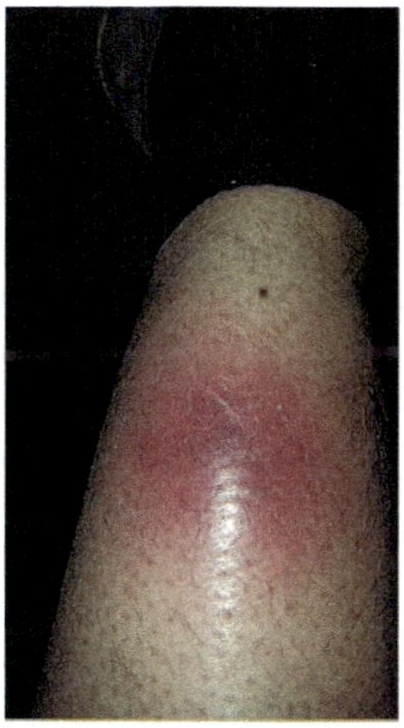

FIGURE 2.19
Early symptom of necrotizing fasciitis. Source*: From Morphx 1982/Own work/; https://creativecommons.org/licenses/by-sa/4.0/deed.en; https://commons.wikimedia.org/wiki/File:Very_early_symptom_of_NF.jpg.*

shock and multiorgan failure. Necrotizing fasciitis and cellulitis (Fig. 2.18) are examples of such invasive infections of the subcutaneous tissue responsible for the rapid destruction of fascia and fat. Incompetence of the immune system, or exposure to the organism due to compromised asepsis during surgical procedures, burns, blunt trauma, minor laceration, and childbirth can result in contraction of the infection. Necrotizing fasciitis is an aggressive and rapidly progressing condition that manifests as localized pain, necrosis of the infected skin lesion, swelling, redness, edema, increased heart rate, and fever with indications of septic shock in severe cases[30] (Figs. 2.19 and 2.20).

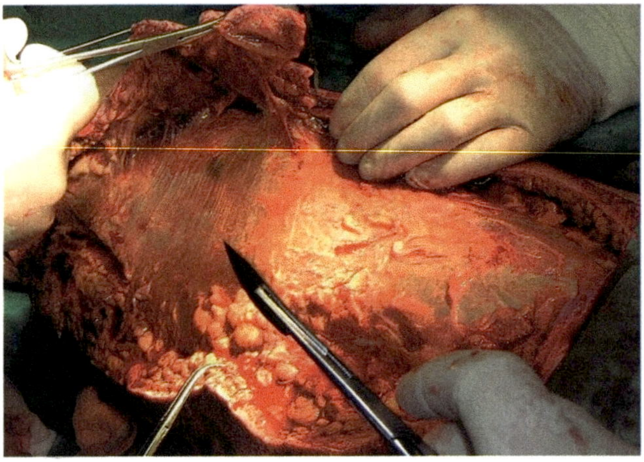

FIGURE 2.20

Surgical debridement of a leg affected by necrotizing fasciitis. Source*: From Piotr Smuszkiewicz, Iwona Trojanowska and Hanna Tomczak/; https://casesjournal.biomedcentral.com/articles/10.1186/1757-1626-1-125/; https://creativecommons.org/licenses/by/2.0/deed.en; https://commons.wikimedia.org/wiki/File:Necrotizing_fasciitis_left_leg_debridement.JPEG.*

S. pyogenes infections can also manifest as genital infections, such as puerperal sepsis. The rupture of membranes involved in labor and parturition result in increased vulnerability to endogenous streptococcal infections, although exogenous sources of infection can be prevented by implementing appropriate asepsis protocols.[31] Puerperal sepsis manifests with a fever, rigors, and tachycardia within 48 hours to up to 7 days in some cases.

Laboratory diagnosis

The diagnosis of subdural empyema caused by *S. pyogenes* infection can be suspected based on collated information, including patient history and physical examination. The confirmation of the diagnosis of subdural empyema, however, relies on cranial imaging.[32,33] A hypodense area on the cerebral hemisphere, or along the falx with well-delineated collection margins seen in a computer tomography scan of the brain performed with an intravenous infusion of a contrast dye, is indicative of a subdural empyema[32] (Fig. 2.21). Such scans also facilitate the analysis of the extent and size of the subdural empyema, indicated by the shifting of midline structures. Magnetic resonance imaging with intravenous gadolinium enhancement is the diagnostic method of choice for intracranial subdural empyema as it provides better visibility of interhemispheric small collections. Hypointense areas that appear crescent-shaped or elliptical underneath the skull or adjacent to the falx cerebri indicate a subdural empyema[34] (Fig. 2.22). MRI diffusion-weighted imaging sequences are now preferred over routine MRI sequences due to higher sensitivity profiles.[35]

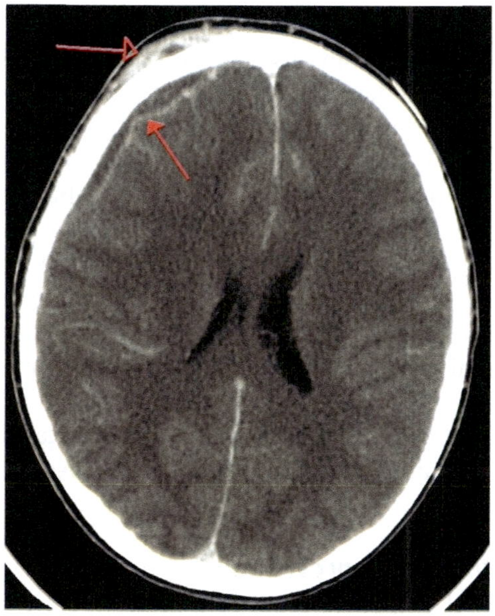

FIGURE 2.21

Intracranial subdural empyema as seen on a CT scan. Source*: From James Heilman/Own work/; https://creativecommons.org/licenses/by-sa/3.0/deed.en; https://commons.wikimedia.org/wiki/File:Subduralempyemaandskinabscess.png.*

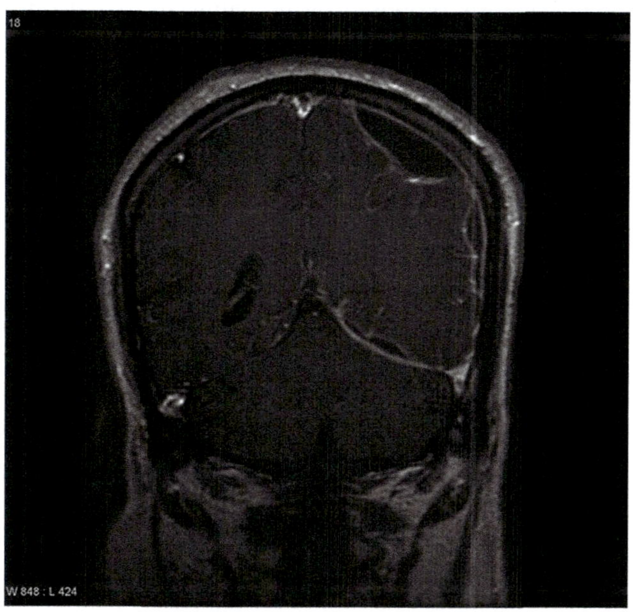

FIGURE 2.22

MRI showing extensive subdural collection with marginal enhancement, indicative of subdural empyema. Source*: From Frank Gaillard/Own work/; https://radiopaedia.org/cases/subdural-empyema/; https://creativecommons.org/licenses/by-sa/3.0/deed.en; https://commons.wikimedia.org/wiki/File:Subdural-empyema.jpg.*

Septic workup can also be used to strengthen the suspected diagnosis. A primary indication of an active infection process could be leukocytosis and neutrophilia on complete blood count analysis, although immunocompromised individuals may not present with these parameters. Serum inflammation markers considered as acute phase reactants, such as erythrocyte sedimentation rate and C-reactive protein could further emphasize on the prevalence of a systemic inflammatory state. Although blood culture samples may be recommended to ensure appropriate antibiotic therapy, the identification of the organism is usually difficult because the infection in cases of subdural empyema is encapsulated.[36] Negative culture results may also be obtained due to the antibiotic treatment prior to the sample collection. Lumbar punctures are usually avoided in cases of subdural empyema due to the possibility of increased intracranial pressure.[21,34]

Treatment and prevention

Subdural empyema requires a multimodal treatment approach that encompasses surgical intervention as well as antibiotic and antiinflammatory therapies. Irrespective of the choice of therapeutic options, a prompt and accurate diagnosis is essential to ensure a good prognosis and a favorable clinical outcome. The time taken for the clinical diagnosis is influenced by the differential diagnosis proposed by the clinician, based on patient history, physical examination, laboratory workup, and imaging findings.

Drainage of the pus, complete evacuation of the subdural purulent collection, and debridement of infected tissue through multiple burr holes, or a craniotomy preceded by preoperative imaging, may be indicated.[32,37] Samples should be collected during the procedure and cultured for aerobic and anaerobic organisms.[32] Antibiotic therapy based on the results of Gram staining should be initiated immediately after the surgical intervention without waiting for the culture reports. Adjustments to the antibiotic therapy can be made later based on the culture reports. Broad-spectrum antibiotic therapy is recommended for cases wherein the results of Gram staining is negative, which could be due to the administration of an antibiotic cover prior to the surgical intervention.[34]

Patients usually require 3–6 weeks of intravenous antibiotic therapy, although cases of osteomyelitis often require longer periods of therapy. Subdural empyema preceded by meningitis and develops as a secondary complication to sinusitis may require 6–8 weeks of antibiotic therapy.[38] Surgical intervention for complete evacuation of subdural empyema is often associated with postoperative seizures, thereby requiring a prescription of antiepileptic drugs.[37]

Follow-ups through laboratory workups on an outpatient basis should be recommended. Repeat CT and MRI scans may also be required in some cases.

Student Notes

Key take-aways

- Subdural empyema is an intracranial condition, resulting from the collection of pus and infected cell and tissue debris in the subdural space between the dura mater and the arachnoid.
- Bacterial infections of the sinuses or the meninges surrounding the brain and the spinal cord can cause subdural empyema.
- *S. pyogenes* is a bacterial species belonging to the *Streptococcus* genus, and is one of the etiological agents responsible for subdural empyema.
- Subdural empyema can lead to severe complications and can even prove to be fatal due to the confined intracranial space.
- Prompt diagnosis and accurate treatment is essential to ensure a good prognosis and outcome in cases of subdural empyema.
- Imaging studies, such as MRI and CT scans, are considered to be the gold standard for confirmation of clinical diagnosis of subdural empyema.
- Culture studies may be inconclusive or negative due to the encapsulated nature of the infection and the administration of preoperative antibiotics.
- Treatment of subdural empyema generally includes both surgical intervention and antibiotic and antiinflammatory treatment.

References

1. Afshar M, Raju M, Ansell D, Bleck TP. Narrative review: tetanus—a health threat after natural disasters in developing countries. *Ann Intern Med.* 2011;154(5):329. Rush University Medical Center, Chicago, Illinois 60612, USA.
2. ACIP. Diphtheria, tetanus, and pertussis: Recommendations for vaccine use and other preventive measures. *Recommendations Advisory Comm Immun Pract (ACIP) MMWR.* 1991;40(No. RR-10):1–28.
3. ACIP. Pertussis vaccination: use of acellular pertussis vaccines among infants and young children. Recommendations of the Advisory Committee on Immunization Practices (ACIP). *MMWR.* 1997;46(No. RR-7):1–25.
4. ACIP. Updated recommendations for use of tetanus toxoid, reduced diphtheria toxoid and acellular pertussis (Tdap) vaccine from the Advisory Committee on Immunization Practices, 2010. *MMWR.* 2011;60(1):13–15.
5. Ananthanarayan R, Paniker CK. 8th ed. *Textbook of Microbiology. Clostridium.* 28. India: Universities Press; 2009:256–261.
6. George EK, De Jesus O, Vivekanandan R. *Clostridium tetani* [Updated February 7, 2021] *StatPearls [Internet]*. Treasure Island (FL): StatPearls Publishing; 2021, January. Available from: https://www.ncbi.nlm.nih.gov/books/NBK482484/.
7. Mueller NH, Gilden DH, Cohrs RJ, Mahalingam R, Nagel MA. Varicella zoster virus infection: clinical features, molecular pathogenesis of disease, and latency. *Neurol Clin.* 2008;26(3):675. Available from: https://doi.org/10.1016/j.ncl.2008.03.011. viii.
8. Cvjetković D, Jovanović J, Hrnjaković-Cvjetković I, Brkić S, Bogdanović M. Reaktivacija herpes zoster infekcije varicela-zoster virusom. Reactivation of herpes zoster infection by varicella-zoster virus. *Med Pregl.* 1999;52(3–5):125–128.

9. Oxman MN. Herpes zoster pathogenesis and cell-mediated immunity and immunosenescence. *J Am Osteopath Assoc.* 2009;109(6 Suppl 2):S13–S17.

10. Nair PA, Patel BC. *Herpes zoster (shingles)* [Updated August 8, 2020] *StatPearls [Internet]*. Treasure Island (FL): StatPearls Publishing; 2020, January. Available from: https://www.ncbi.nlm.nih.gov/books/NBK441824/.

11. Arvin AM. Humoral and cellular immunity to varicella-zoster virus: an overview. *J Infect Dis.* 2008;197(Issue Supplement_2):S58–S60. Available from: https://doi.org/10.1086/522123.

12. Sampathkumar P, Drage LA, Martin DP. Herpes zoster (shingles) and postherpetic neuralgia. *Mayo Clin Proc.* 2009;84(3):274–280. Available from: https://doi.org/10.1016/S0025-6196(11)61146-4.

13. Syed YY. Recombinant zoster vaccine (Shingrix®): a review in herpes zoster. *Drugs Aging.* 2018;35(12):1031–1040. Available from: https://doi.org/10.1007/s40266-018-0603-x.

14. Yee ME, Bakshi N, Graciaa SH, et al. Incidence of invasive *Haemophilus influenzae* infections in children with sickle cell disease. *Pediatr Blood Cancer.* 2019;66(6):e27642. Available from: https://doi.org/10.1002/pbc.27642.

15. Romaneli MTDN, Tresoldi AT, Pereira RM, Garcia MT, Hofling CC, Resende MR. Invasive non-type b *Haemophilus influenzae* disease: report of eight cases. *Rev Paul Pediatr.* 2019;37(2):156–160. Available from: https://doi.org/10.1590/1984-0462/;2019;37;2;00006.

16. Erwin AL, Smith AL. Nontypeable *Haemophilus influenzae*: understanding virulence and commensal behavior. *Trends Microbiol.* 2007;15(8):355–362. Available from: https://doi.org/10.1016/j.tim.2007.06.004. Epub June 27, 2007.

17. Ananthanarayan R, Paniker CK. 8th ed. *Textbook of Microbiology*. 24. India: Universities Press; 2009:218–223.

18. Khattak ZE, Anjum F. *Haemophilus influenzae* [Updated September 15, 2020] *StatPearls [Internet]*. Treasure Island (FL): StatPearls Publishing; 2020, January. Available from: https://www.ncbi.nlm.nih.gov/books/NBK562176/.

19. Bae C, Bourget D. *Periorbital cellulitis. StatPearls [Internet]*. Treasure Island (FL): StatPearls Publishing; 2020, January. [Updated July 21, 2020]. Available from: https://www.ncbi.nlm.nih.gov/books/NBK470408/.

20. De Bonis P, Anile C, Pompucci A, Labonia M, Lucantoni C, Mangiola A. Cranial and spinal subdural empyema. *Br J Neurosurg.* 2009;23(3):335–340. Available from: https://doi.org/10.1080/02688690902939902.

21. Greenlee JE. Subdural empyema. *Curr Treat Options Neurol.* 2003;5(1):13–22. Available from: https://doi.org/10.1007/s11940-003-0019-7.

22. Cunningham MW. Pathogenesis of group A streptococcal infections. *Clin Microbiol Rev.* 2000;13(3):470–511. Available from: https://doi.org/10.1128/cmr.13.3.470-511.2000.

23. Ananthanarayan R, Paniker CK. 8th ed. *Textbook of Microbiology*. 23. India: Universities Press; 2009:205–212.

24. Maxted WR. The indirect bactericidal test as a means of identifying antibody to the M antigen of *Streptococcus pyogenes*. *Br J Exp Pathol.* 1956;37(4):415–422.

25. Musser JM, Hauser AR, Kim MH, Schlievert PM, Nelson K, Selander RK. *Streptococcus pyogenes* causing toxic-shock-like syndrome and other invasive diseases: clonal diversity and pyrogenic exotoxin expression. *Proc Natl Acad Sci USA.* 1991;88(7):2668–2672. Available from: https://doi.org/10.1073/pnas.88.7.2668.

26. Hynes W, Sloan M. Secreted extracellular virulence factors, February 10, 2016 [Updated March 25, 2016] In: Ferretti JJ, Stevens DL, Fischetti VA, eds. Streptococcus pyogenes: *Basic Biology to Clinical Manifestations [Internet]*. Oklahoma City (OK): University of Oklahoma Health Sciences Center; 2016. Available from: https://www.ncbi.nlm.nih.gov/books/NBK333411/.

27. Thaper D, Prabha V. Molecular mimicry: an explanation for autoimmune diseases and infertility. *Scand J Immunol*. 2018;88:e12697. Available from: https://doi.org/10.1111/sji.12697.
28. Ebell MH, Smith MA, Barry HC, Ives K, Carey M. The rational clinical examination. Does this patient have strep throat? *JAMA*. 2000;284(22):2912–2918. Available from: https://doi.org/10.1001/jama.284.22.2912.
29. Basetti S, Hodgson J, Rawson TM, Majeed A. Scarlet fever: a guide for general practitioners. *Lond J Prim Care (Abingdon)*. 2017;9(5):77–79. Available from: https://doi.org/10.1080/17571472.2017.1365677. Published 2017 Aug 11.
30. Misiakos EP, Bagias G, Patapis P, Sotiropoulos D, Kanavidis P, Machairas A. Current concepts in the management of necrotizing fasciitis. *Front Surg*. 2014;1:36. Available from: https://doi.org/10.3389/fsurg.2014.00036.
31. Busowski MT, Lee M, Busowski JD, Akhter K, Wallace MR. Puerperal group a streptococcal infections: a case series and discussion. *Case Rep Med*. 2013;2013:751329. Available from: https://doi.org/10.1155/2013/751329.
32. Suthar R, Sankhyan N. Bacterial infections of the central nervous system. *Indian J Pediatr*. 2019;86(1):60–69. Available from: https://doi.org/10.1007/s12098-017-2477-z. Epub January 3, 2018.
33. Brouwer MC, Tunkel AR, McKhann 2nd GM, van de Beek D. Brain abscess. *N Engl J Med*. 2014;371(5):447–456. Available from: https://doi.org/10.1056/NEJMra1301635.
34. Osborn MK, Steinberg JP. Subdural empyema and other suppurative complications of paranasal sinusitis. *Lancet Infect Dis*. 2007;7(1):62–67. Available from: https://doi.org/10.1016/S1473-3099(06)70688-0.
35. Wong AM, Zimmerman RA, Simon EM, Pollock AN, Bilaniuk LT. Diffusion-weighted MR imaging of subdural empyemas in children. *AJNR Am J Neuroradiol*. 2004;25(6):1016–1021.
36. Madhugiri VS, Sastri BV, Bhagavatula ID, Sampath S, Chandramouli BA, Pandey P. Posterior fossa subdural empyema in children—management and outcome. *Childs Nerv Syst*. 2011;27(1):137–144. Available from: https://doi.org/10.1007/s00381-010-1169-z. Epub May 23, 2010.
37. French H, Schaefer N, Keijzers G, Barison D, Olson S. Intracranial subdural empyema: a 10-year case series. *Ochsner J*. 2014;14(2):188–194.
38. Jim KK, Brouwer MC, van der Ende A, van de Beek D. Subdural empyema in bacterial meningitis. *Neurology*. 2012;79(21):2133–2139. Available from: https://doi.org/10.1212/WNL.0b013e3182752d0e. Epub November 7, 2012.

CHAPTER 3

Infections of the gastrointestinal and hepatobiliary system

Problem number 3.1: *Klebsiella pneumoniae* liver abscess caused due to *Klebsiella pneumoniae* infection

A 30-year-old male presented with shortness of breath, a cough, fever, and redness in the right eye for a duration of 10 days. The patient was traveling when he gradually started developing increasing upper respiratory symptoms that progressed to shortness of breath and a fever with chills. A prescription of oral antibiotics from a local healthcare establishment did not provide any relief from the symptoms. This was followed by swelling, pain, and redness of the right eye along with intermittent episodes of headaches. On returning from his trip, he developed significant photophobia and loss of vision in the right eye, in addition to the symptoms mentioned above. The patient also reported noticeably dark urine with urethral discharge but denied any history of joint pain or genital ulcers.

The following are the reports of the investigations conducted:

1. Vitals
   Temperature: 39.2°C
   Blood pressure: Normal
   Oxygen saturation on room air: Normal
2. Physical examination
   Conscious, oriented, and not distressed at presentation.
   Respiratory system: Reduced air entry bilaterally with a diastolic murmur in the mitral valve area.
   Abdomen: Normal
   Neurological: Normal
   Eye: Severe conjunctival chemosis, clear cornea, exudative membrane over the lens and posterior synechia of the right eye.
3. Serology
   CBC: Leukocytosis with left shift
   ESR: Elevated
   Renal and liver function tests: Normal

CONTENTS

Problem-Based Learning Approach in Microbiology. DOI: https://doi.org/10.1016/B978-0-323-95092-3.00009-3

4. Radiological studies
   Computed tomography (CT) study
   Multiple variable size nodules were observed in both lungs in the upper and lower lobes. Some of the nodules showed cavitation with gas. The largest nodule measured 3 cm with minimal pleural effusion. The eighth segment of the right hepatic lobe displayed well-defined irregular hypodense lesions along the hepatic wall. A density value of 10–20 HU was obtained within the lesion. The lesion was most consistent with a liver abscess. The rest of the study, however, yielded insignificant and nonconclusive results (Fig. 3.1).
   Chest X-ray: Bilateral basal infiltrate (Fig. 3.2)
   Trans-esophageal echocardiogram: Normal left ventricular systolic function with a 60% ejection fraction; mild mitral and tricuspid regurgitation with no evidence of vegetation or clots.
   B-scan ultrasound of the right eye: Severe uveitis, superior posterior vitreous detachment with subretinal abscess extending temporally, suggestive of endophthalmitis.
5. Laboratory and culture studies
   Blood, eye swab, deep tracheal aspiration, and urine cultures: Sterile
   MERS-CoV PCR test: Negative
   Influenza PCR test: Negative
   Respiratory profile (atypical organisms and viruses): Negative
   Mantoux test and vasculitis screen: Negative
   Mycobacterium tuberculosis polymerase chain reaction (PCR) (tracheal aspirate): Negative
   Culture and Gram staining (fluid aspirated from the liver abscess using ultrasound guidance): Growth of pan-sensitive *Klebsiella pneumonia*.
6. Magnetic Resonance Imaging (MRI) studies (brain and orbits)
   Extensive infective etiology involving the right eyeball with possible choroiditis.
   Choroidal detachment associated with periorbital inflammatory soft tissue thickening and small pockets of fluid/abscess similar to those observed in the CT study.
   Mild peripheral enhancement of the right optic nerve with attachment of the eyeball.
   Rest of the optic nerve was unremarkable (Fig. 3.3).
   Multiple rings that enhanced small lesions involving both cerebral hemispheres and posterior fossa—most likely due to septic emboli.

Diagnosis

The final diagnosis concluded a Klebsiella pneumonia liver abscess (KPLA) with septic emboli in the brain and right eye caused by *Klebsiella pneumoniae* infection.

Treatment

1. IV meropenem and moxifloxacin
2. Oral Oseltamivir
3. Vancomycin eye drops
4. Multiple intravitreal injections of vancomycin and ceftazidime.
   After 48 hours of antibiotics, the patient did not show any clinical improvement with continuous spikes in fever up to 39°C.
   Based on culture study findings:
5. Oseltamivir and vancomycin were stopped and meropenem and moxifloxacin were continued.

The patient responded to this course of treatment and the temperature gradually subsided with these antibiotics.

Question and answer

1. What is the difference between nosocomial and community-acquired *K. pneumoniae* infections?
   Community-acquired pneumonia is an acute infection contracted by the affected individual from the community, while hospital-acquired pneumonia generally manifests more than 48 hours after hospital admission.
2. Which are the factors responsible for bacterial virulence of *K. pneumoniae*?
   The polysaccharide capsule, lipopolysaccharides, fimbriae, and siderophores that are iron-chelating compounds secreted by bacterial cells are factors responsible for bacterial virulence of *K. pneumoniae*.
3. What is the characteristic that distinguishes the physiological parameter between pneumonia caused by *Streptococcus pneumoniae* and *Klebsiella pneumoniae*?
   The type of sputum produced in community-acquired pneumonia is "blood-tinged" or "rust-colored," while the sputum produced in Klebsiella pneumonia has a "currant jelly" appearance, resulting from the significant inflammation and necrosis of the surrounding tissue caused by infection with *Klebsiella pneumoniae*.

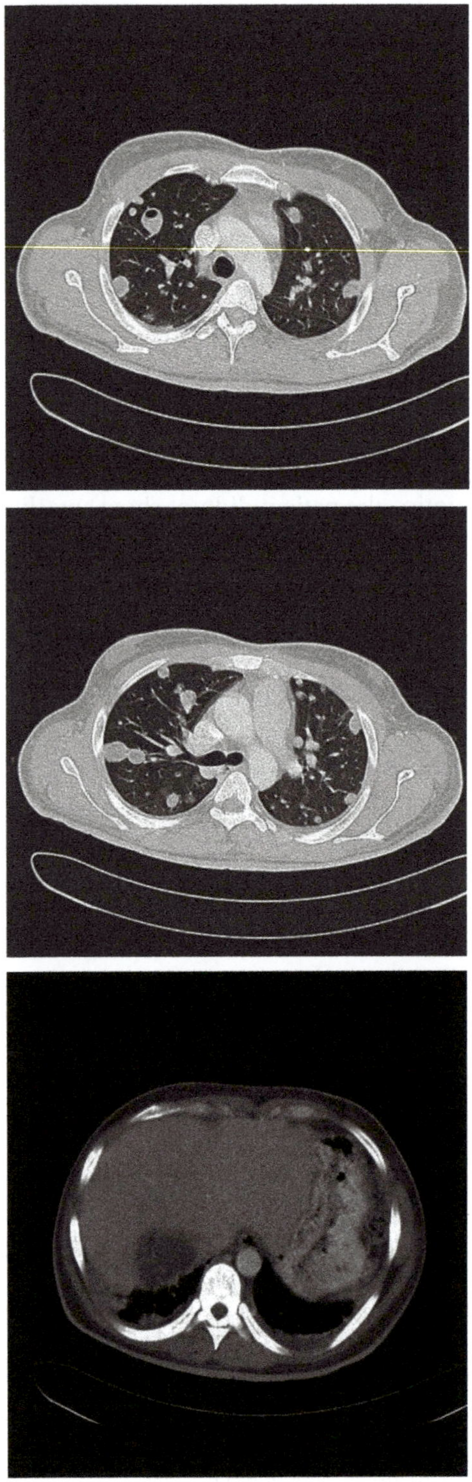

FIGURE 3.1

The computed tomography (CT) film of the abdomen showing the hypodense lesion in the right lobe of the liver. The chest cuts of the CT display the dispersed nodules in both lung fields and the cavitary lesion in the right lung.

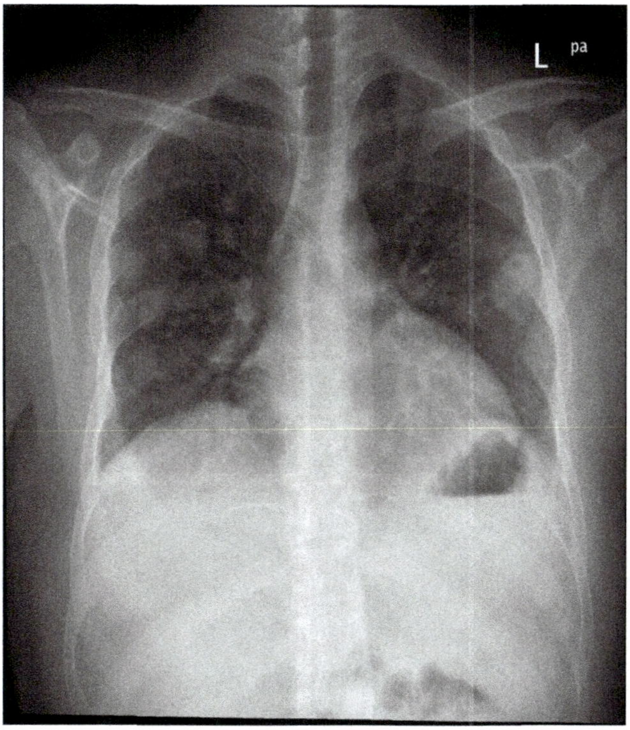

FIGURE 3.2
The chest X-ray on admission to the hospital showing bilateral infiltrate along with multiple nodules in the right lung, which are more numerous than those in the left lung.

4. What are the observations of primary physical investigations in cases of Klebsiella pneumonia?

   Physical examination of the respiratory system in Klebsiella pneumonia cases reveals pronounced signs of unilateral consolidation involving the upper lobes of the lungs, such as crepitation, bronchial breathing, and increased vocal resonance.

5. Which diagnostic parameter enables differentiation between pneumonia caused by *S. pneumoniae* and *Klebsiella pneumoniae*?

   A chest radiograph showing a lobar infiltrate, particularly in the posterior aspect of the right upper lung, differentiates pneumonia caused by *Klebsiella pneumoniae* from that caused by *S. pneumoniae*. Signs of bulging fissures caused by the large extent of infection and inflammation attributable to Klebsiella pneumonia infection is another assertive characteristic observation on chest radiographs. The final diagnosis of Klebsiella pneumonia, however, relies on

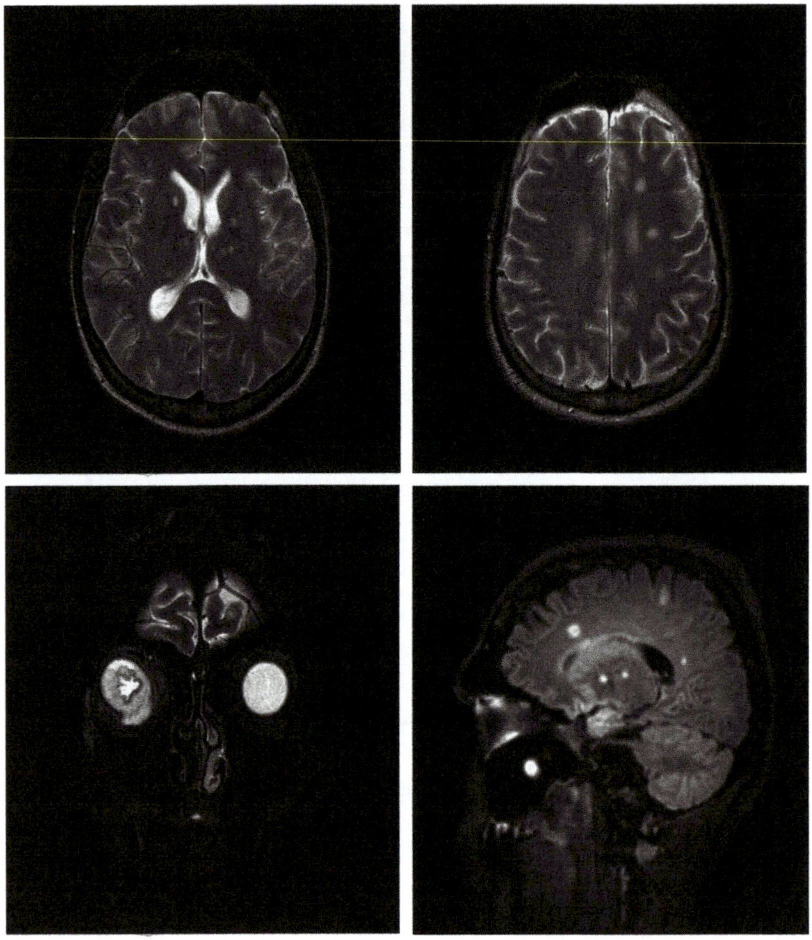

FIGURE 3.3

Magnetic Resonance Imaging (MRI) brain (different views) and orbit showing ring-enhancing lesions suggestive of multiple septic emboli with right orbit collection.

confirmation by either sputum culture analysis or blood culture analysis.

6. How are antibiotic-resistant cases of *Klebsiella pneumoniae* treated?
   Penicillin-resistant *Klebsiella pneumoniae* cases are usually treated using a course of aztreonam or a respiratory quinolone, although carbapenem-susceptible nosocomial infections are known to benefit better from monotherapies comprising of carbapenem-based antibiotics. Carbapenem-resistant cases warrant the use of tigecycline,

fosfomycin, aminoglycosides, dual therapy carbapenems, or antibiotics from the polymyxin class.

Learning objectives

1. Determine the logical flow of differential diagnosis based on patient history, clinical manifestations, physical examination, primary serological investigations, and chest radiographs.
2. Analyze the difference between nosocomial and community-acquired *K. pneumoniae* infection.
3. Describe the characteristics of *K. pneumoniae*, the etiological agent of Klebsiella pneumonia.
4. Predict the prognosis of the disease based on the health and immune status of the patient and the extent of pathology presented.
5. Interpret the correlation between the clinical manifestations and the pathogenicity of *K. pneumoniae*.
6. Summarize the various diagnostic techniques used for the diagnosis of the KPLA based on clinical manifestations of the disease, physical examination, and primary serological investigations.
7. Determine the appropriate treatment regimen based on the patient history and culture sensitivity reports.

Discussion

Differential diagnosis

Etiology	Rationale
Viral hepatitis	Viral hepatitis presents itself with a fever, fatigue, jaundice, loss of appetite, nausea, vomiting, abdominal pain, dark urine, light-colored/clay-colored stools, and joint pain.
Appendicitis	Appendicitis is characterized by pain on the right side of the lower abdomen, nausea, vomiting, loss of appetite, sharp pain originating at the naval and radiating to the lower right abdomen, and pain that increases while coughing and any other jarring movements.
Cholangitis	Cholangitis develops with pain in the upper right side of the abdomen, jaundice, fever, chills, clay-colored stools, dark urine, and low blood pressure.
Necrotic liver mass	Necrotic liver masses or inflammatory pseudo tumors commonly manifest as abdominal pain, fever, weight loss, jaundice, nausea, vomiting, dark-colored urine, and clay-colored stools.
Cholecystitis	Cholecystitis is characterized by severe pain in the upper right or central abdomen, radiating pain affecting the right shoulder and back, abdominal tenderness, nausea, vomiting, and fever.

Discussion

Background

Pyogenic liver abscess is a life-threating condition, which can be caused by bacterial or fungal infections. *Escherichia coli* has long been known as the leading cause of liver abscesses, but recently, *K. pneumoniae* is emerging as the new cause, particularly in Asia and Western countries.[1] Community-acquired KPLA syndrome has been recognized in Asia for more than 20 years, with more than 1000 reported cases by 2008.[2] The reasons for the geographic differences in the incidence of KPLA syndrome are unknown, but hypotheses should address the differences in socioeconomic factors, possible occupational exposures (e.g., food handling), deficits in host immune defenses caused by diabetes mellitus or alcoholism, and possibly differences in genetic and immunologic susceptibilities in different ethnic groups.[3]

KPLA is a community-acquired disease, with increased predisposition toward patients with diabetes and one or more septic emboli. Antibiotic resistance of the organism also affects the effectivity of the treatment, thereby adversely affecting a possible good prognosis and outcome. This warrants the need for awareness about this insidious syndrome and the ability to identify early signs of septic emboli (meningitis or endophthalmitis), as rapid initiation of effective treatment and appropriate time management remains critical for a good outcome in KPLA cases.

Etiological agent

Klebsiella pneumoniae

Classification and morphology

K. pneumoniae is a Gram-negative, encapsulated, nonmotile, rod-shaped bacterium (Figs. 3.4 and 3.5) belonging to the *Enterobacteriaceae* family. It is a facultative anaerobe. Seventy-seven different capsular types of the bacterium have been studied till date.[4]

Epidemiology

The primary reservoir for *K. pneumoniae* has been found to be humans. It is found as a normal inhabitant in the skin, intestine, and mouth of humans but can cause pathological changes to the lung tissue if aspirated. The stool and the nasopharynx of a considerable percentage of the general population show the presence of this bacterium. *K. pneumoniae* is infamous for being one of the leading causes of nosocomial infections as the gastrointestinal tracts of patients and inappropriate sanitized hands of hospital personnel serve as the breeding ground for infections. This bacterium can be responsible for community-acquired or hospital-acquired forms of pneumonia, although community-acquired pneumonia due to infection with

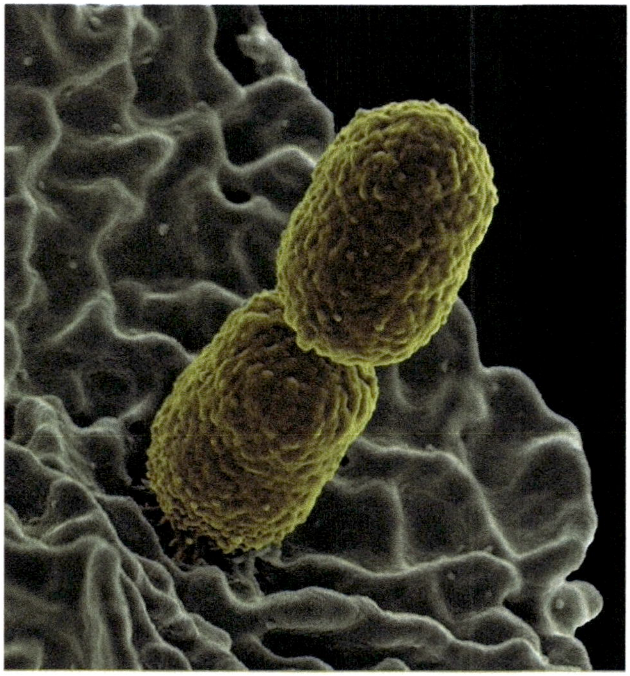

FIGURE 3.4

Scanning electron micrograph showing carbapenem-resistant *Klebsiella pneumoniae* interacting with a human neutrophil. Source*: From NIAID/; https://www.flickr.com/photos/niaid/13383411493/; https://creativecommons.org/licenses/by/2.0/deed.en; https://commons.wikimedia.org/wiki/File:Klebsiella_pneumoniae_Bacterium_(13383411493).jpg.*

K. pneumoniae is relatively uncommon. Prognosis of the disease is affected by the presence of comorbidities, such as diabetes mellitus and alcoholism. Development of septicemia is largely responsible for the high mortality rates associated with Klebsiella pneumonia.

Virulence and resistance

Many factors are responsible for the virulence of the bacterium, resulting in infection and antibiotic resistance. The polysaccharide capsule encapsulating the organism is a significant contributor to the virulence of the organism, which allows the bacteria to evade certain defensive processes of the immune system. The role of the capsule in the virulence of the organism is indicated in the fact that *Klebsiella* species without a capsule tend to be less virulent. The lipopolysaccharides that coat the outer surface of the bacteria is another important contributor to virulence, resulting in inflammation,[5] which is known to be the cause of sepsis and septic shock. Fimbriae, a component of the bacterial cell that allows the organism to attach itself to host cells, also

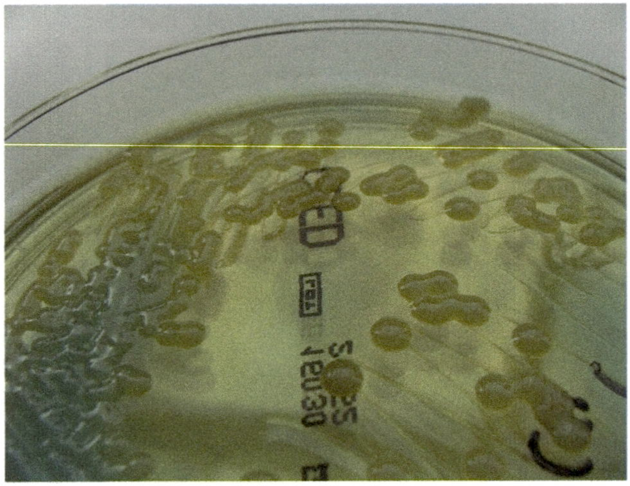

FIGURE 3.5

Klebsiella pneumoniae on cystine–lactose–electrolyte-deficient agar (CLED agar). Source*: From Nathan Reading/Own work/; https://creativecommons.org/licenses/by-nc-nd/2.0/; https://www.flickr.com/photos/nathanreading/6142505941.*

acts as an important virulence factor. Iron-chelating compounds, such as siderophores secreted by bacterial cells, which enable propagation of the organism by acquiring iron from the host, also play a significant role as a virulence factor for the organism.

Increasing antibiotic resistance is threatening to worsen the prognosis and outcome of diseases caused by *K. pneumoniae. Klebsiella pneumoniae* has long been shown to be resistant to antibiotics comprising of the beta-lactam ring, such as penicillin. Extended spectrum beta-lactamase *K. pneumoniae* resistant to third-generation antibiotics, such as cephalosporins, have also been reported,[6] thus increasing the challenge involved in effective clinical treatment of Klebsiella pneumonia infection. Carbapenem-resistant *K. pneumoniae* reported by the Centers for Disease Control and Prevention in 2013 has left the clinical world grappling for effective antibiotic options.[7]

Pathogenicity

The most common condition caused by *K. pneumoniae* is pneumonia, and more typically bronchopneumonia and bronchitis. The mortality with this infection can reach up to 50% despite treatment and up to 100% in cases wherein bacteremia develops.[8] However, it can cause a range of diseases, including urinary tract infection, soft tissue infection, diarrhea, and meningitis.

Klebsiella is mostly seen in immunocompromised people or those with impaired respiratory defense due to underlying conditions, including diabetes, chronic obstructive pulmonary disease, and malignancy. It usually affects the elderly and middle-aged adults.[9]

KPLA is a more recently emerging form of pneumonia, which has been reported with increasing frequency in the far eastern countries in the last three decades, especially in Taiwan and Korea, but is gradually emerging as a global problem. Diabetes is known to predispose patients to this infection and its complications, such as meningitis, bacteremia, endophthalmitis, and necrotizing fasciitis. The organism tends to respond to a prolonged course of antibiotics, particularly cephalosporin, along with drainage of the abscess.

Immune response

Polymorphonuclear granulocytes, which phagocytose the bacteria and bactericidal serum complement proteins, are the two most important components of immune response against Klebsiella pneumonia infection. The alternate pathway of complement activation is also found to be very active in Klebsiella pneumonia infection. Neutrophil myeloperoxidase and lipopolysaccharide-binding protein facilitate in defense against Klebsiella pneumonia infection, although the capsule protects the bacteria from phagocytosis and serum bactericidal proteins.[9]

Clinical manifestations

Klebsiella pneumonia infection causes clinical manifestations that resemble those seen in cases of community-acquired pneumonia, caused by *S. pneumoniae*. Patients are generally febrile and present with a cough, shortness of breath, and in some cases, pleuritic chest pain. However, pneumonia caused by *S. pneumoniae* and *Klebsiella pneumoniae* can be distinguished based on the type of sputum produced. While the sputum produced in community-acquired pneumonia is described as "blood-tinged" or "rust-colored," the sputum produced in Klebsiella pneumonia is described as "currant jelly," resulting from the significant inflammation and necrosis of the surrounding tissue caused by infection with *Klebsiella pneumoniae*.[4]

Klebsiella pneumonia pathology is more pronounced in the upper lobes of the lungs, although in some cases, the lower lobes may also be involved. Primary investigations, therefore, reveal unilateral signs of consolidation, such as crepitation, bronchial breathing, and increased vocal resonance, all of which mostly affect the upper lobe. Cases suspected to be nosocomial infections based on patient history should be investigated for the presence of any wounds, puncture points, surgical stitches etc.

Comorbidities and underlying medical conditions, such as diabetes or alcoholism, that result in an immunocompromised status are known to increase vulnerability to colonization and infection by *Klebsiella pneumoniae*. History of prolonged hospital stays, especially in the intensive care unit that require a prolonged use of invasive medical devices, such as nasogastric tubes, catheters, ventilators, dialyzers etc., predispose patients to Klebsiella pneumonia infection. Elongated use of broad-spectrum antibiotics also increases the risk of acquiring multidrug resistant strains of *Klebsiella pneumoniae*.

Diagnosis

Leukocytosis is one of first serological signs that appear in a laboratory diagnosis of Klebsiella pneumonia infection. Although this, in addition to the clinical manifestations shown by the patient, may indicate a diagnosis for pneumonia, it is not sufficient for deciding whether the pneumonia is caused by *S. pneumoniae* or *Klebsiella pneumoniae*. Klebsiella pneumonia, however, can be considered in the differential diagnosis based on the chest radiograph, if a lobar infiltrate is seen in the posterior aspect of the right upper lung. Although lung abscesses are rare in cases of Klebsiella pneumonia, empyema is commonly observed. Chest radiographs may also show a bulging fissure sign caused by the large extent of infection and inflammation attributable to Klebsiella pneumonia infection. This can further assert the diagnosis, although it is a nonspecific sign of pneumonia caused by *Klebsiella pneumoniae*.[4]

Although the patient history, clinical manifestations, physical examination, and chest radiograph can facilitate the narrowing of the differential diagnosis, the final diagnosis of Klebsiella pneumonia infection should be confirmed by either a sputum culture analysis or blood culture analysis.

Treatment

The occurrence of *Klebsiella pneumoniae* in the general community is considerably low compared to *S. pneumoniae*. Pneumonia should, therefore, usually be treated using standard antibiotic therapy unless infection with *Klebsiella pneumoniae* is either suspected or confirmed, in which case antibiotic treatment should be determined based on culture sensitivities.

Community-acquired pneumonia caused by *Klebsiella pneumoniae* is usually treated using a standard 14-day treatment regimen, either with monotherapies comprising of a third or fourth generation cephalosporin, a respiratory quinolone, or a combination therapy that consists of the monotherapy regimes with an aminoglycoside. Combination therapies have been found to be more effective in reducing mortality compared to monotherapies. Penicillin-allergic cases should be treated using a course of aztreonam or a respiratory quinolone. Nosocomial infections, however, need to be treated differently using an appropriate carbapenem monotherapy unless sensitivities are reported. Carbapenem-resistant *Enterobacteriaceae* cases should be treated using antibiotics from the polymyxin class, tigecycline, fosfomycin, aminoglycosides, or dual therapy carbapenems.[4] Abscess drainage may also be indicated as an effective disease management strategy in some cases, especially in KPLA-related cases.

Student Notes

Key take-aways

- Inclusion of *Klebsiella pneumoniae* as a cause for pneumonia in the differential diagnosis requires a detailed investigation of the patient history, clinical manifestations, physical examination, serological reports, and chest radiographs.
- The final diagnosis of Klebsiella pneumonia requires an appropriate and detailed sputum and blood culture studies. Radiograph findings alone should, therefore, not be used to make a diagnosis of Klebsiella pneumonia infection definitively.
- *Klebsiella pneumoniae* is a Gram-negative bacterium that typically causes nosocomial infections and is notorious for developing antibiotic resistance.
- "Currant jelly" appearance of sputum is a hallmark of infection with *Klebsiella pneumoniae*.
- Comorbidities and underlying medical conditions that cause an immunocompromised state can increase the predisposition of a patient to Klebsiella pneumonia infections.
- Antibiotic resistance contributes significantly to the high level of patient mortality associated with Klebsiella pneumonia infections.
- A prolonged stay in hospitals, especially in intensive care units requiring the use of medical devices, such as catheters, ventilators, nasogastric tubes, dialyzers etc., increases the chances of acquiring nosocomial Klebsiella pneumonia infections.
- Appropriate sterilization and aseptic protocols followed in hospital wards, intensive care units, major or minor surgical procedures etc. can substantially reduce nosocomial infections involving *Klebsiella pneumoniae*.
- Timely and appropriate diagnoses of KPLA and effective treatment using appropriate regimens play an essential role in preventing complications or antibiotic resistance, thereby improving the prognosis and outcome of the disease.

Problem number 3.2: Liver failure caused by cytomegalovirus-induced hepatitis

A 3-year-old male presented with acute liver failure and hepatitis. He was jaundiced and icteric all over his body at presentation but did not have any urinary symptoms.

The patient had been admitted 1 month ago for a full work-up, indicated by the derangement of the liver function tests. At that presentation, the patient was febrile (39°C measured at home), vomiting (large quantities; comprising of food content), unable to take food orally, and complained of blood-stained diarrhea, containing mucous. Iron therapy had been prescribed for 1 month, but stopped after observation of deranged liver function tests.

The patient's birth history was uneventful, with a birth weight of 2.4 kg. There was no history of neonatal intensive care unit (NICU) admissions, and no history of hospitalizations or surgical interventions. Talipes equinovarus diagnosed at birth had been corrected by casting, without the need for any surgical intervention. The patient demonstrated gross motor and language developmental delay. There were no known food or drug allergies, and the immunizations status was up-to-date.

The patient's parents are in a consanguineous marriage, and have two other children who are healthy. There is no history of inherited diseases in the family.

The following are the reports of the investigations conducted:

1. Vitals
   Temperature: 37.2°C
   Blood pressure: 131/68 mmHg while lying
   Respiratory rate: 22 breaths/minute
2. Physical examination
   Icteric (whole body)
   Not cyanosed/no signs of respiratory distress
3. Respiratory examination
   EBAE
   No added sounds
4. Cardiovascular examination
   No murmurs (S1, S2)
5. Abdominal examination
   Distended
   Tympanic with hepatomegaly (3.5 cm below costal margin).
6. Genitalia
   Micro-penis
   Circumcised blood lactate (BL) scrotal testis
7. Musculoskeletal examination
   Normal tone
   Brisk knee reflex
8. Radiological Investigations
   Chest X-ray: Hyperinflation
   Ultrasound (abdomen): Prominent hepatic portal radicles suggesting inflammation.
   Span: 11.8 cm; no focal mass
   CBD and portal vein: normal
   Edematous gall bladder with minimal fluid around liver.

9. Liver function tests

Tests	At presentation	After 8 days hospitalization	Reference values
Protein (Total) (g/L)	65	–	68–80
Albumin (g/L)	33	29.8	35–55
Globulin (g/L)	32	–	15–30
Bilirubin (Total) (μmol/L)	404	95	1.71–20.5
Bilirubin (Direct) (μmol/L)	–	86	–
Alkaline phosphatase (U/L)	896	1138	44–147
Aspartate aminotransferase (U/L)	–	2845	8–33
Alanine aminotransferase (U/L)	369	1250	10–40
G-glutamyl transferase (U/L)	83	416	9–48
Ammonia (μ/dL)	–	60	15–45

10. Coagulation profile

INR	2.74	≤ 1.1
PTT (sec)	67	11–13.5

11. C-reactive protein: 5.43 μg/mL (reference value: <2 μg/mL)
12. Antinuclear antibody: Negative
13. Hepatitis panel
    Anti HCV: Negative
    Anti-hepatitis A (IgM): Negative
    HBS Ag: Negative
14. Epstein – Barr virus (EBV) panel
    EBV IgG Ab (Nuclear Ag): Positive
    EBV IgG Ab (Capsid Ag): Positive
    EBV IgM Ab (Capsid Ag): Negative
15. Cytomegalovirus (CMV) panel
    Anti-CMV IgG: 6.19 IU/mL; repeat test: 7.27 IU/mL (Reference value: <0.9)
    Anti-CMV IgM: Reactive
    Cytomegaloviruc (CMV) viral load (PCR): 208.929 IU/mL (reference value: 227–2270000 IU/mL)
    CMV viral load (LOG): 2.320 log IU/mL (reference value: 2.4–6.4 log IU/mL)

Diagnosis

The final diagnosis was that of CMV hepatitis with acute fulminant liver failure.

Treatment

The patient was admitted to the hospital and kept in an isolation room for 8 days.

Supportive care:

1. IV infusion: ½ NS D5 (KVO)
2. Low fat, high carbohydrate diet
3. Multiple doses of vitamin K and fresh frozen plasma
4. IV dextrose (for recurrent hypoglycemia)

Medications:

1. Valganciclovir—300 mg od
2. Ursodeoxycholic acid—150 mg bd
3. Omeprazole—10 mg od
4. Prednisolone—10 mg od

Question and answer

1. What is the role of the two glycoprotein complexes present on the surface of the CMV?
   The surface glycoproteins of the CMV are important for the structure and function of the virus. A trimeric complex (TC) and a pentameric complex (PC) are the two glycoprotein complexes. While the TC facilitates viral infection in fibroblasts, the PC facilitates viral infection in endothelial and epithelial cells. gB, another surface glycoprotein of the CMV influences the ability of human cytomegalovirus (HCMV) to bind and invade the host cell.
2. Why does the HCMV demonstrate high seroprevalence?
   HCMV infects the majority of the world's population through exposure at an early age and develops into persistent lifelong infections, resulting in a high seroprevalence of HCMV.
3. How do congenital CMV infections occur in fetuses?
   CMV infections transmit vertically from an infected mother to the fetus, manifesting as congenital CMV infections in fetuses.
4. Why is the risk of fetal transmission higher in cases of primary CMV infections in seronegative pregnant women compared to cases of reinfection in seropositive pregnant women?

Seropositive pregnant women have preconceptional immunity to CMV that prevents vertical transmission of the virus in case of reinfection during pregnancy. However, the absence of preconceptional immunity in seronegative pregnant women increases the risk of CMV infection to the fetus in case of a primary infection during pregnancy.

5. How does CMV persist in the host's system despite clearance of the primary infection by the host's immune system?
The ability of the CMV to achieve dormancy and persist in the myeloid cells of the host's system enables persistence of the virus in the system despite clearance of the primary infection by the host's immune system. Primary CMV infections are usually asymptomatic but facilitate the establishment of a persistent infection. Cytotoxic T-cell vigilance by the host's immune system keeps viral replication and spread under control. Periodic reactivation of infection and persistent viral shedding from seropositive individuals is responsible for the spread of the infection to naïve hosts through virus shedding.
6. How do the clinical manifestations seen in certain cases of primary CMV infections differ from those seen in cases of reactivation of the virus?
The clinical manifestations seen in certain cases of primary CMV infections are caused by cytokine production, induced in response to a systemic antiviral state resulting from nonlytic processes. The clinical manifestations demonstrated in cases of reactivation of the virus result from the release of virions from the infected cells into the bloodstream through lytic processes. Reactivation of the virus in immunocompromised individuals is known to precipitate HCMV-induced immuno-pathological responses that cause significant organ damage.
7. Why are solid organ and bone marrow transplant patients prone to CMV reactivation in the posttransplantation period?
The CD8 + responses of the cell-mediated immune system are specific to the CMV. The administration of immunosuppressive drugs in solid organ and bone marrow transplant patients in the posttransplantation period results in a decrease in these responses. This causes reactivation of the CMV during the posttransplant period, resulting in severe infections.
8. How do the clinical manifestations of HCMV-associated mononucleosis in healthy individuals differ from EBV-associated mononucleosis?
The most prevalent clinical manifestations of HCMV-associated mononucleosis include fever, rash, and leukocytosis, while pharyngeal exudate and heterophile antibodies, characteristic of EBV-associated mononucleosis, are absent.
9. What are considered to be hallmark features of a CMV infection in histological investigations?
The presence of large refractile cells (cytomegalic) with intranuclear inclusions, referred to as "owl's eye inclusions," also known as CMV

inclusion bodies in histopathological samples, are considered to be the hallmark features of a CMV infection in histological investigations.

10. How do CMV-antibody tests help the diagnostic process?
The antibody tests for CMV infections comprise of CMV – IgG and CMV – IgM tests. While the CMV – IgG test indicates an active or a past latent CMV infection, the CMV – IgM test enables differentiation between an active and a latent CMV infection.

Learning objectives

1. Determine the appropriate differential diagnosis based on patient history, clinical manifestations, serological investigations, organ profiling investigations, histopathological findings, imaging reports, and PCR tests reports.
2. Differentiate hepatitis caused by CMV from other causes based on patient history, PCR test reports, and histopathological findings.
3. Differentiate between a latent and an active CMV infection based on anti-CMV antibody tests.
4. Describe the characteristics of the CMV.
5. Summarize the pathogenesis of CMV infection.
6. Predict the prognosis of the case based on the clinical manifestations and the extent of pathology presented.
7. Correlate the clinical manifestations with the pathogenicity of the human CMV.
8. Assess the need for antiviral intervention by weighing the risks associated with the toxicity of the therapeutic agents against the benefits of the treatment.

Discussion

Differential diagnosis

Human immunodeficiency virus (HIV) infection	HIV infection presents itself with a broad spectrum of symptoms, varying with the affected organ system. The symptoms can range from mild ones, such as fever, cold, and weakness, to severe symptoms caused by organ failure.
Human herpes virus (HHV 6) infection	HHV 6 infection manifesting in young children usually presents as a fever, rash, and diarrhea, although severe symptoms such as febrile seizures, encephalitis, and intractable seizures are also seen in some cases.

Continued

Continued

EBV infection	Symptomatic cases of EBV infection in children and young adults present as fatigue, fever, rash, sore throat, lack of appetite, swollen glands in the neck, weakness, and sore muscles.
Hepatitis A/B/C infection	Hepatitis A/B/C infections may present themselves as abdominal pain, fever, joint pain, itchy skin, yellowing of skin/nails/conjunctiva, nausea, vomiting, loss of appetite, light-colored stool, and dark orange-yellow colored urine.
Infectious mononucleosis	Infectious mononucleosis can be caused by CMV or EBV infection in young adults and presents itself as fatigue, fever, rash, sore throat, lack of appetite, swollen glands in the neck, weakness, and sore muscles.

Background

CMV is a common human pathogen infecting almost half of the world's population, with almost the entire population being infected in developing countries.[10] CMV infections may manifest as asymptomatic infections or may lead to chronic CMV infection, which could be responsible for severe multiorgan failure in immunocompromised patients.[11,12] While subclinical asymptomatic CMV infections are acquired at an early age, manifestation of acute life-threatening diseases is generally limited to immunocompromised hosts. CMV infections are therefore established as persistent latent infections that can opportunistically reactivate in the event that the host's immune status is compromised.

Etiological agent

Human cytomegalovirus

Classification and morphology

CMV is a genus comprising of double-stranded DNA viruses belonging to the *Betaherpesvirinae* subfamily of the *Herpesviridae* family. Dormancy is a characteristic of all herpes viruses, and HCMV is well-known for its ability to remain dormant in the human host after primary infection. An immunocompromised status of the host triggered by any trauma or medical condition provides an opportunity for reactivation of HCMV.

HCMV is spherical-shaped and comprises of a lipid bilayer envelope composed of various viral glycoproteins surrounding a ribonucleoprotein core. The ribonucleoprotein core comprises of a proteinaceous matrix containing a double-stranded linear DNA within an icosahedral nucleocapsid. The surface glycoproteins are important for the structure and function of the virus. One of the surface glycoproteins, gB, influences the ability of HCMV to bind and invade the host's cell. The viral surface comprises of two glycoprotein complexes, a TC that facilitates viral infection in fibroblasts, and a PC that facilitates viral infection in endothelial and epithelial cells.

Epidemiology

The majority of the world's population is exposed to HCMV at an early age and harbor latent lifelong infection, thereby resulting in high seroprevalence of HCMV. Although HCMV shows marked restriction in host cell tropism and is generally associated with the salivary glands, transmission of infection may occur through transfusion of blood products, organ transplants, breastfeeding, direct contact with contaminated saliva, perinatally, and sexual contact. Congenital CMV infections of fetuses frequently occur by vertical transmission of the virus from the mother to the fetus. A primary CMV infection in seronegative pregnant women is associated with a higher risk of fetal transmission and severe damage to the fetus compared to reinfection in seropositive pregnant women with preconceptional immunity.[13]

Primary infections are usually asymptomatic in seronegative healthy individuals, while symptomatic recurrent infections may be a result of reinfection or reactivation of HCMV in seropositive patients. The morbidity and mortality levels are very high in cases of acute multiorgan failure induced by reactivated CMV infections in immunocompromised individuals.

Periodic reactivation of infection and persistent viral shedding from seropositive individuals facilitates the spread of CMV infections in the population, especially in communities or populations with overcrowded and unhygienic living conditions.

Pathogenicity

Primary CMV infections are usually asymptomatic but result in the establishment of a persistent infection, which is also responsible for the spread of the infection to naïve hosts through virus shedding. Transmission of infection is generally followed by the clearance of the primary infection from the host's system, after which the virus becomes dormant in the myeloid cells. The pathogenesis of CMV infection results from the compromised efficacy of the host's immune system to control viral replication and spread, which is primarily kept under control by cytotoxic T-cell vigilance.

Primary CMV infections can also result in the manifestation of certain chronic symptoms caused by cytokine production induced by a systemic antiviral state through nonlytic processes. Acute pathogenic effects of reactivation, seen predominantly in immunocompromised patients, on the other hand, occur through lytic processes wherein the virions are released from the infected cells into the bloodstream, resulting in the manifestation of symptoms.[14,15] HCMV-induced immunopathological responses result in significant organ damage, which is characteristic of HCMV reactivation in immunocompromised individuals. Although HCMV shows a restriction of host cell tropism for infection, pathogenesis of the infection can manifest in almost all organ systems.

Immune responses

The resolution of the primary infection and control of the latent infection require both the humoral and cell-mediated immune responses. The host's immune system plays a very dominant role in controlling dormant viral infections such as those caused by HCMV. A decrease in either the antibody or the cellular responses as a result of immunosuppressive conditions provides an opportunity for the dormant infection to enter into the lytic phase, causing extensive damage to the host's system.

The CD8 + responses of the cell-mediated immune system are HCMV-specific, and a decrease in these responses in solid organ and bone marrow transplant patients caused by the administration of immunosuppressive drugs is known to be associated with severe CMV infections during the posttransplant period.[16] The correlation of cellular responses to intrauterine HCMV transmission, causing congenital CMV disease, also shows the importance of cellular responses in pregnant women.[17] The absence of cellular responses in AIDS patients due to lymphocyte depletion is responsible for an invasive and severe CMV disease.[18] Decreased cellular responses in congenitally infected children is also associated with acute CMV infections, resulting in fatal multiorgan pathology.[19]

Higher levels of immunogenic proteins specific to HCMV are found in cases of primary CMV infection and reactivation in immunocompromised individuals. HCMV-specific antibody responses therefore play a significant role in immune responses against CMV infections.[20] Antibodies against surface exposed viral glycoproteins are very important in viral neutralization processes. The presence of IgG antibodies specific to the HCMV glycoproteins is also very useful in detecting primary as well as secondary infections, especially in pregnancy.[21]

Clinical manifestations

Primary CMV infections are usually asymptomatic, and therefore go unnoticed. Predominantly, HCMV-associated mononucleosis manifests in healthy individuals who have been exposed to the virus. The most prevalent clinical manifestations of HCMV-associated mononucleosis include fever, rash, and leukocytosis, while pharyngeal exudate and heterophile antibodies, which are characteristic presentations of EBV-associated mononucleosis, are absent. Although organ involvement is rare in immunocompetent hosts, some other manifestations, such as anemia, abnormal liver function tests, thrombocytopenia, positive rheumatoid factor, and positive antinuclear antibody levels are also seen. The aggressiveness of CMV infections is inversely proportional to the immunocompetence of the infected individual. Clinical manifestations of exacerbated CMV infections may include hepatitis, leading to fulminant liver failure, retinitis, esophagitis, colitis, pneumonitis, placentitis, encephalitis, and transverse myelitis, depending on the organ system involved.

Chronic CMV infections have also been implicated in certain rheumatological diseases[22] and genitourinary tract malignancies.[23] The high-risk groups predisposed to exacerbations of CMV infections include patients with acquired immunocompromised conditions due to acquired immunodeficiency syndrome (AIDS), malignancies, organ transplants, and infants with congenital CMV disease due to in utero infection.

Laboratory diagnosis

A diagnosis of CMV infections can be clinically challenging, owing to the varied clinical manifestations and multiorgan involvement. Therefore patient history plays a very important role in facilitating the inclusion of CMV infections in differential diagnosis. The identification of high-risk groups can direct the clinician toward establishing a correlation between the probable etiology and the clinical manifestations of the patient. The need for a definitive clinical diagnosis of a CMV infection is not indicated in immunocompetent patients.

Serological profiling, profiling of the affected organs, and histological observations are the primary investigations undertaken to identify CMV infections as the etiology for the clinical presentation of the case. Large refractile cells (cytomegalic) with so-called owl's eye intranuclear inclusions (Fig. 3.6), known as CMV inclusion bodies, are considered to be the hallmark features of a CMV infection in histological investigations.[10]

The ubiquitous nature of HCMV makes it empirical that a distinction should be made between the detection of HCMV, indicative of a latent CMV infection, and implication of HCMV in correlation with the clinical presentation. While histopathological diagnosis is considered the gold standard for a diagnosis of active CMV infections, viral detection and identification still relies on quantitative PCR assays. A negative PCR test, however, should not be used to rule out the possibility of a CMV infection without correlation with histopathological findings.[24]

Serological testing for anti-CMV IgG antibodies is especially indicated before organ transplantation to ensure better risk evaluation, although distinction between primary and secondary infections may not be possible. Anti-CMV IgM antibodies, on the other hand, may be used to distinguish latent infections from active ones; however, the low sensitivity and specificity of the assay restricts its use in the clinical setting.

Although risks of acute CMV infection from transfusion of blood and blood products are low, CMV-negative blood and blood products are earmarked for transfusion to high-risk patient groups, such as infants and immunocompromised patients.

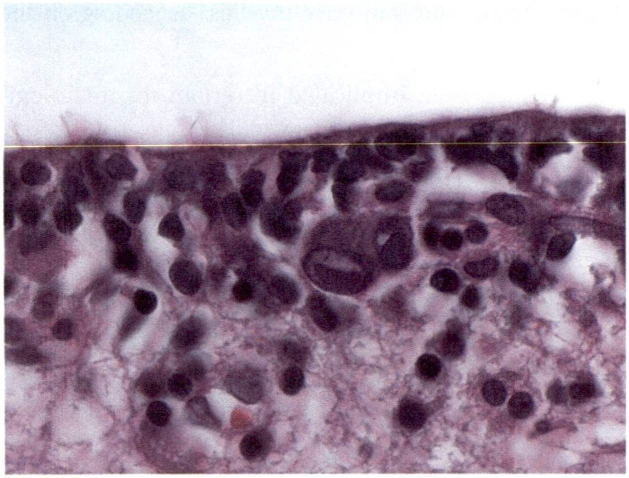

FIGURE 3.6

Encephalitic tissue demonstrating the typical owl eye appearance of CMV infected cells. Source*: From Jensflorian/ Own work/; https://creativecommons.org/licenses/by-sa/4.0/deed.en; https://commons.wikimedia.org/wiki/File:CMV_encephalitis_owl_eye_inclusions_HE_stain.jpg.*

Treatment and prevention

The advent of several efficient antiviral agents, such as cidofovir, foscarnet, ganciclovir, and valganciclovir, has improved the prognosis and outcome in cases of acute CMV infections to a large extent.[25–27]

Immunocompetent patients may either be asymptomatic or may present with mild symptoms, which may be self-limiting. Although such cases usually do not require specific therapy, symptomatic management may be indicated. Immunocompromised patients, on the other hand, should be diagnosed quickly, and the need for antiviral intervention should be evaluated and initiated immediately after diagnosis. The risks associated with possible toxicity caused by antiviral agents should be considered and evaluated in relation with the clinical presentation of the case.

Organ transplantations, involving naïve receivers and infected donors, should be followed up with valganciclovir or ganciclovir prophylaxis and regular serological monitoring.

Several vaccine candidates are being considered for the prevention of CMV infections in high-risk patient groups. The HCMV PC has been the focus of HCMV vaccine development for a long time.[28] However, the challenges involved in the development of an effective vaccine against HCMV has hampered the progress of the process, although several promising vaccine candidates are in clinical trials now.

Student Notes

Key take-aways

- HCMVs infect the majority of the world's population, and are therefore, ubiquitous.
- Primary infection with HCMV usually occurs early in life followed by a latency period during which the immune system keeps the infection under control. Reinfections and reactivations of primary infections may occur due to acquired immunocompromised conditions.
- HCMV usually causes asymptomatic infections in immunocompetent individuals, although mild self-limiting symptoms may be present in some cases.
- Immunocompromised status triggers the opportunistic nature of HCMV, causing it to manifest as severe conditions involving different organ systems. Multiorgan involvement and failure caused by CMV infections in immunocompromised individuals is responsible for poor prognosis and outcome.
- Vertical transmission from the infected mother to the fetus results in congenital CMV diseases, responsible for several developmental and physical anomalies in newborn infants.
- Inclusion of CMV infections as an etiology in differential diagnosis relies on accurate patient history and clinical manifestations.
- Diagnosis of CMV infections requires serological, histopathological, and viral detection and identification tests such as PCR.
- While the risk of CMV infections through transfusions in immunocompetent individuals is very low, CMV-negative blood and blood products should be earmarked for use in high-risk patients.
- Antiviral treatments should be administered only after weighing the risks associated with the toxicity of the therapeutic agents against the benefits of the treatment as per the clinical presentation of the case.
- Antiviral prophylaxis is recommended in case of high-risk organ transplantations, wherein the receiver is naïve while the donor is infected.

Problem number 3.3: Ulcerative colitis caused by *rotavirus* infection

A 30-year-old man presented with complaints of intermittent lower abdominal cramps, bloody mucus, containing diarrhea (4–5 times/day), and tenesmus since the past week. The patient denied any history of recent travel or consumption of outside food within the past 15 days. However, the patient mentioned a history of similar occurrences recurring in the past intermittently since a childhood *rotavirus* infection.

The following are the reports of the investigations carried out:

1. Physical examination
   a. Appearance: Pale and dehydrated
   b. Abdomen: Slightly distended with lower abdominal tenderness
   c. Anus: Appearance—Normal
   d. Digital rectal examination: Blood and mucus; no masses/strictures
   e. Prostate: Normal
2. CBC

Test	Values	Reference range
Hb (g/dL)	10.4	14–16
White blood cells (WBC) (/L)	13×10^9	$4–11 \times 10^9$
Neutrophils (%)	54	42.2–75.2
Lymphocytes (%)	42	20–50
Monocytes (%)	3	2–9
Eosinophils (%)	2	1–6
Basophils (%)	2	1–2

3. Microscopic examination of fresh stool sample: RBC, WBC, and mucus present; trophozoites, cysts, or ova absent. *Entamoeba histolytica* infection was ruled out.
4. Stool sample culture: Bacterial pathogens—absent.
5. Proctoscopy and sigmoidoscopy: Mucus and blood found in rectum; few ulcers present.
6. Colonoscopy: Descending colon—multiple small ulcers having congested mucosa with granular surface prone to bleeding; mucosal surface covered with straw-colored exudate.
7. Colon biopsy: Ulcerative colitis resulting in inflammatory bowel disease.
8. Abdominal radiograph: Colonic dilation.

Diagnosis

The final diagnosis was that of inflammatory bowel disease elicited by ulcerative colitis. The history of childhood *rotavirus* infection followed by intermittently recurring bouts of similar episodes of abdominal cramps and bloody diarrhea was assumed to have elicited chronic ulcerative colitis resulting in inflammatory bowel disease over the years.

Treatment

1. Mesalamine
2. Corticosteroid

Follow-up and outcome

The patient reported back for a follow-up after 7 weeks and reported having two normal bowel movements daily comprising of slight mucus discharge, but without any bleeding. As the patient had attained remission with medical therapy, there was no need for surgical intervention. The patient was counseled regarding the importance of incorporating healthy dietary and lifestyle changes to ensure maintenance of the remission state, and his fears related to colon cancer were addressed appropriately.

An endoscopy was advised after 3 months to get a better insight into the condition posttreatment.

Question and answer

1. What are the causative factors for diarrhea?
   Diarrhea can be caused due to an infection by entero-pathogens including bacteria, viruses, protozoa, and helminths. The disturbance to the normal intestinal flora caused by certain antibiotics can also result in diarrhea. Diarrhea can also manifest as a symptom of inflammation caused by conditions such as irritable bowel, thyrotoxicosis, diabetes mellitus, and certain allergies such as lactose and gluten intolerance.
2. What are the causative factors for bleeding and mucus discharge in the stools?
   Inflammatory conditions, the presence of ulcers, infections, and neoplastic conditions can cause bleeding and mucus discharge in stools.
3. What is the relevance of recent travel in the patient's clinical manifestations?
   Entero-pathogens can enter the host system through contaminated food and water, especially in regions where such entero-pathogens are endemic. As the patient's condition is associated with the gastrointestinal system, a history of recent travel to a region where entero-pathogens are endemic would indicate an infection, thus aiding the process of diagnosis.

4. Why is the patient not consuming food from outside or traveling anywhere within the last 15 days relevant to the case?
   The nonconsumption of outside food and no history of travel within the last 15 days helps in ruling out infection by entero-pathogens, as the incubation period of the pathogens before clinical manifestations are seen is completed within 15 days.
5. Why is it important to conduct the microscopic examination of fresh stool samples immediately after they are obtained?
   Fresh stool samples can reveal the presence of active trophozoite forms of parasites that may be present in the sample. The trophozoite form of the parasite does not survive in stool samples that are not fresh, hence resulting in the nondetection of the parasite.
6. What is an ameba?
   An ameba is a unicellular protozoon consisting of cytoplasm, ectoplasm, and nucleus. It is a parasite and is a clinically significant pathogen in humans.
7. What are trophozoites, cysts, and ova?
   Trophozoites and cysts are stages in the life cycle of protozoa, while ova is a stage in the life cycle of helminths such as ascaris and hookworms. Trophozoites are detected in acute manifestations of disease, and their presence is often considered indicative of the active clinical stage of the disease. Cysts, on the other hand, are detected in patients with chronic or subacute conditions. Asymptomatic carriers of parasitic protozoa often release cysts in stools, which can become potential sources of infection. Ova are produced exclusively by helminths and play an important role in the identification of the etiological agent.
8. Which are the most common protozoa responsible for causing gastrointestinal infections?
   Amoeba and *Giardia lamblia* are the most common protozoa responsible for causing gastrointestinal infections. These protozoa can be differentiated based on the cysts present in stool samples.
9. Which bacteria and viruses can cause clinical manifestations such as those seen in this case?
   Bacteria belonging to the *Salmonella, Shigella,* and the *Campylobacter* species and viruses such as *adenovirus* and *rotavirus* are known to present clinical manifestations such as those seen in this case.
10. What is microbiota? List the bacteria that form a part of the intestinal microbiota.
    Microbiota comprises the microorganisms that form the normal flora of the human body. *E. coli, Enterococci, Lactobacilli, Clostridia,*

Bacteroides comprise the most significant part of the human intestinal microbiota.

11. What are the immune mechanisms responsible for protecting the colon from the normal flora of the human intestines?
    The mucosal layer that secretes mucus, antibacterial substances secreted by epithelial cells in the gastrointestinal system, and immune mechanisms elicited by the Gut Associated Lymphoid Tissue acting as a barrier ensures tolerance against the pathogenic effects of normal intestinal flora are responsible for protecting the colon from the normal flora of the human intestine.
12. What causes the clinical manifestations of ulcerative colitis?
    Ulcerative colitis manifests as inflammation and ulcerations in the upper lining of the large intestine. The condition presents with a broad spectrum of symptoms, which overlap with those of infectious colitis and Crohn's disease, making it difficult to diagnose. An immune response to a viral or bacterial infection of the gastrointestinal system, which continues to be elicited even after the pathogen has been eliminated, results in an ongoing inflammation of the intestinal membranes eventually causing ulcerative colitis.

Learning objectives

1. List the causes of diarrhea.
2. Describe the fecal – oral route of disease transmission.
3. Explain the significance of the incubation period.
4. Describe stool analysis for detection of ova and cysts of parasites infecting the gastrointestinal tract.
5. Distinguish between the normal microbiota and pathogens of the gastrointestinal tract.
6. Describe the microbiological, clinical aspects, and complications of amebic infections.
7. Enumerate the various bacterial and viral pathogens responsible for causing diarrhea.
8. Describe the treatment for amebiasis.
9. Describe the immunological barrier present in the intestines.
10. Explain the differential histopathological diagnosis of inflammatory bowel disease.
11. Describe the clinical aspects of inflammatory bowel disease.
12. Explain the complications that may be associated with ulcerative colitis.

Discussion

Differential diagnosis

Parasitic colitis	Parasitic colitis generally presents itself with frequent nonbloody, watery diarrhea, abdominal cramps, nausea, sweating, and loss of appetite.
Crohn's disease	Crohn's disease causes severe diarrhea, abdominal pain, fatigue, unexplained weight loss, loss of appetite, and malnutrition.
Intestinal tuberculosis	Intestinal tuberculosis results in the thickening of the intestines, thereby presenting with symptoms of intestinal obstruction, pain in the right iliac fossa, abdominal pain, weight loss, loss of appetite, anemia, fever, night sweats, fatigue, and the possibility of a palpable mass in the right iliac fossa.
Colon cancer	Colon cancer can manifest as very nonpeculiar symptoms such as a sudden and persistent change in bowel habits including unexplained diarrhea or constipation, abdominal cramps, bloating, flatulence, rectal bleeding, feeling of incomplete bowel emptying, extreme fatigue, loss of appetite, and unexplained weight loss.
Radiation colitis	Radiation therapy for prostate or cervical cancer often results in radiation colitis, which presents itself with wave-like abdominal cramps, frequent watery diarrhea, mucus discharge from the rectum, nausea, vomiting, frequent urges to urinate, rectal pain, weight loss, and rectal bleeding.
Toxic megacolon	Toxic megacolon, often a complication of inflammatory bowel disease, leads to intense bloody watery diarrhea, abdominal swelling, fever, dizziness, palpitations, and racing heart beats.
Bacterial/viral gastroenteritis	Gastroenteritis caused by infection with a bacterial or viral pathogen usually presents itself with nonbloody watery diarrhea, nausea, vomiting, abdominal cramps, low-grade fever, dizziness, fatigue, myalgia, and headaches.

Background

Rotaviruses are known to cause a vast majority of childhood gastroenteritis cases in humans and are also responsible for acute diarrhea in the young of many other mammalian species. *Rotavirus* is responsible for causing a food- and water-borne disease that is transmitted through the fecal – oral route. Although *rotavirus* infections can be prevented by vaccination against the pathogen, the vaccine being nonmandatory is responsible for a large number of rotavirus-associated mortalities worldwide.

While *rotavirus* generally infects children below 5 years of age, infection in immunocompromised adults or those on regular steroid treatment has also been reported. *Rotavirus* infections are manageable and rarely lead to complications resulting in morbidity and mortality; however, untreated infections can result in hospitalization and even death due to severe dehydration and lack of oral intake.[10] Complications related to the nervous system such as convulsions, encephalitis, encephalopathy,[11] and immune system elicited complications of the gastrointestinal tract such as inflammatory bowel disease and ulcerative colitis are also known to be associated with untreated or inappropriately managed cases of *rotavirus* infections.[12]

Etiological agent

Rotavirus

Classification and morphology

The *rotavirus* genus consists of segmented double-stranded RNA virus species belonging to the *reoviridae* family. There are 10 known viral species belonging to the *rotavirus* genus referred to as *rotaviruses* A – J.[13] While the other species of the genus are responsible for only about 10% of the reported *rotavirus* infections, 90% of the infections are caused by *rotavirus A*. The *rotavirus* genus gets its name from the characteristic "wheel-like" appearance of the double-shelled virus particles seen under the electron microscope[10] (Fig. 3.6).

The rotavirus genome consists of 11 RNA segments that encode six structural proteins (VP1, VP2, VP3, VP4, VP6, VP7) and five nonstructural proteins (NSP1–NSP5). The core structural proteins are VP1–VP3, while VP6 comprises the inner shell, and VP4 (a protease-sensitive protein known as a P-type antigen) and VP7 (a glycoprotein known as a G-type antigen) form the outer shell (Fig. 3.7). Different permutations and combinations of the P and G-type antigens result in variations in the circulating strains of the *rotavirus A* species.

Epidemiology

Rotaviruses are entero-pathogens infecting young children and the young ones of certain other mammalian species, throughout the world. Asymptomatic infections are generally seen in neonates owing to the prevalence of maternal antibodies, while infections in young children are usually symptomatic. The primary route of transmission of the disease is the fecal – oral route, although indirect transmission through fomites also occurs. Although the occurrence of rotavirus infections is common globally, incidences of complications and mortalities due to the infections are generally seen to be higher in underdeveloped and developing countries owing to the prevalence of low standards of living, lack of hygiene and sanitation, inadequate healthcare facilities, and higher prevalence of malnutrition.[10]

While incidences of rotavirus infections in tropical climates occur all year round, temperate regions tend to show an increasing trend of rotavirus infections in the winter months and the dry months when conditions of low

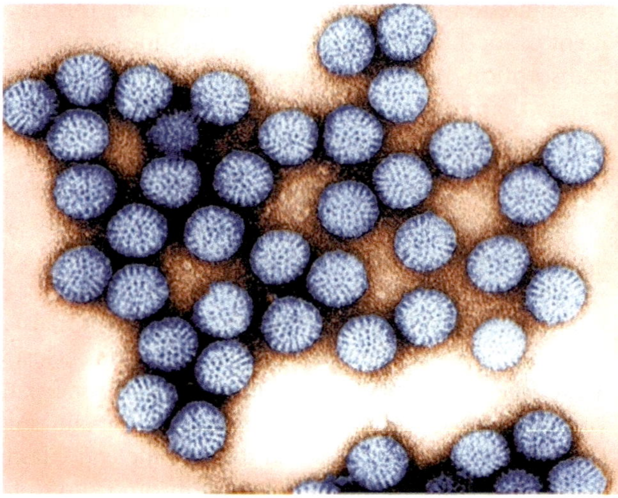

FIGURE 3.7

Microscopic image of intact double-shelled rotavirus particles. Source*: From Dr. Erskine Palmer/Own work/; https://creativecommons.org/licenses/publicdomain/; https://pixnio.com/science/microscopy-images/ultrastructural-morphology-of-a-number-of-intact-rotavirus-double-shelled-particles.*

humidity prevail.[15,16] Geographical influences result in strain variations in *rotavirus A* species circulating globally, which are responsible for the majority of the cases of rotavirus infections occurring worldwide.[10]

The advent of rotavirus vaccines has been instrumental in changing the epidemiologic scenario of rotavirus infections. There has been a steep decrease in the incidence of rotavirus disease in developed countries.[17] Rotavirus vaccines have also resulted in the trend of infections shifting toward older children and aged individuals who cannot be vaccinated due to age-associated limitations of the vaccines.

Pathogenicity and pathogenesis

Oral ingestion of food and/or water contaminated with rotavirus particles causes the viral particles to enter the small intestine. Rotavirus particles, like most other entero-pathogens, are resistant to the hostile environment created by the gastric juices and stomach acid in the digestive tract. Invasion of the small intestine is followed by replication of the virus in the host cells, which is required for the establishment of an active infection. The VP4 protein present on the outer shell of the double-shelled viral particles facilitates the attachment of the viral particles to the mature enterocytes present at the tips of the villi throughout the lumen of the small intestine.[14]

The alteration in the enterocytes triggered by this invasion changes the movement of food bolus into the large intestine, subsequently impairing the water reabsorption process in the large intestine. The clinical manifestation of watery diarrhea seen in rotavirus infections is a direct result of the impaired water reabsorption process.[18]

The cleavage of the VP4 protein into VP5 and VP8 is essential for the entry of the viral particles into the cytoplasm of the cells, which is the exclusive site for the process of viral replication. RNA-dependent RNA polymerase associated with the virus is activated in the cytoplasm of the host cell, thereby triggering the synthesis of viral mRNAs. This is followed by a translation of the viral mRNAs into viral proteins. The viral proteins then form inclusion bodies, thus providing a site for the assembly of the newly formed viral particles. The newly formed viral particles enclose 11 RNA segments but are single-shelled particles. The single-shelled viral particles bud through the endoplasmic reticulum after attachment of VP4 and is followed by the incorporation of VP7. VP4 and VP7 then form a highly organized outer shell structure, thus converting the single-shelled viral particles into double-shelled viruses. This is followed by cell lysis, which releases the fully functional rotaviruses into the lumen of the small intestine where the double-shelled viral particles now infect other healthy mature enterocytes, thus establishing the infection.[14]

The release of the newly formed viral particles requires cell lysis, which causes cell death and desquamation of the mature enterocytes (Fig. 3.8) in

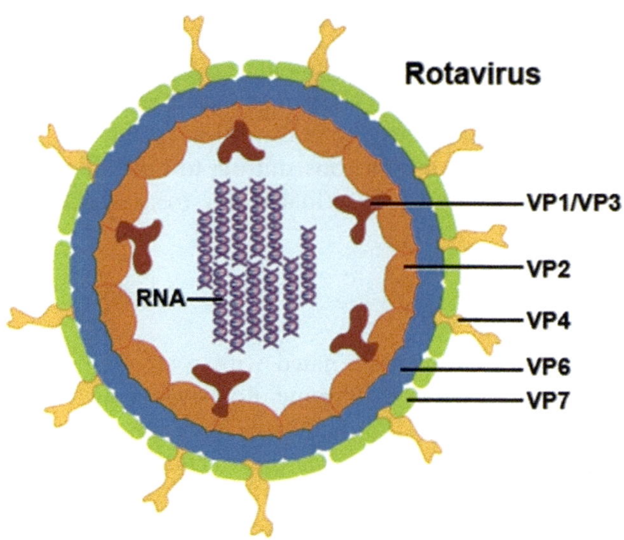

FIGURE 3.8
Structure and composition of the rotavirus. Source: *From NIAID/; https://www.flickr.com/people/54591706@N02/https://creativecommons.org/licenses/by/2.0/deed.en; https://commons.wikimedia.org/wiki/File:Rotavirus_(5102281461).jpg.*

the small intestine resulting in villous atrophy[19] (Fig. 3.9). This hampers the process of digestion, thereby compromising nutrient absorption. The expression of the absorptive enzymes instrumental in the process of digestion and nutrient absorption are also downregulated due to the viral infection. Crypt cell hyperplasia triggers increased secretion, further contributing to the severity of diarrhea. The VP4 and VP7 proteins are known to be the major determinants of viral pathogenicity in rotaviruses, although VP3 coding for the mRNA capping enzyme, NSP1 is responsible for the evasion of the host immune system by interferon antagonism, NSP2 is responsible for the assembly of the newly replicated viral particles, and NSP4-virus-encoded enterotoxin that stimulates the enteric nervous system to contribute to the regulation of viral replication and calcium homeostasis, and have been associated with rotavirus-associated pathogenicity.[18]

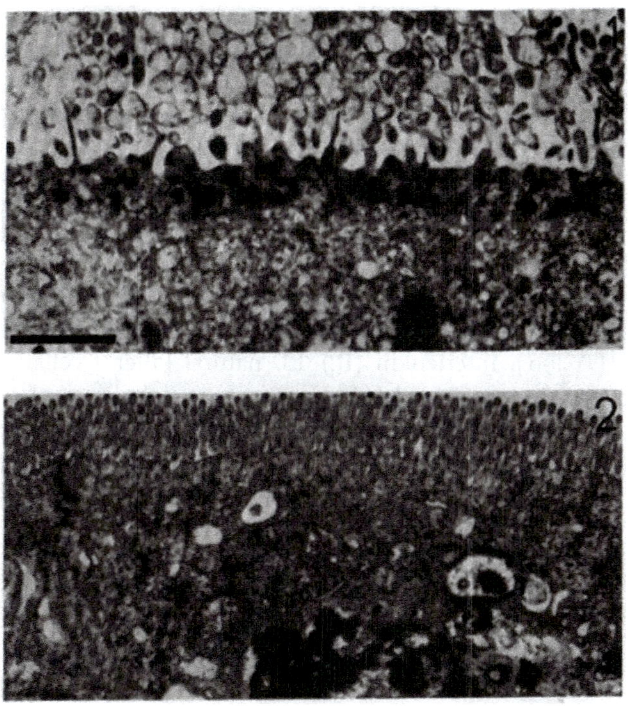

FIGURE 3.9
(A) Transmission electron micrograph of a rotavirus-infected enterocyte showing the destruction of the microvilli (B) Uninfected enterocyte for comparison. Source: *From Graham Beards/Own work/; https://creativecommons.org/licenses/by/3.0/deed.en; https://commons.wikimedia.org/wiki/File:Rotavirus_infected_gut.jpg.*

Immune responses

Rotavirus infections elicit both innate as well as acquired immune responses. The innate immune responses that are elicited rapidly after rotavirus infection consists of interferon (types I and III) secretion stimulated by the toll-like receptors (TLRs) activated by the viral RNA and Th1 and Th2 cytokines that recruit macrophages and natural killer cells. The viral NSP1 protein, however, is an interferon antagonist and thereby suppresses the antiviral activity of certain interferon regulatory proteins. The acquired immune responses triggered by rotavirus infections consist of both B- and T-cell immune responses. The VP4 and VP7 proteins of the rotavirus are known to stimulate the production of IgG, IgM, and IgA classes of antibodies, which can effectively neutralize the virus particles.

Although chronic conditions elicited by rotavirus infections are rare, the possibility of immune hyperactivity leading to an ongoing assault of the gastrointestinal system manifesting as chronic relapsing clinicopathologies of inflammatory bowel disease such as ulcerative colitis cannot be undermined.[12] Defects in the intestinal epithelial layer caused by a viral or bacterial infection, uncontrolled immune responses, and altered intestinal microflora are the major players in the pathophysiology of ulcerative colitis.[20]

The ongoing immune responses in the intestines result in an increased number of activated and mature dendritic cells including TLR2 and TLR4, while the Th cell responses were typically seen in ulcerative colitis exert a cytotoxic effect on the intestinal epithelial cells. The pathophysiology of ulcerative colitis is also dictated by the elevated levels of tumor necrosis factor-alpha (TNF-α), interleukin (IL) 13, natural killer T-cells, and neutralizing antibodies IgM, IgG, and IgA. Homeostatic imbalance of the host immune system to the intestinal microflora is also responsible for the vulnerability of the gastrointestinal system to entero-pathogens.[20] The characteristic inflammation of the mucosa and submucosa of the colon associated with ulcerative colitis is a result of the uncontrolled immunological onslaught of the mucosal surfaces of the colon.

Clinical manifestations and complications

Rotavirus infections are typically known for the sudden onset of severe gastrointestinal manifestations such as vomiting, abdominal cramps, watery diarrhea, and rapid dehydration, often accompanied by mild fever within 1 – days of exposure to the pathogen. The severity of the disease depends on the age and the immune status of the individual, and most cases resolve without any complications within 5–7 days from the onset of the disease.

However, certain cases may present with signs of severe dehydration such as decreased urine output, tachycardia, dry mucous membranes, prolonged capillary refill, electrolyte imbalances, and decreased skin turgor. Such cases may necessitate hospital admissions and measures to effectively restore the electrolyte balance.[13]

Systemic infections and infections of the nervous system such as meningitis, encephalitis, and encephalopathy caused by rotaviruses are rare and are usually seen only in immunocompromised individuals. Chronic relapsing gastrointestinal pathologies such as ulcerative colitis are also possible, albeit rare complications of rotavirus infections.

The onset of ulcerative colitis (Fig. 3.10) is gradual, triggered by the uncontrolled immunological onslaught of the colon following an enteropathogenic infection. Intermittent periods of spontaneous remissions followed by acute flares are characteristic of ulcerative colitis. During an acute flare episode ulcerative colitis presents with abdominal pain, tenesmus, bloody diarrhea, mucus discharge from the rectum, rectal bleeding, fever, and fatigue. The severity of the presenting symptoms is directly proportional to the severity of the particular flare. While the reasons for remissions are not completely understood, certain factors known as "triggers" varying from individual to individual such as the use of nonsteroidal antiinflammatory drugs are known

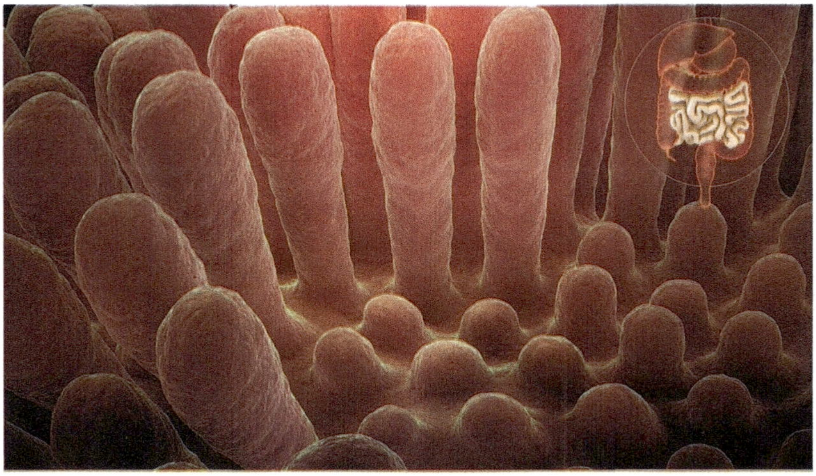

FIGURE 3.10

Inflammation of the intestinal mucosa resulting in villous atrophy in the small intestine. Source*: From http://www.scientificanimations.com/ http://www.scientificanimations.com/wiki-images/; https://creativecommons.org/licenses/by-sa/4.0/deed.en; https://commons.wikimedia.org/wiki/File:Inflammed_mucous_layer_of_the_intestinal_villi_depicting_Celiac_disease.jpg.*

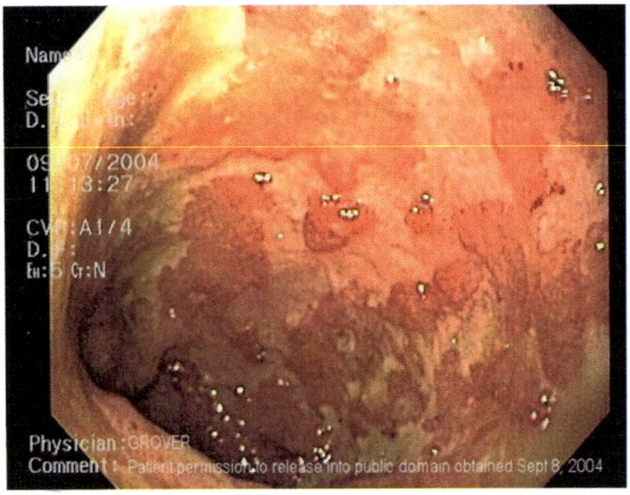

FIGURE 3.11

Endoscopic image of ulcerative colitis affecting the left side of the colon. Source*: From Samir at English Wikipedia/Own work/; https://creativecommons.org/licenses/by-sa/3.0/deed.en; https://commons.wikimedia.org/wiki/File:Ulcerative_colitis.jpg.*

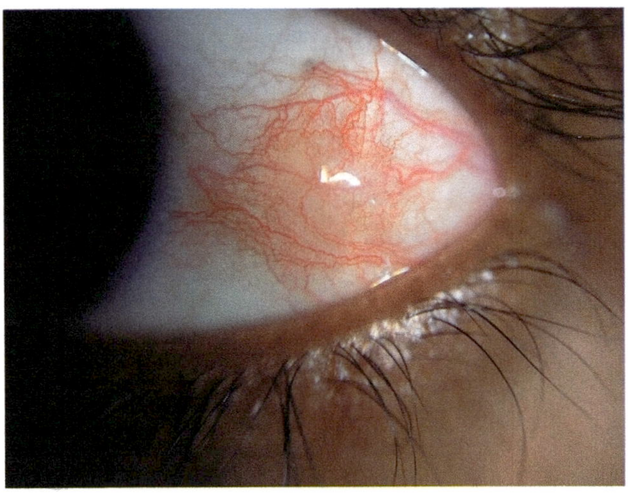

FIGURE 3.12

Episcleritis. Source*: From Imrankabirhossain/Own work/; https://creativecommons.org/licenses/by-sa/4.0/deed.en; https://commons.wikimedia.org/wiki/File:Episcleritis.jpg.*

to be responsible for the exacerbation of the symptoms. Some cases of ulcerative colitis may also present with extraintestinal manifestations such as peripheral arthropathies, episcleritis (Fig. 3.11), uveitis (Fig. 3.12), erythema nodosum (Fig. 3.13), scleritis, and pyoderma gangrenosum (Fig. 3.14)

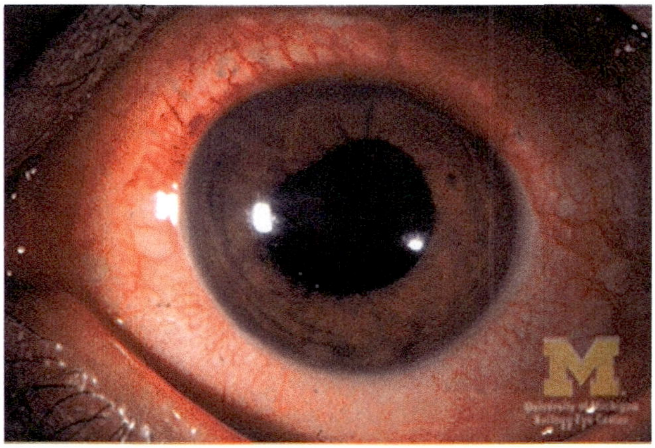

FIGURE 3.13

Anterior uveitis. Source*: From Jonathan Trobe, MD/Own work/; http://kellogg.umich.edu/theeyeshaveit/about.html/; http://kellogg.umich.edu/theeyeshaveit/redeye/anterior_uveitis.html/; https://creativecommons.org/licenses/by/3.0/deed.en; https://commons.wikimedia.org/wiki/File:Anterior-uveitis.jpg.*

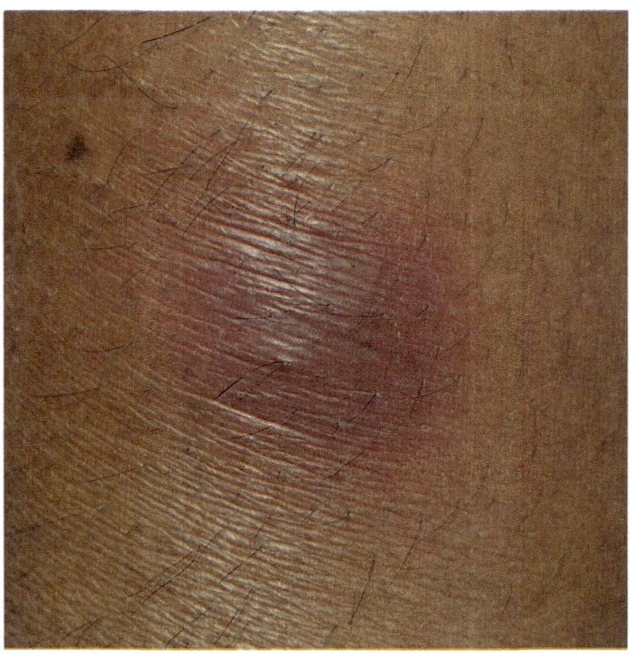

FIGURE 3.14

Erythema nodosum on the shin. Source*: From Biswarup Ganguly/Own work/; https://creativecommons.org/licenses/by/3.0/deed.en; https://commons.wikimedia.org/wiki/File:Erythema_nodosum_-_Kolkata_2012-01-03_7753.JPG.*

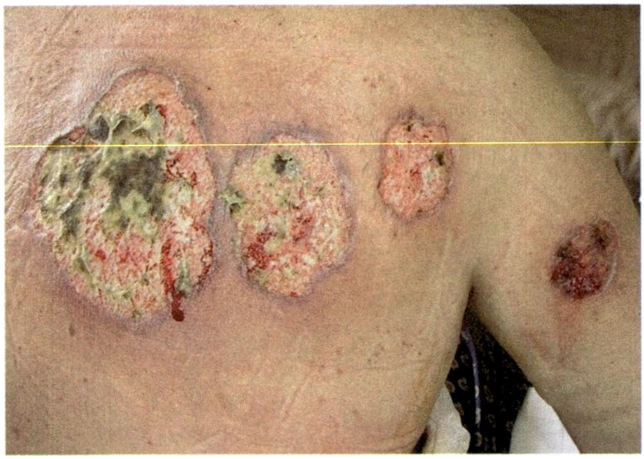

FIGURE 3.15
Pyoderma gangrenosum. Source: *From Monopol/Own work/; https://creativecommons.org/licenses/by-sa/3.0/deed.en; https://commons.wikimedia.org/wiki/File:Pyoderma_gangrenosum_01.jpg.*

during an active flare of the disease. Ankylosing spondylitis (Fig. 3.15), sacroiliitis (Fig. 3.16), axial arthropathies (Fig. 3.17), and primary sclerosing cholangitis (Figs. 3.16 and 3.18) are extraintestinal manifestations that may present even during the remission phase in a known case of ulcerative colitis. The association between ulcerative colitis and increased risk of colon cancer is well-established, especially in cases that present with hepatic extraintestinal manifestations such as primary sclerosing cholangitis (Fig. 3.19).[20]

Diagnosis

Clinical diagnosis of rotavirus-induced gastroenteritis can be a challenge due to the nonspecific nature of the symptoms. However, as a thumb rule, a history of possible exposure within 1–3 days of the onset of symptoms, low-grade fever, watery diarrhea without blood, and vomiting are considered to be a congregation of symptoms indicative of rotavirus-induced gastroenteritis. The presence of acid-reducing agents in stool samples and low levels of serum bicarbonate are also considered diagnostic indicators of rotavirus-induced gastroenteritis.[10] While laboratory testing may not be required, confirmation of diagnosis when required in critical unmanageable cases warrants the need for reliable quantifiable tests such as enzyme-linked immunosorbent assay (ELISA) and reverse transcriptase- PCR. Detection of the viral etiological agent may also be done by antigen detection assays,

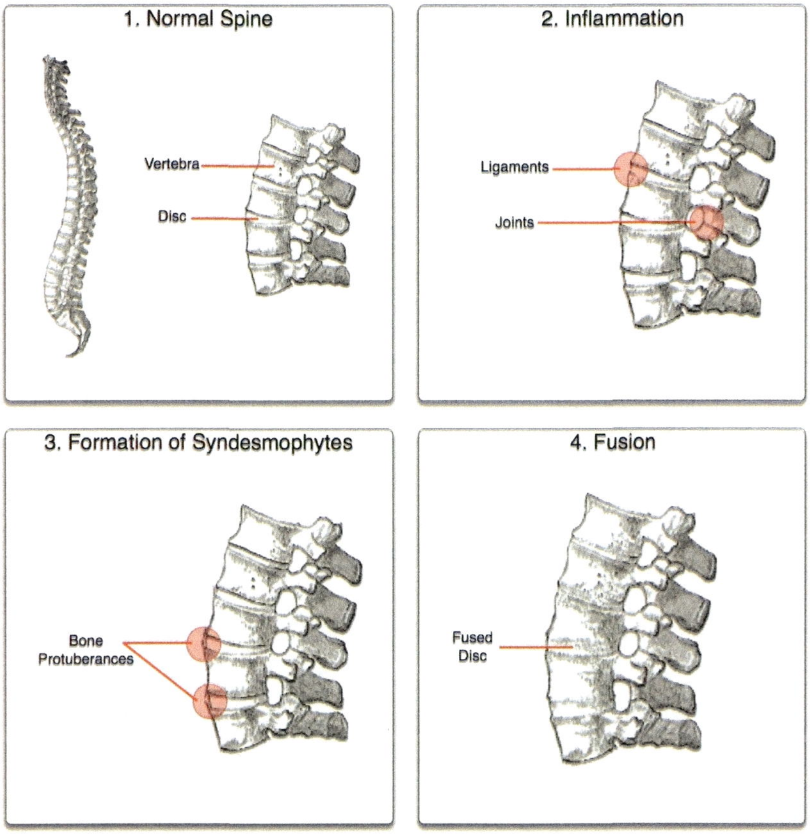

FIGURE 3.16

Ankylosing spondylitis process. Source: *From http://en.wikipedia.org/wiki/User:Senseiwa/; Own work (with an image of the spine from Gray's anatomy)/; https://creativecommons.org/licenses/by-sa/3.0/deed.en; https://commons.wikimedia.org/wiki/File:Ankylosing_process.jpg.*

polyacrylamide gel electrophoresis, electron microscopy, and virus isolation through culturing.[21]

The diagnosis of ulcerative colitis encompasses a multitude of tests including physical examination, culture studies, radiologic examinations, histopathological studies, and laboratory evaluations. Detailed and accurate patient history also plays a very important role in indicating the possibility of a diagnosis of ulcerative colitis. A history of intermittent episodes of bloody diarrhea, abdominal cramps, and tenesmus without fever with or without a history of infectious gastroenteritis is suggestive of relapsing inflammatory bowel disease clinicopathologies such as ulcerative colitis. Culture studies of

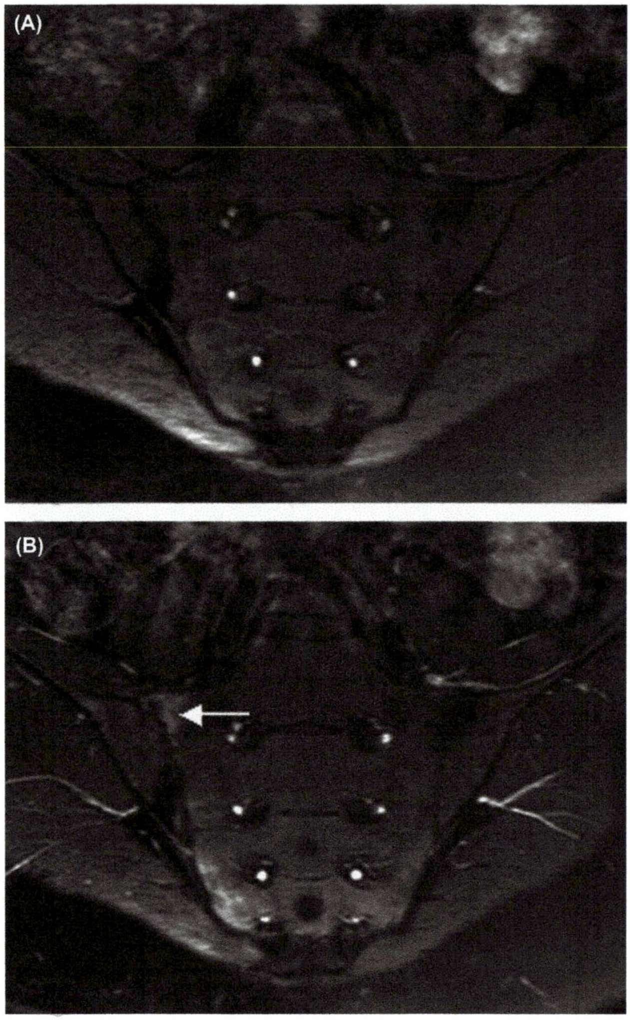

FIGURE 3.17

MRI of sacroiliac joints (A) without contrast agent (B) with contrast showing enhancement at the right sacroiliac joint (arrow), indicative of active sacroiliitis. Source*: From Fiona McQueen, Marissa Lassere and Mikkel Østergaard/Own work/; https://creativecommons.org/licenses/by/2.0/deed.en; https://arthritis-research.biomedcentral.com/articles/10.1186/ar1934/.*

stool samples are recommended to rule out infectious causes for the clinical manifestations.

Radiologic examinations such as endoscopy and colonoscopy are required to get an insight into the extent of the disease. Endoscopies can reveal the presence of fragile and/or granular mucosa, loss of vascular pattern, signs

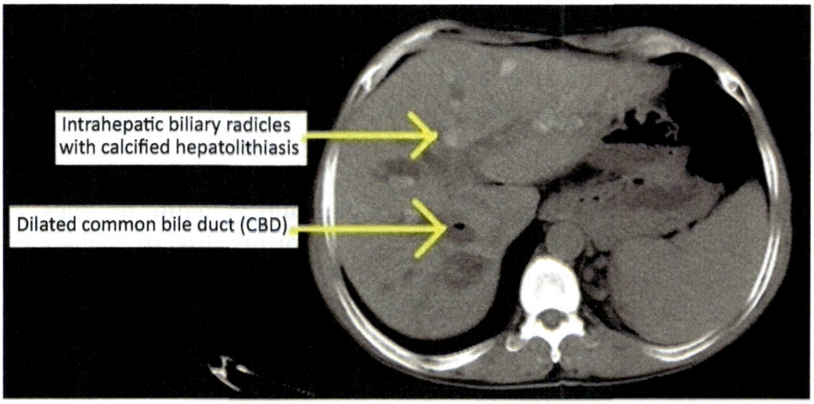

FIGURE 3.18

CT scan of primary sclerosing cholangitis. Source: *From Dima Nimri (https://www.wikidoc.org/index.php/User:Dima_Nimri) (image) and Akshun Kalia (https://www.wikidoc.org/index.php/User:Akshun_Kalia) (annotations)/Own work/; https://creativecommons.org/licenses/by-sa/3.0/deed.en; https://www.wikidoc.org/index.php/File:CT_PSC.jpg.*

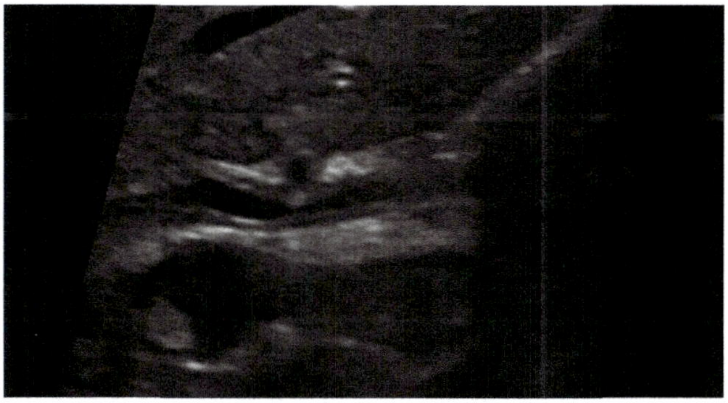

FIGURE 3.19

Ultrasound of sclerosing cholangitis. Source: *From Mme Mi/Own work/; https://creativecommons.org/licenses/by-sa/4.0/deed.en; https://commons.wikimedia.org/wiki/File:Sclerosing_cholangitis_0001.jpg.*

of erosions and pseudopolyposis, providing a clinical insight into the severity of the disease. Colonoscopies, on the other hand, are instrumental in determining the extent of the disease; proctitis indicative that the disease is localized to the rectum and sigmoid colon, while pancolitis is indicative that the disease has affected the entire colon. Colonoscopies can also reveal the extent of damage caused by the disease via the revelation of granularity, friability, loss of typical vasculature, and presence of ulceration involving the distal part of the rectum and the proximal part of the colon. The Montreal classification system is commonly used to determine the extent (E) and severity (S) of ulcerative colitis in confirmed cases. Endoscopy classifies the extent (E) between E1 and E; E1 being indicative of proctitis, E2 being indicative of distal/left-sided colitis, and E3 being indicative of pancolitis. Severity (S) ranging from S0 to S3 is classified based on clinical manifestations and systemic findings; S0 is considered suggestive of the remission phase of the disease, while S3 is considered to be suggestive of a severe form of the disease.[20] Double-contrast barium enema often reveals the characteristic "lead pipe" sign suggestive of chronic long-term ulcerative colitis.

Biopsy samples are recommended for histopathological studies, which are also very important in confirming the diagnosis of ulcerative colitis, and in understanding the severity of the disease. Macroscopically, ulcerative colitis is characterized by a continuous and symmetrical pattern of inflammation, which is restricted to the colonic mucosa. The biopsy sample usually shows edema, erythema, mucosal granularity, while ulceration may or may not be present. The presence of mucosal polyps, a slight narrowing of the distal colon, and a "cobblestoning" pattern, without any strictures or fistula, are often indicative of chronic long-term cases of ulcerative colitis. Microscopically, the presence of neutrophil-mediated epithelial injury demonstrated as the infiltration of the crypt or surface epithelium by neutrophils and neutrophil aggregates present in the crypt lumen, with or without ulcerations is considered to be a sign of ulcerative colitis. Chronic ulcerative colitis presents with shortening and distortional branching of the crypt and collections of lymphoplasmacytic infiltrates within the crypt bases and the mucosa. Metaplasia of the pyloric gland and the presence of Paneth cells in the left colon, indicative of metaplasia of the cells from the right colon where they are usually present, are also strong histopathological indicators of ulcerative colitis[22] (Fig. 3.20A,B, and C).

Laboratory evaluation parameters for the diagnosis of ulcerative colitis include erythrocyte sedimentation rate (ESR), C-reactive protein (CRP), and WBC count, which demonstrate elevated levels, especially during an acute flare of the disease. Perinuclear antineutrophil cytoplasmic antibodies (P-ANCA) are considered to be a reliable marker for the diagnosis of ulcerative colitis, irrespective of the stage of the disease, although certain cases of Crohn's disease may also be positive for P-ANCA. The antisaccharomyces cerevisiae antibodies (ASCA) test, on the other hand, tests positive more commonly in Crohn's disease than in ulcerative colitis, and therefore, testing for both P-ANCA and ASCA can enable distinction between the two conditions.[20]

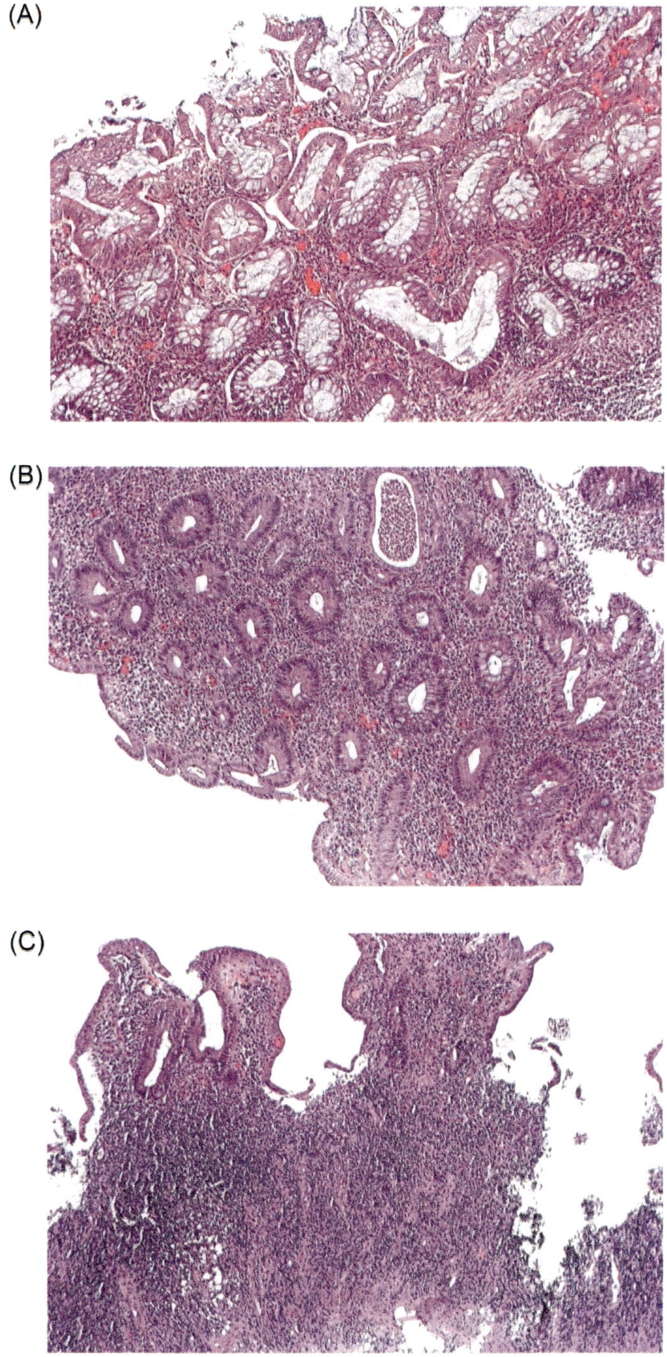

(*Continued*)

Treatment and management

Rotavirus infections often need management of the symptoms and quick and effective treatment of associated dehydration. Oral, subcutaneous, or intravenous rehydration using WHO-approved oral rehydration salts fortified with added glucose is used for dehydration treatment, depending on the severity of the condition. While most cases can be resolved on an outpatient basis, certain complicated cases such as pregnant women, electrolyte disturbances, renal failure, severe dehydration, persistent diarrhea and vomiting, ileus, and abdominal cramps may warrant the need for hospitalization.[23] Diphenoxylate, loperamide, codeine, and antiemetic medications may be used in adults to manage diarrhea and vomiting. Bismuth salicylate, although useful in treating rotavirus infections, is contraindicatory in infections caused by other enteropathogens, and therefore should be used only in confirmed cases of rotavirus infections. Studies have also demonstrated the efficacy of human serum immunoglobulins[24] and the use of zinc, ondansetron, and probiotics[21] in the treatment of acute cases of rotavirus-induced gastroenteritis.

The choice of treatment for ulcerative colitis is decided based on the extent and severity of the disease. Cases that are diagnosed early in the disease course generally have a good prognosis and usually go into remission. Suppository- or enema-based therapies are usually effective for conditions limited to proctitis, while a combination of rectal and systemic therapies may be required to treat inflammation in the distal colon, also managing the tenesmus in the process.[25] Oral or rectal administration of sulfasalazine and 5-aminosalicylates is considered the first-line treatment for ulcerative colitis. Remission is expected in most cases within 2 weeks, without which glucocorticoids are added to the treatment regime. The treatment regime, excluding glucocorticoids, can also be prescribed for the maintenance of the remission state. Immunosuppressants in the form of thiopurines such as azathioprine and 6-mercaptopurine are recommended only in cases that do not respond to glucocorticoids.

FIGURE 3.20

(A) Chronic inactive ulcerative colitis showing chronic inflammation and crypt architectural distortion. (B) Moderately active chronic ulcerative colitis showing cryptitis, crypt abscess, and erosion of goblet cells. (C) Severe chronic ulcerative colitis showing high levels of inflammation, mucosal ulceration, marked crypt architectural distortion, and loss of goblet cells. Source*: From (A) CoRus13/Own work/; https://creativecommons.org/licenses/by-sa/4.0/deed.en; https://commons.wikimedia.org/wiki/File:Chronic_inactive_ulcerative_colitis_intermed._mag.jpg. (B) CoRus13/Own work/; https://creativecommons.org/licenses/by-sa/4.0/deed.en; https://commons.wikimedia.org/wiki/File:Moderate_chronic_active_ulcerative_colitis_intermed._mag.jpg. (C) CoRus13/Own work/; https://creativecommons.org/licenses/by-sa/4.0/deed.en; https://commons.wikimedia.org/wiki/File:Severe_chronic_active_ulcerative_colitis,_intermed._mag.jpg.*

Monoclonal antibody therapy uses drugs such as infliximab, which inhibit the effects of proinflammatory cytokines such as TNF-α, and vedolizumab, which inhibit the actions of the adhesion molecules responsible for the influx of lymphocytes,[26] is also used for severe cases of ulcerative colitis. Increased NF-Kß-associated inflammation in ulcerative colitis is attributable to the decreased expression of peroxisome proliferator-activated receptor-gamma (PPAR-γ) in their colonocytes. 5-aminosalicylic acid analogs are novel drugs that have demonstrated PPAR-γ agonistic activity and are therefore being developed as a replacement for the existing cardiotoxic and metabolically toxic PPAR-γ agonists.

Surgical alternatives including proctocolectomy with ileal pouch-anal anastomosis or proctocolectomy with ileostomy may be warranted in cases that do not respond to therapy or demonstrate complications such as perforation of the colon, toxic megacolon, strictures, malignant growth, intolerance to therapy, nonremovable dysplasia, extreme inflammation, and critically high levels of bleeding.

The role of probiotics in attaining and maintaining remission has also been suggested. Reestablishment of healthy gut microbiota through fecal microbiota transplant is also recommended as an effective treatment option. Counseling patients about the importance of dietary and lifestyle changes, in addition to emphasizing the importance of regular dietary and lifestyle changes, with special emphasis on regular colonoscopies due to the predisposition toward colon cancer should form an indispensable part of treatment and management of ulcerative colitis.[20]

Prevention

Vaccines against rotaviruses have been shown to play a very important role in the reduction of incidences of gastroenteritis caused due to rotavirus infections. Although the strain variations pose a formidable challenge for the development of an effective vaccine, the live attenuated pentavalent reassortant vaccines available now have proved to be very effective against rotavirus infections, irrespective of geographical variations in the circulating viral strains. The vaccines may be given in three or two doses between the age of 2 and 6 months, depending on the specific vaccine being used.

Although the use of rotavirus vaccines may not ensure herd immunity to the disease due to the inability to vaccinate older children and elderly individuals, the vaccines have shown definite promise in ensuring a downward trend in incidences of rotavirus-associated gastroenteritis and related complications, especially in the developing and underdeveloped countries.

Student Notes

Key take-aways

- Rotaviruses are known to be one of the major causes of viral gastroenteritis, especially in young children and immunocompromised individuals.
- Rotavirus infections are transmitted through the fecal – oral route and poor hygiene and sanitation conditions existing in developing and underdeveloped nations are responsible for the higher incidences of rotavirus-associated gastroenteritis in these countries.
- While most rotavirus-associated gastroenteritis cases resolve within about a week, complications related to the nervous system, gastrointestinal system, or systemic manifestations are also known, especially among malnourished and immunocompromised individuals.
- Chronic relapsing clinicopathologies associated with the gastrointestinal system such as manifestations of inflammatory bowel disease in the form of ulcerative colitis and Crohn's disease may develop as sequelae to severe, untreated, or ineffectively treated cases of rotavirus-associated gastroenteritis.
- Ulcerative colitis is a chronic inflammatory condition of the colon, characterized by intermittent periods of remissions and flares.
- Although largely manageable, ulcerative colitis is a debilitating condition responsible for a large number of life-threatening conditions related to the colon such as colon cancer and toxic megacolon.
- Early diagnosis of ulcerative colitis is instrumental in ensuring a good prognosis of the disease.
- Chronic ulcerative colitis predisposes individuals to colon cancer, thereby necessitating colonoscopies at regular intervals.
- Rotavirus vaccination within the first 6 months of life is an effective preventive measure against rotavirus-associated gastroenteritis.

Problem number 3.4: Gastric ulcer caused by *Helicobacter pylori*

A 42-year-old man presented with epigastric pain and unintentional weight loss in the past 8 weeks. The patient also complained of chronic back pain that requires frequent self-medication with ibuprofen. He is a heavy smoker (about 15 cigarettes/day) and is professionally stressed.

The following are the reports of the investigations carried out:

1. Physical examination
   a. Epigastric region: Very tender
2. CBC

Test	Values	Reference range
Hb (g/dL)	9.5	14–16
RBC (/L)	4.8×10^9	$4.5–6.2 \times 10^9$
PCV (%)	29	40–52
WBC (/L)	12×10^9	$4–11 \times 10^9$
Neutrophils (%)	67	42.2–75.2
Lymphocytes (%)	29	20–50
Monocytes (%)	2	2–9
Eosinophils (%)	2	1–6
Platelets (/μL)	3,75,000	1,50,000–4,00,000

3. Blood group: O +ve
4. Stool sample: Occult blood present.
5. GI endoscopy: Single gastric ulcer visible; no neoplasm is present.
6. Rapid urease test [or test for campylobacter-like organism (CLO)] from stomach biopsy sample: Positive (indicative of *H. pylori* infection).

Diagnosis

The final diagnosis concluded a gastric ulcer caused by *H. pylori* infection.

Treatment

1. Lansoprazole
2. Clarithromycin (CLA)
3. Amoxicillin (AMO)

Follow-up and outcome

The patient was recommended to follow-up with a urea breath test (UBT) after 1 month.

Question and answer

1. What is *H. pylori*?

   H. pylori is a spiral-shaped, Gram-negative bacillus with multiple flagella, and is microaerophilic. These bacteria colonize the mucus, coating the epithelial cells in the stomach and producing the urease enzyme, detected by the rapid urease test or the UBT.

2. What is the clinical significance of *H. pylori*?
   H. pylori has been identified as the etiological agent responsible for a large majority of duodenal ulcer (DU) and gastric ulcer cases. The organism is also implicated in active chronic gastritis and infection and is known to be a predisposing factor for the development of mucosal-associated lymphoma (MALT) and carcinoma.
3. How is a *H. pylori* infection diagnosed?
   H. pylori infection can be diagnosed by invasive and noninvasive means. The CLO and rapid urea test (RUT) tests are invasive tests that detect the presence of urease in stomach biopsy samples. Microbiological culturing and histopathological analysis are also used for diagnosis, albeit less commonly. Serological tests and the UBT are noninvasive tests used for the diagnosis of *H. pylori* infection.
4. What is the CLO test?
   The CLO test, also known as the RUT, detects the presence of campylobacter-like organisms. *H. pylori* produces an enzyme known as urease, which breaks down urea and releases ammonia and carbon dioxide. A stomach biopsy sample is incubated with urea and the phenol red pH indicator is added. In the presence of *H. pylori*, the urea is broken down by urease, which has been produced by the bacteria to release ammonia, increasing the pH of the solution as indicated by a change in color.
5. What is the principle of the *H. pylori* UBT?
   The UBT is used to detect the presence of gastric *H. pylori*. The bacteria produce the urease enzyme, which breaks down urea to release carbon dioxide and ammonia. The test requires the patient to swallow a radioactive carbon C14 or nonradioactive carbon C13, containing a urea capsule. The isotope-labeled carbon dioxide is detected in the exhaled breath about 30 minutes following the intake of the capsule, indicating that the urea was split by the urease produced by the presence of *H. pylori*.
6. What are the indications for clinical suspicion of *H. pylori* infection?
   Symptoms of gastrointestinal distress characterized by heartburn, flatulence, abdominal pain, cramps, loss of appetite, and unintentional weight loss, along with the diagnosis of active gastric or DUs and peptic ulcer disease, are indications for clinical suspicion of *H. pylori* infection.

Learning objectives

1. Explain the etiology of gastric and DUs.
2. Describe the pathogenetic mechanism of *H. pylori*.
3. Summarize the epidemiology of *H. pylori*.

4. Describe the microbiological and clinical aspects of *H. pylori* infections.
5. Enumerate the clinical manifestations and complications of *H. pylori* infections.
6. Explain the principle of the UBT.
7. Summarize the treatment strategies for *H. pylori* infection.

Discussion

Differential diagnosis

Celiac disease	Celiac disease is an autoimmune condition that causes diarrhea, fatigue, loss of appetite, weight loss, bloating, flatulence, abdominal pain, nausea, vomiting, and constipation.
Peptic ulcer disease	Peptic ulcer disease is characterized by a gnawing or burning pain in the middle or upper part of the stomach that is especially aggravated between meals or at night; abdominal pain that is relieved by antacids; bloating, heartburn, nausea, and vomiting.
Crohn's disease	Crohn's disease is a chronic autoimmune condition that presents itself with severe diarrhea, abdominal pain, fatigue, unexplained weight loss, loss of appetite, and malnutrition.
Gastroesophageal reflux disease	Gastroesophageal reflux disease manifests as a burning sensation in the chest that usually presents itself after eating and worsens at night, chest pain, regurgitation of food or sour-tasting fluid, sensation of a lump in the throat, and difficulty swallowing.
Pancreatitis	Pancreatitis presents itself with upper abdominal pain that radiates to the back and worsens after eating, fever, rapid pulse, nausea, vomiting, and abdominal tenderness.
Lactose intolerance	Lactose intolerance is characterized by abdominal bloating, flatulence, belching, farting, abdominal cramps, stomach ache, general nausea, and stomach rumbling.
Hepatic congestion	Hepatic congestion presents itself with discomfort in the right upper quadrant, dyspnea, hepatomegaly, tenderness in the right upper quadrant, jaundiced appearance, and dry itchy skin.
Biliary colic	Biliary colic presents itself with intermittent periods of sharp pain or cramps in the middle or right upper quadrant of the abdomen that aggravates after a heavy meal or at night.
Eosinophilic esophagitis	Eosinophilic esophagitis presents itself with dysphagia, food chokage in the esophagus, also known as impaction, centrally located chest pain that is refractory to antacids, and regurgitation.
Stomach cancer	Stomach cancer may present itself with poor appetite, unintentional weight loss, abdominal pain or vague discomfort, bloating, feeling of fullness after eating only a small meal, heartburn, indigestion, nausea, blood-stained vomiting, fluid build-up in the abdomen, blood-stained stools, fatigue, malaise, anemia, weakness, and jaundice.

Continued

Continued	
Non-steroidal anti-inflammatory drugs (NSAID)-induced gastritis	NSAID-induced gastritis causes symptoms of gastrointestinal distress, such as dyspepsia, diarrhea, nausea, and vomiting.
Cholecystitis	Cholecystitis presents itself with intense, sudden pain in the upper right quadrant of the abdomen, pain that worsens when taking deep breaths, pain that radiates to the back or below the right shoulder blade, nausea, vomiting, fever, a jaundiced appearance, greyish-colored smelly stools with a loose consistency, and abdominal bloating.
Constipation	Constipation is characterized by fewer than three bowel movements a week, difficulty or pain while passing stools, dry and hard consistency of stools, a stomach ache, abdominal cramps, bloating, nausea, and feeling of incomplete bowel emptying.
Inferior myocardial infarction	Inferior myocardial infarction presents itself with chest pain, feeling of crushing weight or pressure in the chest region, dyspnea, diaphoresis radiating to the jaw or the arms, fatigue, light headedness, and nausea.
Superior mesenteric artery syndrome	Superior mesenteric artery syndrome causes a rapid feeling of fullness while eating, bloating, belching, indigestion, nausea, vomiting, regurgitation of partially digested food or bile-like fluid, small bowel obstruction, weight loss, and mid-abdominal cramps that are relieved by lying in the prone position, the knee-to-chest position, or by lying on the left side.

Background

H. pylori is a pathogenic bacterium that can infect humans. This bacterium colonizes the mucosal membranes of the stomach, thereby affecting the normal functioning of the gastrointestinal system. *H. pylori* is notorious for causing inflammation and ulceration in the stomach. *H. pylori* infections are more prevalent in developing countries due to overcrowding, lower socioeconomic status, poor sanitation, and unhygienic standards of living.[29,30] Early childhood is the most favorable time for acquiring *H. pylori* infections, although symptoms of the infection do not manifest until adulthood.[31] Chronic or atrophic gastritis, peptic ulcer, duodenitis, gastric lymphoma, and gastric carcinoma that manifest only in adulthood are mostly attributed to *H. pylori* infection.

Etiological agent

Helicobacter pylori

Classification and morphology

H. pylori is a helical-shaped, Gram-negative bacterium that is the type species of the *Helicobacter* genus (Fig. 3.21). This bacterium is a microaerophile,

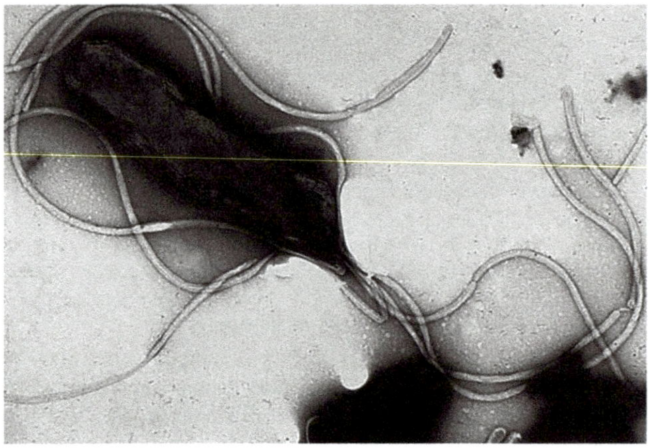

FIGURE 3.21
Helicobacter pylori. Source*: From Yutaka Tsutsumi/Fujita Health University School of Medicine/; https://creativecommons.org/licenses/by-nc-nd/2.0/; https://www.flickr.com/photos/pennstatelive/14227300617.*

requiring low oxygen concentration ranging between 5% and 14% and a capnophile that can survive at high concentrations of carbon dioxide. *H. pylori* is highly motile due to the presence of numerous flagella.[32] The urease-producing ability of these bacteria allow them to survive and even thrive in the highly acidic gastric environment by increasing the pH from an unfavorable range of 2–3 to a more conducive range of 6–7 (Fig. 3.22).

Epidemiology

The prevalence of *H. pylori* infections is higher in developing countries and is one of the most common childhood infections, although manifestations of the infection are usually seen only in adulthood. This higher prevalence is attributed to lower socioeconomic status and compromised standards of living in developing countries. *H. pylori* infections have also been seen to show a higher prevalence in Hispanic and African American populations rather than in White American populations.[29,33,34] Contaminated sources of food and water are the most common contributors in the environmental transmission of *H. pylori*. The fecal – oral and gastric – oral, oral – oral, and sexual routes are the most common factors for the spread of the infection.[35]

Pathogenicity and pathogenesis

The pathogenicity of *H. pylori* is a function of four important etiological attributes namely the urease-producing ability, the flagella-mediated motility, the bacterial adhesins, and the effector proteins or toxins encoded by the bacteria. Urease breaks down urea, which releases carbon dioxide and ammonia. The ammonia generated in the process causes a shift in the hostile acidic environment of the

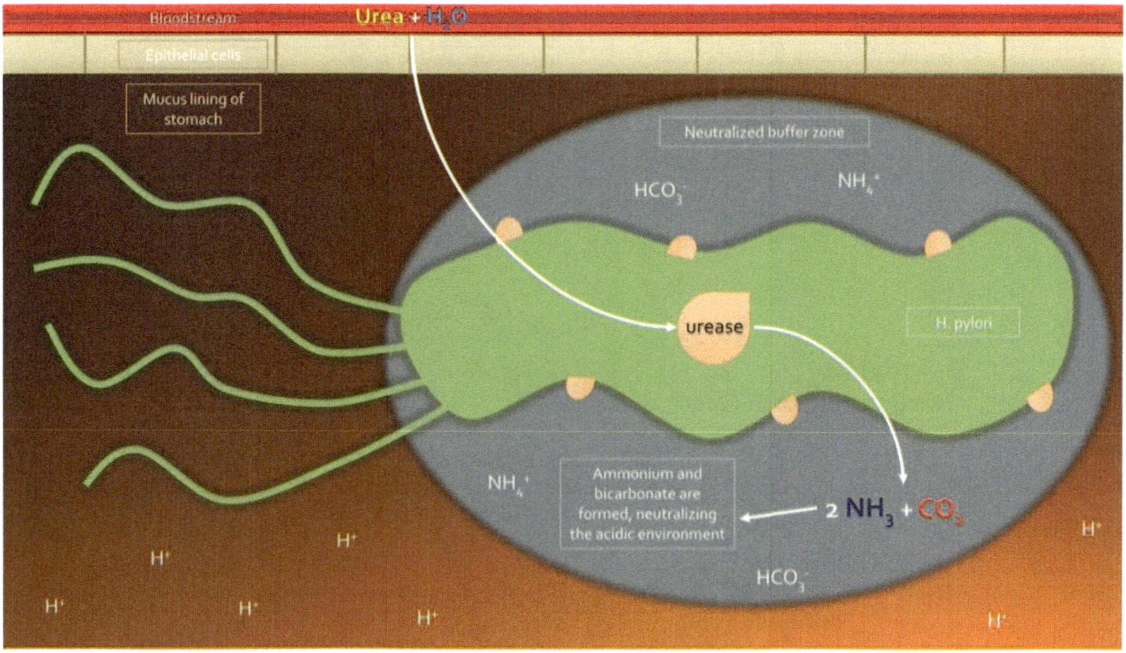

FIGURE 3.22

Production of urease enzyme production by *H. pylori*. Source*: From Olivia Child/Own work/; https://creativecommons.org/licenses/by-sa/4.0/deed.en; https://commons.wikimedia.org/wiki/File:H._pylori_urease_enzyme_diagram.svg.*

stomach to a more neutral pH range, facilitating bacterial survival. The flagella-mediated motility enables the bacterial cells to migrate effectively toward the gastric epithelial cells, while the interaction of the bacterial adhesins with the host cell receptors mediate colonization and infection. The bacterial adhesins also play an important role in facilitating persistent *H. pylori* infection in the host. The effector proteins, such as cytotoxin-associated gene A (Cag A) and vacuolating cytotoxin A (VacA) encoded by the bacterial cells, are toxins that mediate tissue damage in the host (Fig. 3.23).

H. pylori invade the gastrointestinal system of the host and mobilize toward the epithelial cells of the gastric mucosa, aided by flagella-mediated motility. The bacteria then penetrate the gastric mucosa and the bacterial adhesins enable adherence to the surface of the mucosal epithelial cells. Urease activity mediates the production of alkalizing ammonia from urea, which helps the bacteria combat the hostile acidic gastric environment. This confers resistance to elimination and enables the survival and proliferation of *H. pylori*. The progression of the infection results in inflammation, mucosal destruction, and cell death (Fig. 3.24). Histopathological analysis of *H. pylori*-infected gastric tissues typically demonstrate visible signs of inflammation[36] (Fig. 3.25).

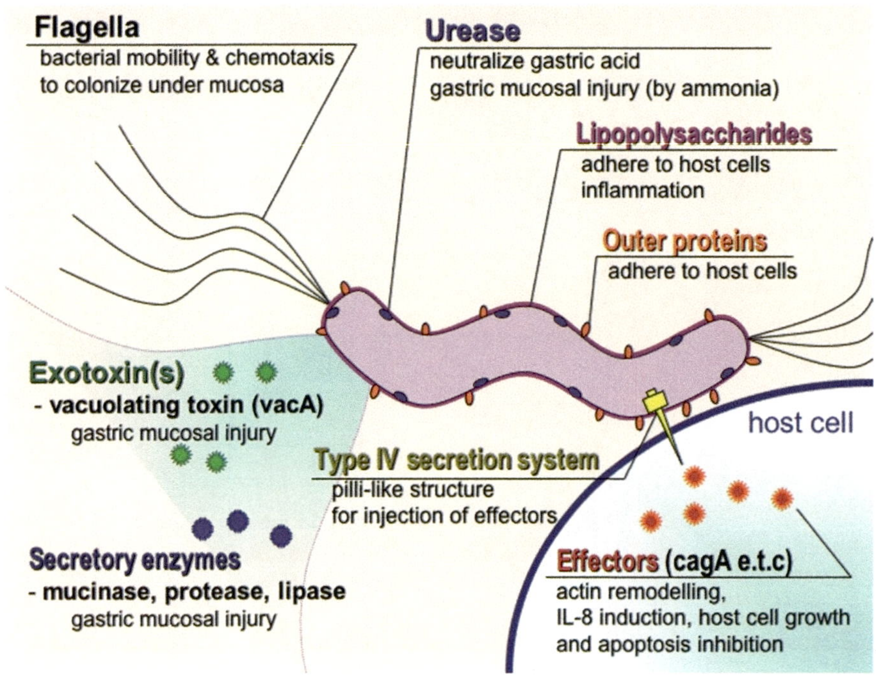

FIGURE 3.23

Schematic diagram of virulence factors of *H. pylori*. Source*: From Y_tambe/Y_tambe's file/; https://creativecommons.org/licenses/by-sa/3.0/deed.en; https://commons.wikimedia.org/wiki/File:H_pylori_virulence_factors_en.png.*

Infection usually results in acute neutrophilic gastritis in the initial stages and may potentially progress to active chronic gastritis. As *H. pylori* infects almost half the population, especially in developing countries, it is imperative to distinguish between a benign presence and a pathogenic role of the bacteria. Products of bacterial metabolism, such as ammonia or ammonium products, cytotoxins, and phospholipases, in addition to the generation of some proinflammatory products, like lipopolysaccharide and platelet-activating factor, are implicated in direct tissue injury, consequentially leading to surface epithelial degeneration. The host response to the infection typically results in wide-spread inflammation akin to pangastritis. Subsequently, pathologies such as multifocal atrophy and intestinal metaplasia develop. This weakens mucosal defenses and mediates peptic ulceration. Certain physiological attributes, such as increased parietal cell mass that mediates a high level of acid output, can prevent gastritis in the antral region of the stomach by restricting bacterial adhesion and inflammatory responses. However, the proximal duodenum may demonstrate acid-induced gastric metaplasia in such cases.[37]

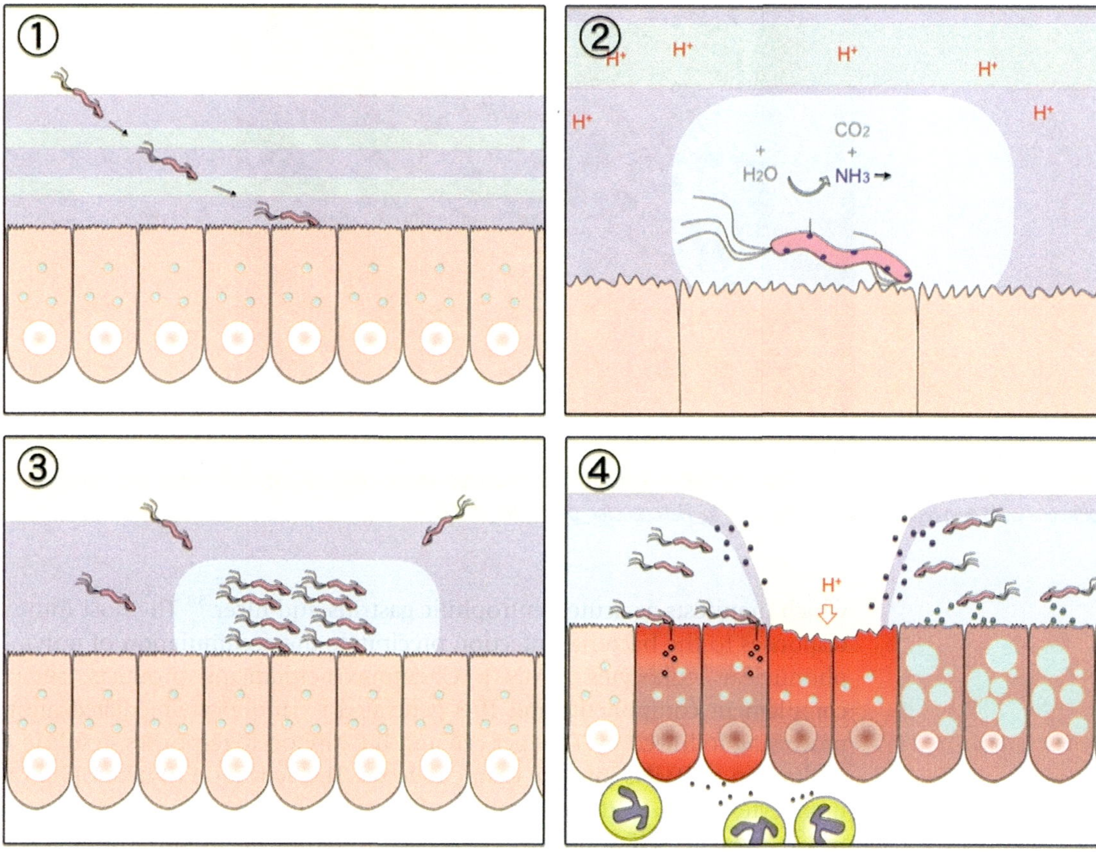

FIGURE 3.24

H. pylori infection-induced gastric ulceration. Source: *From Y_tambe/ Y_tambe's file/; https://creativecommons.org/licenses/by-sa/3.0/deed.en; https://commons.wikimedia.org/wiki/File:H_pylori_ulcer_diagram.png.*

H. pylori has been classified as a group 1 carcinogen, and chronic untreated infections are known to result in gastric adenocarcinoma. The etiology of carcinoma because of *H. pylori* infection follows a general sequence initiated with gastritis, followed by atrophy that results in intestinal metaplasia, leading to dysplasia. Carcinoma is a common consequence that follows dysplasia.[33]

Immune responses

Eosinophils, neutrophils, mast cells, and dendritic cells are stimulated in *H. pylori*-mediated gastritis.[33] The cells of the epithelial layer in the gastric cavity secrete chemicals that signal the elicitation of innate immune responses. Subsequently, the activation of neutrophils causes damage to the host tissue,

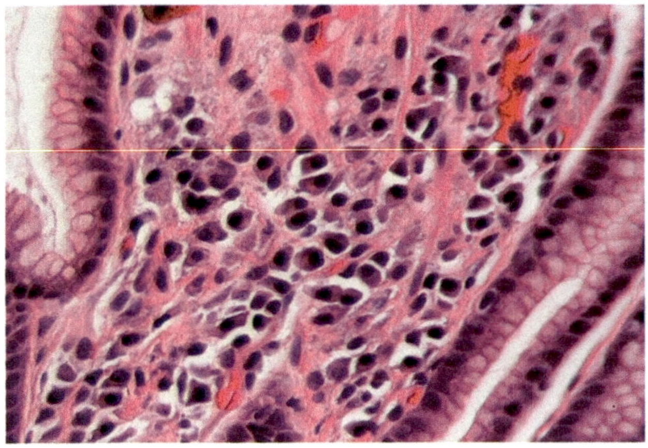

FIGURE 3.25

Gastritis in *H. pylori* infection. Source: *From Patho/Own work/; https://creativecommons.org/licenses/by-sa/3.0/deed.en; https://commons.wikimedia.org/wiki/File:Gastritis_due_to_Helicobacter_pylori,_HE_4.JPG.*

which manifests as acute neutrophilic gastritis and ulcer.[38] The host immune responses to the bacterial infection precipitate the accumulation of polymorphonuclear leukocytes (PMNs). Chemotaxis-enhancing products, such as complement components and IL-8 generated by monocyte-mediated antigen processing followed by T-helper cell and B-lymphocyte responses, result in the infiltration of PMNs into the gastric mucosa. The activation of neutrophils and monocytes leads to the generation of reactive oxygen species and proteases, which mediate self-destructive processes, resulting in atrophy. Although the innate immune responses dominate the initial host responses to *H. pylori* infection, adaptive immune responses also have a significant impact on the pathophysiology of the disease. Intestinal metaplasia is a manifestation of adaptive responses to the infection. The pathology of intestinal metaplasia is further exacerbated due to bile reflux and exposure to dietary irritants.[37]

Clinical manifestations and complications

H. pylori infections are implicated in acute and chronic gastritis, peptic ulcer disease, nonulcer dyspepsia, DUs, duodenitis, gastric cancers, and MALT-lymphoma.[38] *H. pylori* infection in children is typically asymptomatic, although some clinical manifestations of gastritis or peptic ulcer disease may be present on rare occasions. Gastrointestinal symptoms, such as abdominal pain, nausea, vomiting, dyspepsia, regurgitation of undigested food or sour-tasting fluid, loss of appetite, and unintentional weight loss, warrant the need for clinical investigations aimed at identification of the underlying etiology. Chronic extra-gastrointestinal manifestations, such as unexplained iron deficiency anemia and chronic immune thrombocytopenia (cITP), may also be indicative of pathologies associated with *H. pylori* infection. Therefore noninvasive detection of *H. pylori* infection is recommended in children manifesting cITP or idiopathic refractory iron deficiency anemia.[39] Clinical manifestations of *H. pylori* infection are generally more prominent in adulthood and may manifest demonstrable symptoms only after the progression to gastritis and ulcers. Typically, an individual may present with a stomach ache, burning sensation in the abdomen, abdominal pain that worsens on an empty stomach, nausea, loss of appetite, unintentional weight loss, frequent burping, and bloating.

Diagnosis

The diagnosis of *H. pylori* infection may be done either by invasive or noninvasive methods. An invasive endoscopic assessment of the disease pathology is reliable for gaining a clinical understanding of the disease progression. A biopsy sample of the gastric tissue can also be obtained during an upper gastrointestinal endoscopy. The gastric sample is used for detection of *H. pylori* by culture, a method that has 100% specificity albeit the low sensitivity. RUT, also known as the CLO test, is the most recommended invasive diagnostic test. However, due to the variability of the sensitivity and specificity levels, it is recommended that the diagnosis of *H. pylori* infection is based on concordant results of at least two tests. Histopathology, PCR, and fluorescent in situ hybridization are other frequently used tests that require gastric samples. A positive histopathology report, combined with either a positive *H. pylori* culture or a positive RUT report, is usually required for a confirmed diagnosis of *H. pylori* infection.

The UBT is the most recommended noninvasive test for the diagnosis of *H. pylori* infections, owing to its high sensitivity and specificity levels. Antigen and antibody detection tests are also effective noninvasive means of diagnosing *H. pylori* infections. These tests detect antigens in stool samples and *H. pylori*-specific antibodies in serum, urine, and buccal cavity samples. Similar to the UBT, the ELISA-based stool antigen test has high levels of sensitivity and is specific comparable to invasive diagnostic methods.[40] The low sensitivity of IgG antibody detection tests, on the other hand, makes them unreliable for a clinical diagnosis of *H. pylori* infections.[39] UBT and stool antigen tests are recommended for posttreatment determination of *H. pylori* elimination.[34]

Treatment and management

The complete eradication of *H. pylori* relies on the widely recommended triple therapy comprising of a proton pump inhibitor (PPI) (1–2 mg/kg/day), AMO (50 mg/kg/day), and CLA (20 mg/kg/day). The treatment regimen is prescribed for 14 days. However, for strains that are resistant to CLA, metronidazole (MET) (20 mg/kg/day) can also be prescribed. The triple therapy can be taken together or as a sequential therapy for 10 days. The sequential therapy recommends that the PPI and AMO are taken simultaneously for 5 days followed by PPI and MET/CLA together for 5 days. The sequential therapy is most commonly prescribed for pediatric patients infected with susceptible bacterial strains and is not recommended for patients infected with strains resistant to MET or CLA. Children below the age of 8 years are given AMO, while those above 8 years of age are given tetracycline. The use of three different high dose antibiotics in the triple therapy warrants the need for body weight-based calculations for determination of medication doses. Pediatric patients treated with the triple therapy are prescribed higher doses of PPI per kg body weight to be given 15–30 minutes before meals to counter the increased acid production caused using multiple antibiotics. The bismuth quadruple therapy comprising of bismuth salts (8 mg/kg/day), PPI, AMO/tetracycline, and MET is effective in cases that are resistant to both MET and CLA.[34]

CLA-resistant strains of *H. pylori* have increasingly emerged in recent years due to the irrational use of macrolide group antibiotics for respiratory tract infections. The threat of treatment failure due to antibiotic-resistant strains warrants the determination and prescription of individualized treatment therapies based on antibiotic susceptibility reports.[41] The appropriate choice of PPIs also plays an important role in enhancing the effectiveness of the treatment regimen. The use of probiotics has also been contemplated, but there is a lack of evidence indicative of any benefits.[34]

Prevention

The prevalence of *H. pylori* infections is found to be higher in developing countries due to the low standards of hygiene and sanitation, crowded living conditions, and contaminated water supply. Most cases of infection are known to occur in early childhood, although these cases usually remain asymptomatic. The bacteria primarily reside in the stomach of the infected individual, and therefore, person-to-person spread through fecal – oral or oral – oral routes are the most common.

Transmission from *H. pylori*-infected mothers and siblings is known to be predominant. Although breastfeeding is not contraindicated in *H. pylori*-infected mothers, appropriate hand washing practices should precede breastfeeding. Individuals sharing intimate contact are also prone to infect each other. Awareness about good hygiene and sanitation practices among parents, grandparents, or caregivers of children is important for prevention of childhood infections. Symptomatic adults should be particular about the diagnosis, treatment, and follow-up of their condition. Practicing good hand hygiene, clean and hygienic food preparation practices, and ensuring safe and clean sources of drinking water are important steps for prevention of *H. pylori* infection in children.[42]

Student Notes

Key take-aways

- *H. pylori* is a pathogenic bacterium that infects humans.
- The prevalence of *H. pylori* infections is found to be higher in developing countries due to the low standards of hygiene and sanitation, crowded living conditions, and contaminated water supply.
- Most cases of infection are known to occur in early childhood, although these cases usually remain asymptomatic.
- The bacteria primarily reside in the stomach of the infected individual, and therefore, person-to-person spread through fecal – oral or oral – oral routes are the most common.
- *H. pylori* is notorious for causing inflammation and ulceration in the stomach.
- Chronic or atrophic gastritis, peptic ulcer, duodenitis, gastric lymphoma, and gastric carcinoma that manifest only in adulthood are mostly attributed to *H. pylori* infection.
- Typically, an individual may develop a stomach ache, burning sensation in the abdomen, abdominal pain that worsens on empty stomach, nausea, loss of appetite, unintentional weight loss, frequent burping, and bloating.
- The diagnosis of *H. pylori* infection may be done either by invasive or noninvasive methods.
- The complete eradication of *H. pylori* relies on the widely recommended triple therapy, comprising of a PPI (1–2 mg/kg/day), AMO (50 mg/kg/day), and CLA (20 mg/kg/day).
- Awareness about good hygiene and sanitation practices among parents, grandparents, or caregivers of children is important for the prevention of childhood infections.

References

1. Rahimian J, Wilson T, Oram V, Holzman RS. Pyogenic liver abscess: recent trends in etiology and mortality. *Clin Infect Dis*. 2004;39(11):1654–1659. Available from: https://doi.org/10.1086/425616. Epub November 9, 2004.
2. Mciver C, Janda JM. Pathogenesis and laboratory identification of emerging hepatovirulent *Klebsiella pneumoniae*. *Clin Microbiol News*. 2008;30:127–131.
3. Habib AG, Tambyah PA. Community-acquired *Klebsiella pneumoniae* central nervous system infections in adults in Singapore. *Eur J Clin Microbiol Infect Dis*. 2003;22(8):486–488. Available from: https://doi.org/10.1007/s10096-003-0963-x.
4. Ashurst JV, Dawson A. *Klebsiella pneumonia* [Updated March 25, 2020]*StatPearls [Internet]*. Treasure Island (FL): StatPearls Publishing; 2020. Available from: https://www.ncbi.nlm.nih.gov/books/NBK519004/.
5. Siu LK, Yeh KM, Lin JC, Fung CP, Chang FY. *Klebsiella pneumoniae* liver abscess: a new invasive syndrome. *Lancet Infect Dis*. 2012;12(11):881–887. Available from: https://doi.org/10.1016/S1473-3099(12)70205-0.

6. Garbati MA, Al Godhair AI. The growing resistance of *Klebsiella pneumoniae*; the need to expand our antibiogram: case report and review of the literature. *Afr J Infect Dis*. 2013;7(1):8–10. Available from: https://doi.org/10.4314/ajid.v7i1.
7. Centers for Disease Control and Prevention (CDC). Vital signs: carbapenem-resistant Enterobacteriaceae. *MMWR Morb Mortal Wkly Rep*. 2013;62(9):165–170.
8. Ryan KJ, Ray CG, eds. *Sherris Medical Microbiology*. 4th ed. McGraw Hill; 2004. ISBN 0-8385-8529-9.
9. Podschun R, Ullmann U. Klebsiella spp. as nosocomial pathogens: epidemiology, taxonomy, typing methods, and pathogenicity factors. *Clin Microbiol Rev*. 1998;11(4):589–603.
10. Gupta M, Shorman M. *Cytomegalovirus. StatPearls [Internet]*. Treasure Island (FL): StatPearls Publishing; 2020 [Updated August 10, 2020]. Available from: https://www.ncbi.nlm.nih.gov/books/NBK459185/.
11. Ngai JJ, Chong KL, Oli Mohamed S. Cytomegalovirus retinitis in primary immune deficiency disease. *Case Rep Ophthalmol Med*. 2018;2018:8125806. Available from: https://doi.org/10.1155/2018/8125806.
12. Kim J, Lee S. Ocular ischemic syndrome as the initial presenting feature of cytomegalovirus retinitis. *Korean J Ophthalmol*. 2018;32(5):428–429. Available from: https://doi.org/10.3341/kjo.2018.0017.
13. Stagno S, Cloud G, Pass RF, Britt WJ, Alford CA. Factors associated with primary cytomegalovirus infection during pregnancy. *J Med Virol*. 1984;13:347–353. Available from: https://doi.org/10.1002/jmv.1890130405.
14. Nosotti M, Tarsia P, Morlacchi LC. Infections after lung transplantation. *J Thorac Dis*. 2018;10(6):3849–3868. Available from: https://doi.org/10.21037/jtd.2018.05.204.
15. Faith SC, Durrani AF, Jhanji V. Cytomegalovirus keratitis. *Curr Opin Ophthalmol*. 2018;29(4):373–377. Available from: https://doi.org/10.1097/ICU.0000000000000481.
16. Bunde T, Kirchner A, Hoffmeister B, et al. Protection from cytomegalovirus after transplantation is correlated with immediate early 1-specific CD8 T cells. *J Exp Med*. 2005;201(7):1031–1036. Available from: https://doi.org/10.1084/jem.20042384.
17. Brizić I, Hiršl L, Britt WJ, Krmpotić A, Jonjić S. Immune responses to congenital cytomegalovirus infection. *Microbes Infect*. 2018;20(9–10):543–551. Available from: https://doi.org/10.1016/j.micinf.2017.12.010.
18. Slyker JA, Rowland-Jones SL, Dong T, et al. Acute cytomegalovirus infection is associated with increased frequencies of activated and apoptosis-vulnerable T cells in HIV-1-infected infants. *J Virol*. 2012;86(20):11373–11379. Available from: https://doi.org/10.1128/JVI.00790-12.
19. Schleiss MR. Congenital cytomegalovirus infection: molecular mechanisms mediating viral pathogenesis. *Infect Disord Drug Targets*. 2011;11(5):449–465. Available from: https://doi.org/10.2174/187152611797636721.
20. La Rosa C, Diamond DJ. The immune response to human CMV. *Future Virol*. 2012;7(3):279–293. Available from: https://doi.org/10.2217/fvl.12.8.
21. Prince HE, Lapé-Nixon M. Role of cytomegalovirus (CMV) IgG avidity testing in diagnosing primary CMV infection during pregnancy. *Clin Vaccine Immunol*. 2014;21(10):1377–1384. Available from: https://doi.org/10.1128/CVI.00487-14.
22. Eisenstein EM, Wolf DG. Cytomegalovirus infection in pediatric rheumatic diseases: a review. *Pediatr Rheumatol Online J*. 2010;8:17. Available from: https://doi.org/10.1186/1546-0096-8-17.
23. Alibek K, Karatayeva N, Bekniyazov I. The role of infectious agents in urogenital cancers. *Infect Agent Cancer*. 2012;7(1):35. Available from: https://doi.org/10.1186/1750-9378-7-35.
24. Ross SA, Novak Z, Pati S, Boppana SB. Overview of the diagnosis of cytomegalovirus infection. *Infect Disord Drug Targets*. 2011;11(5):466–474. Available from: https://doi.org/10.2174/187152611797636703.

25. Ngai JJ, Chong KL, Oli Mohamed S. Cytomegalovirus retinitis in primary immune deficiency disease. *Case Rep Ophthalmol Med.* 2018;2018:8125806. Available from: https://doi.org/10.1155/2018/8125806.

26. Mozaffar M, Shahidi S, Mansourian M, Badri S. Optimal use of ganciclovir and valganciclovir in transplanted patients: how does it relate to the outcome. *J Transpl.* 2018;2018:8414385. Available from: https://doi.org/10.1155/2018/8414385.

27. Bartlett AW, Hall BM, Palasanthiran P, McMullan B, Shand AW, Rawlinson WD. Recognition, treatment, and sequelae of congenital cytomegalovirus in Australia: an observational study. *J Clin Virol.* 2018;108:121–125. Available from: https://doi.org/10.1016/j.jcv.2018.09.017.

28. Anderholm KM, Bierle CJ, Schleiss MR. Cytomegalovirus vaccines: current status and future prospects. *Drugs.* 2016;76(17):1625–1645. Available from: https://doi.org/10.1007/s40265-016-0653-5.

29. Hooi JKY, Lai WY, Ng WK, et al. Global prevalence of *Helicobacter pylori* infection: systematic review and *meta*-analysis. *Gastroenterology.* 2017;153(2):420–429. Available from: https://doi.org/10.1053/j.gastro.2017.04.022.

30. Graham DY, Adam E, Reddy GT, et al. Seroepidemiology of *Helicobacter pylori* infection in India. Comparison of developing and developed countries. *Dig Dis Sci.* 1991;36(8):1084–1088. Available from: https://doi.org/10.1007/BF01297451.

31. Jones NL, Sherman PM. *Helicobacter pylori* infection in children. *Curr Opin Pediatr.* 1998;10(1):19–23. Available from: https://doi.org/10.1097/00008480-199802000-00005.

32. Dunn BE, Cohen H, Blaser MJ. Helicobacter pylori. *Clin Microbiol Rev.* 1997;10(4):720–741. Available from: https://doi.org/10.1128/CMR.10.4.720.

33. Diaconu S, Predescu A, Moldoveanu A, Pop CS, Fierbințeanu-Braticevici C. *Helicobacter pylori* infection: old and new. *J Med Life.* 2017;10(2):112–117.

34. Parikh NS, Ahlawat R. *Helicobacter pylori* [Updated August 10, 2020]*StatPearls [Internet]*. Treasure Island (FL): StatPearls Publishing; 2021. Available from: https://www.ncbi.nlm.nih.gov/books/NBK534233/.

35. Zamani M, Vahedi A, Maghdouri Z, Shokri-Shirvani J. Role of food in environmental transmission of *Helicobacter pylori*. *Casp J Intern Med.* 2017;8(3):146–152. Available from: https://doi.org/10.22088/cjim.8.3.146.

36. Ricci C, Holton J, Vaira D. Diagnosis of *Helicobacter pylori*: invasive and non-invasive tests. *Best Pract Res Clin Gastroenterol.* 2007;21(2):299–313. Available from: https://doi.org/10.1016/j.bpg.2006.11.002.

37. Dixon MF. Pathophysiology of *Helicobacter pylori* infection. *Scand J Gastroenterol Suppl.* 1994;201:7–10.

38. Aleksić S. *Glas Srp Akad Nauka Med.* 2005;48:41–54.

39. Rajindrajith S, Devanarayana NM, de Silva HJ. *Helicobacter pylori* infection in children. *Saudi J Gastroenterol.* 2009;15(2):86–94. Available from: https://doi.org/10.4103/1319-3767.48964.

40. Fischbach W, Malfertheiner P. *Helicobacter pylori* infection. *Dtsch Arztebl Int.* 2018;115(25):429–436. Available from: https://doi.org/10.3238/arztebl.2018.0429.

41. Mégraud F. *H. pylori* antibiotic resistance: prevalence, importance, and advances in testing. *Gut.* 2004;53(9):1374–1384. Available from: https://doi.org/10.1136/gut.2003.022111.

42. Yucel O. Prevention of *Helicobacter pylori* infection in childhood. *World J Gastroenterol.* 2014;20(30):10348–10354. Available from: https://doi.org/10.3748/wjg.v20.i30.10348.

CHAPTER 4

Infections of the respiratory system

Problem number 4.1: Community-acquired pneumonia caused by *Streptococcus pneumoniae*

An 89-year-old male presented with a fever and cough, which had manifested 2 days ago. The cough was intermittent and nonproductive, while the fever developed suddenly and remained at a constant temperature of 38°C. His condition was associated with chills, chest pain while breathing, and myalgia. There were no other complaints, such as shortness of breath, wheezing, hemoptysis, night sweats, headache, sore throat, dyspnea on exertion, or orthopnea. Although the patient mentioned having some loss of appetite, he did not report any weight loss. His sleep cycle was unaffected, and he did not have any recent history of trauma.

The patient was a nonsmoker and did not abuse alcohol or drugs, but he had a 20-year-long history of hypertension, dyslipidemia, and ischemic heart disease. He was on medication for the known medical conditions and did well until he developed a fever and cough 2 days prior to the presentation. The patient history did not reveal any aggravating factors, or any history of recent travel. The patient did not have any known food or drug allergies, and denied any specific memory of childhood illnesses, and his vaccination profile was complete.

There was a history of cardiac catheterization 10 years ago to open clogged heart arteries using coronary angioplasty and stents, and two units of blood transfusion given after cardiac catheterization with no complications. The patient had also been operated successfully for a femoral fracture 18 years ago with no complications.

The patient was in a consanguineous marriage and had nine children. His spouse and children were devoid of any medical conditions other than type 2 diabetes and hypertension. He did not mention any history of genetic abnormalities in the family. The patient is capable of independent movement and can efficiently take care of his day-to-day activities.

CONTENTS

Problem-Based Learning Approach in Microbiology. DOI: https://doi.org/10.1016/B978-0-323-95092-3.00006-8

Current (on-going) medications

1. Angiotensin Receptor Blockers (Atacand): 16 mg tablet to be taken once daily for hypertension.
2. B-blockers (Tenormin): 100 mg tablet to be taken once daily for hypertension.
3. Ca channel blockers (Plendil): 10 mg tablet to be taken once daily for hypertension.
4. Rosuvastatin (Creston): 20 mg tablet to be taken every day at bedtime for dyslipidemia.
5. Isosorbide mononitrate (Imdur): 60 mg tablet to be taken once daily to prevent angina due to history of coronary artery disease.
6. Acetylsalicylic acid (Aspirin): 81 mg tablet to be taken once daily as a platelet aggregation inhibitor (anticlotting agent).

The following are the reports of the investigations conducted:

1. Vitals
   Temperature: 39.9°C
   Pulse: 109 beats/minute, regular and with normal volume, normal in character, symmetrical and vessel wall not palpable, no signs of collapsing pulse.
   Blood pressure: 131/68 mmHg while lying.
   Respiratory rate: 22 breaths/minute.
2. Physical examination
   Well-built; weight: 80 kg; height:168 cm; BMI: 28 (overweight)
   Comfortable in supine position
   Conscious, alert, and oriented but tachypneic
   No muscle wasting/deformities/missing limbs
   No radio-radial/radio-femoral delay
   No swollen lymph nodes
   No visible rashes, bruises, scars, lesions, or ulcers
   No evidence of pallor or jaundice
   No parotid swelling
3. Eye examination
   Arcus senilis
   No conjunctival pallor
   No squint
   No conjunctivitis or secretions
   Pupils: 3 mm equally round and reactive to light
4. ENT and oral examination
   External auditory canals normal
   Tympanic membranes intact
   Nasal mucosa pink, no discharge, or polyps

No glossitis, angular stomatitis, or ulcers in the mouth
Mucous membrane is moist
Good dental hygiene
No sore throat

5. Examination of the neck
   No abnormal masses seen
   No neck vein distension (no raised Jugular venous pressure (JVP))
   No visible swelling of lymph nodes
   Not using accessory respiratory muscle
6. Cardiovascular examination
   Tachycardia
   Chest pain
   No palpitations
   No dizzy spells
   No swollen ankles
   No dyspnea
   No orthopnea
7. Respiratory examination
   Chest pain while breathing
   No shortness of breath/dyspnea on exertion/orthopnea
   No wheezing
   No hemoptysis
8. Gastrointestinal examination
   No abdominal pain/nausea/vomiting
   No hematemesis
   Some loss of appetite
   No heartburn
   No constipation
   No melena
   No diarrhea
   No dysphagia
9. Genitourinary examination
   No difficulty/pain during urination
   No history of hematuria
   No urgency or incontinence
   No history of sexual transmitted diseases or prostate diseases
10. Neurological examination
    Normal memory and speech
    No syncope
    No dizziness
    No headache
    No seizures
    No altered sensation or weakness

No difficulty in walking
No visual disturbances
No hearing loss
No memory/concentration changes

11. Locomotor examination
Occasional complaints of joint and back pain
Occasional complaints of joint stiffness and swelling
No diurnal variation
No functional deficit
Normal spine structure and function
Normal movement of limbs

12. Endocrinological examination
No polyuria
No polydipsia
No excessive sweating
No heat or cold intolerance
No irritability
No dry skin or hair changes

13. Psychiatric examination
Feels healthy and happy
No history of psychiatric disorders
No ongoing mental health medications

14. Arterial blood gas profile

pH	7.43	7.35–7.45
PaO_2 (mmHg)	89	75–100
$PaCO_2$ (mmHg)	41.9	38–42
HCO_3 (mmol/L)	28.2	22–26

15. Hematological profile

WBC (/L)	4.01×10^9	$4.3–10.8 \times 10^9$
RBC (/L)	4.39×10^{12}	$4.2–5.9 \times 10^{12}$
Hb (g/dL)	12.5	13.5–17.5
Hematocrit (%)	38.50	33–40
Mean corpuscular volume (MCV) (fL)	89.6	80–100
Mean corpuscular Hb (MCH) (pg)	32.4	27–32
Mean corpuscular Hb concentration (MCHC) (%)	33.40	33.5–35.5
RDW (%)	13.5	11.8–14.5
Platelets (/L)	146×10^9	$150–400 \times 10^9$
Mean platelet volume (MPV) (fL)	9.5	7.5–12.0

Continued

Continued

Plateletcrit (%)	0.1	0.20–0.36
Platelet size distribution (%)	44	8.3–56.6
Neutrophils (%)	80	54–62
Lymphocytes (%)	10.80	25–33
Monocyte (%)	7.5	3–7
Eosinophils (%)	2.1	1–3
Basophils (%)	2.0	0.5–1
Neutrophils absolute count (/L)	3×10^9	$1.7–7.0 \times 10^9$
Lymphocyte absolute count (/L)	0.35×10^9	$0.90–2.90 \times 10^9$
Monocytes absolute count (/L)	0.30×10^9	$0.30–0.90 \times 10^9$
Eosinophils absolute count (/L)	0.09×10^9	$0.05–0.50 \times 10^9$
Basophils absolute count (/L)	0.1×10^9	$0.00–0.30 \times 10^9$
Nucleated RBC (%)	0	0
Reticulocyte count (%)	1.7	0.5–2.5
Reticulocyte absolute count (/L)	73.40×10^9	$50–100 \times 10^9$

16. Coagulation profile

Prothrombin time (seconds)	13.30	11–13.5
INR	1.11	≤1.1
Activated partial thromboplastin time (aPTT) (seconds)	19.10	30–40
aPTT ratio (seconds)	0.68	1.8–3.3
D-dimer	6 mg/L	<0.50

17. Biochemical profile

Urea (mmol/L)	6.8	3–6.3
Ketone bodies	Negative	Negative
Amylase (U/L)	74	30–110
Glucose (random) (mmol/L)	10.9	4.4–7.8
Creatinine (μmol/L)	97	61.9–114.9
Calcium (mmol/L)	2.05	2.2–2.7
Phosphorous (mmol/L)	0.9	1.12–1.45
Magnesium (mmol/L)	0.7	1.7–2.2
Creatinine kinase (U/L)	197	55–170
Potassium (mmol/L)	3.8	3.5–5.3
Sodium (mmol/L)	133	135–145
Chlorine (mmol/L)	101	95–110
Bicarbonate (mmol/L)	28	23–32

18. Liver function test

Protein (total) (g/L)	65	68–80
Albumin (g/L)	39	35–55
Globulin (g/L)	27	15–30
Total bilirubin (μmol/L)	13	1.71–20.5
Alkaline phosphatase (U/L)	84	44–147
Alanine aminotransferase (U/L)	13	10–40
G-glutamyl transferase (U/L)	29	9–48

19. Culture and sensitivity
    Blood: Sterile
    Urine (midstream): Sterile
20. Radiological investigations
    Chest X-ray: Right middle lobe infiltration
    Chest CT: No signs of pulmonary embolism
    Electrocardiogram: Sinus tachycardia
21. Urine analysis

pH	8
Protein	Negative
Glucose	Normal
Ketones	+1
Blood hemoglobin	Negative
Leukocyte esterase	Negative
Nitrite	Negative
Bilirubin	Negative
Bilirubinogen	Normal
RBC (cells/hpf)	0–2
WBC (cells/hpf)	0–2
Crystals	Not seen
Casts	Not seen
Epithelial cells	Not seen
Yeast cells	Not seen
Parasite	Not seen
Triple phosphates	Numerous/hpf

Diagnosis

The final diagnosis involved a community-acquired pneumonia (CAP) caused by *S. pneumoniae* infection.

Treatment

1. Penicillin intravenous (IV)—1 million units (every 4 hours)
2. Ceftriaxone—1 g (every 24 hours)

Question and answer

1. What forms the basis for the classification of *S. pneumoniae*?
   S. pneumoniae is an encapsulated extracellular pathogen in humans. The capsular polysaccharide comprising the capsule that encases the bacterial diplococci pair is responsible for bacterial virulence and is also the basis for the classification of the organism.
2. Why does asplenia predispose individuals to *S. pneumoniae* infections?
   The spleen plays an instrumental role in the elimination of polysaccharide-encapsulated bacteria, such as *S. pneumoniae*. Therefore the host's immune responses to *S. pneumoniae* infection largely rely on the pathogen clearance from the bloodstream and rapid antigen production mediated by the spleen. Asplenia is associated with severe consequences of pneumococcal infections, such as bacteremia and meningitis, due to increased predisposition to the development of invasive pneumococcal disease and an overwhelming postsplenectomy infection.
3. What is the mode of beta-lactam resistance exhibited by *S. pneumoniae*?
   The mode of beta-lactam resistance exhibited by *S. pneumoniae* relies on an alteration in the penicillin binding proteins on the bacterial surface rather than an alteration affecting the production of beta lactamase. This mode of resistance confers bacterial resistance to multiple classes of antibiotics.
4. How are pneumococci differentiated from other streptococci in culture?
   The sensitivity exhibited by pneumococci to optochin (ethyl hydrocuprein) enables their differentiation from other streptococci in culture.
5. Why is the polysaccharide capsule of *S. pneumoniae* considered to be the primary virulence factor?
   Encapsulated *S. pneumoniae* is implicated in pathogenic diseases in humans, while nonencapsulated *S. pneumoniae* is considered nonpathogenic to humans. This difference in virulence mediated by the presence or absence of the polysaccharide capsule is indicative of its role in facilitating bacterial virulence. Therefore the polysaccharide capsule of *S. pneumoniae* is considered the primary virulence factor.
6. Why does *S. pneumoniae* usually cause lower respiratory tract infections, although the bacteria are normal inhabitants of the upper respiratory tract?
   The anatomical barriers of the innate immune system in the nasopharyngeal mucosa, such as entrapment, expulsion, and cough reflex, aided by the ciliary escalator effect, prevent establishment of infection in the upper respiratory tract. However, the bacteria may establish infection in the lower respiratory tract due to aspiration of bacterial colonies.
7. What are the host responses responsible for the extensive cytopathic effects seen in CAP?
   The cytopathic effects seen in CAP are a consequence of the hyperinflammatory responses caused by the pneumolysin-mediated

activation of the complement cascade and the cytokine release triggered by the capsular polysaccharides.

8. What are the classical red flag signs of pneumonia?
   Egophony, rales, whispered pectoriloquy, dullness to percussion, and bronchial breath sounds are the classical red flag signs of pneumonia.

Learning objectives

1. Determine the logical flow of differential diagnosis based on patient history, clinical manifestations, physical examination, serological investigations, and radiological investigations.
2. Describe the characteristics of *S. pneumoniae*.
3. Summarize the pathogenesis of *S. pneumoniae* infection.
4. Predict the prognosis of the disease based on the clinical manifestations and the extent of the pathology presented.
5. Enumerate the methods involved in the diagnosis of CAP.
6. Interpret the correlation between the clinical manifestations shown by the patient and the pathogenicity of *S. pneumoniae*.
7. Determine the appropriate antibiotic-based treatment regimen for the patient based on a history of allergies, underlying medical conditions, and culture sensitivity reports.

Discussion

Differential diagnosis

Etiology	Rationale
Viral pneumonia	Viral pneumonia presents itself with a productive cough, fever, chills, sweating, rapid shallow breaths, stabbing chest pain that worsens on breathing, loss of appetite, and fatigue.
Klebsiella pneumonia	Klebsiella pneumonia presents itself with symptoms related to the lower respiratory tract similar to those seen in viral pneumonia, but has an increased tendency to develop lung abscess, cavitation, empyema, and pleural adhesions.
Legionnaire's disease	Legionnaire's disease is an atypical form of pneumonia, and the disease manifests as symptoms related to the lower respiratory tract.
Pneumocystis pneumonia	Pneumocystis pneumonia is an opportunistic fungal infection, which presents itself with symptoms related to the lower respiratory tract and can be fatal for immunocompromised patients, such as those with an human immunodeficiency virus (HIV) infection.
Lung abscess	Lung abscesses are characterized by blood-streaked, sour-tasting, and foul-smelling sputum, chest pain that worsens on breathing, fever, loss of appetite, weight loss, fatigue, night sweats, and a cough.

Continued

Continued	
Etiology	**Rationale**
Influenza	Influenza presents itself with symptoms related to the upper respiratory tract, and pneumonia is a well-known secondary complication of influenza infection. The infection presents itself with a sore throat, high grade fever, runny nose, chills, body aches, cough, malaise, and fatigue.
Pleural effusion	Pleural effusion is characterized by a dry cough, tightness in the chest, dyspnea, chest pain, inability to lie in the supine position, and malaise.

Background

CAP is one of the leading causes of morbidity and mortality among all the infectious diseases worldwide, especially in elderly patients[1] and in children below 5 years of age. The morbidity associated with the disease results in hospitalizations, costing approximately billions of dollars each year.

S. pneumoniae is a bacterium known to be the most common pathogen responsible for causing CAP worldwide. The advent of antibiotics has brought about a substantial decrease in the percentage of pneumonia cases caused by *S. pneumoniae*.[2] The spectrum of diseases caused by *S. pneumoniae* ranges from mild respiratory tract mucosal infections, such as otitis media and sinusitis, to more severe diseases, such as pneumonia, septicemia, and meningitis. The clinical diagnosis of pneumonia caused by *S. pneumonia* is challenging due to the elusive nature of the bacterial growth in blood cultures. The numerous serotypes of the pneumococcal bacterium is the main challenge in developing strategies to control this global pathogen. Although pneumococcus can cause lethal diseases, it commonly colonizes the upper respiratory tract, resulting in an asymptomatic carriage state in most children.[3] Such asymptomatic carriers are infectious and can spread the disease to their contacts without showing any clinical manifestations of the disease.

Etiological agent

Streptococcus pneumoniae

Classification and morphology

S. pneumoniae, belonging to the genus *Streptococcus*, is a lancet-shaped, Gram-positive, catalase-negative, facultative anaerobic bacterium that typically occurs as diplococci or as short chains (Fig. 4.1). These are nonmotile and nonsporing bacteria. *S. pneumoniae* is an encapsulated extracellular pathogen in humans, and the capsule encasing each diplococci pair is composed of the capsular polysaccharide, responsible for the bacterial virulence and the basis for the classification of the organism. As of 2011, a total of 92 separate

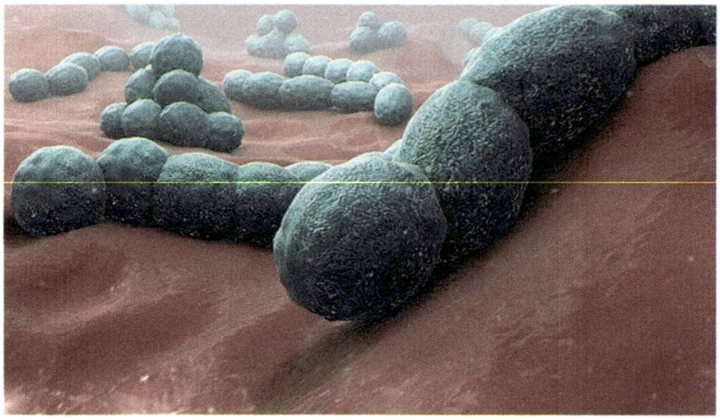

FIGURE 4.1
Streptococcus pneumoniae. Source: *http://www.scientificanimations.com/; http://www.scientificanimations.com/wiki-images/; https://creativecommons.org/licenses/by-sa/4.0/deed.en; https://commons.wikimedia.org/wiki/File:Streptococcus_pneumoniae_-_A_causative_bacteria_of_meningitis.jpg.*

serotypes of *S. pneumoniae* have been isolated,[4] all of which differ in virulence, prevalence, and extent of drug resistance.

The *S. pneumoniae* cocci are slightly elongated with one broad end and another pointed end, giving it the flame-shaped or lancet-shaped appearance. However, in culture, the typical morphology may not be apparent as the cocci are more rounded, tending to occur in short chains (Fig. 4.2). The growth of these bacteria requires media enriched with very specific and complex growth requirements.

The colonizing and infecting strains can be differentiated on culture media by their appearance, as opaque and transparent types. While the transparent type usually colonizes the nasopharynx, the opaque type is responsible for lung, brain, and bloodstream infection.[4]

Epidemiology

S. pneumoniae can colonize the human nasopharynx without causing any infection, resulting in asymptomatic carriers spreading the disease to their unsuspecting contacts through contaminated respiratory droplets, thereby largely contributing to its worldwide prevalence. Pneumococcal infections are usually seasonal, occurring most commonly during the winter and early spring months. The development of childhood conjugate vaccines for *S. pneumoniae* has resulted in decreased colonization frequency.

S. pneumoniae infections are particularly serious in patients older than 65 years, younger than 5 years, smokers, alcoholics, those who have underlying respiratory conditions, such as asthma or chronic obstructive pulmonary disease (COPD), and in immunocompromised patients, such as those with HIV or who are asplenic. CAP is also a common coinfection in influenza patients and affects the morbidity and mortality in such patients.

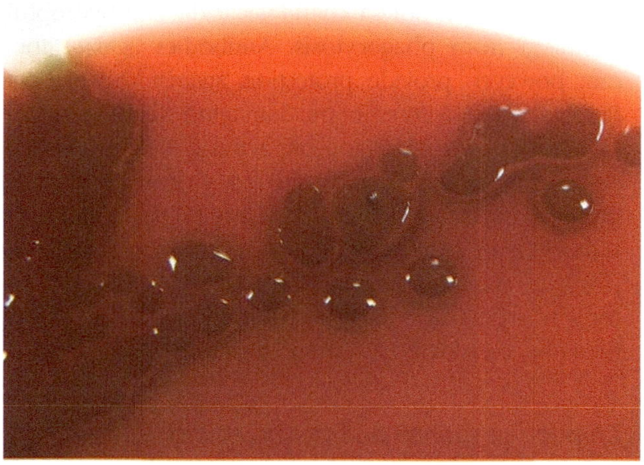

FIGURE 4.2

Streptococcus pneumoniae on Columbia Horse Agar. Source: *Nathan Reading on Flickr/; https://www.flickr.com/photos/nathanreading/6751940871/; https://creativecommons.org/licenses/by/2.0/deed.en; https://commons.wikimedia.org/wiki/File:Streptococcus_pneumoniae_on_Columbia_Horse_Blood_Agar_-_Detail.jpg.*

Resistance

S. pneumoniae have very low resistance to heat and chemicals present in antiseptics. Until the late 1960s, beta-lactams such as penicillin were considered the antibiotics of choice for treatment of CAP. However, as beta-lactam-resistant strains began to appear, the hunt for more effective antibiotics commenced. A peculiar characteristic of the beta-lactam resistance, shown by *S. pneumoniae,* is that the mode of resistance is an alteration in the penicillin binding proteins on the bacterial surface rather than the production of beta lactamase. This characteristic also confers resistance to multiple classes of antibiotics,[5] thereby posing as a challenge in effective clinical treatment.

Pneumococci are sensitive to optochin (ethyl hydrocuprein), which enables their differentiation from other streptococci in culture.

Virulence

Gram-positive *S. pneumoniae* bacteria express certain proteins and enzymes on their surface, which are the virulence factors in these organisms and are indispensable for the pathogenicity attributed to these pathogens. Some of these proteins interact directly with host tissues, while others enable the bacteria to evade the host's immune system. As encapsulated *S. pneumoniae* have been implicated in pathogenic diseases, while the nonencapsulated bacteria are nonpathogenic to humans, the polysaccharide capsule encasing the encapsulated *S. pneumoniae* bacteria is considered the primary virulence factor. The capsular polysaccharide is acidic and hydrophilic, thereby protecting the organism from phagocytosis. Antibodies against the capsular polysaccharide can, however, provide protection against the pathogenicity of the organism. Pneumolysin, a membrane-damaging toxin produced by pneumococci, is known to be cytotoxic and can activate the complement cascade of the immune system and is therefore considered a virulence factor. In addition, some other proteins, including hyaluronate lyase (Hyl),[6] pneumolysin (Ply),[7] two neuraminidases (NanA and NanB),[6] major autolysin (LytA),[6] choline binding protein A (CbpA),[8] pneumococcal surface antigen A (PsaA),[9] and pneumococcal surface protein A (PspA),[10] have been found to have immunogenic properties and are, therefore, considered as potential vaccine candidates.

Immune response and pathogenicity

S. pneumoniae is a normal inhabitant of the upper respiratory tract in humans, capable of colonization without causing infection. While anatomical barriers of the innate immune system in the nasopharyngeal mucosa, such as entrapment, expulsion, and cough reflex, aided by the ciliary escalator effect prevent establishment of infection, the lower respiratory tract can get infected by aspiration of the bacterial colonies. Compromised immune defenses give the pneumococci an opportunity to multiply, penetrate the bronchial mucosa, and spread to the lung through the peri-bronchial tissues and lymphatics.[11]

The bacteria can cause infections in the middle ear, paranasal sinus, and respiratory tract, in addition to infecting the brain, heart, and blood stream in severe cases. Certain risk factors, such as extremities of age, underlying respiratory conditions, immunocompromised status, smoking, and alcoholism, predispose an individual to infection. Although usually endogenous, infections can also be exogenous, especially when a particularly virulent strain is involved.[11] Exogenous infections with extremely virulent strains of the bacteria can also lead to serious infections in healthy individuals.

The capsular polysaccharide and pneumolysin, a toxin produced by pneumococci, are mainly responsible for the pathogenicity attributed to *S. pneumoniae*. The ability of *S. pneumoniae* to adhere to and invade the respiratory epithelium is another contributing factor to the pathogenicity of the bacteria.

Pneumonia, both lobar and bronchopneumonia, are mainly caused by *S. pneumoniae*. These organisms are also responsible for empyema and acute tracheobronchitis. Chronic bronchitis can also result from infection with *S. pneumoniae*. Bronchopneumonia, considered a very grave event in elderly and debilitated patients, is a very common secondary infection resulting from the pneumococci invading the bronchial tree, which is a direct consequence of the severe damage to the respiratory epithelium caused by the primary infection. Bacteremia and toxemia, caused by the diffusion of the capsular polysaccharide into the blood and tissues, can occur during the early stages of lobar pneumonia.

Meningitis, a serious infection of the brain, can occur as a secondary infection to any of the other pneumococcal infections, such as pneumonia, otitis media, sinusitis, empyema, pericarditis, and conjunctivitis. Fatality rates of meningitis are extremely high even when treated with antibiotic therapy, therefore demanding rapid diagnosis and effective treatment.

The activation of the complement cascade by pneumolysin and the release of cytokines triggered by the capsular polysaccharides results in a significant inflammatory response, eventually responsible for the extensive cytopathic effects seen in CAP.

Clinical manifestations

Pneumonia patients may develop a wide array of symptoms, including dyspnea, cough, pleuritic pain, sputum production, and fever. However, these symptoms are not considered specific for the diagnosis of pneumonia. Extremities of age can result in atypical clinical manifestations, leading to a delay in diagnosis and increased mortality. Generic presentations, such as generalized weakness, altered mental status, and malaise could also be the first signs of CAP infection in certain patients, thereby increasing the challenge of an accurate clinical diagnosis.

Physical examination focusing on the patient's vital signs is therefore considered the most dependable factor in diagnosing those with CAP. Tachypnea, hypoxia, and hyperthermia should all be considered as indications, including CAP in the differential diagnosis, and immediate further diagnostic evaluation should be undertaken. Classical signs of pneumonia, such as egophony, rales, whispered pectoriloquy, dullness to percussion, and bronchial breath sounds should be considered as red flags, and should result in immediate diagnostic and therapeutic action to ensure a good prognosis and outcome.

Diagnosis

The generic nature of the clinical manifestations of CAP makes it a challenge for clinicians to diagnose the disease based on physical examination. Although routine laboratory evaluation is also not specific enough for determining the exact etiological agent involved in pneumonia, it can enable the clinicians to strategize the treatment and management of the disease better. Chest radiography has been considered the mainstay for diagnosis of CAP (Figs. 4.3 and 4.4), although a radiograph is not reliable for the determination of the causative organism. Computed tomography (CT) has better sensitivity and accuracy in the diagnosis of pneumonia compared to chest radiography; however, relatively high cost and radiation exposure limits its use in hospitals.

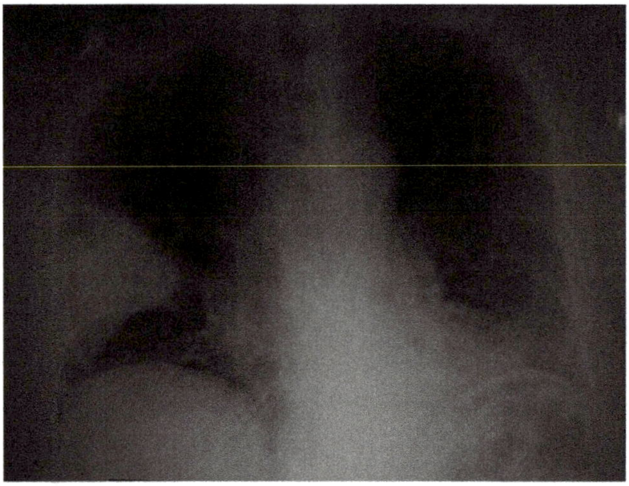

FIGURE 4.3

Chest X-ray showing pneumonia infiltrating the right middle lobe. Source *James Heilman, MD/; https://commons.wikimedia.org/wiki/File: PneumonisWedge09.JPG/; https://creativecommons.org/licenses/by-sa/4.0/deed.en; https://commons.wikimedia.org/wiki/File: Pneumonia_of_Right_Lung_Middle_Lobe.jpg.*

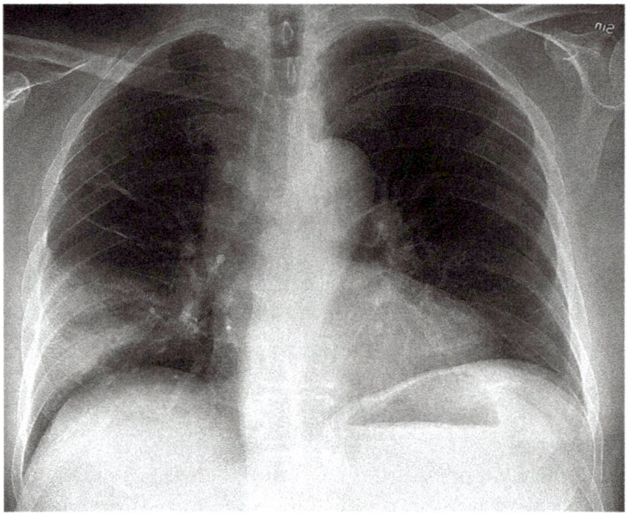

FIGURE 4.4

Chest X-ray showing lobar pneumonia affecting the right middle lobe. Source: *Mikael Haggstrom, MD/Own work/; https:// creativecommons.org/publicdomain/zero/1.0/deed.en. https://commons.wikimedia.org/wiki/File:X-ray_of_lobar_pneumonia.jpg.*

The causative organism for pneumonia can be determined through blood cultures, sputum analysis, and urine antigen test. However, it is recommended that blood cultures be avoided in cases other than the serious ones as positive blood cultures largely show the presence of *S. pneumonia*, which changes the clinical management of the disease. Although sputum cultures are considered to have low sensitivity and specificity in detecting the causative organism for pneumonia, it is recommended to obtain sputum cultures in cases suspected of having a drug-resistant organism. Urine antigen tests are highly sensitive and specific for the diagnosis of pneumonia cases caused by *S. pneumoniae*, although its usefulness in children is debatable.[4]

Treatment

Antibiotic therapy and symptomatic relief are the mainstays of treatment in CAP cases caused by *S. pneumoniae*. Supportive care involving tube thoracostomy and mechanical ventilation may be indicated by parapneumonic effusion in acute cases of complicated pneumonia. Cases that progress to empyema and are resistant to drainage by tube thoracostomy, may require decortication with video-assisted thoracoscopic surgery.[4]

Low-risk CAP is generally treated with parenteral penicillin monotherapy, and admission to a hospital may not be required. Erythromycin and tetracycline can be used for penicillin-resistant cases, provided the strains are sensitive to the antibiotics. Third-generation cephalosporins are recommended in cases resistant to first- and second-line antibiotics. Vancomycin should, however, be used very diligently only in high-risk cases infected with highly resistant strains of *S. pneumoniae*. High-risk cases may also require the use of respiratory fluoroquinolones. Some acute cases may also indicate the need for dual therapy with a beta-lactam-based antibiotic and a macrolide or a beta-lactam-based antibiotic and a respiratory fluoroquinolone.[12] The administration of the first dose of the antibiotics of choice should be given as soon as a definitive diagnosis is made.

Prevention

The production of vaccines against *S. pneumoniae* have been challenging, owing to the existence of more than 90 serotypes of the organism. Immunity to CAP is serotype-specific and depends on the antibodies against the capsular polysaccharide. A conjugate pneumococcal vaccine is now included universally in childhood immunization regimes. In addition, all high-risk individuals, such as those above 65 years of age, those with underlying respiratory disease, alcoholics, smokers, those with chronic kidney and liver diseases, individuals who are immunocompromised due to asplenia, HIV, cancer, or any other medical conditions should be vaccinated.

In addition to appropriate vaccination, prevention of exogenous infection should also be prioritized by maintaining good nutrition, good hygiene and sanitation practices, avoiding crowded unhygienic spaces, and smoking cessation.

Student Notes

Key take-aways

- CAP is a disease with a high risk of morbidity and mortality, especially in geriatric and pediatric patients.
- *S. pneumoniae,* the bacteria usually responsible for CAP, can colonize the human nasopharynx without causing an infection in healthy individuals.
- Underlying respiratory conditions, chronic illnesses, smoking, alcoholism, and an immunocompromised state also predisposes an individual to infection with *S. pneumoniae* and to exacerbation of the disease.
- Rapid diagnosis and timely treatment with appropriate antibiotics are very important in ensuring a good prognosis and outcome of the disease.
- Inclusion of the conjugate pneumococcal vaccine in childhood immunization regime substantially reduces the spread of CAP. High-risk individuals should also be vaccinated against the disease.

Problem number 4.2: Cystic fibrosis exacerbated by *Pseudomonas aeruginosa* infection

An 11-year-old male with a known history of glucose-6-phosphate dehydrogenase (G6PD) deficiency and cystic fibrosis developed an exacerbated cough the previous night. The patient had been experiencing a mild productive cough with greenish sputum, devoid of blood for the last 2 weeks. The symptoms were exacerbated the previous night, resulting in continuous coughing and the inability to sleep. The patient was afebrile and did not complain of any earache, nausea, vomiting, or change in bowel habits. There was no recent history of travel/exposure to sick individuals.

The patient's birth history was uneventful; normal delivery without any antenatal or postnatal complications and had a normal birth weight. The developmental milestones had been achieved on an appropriate timeline. The patient had been diagnosed with cystic fibrosis around 5 months of age, and had a history of multiple hospital admissions since, although there was no history of pediatric intensive care unit (PICU) or recovery room (RR) admissions. The patient had undergone a chest tube insertion 3 years ago. The last hospital admission had been about 6 months ago, which had been precipitated by an exacerbated cough. The patient had tested positive (sputum culture) for *Staphylococcus aureus* and *P. aeruginosa* during this hospitalization. The infection had been treated with an 18-day course of meropenem, gentamycin, and vancomycin. There was no known history of food or drug allergies, and his immunizations were updated, except the yearly seasonal flu vaccine.

The patient's parents are in a consanguineous marriage, and both parents have G6PD deficiency, and the mother also has sickle cell traits. The patient has two younger sisters, both of whom are healthy, and there is no family history of cystic fibrosis.

The following are the reports of the investigations conducted:

1. Vitals
   Temperature: 37.3°C
   Heart rate: 121 beats/min
   SpO2: 90%
2. Physical examination
   Comfortable in supine position
   Conscious, alert, and oriented
   No muscle wasting/deformities/missing limbs
   Well-hydrated
   Pink, moist lips
   Not pale/jaundiced
   Normal activity levels
   Able to maintain oral intake of food and water
   Clubbed fingers
3. Respiratory examination
   Subcostal retractions are present
   Runny nose with greenish discharge (1 week)
   Throat and chest pain associated with cough
   Good accident & emergency (A/E) sounds bilaterally
   Bilateral crepitations
4. Cardiovascular examination
   No murmur (S1, S2)
5. Abdominal examination
   Soft and lax
6. Neurological examination
   Awake and alert
   Fully responsive
   Oriented to person, place, and time
7. Radiological investigations
   Chest X-ray:
   Hyperinflation
   Atelectatic changes with collapse consolidation
   Cystic fibrosis-associated changes

8. Hematological profile

WBC (/L)	18.45×10^9	$4.3–10.8 \times 10^9$
Hb (g/dL)	12.8	13.5–17.5
Platelets (/L)	550×10^9	$150–400 \times 10^9$
PMN (%)	62.8	40–60
Lymphocytes (%)	31	25–33

9. Biochemical profile

Sodium (mmol/L)	133	135–145
Potassium (mmol/L)	4.6	3.5–5.3
Chlorine (mmol/L)	98	95–110
Glucose (mmol/L)	6.7	4.4–7.8
Bicarbonate (mmol/L)	23	23–32
Calcium (mmol/L)	2.19	2.2–2.7
Phosphate (mmol/L)	1.3	1.12–1.45
Magnesium (mmol/L)	0.8	1.7–2.2

10. Sputum analysis (culture and sensitivity)
    WBC: Abundant
    Epithelial cells: Few
    Organisms: Mixed growth comprising of clusters of Gram-positive cocci and three types of Gram-negative bacilli including *P. aeruginosa* was isolated.
11. Blood culture and sensitivity: Sterile

Diagnosis

The final diagnosis was that of *P. aeruginosa* infection, exacerbating the existing cystic fibrosis condition.

Treatment

1. IV infusion: 500 mL/24 hours; ½ NS D5
2. Antibiotics:
   Vancomycin 900 mg Q8
   Gentamycin 60 mg Q8
   Clindamycin 300 mg Q8

3. Medicines to be continued:
   Tobramycin—300 mg BD
   Seretide inhalation—BD
   Creon—15 capsules/day
   Ventolin dornase alfa inhalation—OD

Question and answer

1. Why is *P. aeruginosa* the most common organism responsible for the exacerbation of symptoms in cystic fibrosis patients?
   P. aeruginosa is an opportunistic pathogen. The ability to produce various pigments, such as pyocyanin (bluish green) and fluorescein (greenish yellow), is characteristic of *P. aeruginosa,* and these are known to inhibit the growth of other bacteria and contribute to the persistence of *P. aeruginosa* in the lungs of cystic fibrosis patients. In addition, the bacterial strains isolated from cystic fibrosis patients have a mucoidal slimy layer and an abundance of extracellular polysaccharides composed of alginate polymers, forming a loose capsule called glycocalyx, which encapsulates microcolonies of the bacillus, thus protecting the bacterium from host defences. The compromised health of a respiratory system affected by cystic fibrosis in combination with these beneficial attributes of the infectious bacterial strains causes *P. aeruginosa* to emerge as the predominant bacterial strain in cystic fibrosis cases and result in the exacerbation of symptoms.
2. What gives *P. aeruginosa* a competitive advantage over other microorganisms and facilitates the establishment of persistent infections?
   P. aeruginosa produce various pigments, such as pyocyanin (bluish green) and fluorescein (greenish yellow), which facilitate their existence in the host system and provide a competitive advantage over other microorganisms. These pigments also enable the establishment of persistent *P. aeruginosa* infections in the host system.
3. Which attribute facilitates *P. aeruginosa* to cause nosocomial infections associated with invasive medical devices?
   P. aeruginosa cause nosocomial infections associated with various invasive medical devices, such as ventilators, in-dwelling catheters, endotracheal tubes, nasogastric tubes, and many others[3,4] because of its ability to form undetectable biofilms.
4. What causes the broad-spectrum antibiotic resistance associated with *P. aeruginosa*?

The broad-spectrum antibiotic resistance seen in *P. aeruginosa* is attributed to multiple mechanisms, such as intrinsic antibiotic resistance, which prevents antimicrobials from permeating the membrane; efflux systems, which facilitate the transportation of toxic compounds out of the bacterial cell; and antibiotic-inactivating enzymes, such as beta-lactamases and extended-spectrum beta-lactamases (ESBLs).

5. How does the mucoidal slimy capsule of certain *P. aeruginosa* bacterial strains facilitate the pathogenicity of the bacteria?

   The mucoidal slimy capsule facilitates evasion of the host immune responses and enables the bacteria to establish chronic infections in the host system, especially if the system is compromised due to an underlying respiratory condition, such as cystic fibrosis.

6. Why is the isolation of *P. aeruginosa* in microbial cultures not considered a confirmation of *P. aeruginosa* infection?

   P. aeruginosa is ubiquitously present in the human body. Therefore isolation of the bacteria in culture may not be indicative of an established infection in the system. Hence, positive microbial cultures need to be combined with patient history, clinical manifestations, physical examinations, serological investigations, radiological examinations, and biochemical investigations, such as oxidase reaction and arginine hydrolysis, for a diagnostic confirmation of *P. aeruginosa* infections.

7. What is the significance of local antibiograms in the treatment of *P. aeruginosa* infections?

   Antibiotic resistance is a matter of primary concern in *P. aeruginosa* infections, thereby emphasizing on the need for culture sensitivity results to ensure the right course of treatment. However, culture sensitivity tests require time, and timely initiation of treatment is important to ensure good prognosis and outcome in *P. aeruginosa* infections. Therefore local antibiograms are significant in treatment initiation until sensitivity reports are available.

Learning objectives

1. Identify the etiological agent responsible for the exacerbation of cystic fibrosis in the patient.
2. Describe the characteristics of *P. aeruginosa*.
3. Summarize the pathogenesis of *P. aeruginosa* infection.
4. Predict the prognosis of the infection and the outcome based on the clinical manifestations and the extent of pathology presented.

5. Enumerate the methods involved in the diagnosis of *P. aeruginosa* infections.
6. Interpret the correlation between the clinical manifestations shown by the patient and the pathogenicity of *P. aeruginosa*.
7. Analyze the predisposition of cystic fibrosis patients to respiratory infections.
8. Determine the appropriate antibiotic-based treatment regimen for the patient based on patient history, the clinical manifestations, and the culture sensitivity reports.

Discussion

Differential diagnosis

Etiology	Rationale
Hot tub folliculitis	Hot tub folliculitis is a skin infection caused by *Pseudomonas* that develops within a few hours of coming into contact with contaminated water in community spaces, such as hot tubs, jacuzzies, water slides, etc. The infection presents itself as a rash that may be painful and/or itchy, which may be self-limiting or may require medical intervention if lesions are formed.
Malignant otitis externa	Malignant otitis externa is an invasive bacterial infection involving the external auditory canal and skull base. This is a complication of otitis externa, which is seen usually in immunocompromised patients. The symptoms may consist of persistent foul-smelling yellow/green discharge from the ear, ear pain that worsens on head movement, itchy ear canals, hearing loss, fever, difficulty swallowing/talking, swelling/redness around the ear, and inflammation of the throat.
Staphylococcal infection/ sepsis	Staphylococcal infections usually affect immunocompromised individuals and manifest with symptoms involving multiple organ systems, most commonly the skin. Infections may present as pus-filled pimples/boils, cellulitis, chills and fever, swelling, warmth, redness, joint pains; endocarditis symptoms such as a rapid heartbeat, shortness of breath; and food poisoning symptoms such as abdominal pain, vomiting, and diarrhea. Untreated streptococcus infections can result in multiorgan systems being affected, which in some extreme cases can lead to sepsis. Sepsis usually manifests as a high fever, sudden low blood pressure, vomiting, diarrhea, and confusion.
Streptococcal infection/ sepsis	Streptococcal infections often affect the throat, resulting in a sudden onset of throat pain, difficulty swallowing, red and swollen tonsils with white patches/pus streaks, tiny red

Continued

Continued

Etiology	Rationale
	spots at the back of the throat, swollen/tender lymph nodes, fever sometimes accompanied by chills, headache, and a rash. Untreated streptococcus infections can result in multiorgan systems being affected, which in some extreme cases can lead to sepsis. Sepsis usually manifests as high fever, sudden low blood pressure, vomiting, diarrhea, and confusion.
Community-acquired pneumonia (CAP)	CAP is a bacterial disease usually caused by *S. pneumoniae*. The disease manifests with mild or severe symptoms related to the lower respiratory tract.
Viral pneumonia	Viral pneumonia develops similar symptoms to bacterial pneumonia along with some additional generalized symptoms.
Klebsiella pneumonia	Klebsiella pneumonia is a well-known nosocomial infection, although community-acquired forms of the infection are also known. The disease can develop symptoms related to the lower respiratory tract, but has an increased tendency to develop lung abscesses, cavitation, empyema, and pleural adhesions.

Background

P. aeruginosa is a bacterium notorious as an opportunistic organism that causes serious infections in immunocompromised hosts. The organism is especially implicated in nosocomial infections, and although it usually causes severe infections in immunocompromised individuals, immunocompetent individuals can also get infected. The treatment of infections caused by this organism poses a formidable challenge for clinicians, owing to the extremely versatile types of infections, antibiotic resistance, and mechanisms of immune evasion. *P. aeruginosa* is the most prevalent organism causing infections in individuals with underlying respiratory conditions, such as cystic fibrosis.

Etiological agent

Pseudomonas aeruginosa

Classification and morphology

P. aeruginosa, belonging to the *Pseudomonas* genus, are Gram-negative, aerobic, nonspore forming, motile, noncapsulated, slender, rod-shaped bacteria (Figs. 4.5 and 4.6) known for its widespread prevalence and its ability to cause opportunistic infections in humans. The ability of these organisms to cause infections in almost all organ systems in the human body is astounding.

FIGURE 4.5
Pseudomonas aeruginosa. Source: DataBase Center for Life Science (DBCLS)/; *https://doi.org/10.7875/togopic.2017.38/; https://creativecommons.org/licenses/by/4.0/deed.en.https://commons.wikimedia.org/wiki/File:201708_Pseudomonas_aeruginosa.svg.*

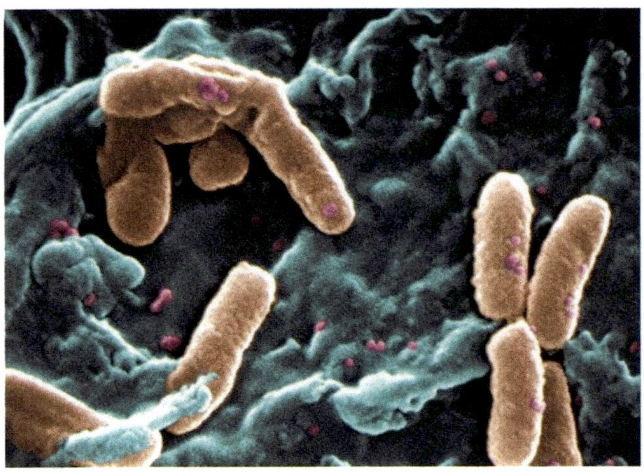

FIGURE 4.6
Scanning electron micrograph of *Pseudomonas aeruginosa*. Source: Janice Haney Carr, USCDCP/Own work/; *https://creativecommons.org/licenses/publicdomain/. https://pixnio.com/science/microscopy-images/scanning-electron-micrograph-sem-of-a-number-of-pseudomonas-aeruginosa-bacteria.*

A polar flagellum allows for active motility of the organism (Fig. 4.7). The bacterial strains isolated from cystic fibrosis patients are an exception in that they have a mucoidal slimy layer and an abundance of extracellular polysaccharides composed of alginate polymers, forming a loose capsule called glycocalyx, which encapsulates microcolonies of the bacillus, thus protecting the bacterium from host defenses.

P. aeruginosa are obligate aerobes with an oxidative and nonfermentative metabolism, but can also survive in anaerobic conditions in the presence of

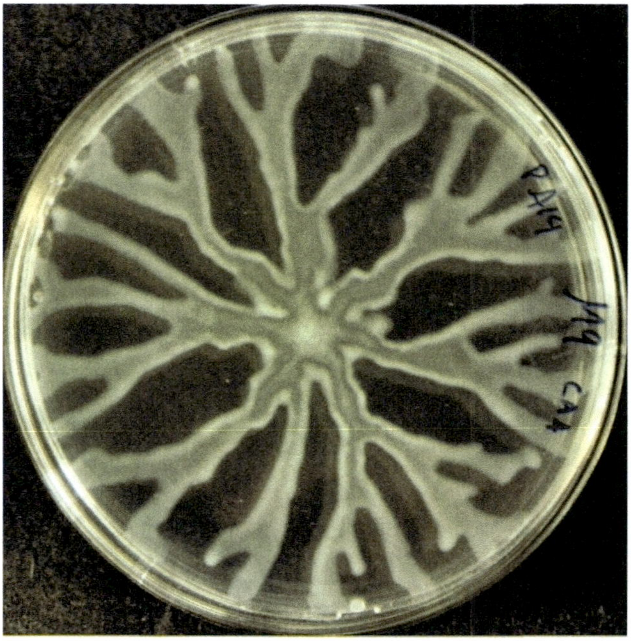

FIGURE 4.7

Swarming motility of *Pseudomonas aeruginosa*. Source: Julio514/Transferred from en.wikipedia to Commons by Jacopo Werther/; *https://creativecommons.org/licenses/by-sa/2.5/deed.en; https://creativecommons.org/licenses/by-sa/3.0/deed.en; https://creativecommons.org/licenses/by-sa/2.0/deed.en; https://creativecommons.org/licenses/by-sa/1.0/deed.en; https://commons.wikimedia.org/wiki/File:Swarming_motility_of_Pseudomonas_aeruginosa.jpg.*

nitrate. Thus biochemical tests using nitrates show a reduction to nitrites and further to gaseous nitrogen in the presence of *P. aeruginosa*. The bacteria are catalase-, arginine dihydrolase-, and oxidase-negative (Fig. 4.8). Although the optimum growth temperature is 37°C, the bacteria can grow at temperatures ranging from 6°C to 42°C.

The ability to produce various pigments, such as pyocyanin (bluish green) and fluorescein (greenish yellow), is characteristic of *P. aeruginosa* (Fig. 4.9). Although the precise mechanism by which these pigments enhance the pathogenicity of *P. aeruginosa* is unknown, they are known to inhibit the growth of other bacteria and contribute to the persistence of *P. aeruginosa* in the lungs of cystic fibrosis patients, thereby facilitating the emergence of *P. aeruginosa* as the dominant bacterium in mixed infections.[1,2]

Epidemiology

P. aeruginosa is ubiquitously present in the environment and shows special preference for moist conditions. While public spaces, such as hot tubs,

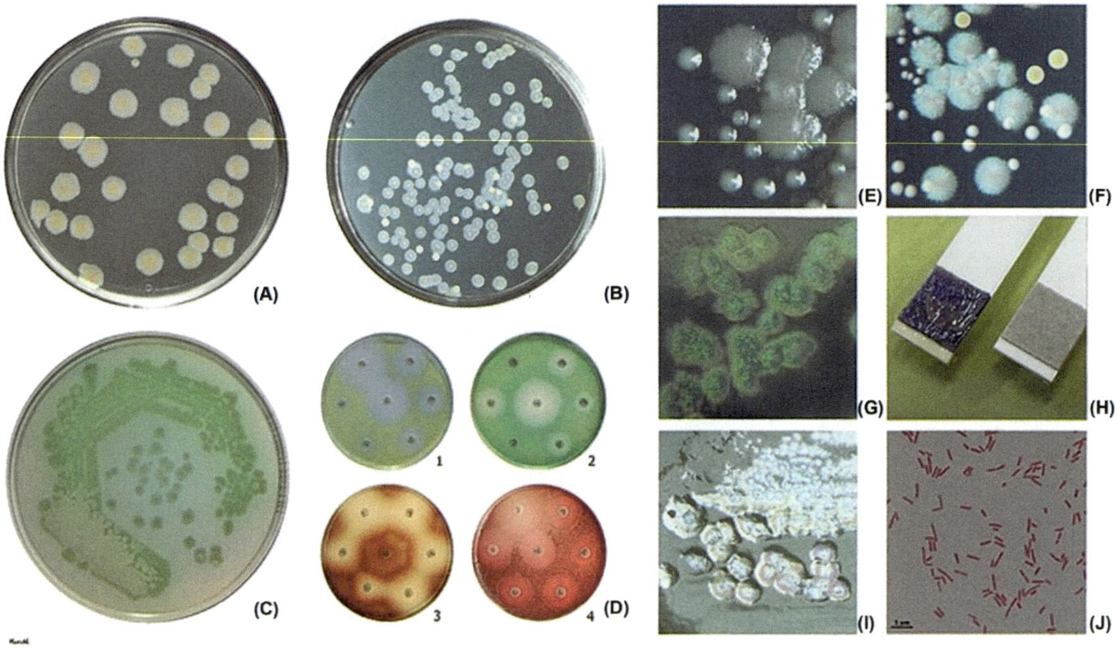

FIGURE 4.8

Pseudomonas aeruginosa pigment production, growth on cetrimide agar, oxidase test, plaque formation, and Gram stain. Source: HansN/Own work/ *https://creativecommons.org/licenses/by-sa/4.0/deed.en; https://commons.wikimedia.org/wiki/File:Pseudomonas_aeruginosa_pigment_production.jpg.*

FIGURE 4.9

Production of pyocyanin, water-soluble green pigment of *Pseudomonas aeruginosa*. Source: Y tambe/Own work/; *https://creativecommons.org/licenses/by-sa/3.0/deed.en; https://commons.wikimedia.org/wiki/File:Pseudomonas_aeruginosa_pyocyanin.jpg.*

jacuzzies, and swimming pools, act as reservoirs in urban communities, potable water, taps, sinks, toothbrushes, icemakers, disinfecting solutions, sanitizers, soap bars, respiratory therapy equipment, endoscopes, and endoscope washers are the reservoirs in hospital and healthcare settings. *P. aeruginosa* can also cause a wide array of community-acquired infections like folliculitis, puncture wounds leading to osteomyelitis, pneumonia, otitis externa, and many others. The bacteria acts as opportunistic pathogens responsible for nosocomial infections associated with various invasive medical devices, such as ventilators, in-dwelling catheters, endotracheal tubes, nasogastric tubes, and many others[3,4] because of its ability to form undetectable biofilms. *P. aeruginosa* infections are common in individuals with underlying medical and immunocompromised conditions, such as cystic fibrosis, bronchiectasis, neutropenia, burns, cancer, acquired immunodeficiency syndrome (AIDS), abnormalities of the genital or the urinary tract, organ transplant, and uncontrolled diabetes mellitus.

Pathogenicity

P. aeruginosa has been well-established as one of the most notorious pathogens responsible for nosocomial infections. While community-acquired infections may be chronic and difficult to treat, they are usually not responsible for debilitating conditions. On the other hand, nosocomial infections may be localized, such as bedsores, eye infections, and urinary infections or systemic, such as bacteremia and sepsis. *P. aeruginosa*, the most dreaded pathogen causing severe infections in burn victims, is responsible for posttracheostomy pulmonary infections, and is the culprit in iatrogenic meningitis following lumbar puncture. Concomitant infections, malignancies, immunosuppressive therapies, and certain congenital abnormalities increase the predisposition to septicemia, ecthyma gangrenosum, and endocarditis caused by *P. aeruginosa*. Excessive exposure of hands to detergents and water can result in the infection of the nail bed caused by *P. aeruginosa*.

The extreme adaptability of the organism allows it to survive and multiply even in nutrient-deficient conditions in presence of moisture. High levels of resistance to common antibiotics, disinfectants, and antiseptics, in addition to its ability to establish itself widely in hospitals, makes the eradication of *P. aeruginosa* from hospital and healthcare settings very challenging. Medical equipment and devices, such as respirators, endoscopes, and bed pans may be contaminated. *P. aeruginosa* can also contaminate lotions, ointments, eye drops, and stocks of distilled water.

Although the mechanisms of pathogenesis are yet to be well-understood, several toxic extracellular products, such as toxins A and S, have been found to have a role in the pathogenicity of *P. aeruginosa*. Antibody responses to exotoxin A have been shown to indicate a good prognosis in severe *P. aeruginosa* infections. Proteases, elastases, hemolysins, and enterotoxins have also been found to be responsible for pathogenesis in *P. aeruginosa* infections.[4]

Resistance, virulence, and immune response

The *P. aeruginosa* bacillus is heat labile, and can be killed at a temperature of 55°C within 1 hour. However, the bacteria exhibits a high degree of resistance to chemical agents, such as common antiseptics and disinfectants, but is susceptible to acids, beta glutaraldehyde, silver salts, and strong phenolic disinfectants. Therefore topical creams containing sulfonamide compounds are used for application in burn victims. Although *P. aeruginosa* is naturally resistant to many antibiotics, some antibiotic groups, such as aminoglycosides, cephalosporins, ofloxacin, fluoroquinolones, and penicillins may be clinically effective. Localized infections have also been found to respond well to topical colistin, polymyxin B, and 1% acetic acid.

The broad-spectrum antibiotic resistance seen in *P. aeruginosa* is attributed to multiple mechanisms, such as intrinsic antibiotic resistance, which prevents antimicrobials from permeating the membrane; efflux systems, which facilitate the transportation of toxic compounds out of the bacterial cell; and antibiotic-inactivating enzymes, such as beta-lactamases and ESBLs.[5] Resistance to antibiotics and the host immune system is further enhanced by the biofilm-forming ability of *P. aeruginosa*[2,5] (Figs. 4.10 and 4.11).

The ability of *P. aeruginosa* to form biofilms facilitates persistence of the organism in the host system. Although the invasion of the host system results in simultaneous activation of both, the innate and adaptive immune systems of the host, the evasive mechanisms of *P. aeruginosa* prevent elimination of the pathogen. The simultaneous activation of both types of immunity, however, proves detrimental to the host because of collateral tissue damage.[6]

Cystic fibrosis patients generally acquire community-associated or nosocomial infections caused by encapsulated *P. aeruginosa* strains comprising of a mucoidal slimy capsule within the first year of life. The capsule facilitates evasion of the host immune system and also enables the infection to become chronic in some cases.

As immunopathology in *P. aeruginosa* biofilm infections is attributable to host immune responses and bacterial defence mechanisms, a better understanding of both would facilitate the development of better therapeutic measures capable of treating otherwise immunotolerant *P. aeruginosa* infections.

Clinical manifestations

P. aeruginosa infections can manifest with a broad spectrum of symptoms, varying with the site of infection, severity of infection, and health and immune status of the patient. Local tenderness, purulent discharge, pain,

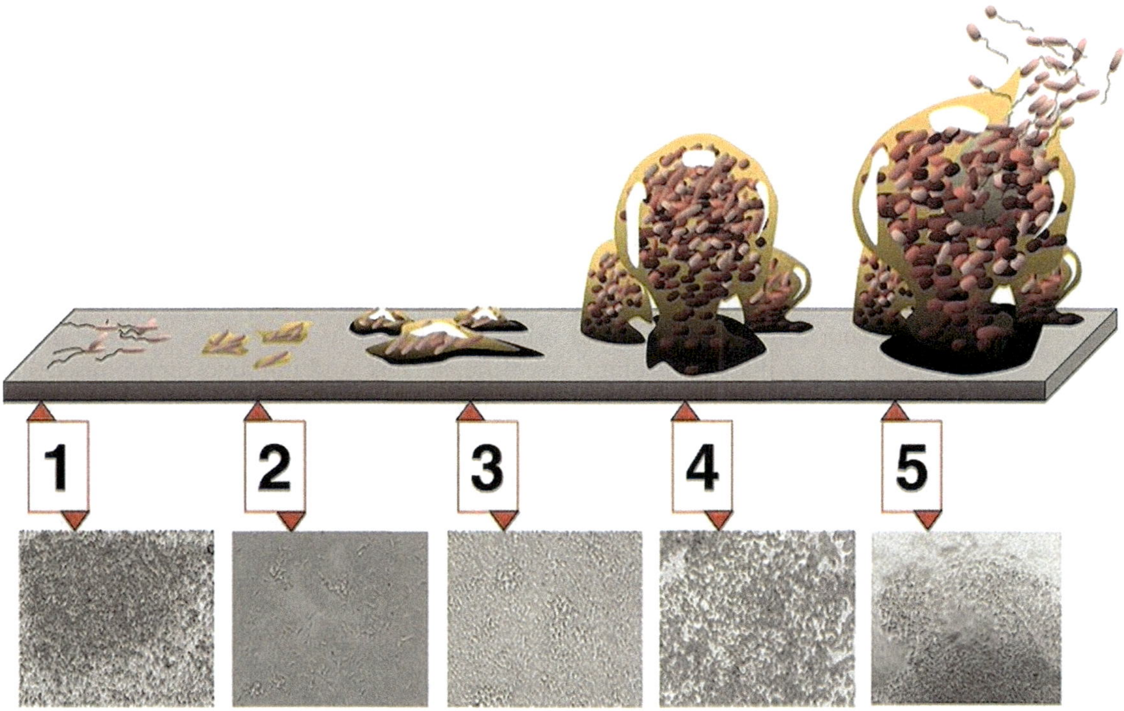

FIGURE 4.10

Representative image demonstrating the five stages of biofilm development of *Pseudomonas aeruginosa*. Stage 1: Initial attachment; Stage 2: Irreversible attachment; Stage 3: Maturation I; Stage 4: Maturation II; Stage 5: Dispersion. Source: Davis D./From: Monroe D. *Looking for chinks in the armor of bacterial biofilms*. PLoS Biol. *5(11):e307. DOI: 10.1371/journal.pbio.0050307/; https://creativecommons.org/licenses/by/2.5/deed.en; https://commons.wikimedia.org/wiki/File:Biofilm.jpg.*

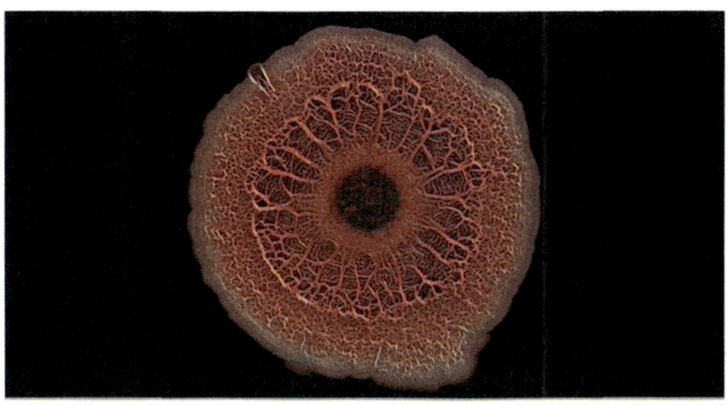

FIGURE 4.11

Intricate colony of millions of the single-celled *Pseudomonas aeruginosa* bacteria, self-organized into a sticky, mat-like colony known as a biofilm. Source: Scott Chimileski and Roberto Kolter, Harvard Medical School, Boston/; *https://directorsblog.nih.gov/2017/01/19/snapshots-of-life-portrait-of-a-bacterial-biofilm/; https://creativecommons.org/licenses/by-nc/2.0/; https://www.flickr.com/photos/nihgov/31846788804.*

reddening, and compromised functioning of the affected region may be seen in case of localized infections, such as those caused by puncture wounds. Infections of specific organ systems, such as the respiratory system as seen in case of cystic fibrosis patients or urinary tract infections, also in patients with underlying medical conditions, such as congenital anomalies of the kidney and urinary tract, may present as exacerbations of symptoms related to the functioning of the affected organ system. While exacerbated infections of the respiratory tract (Fig. 4.12) may present as a fever, productive cough, greenish sputum, and chest pain while breathing and coughing, exacerbated infections of the urinary tract may develop as abdominal pain, pain and burning while micturition, fever, turbid urine, and blood in urine. Burn injuries, uncontrolled diabetes mellitus, organ transplants, and immunocompromised state because of chemotherapy, malignancies or HIV/AIDS are some other factors that predispose individuals to *P. aeruginosa* infection.

Chronic *P. aeruginosa* infections are commonly seen in cystic fibrosis patients, and are associated with high levels of mortality in such patients.[4] Therefore *P. aeruginosa* infection should always be included in the differential diagnosis in cases of cystic fibrosis exacerbations.

Diagnosis

A diagnosis of *P. aeruginosa* infections can be challenging as the presence of the bacillus may not always indicate a pathological role because of its ubiquitous presence. Repetitive isolations may be required to confirm the diagnosis. However, a delay in the diagnosis and initiation of treatment can result in a poor prognosis and outcome, especially in cases of exacerbated

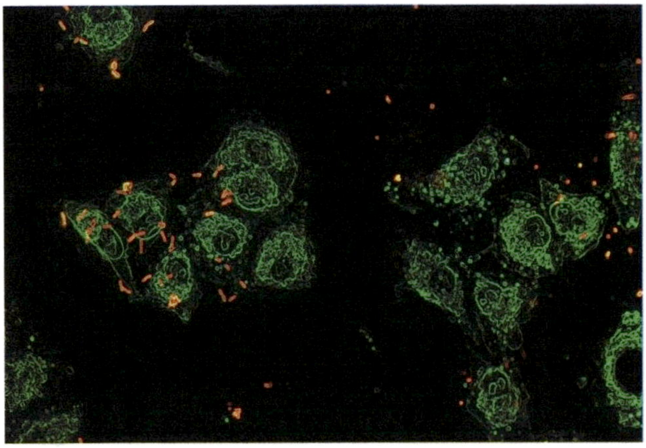

FIGURE 4.12
Early infection of epithelial lung cells with *Pseudomonas aeruginosa*. Source: Laventie Benoit-Joseph, University of Basel/Own work/; https://creativecommons.org/licenses/by-nc-nd/2.0/; https://www.flickr.com/photos/snsf_image_competition/47519246952.

underlying chronic conditions. Chronic *P. aeruginosa* infections are associated with high mortality rates and pose a huge diagnostic challenge, owing to the chronic presence of the bacteria in the patient's system. Differentiating a dormant presence of the bacteria from a pathological role in the presented condition can become challenging for clinicians.

While bacterial culture is the gold standard for a diagnosis of *P. aeruginosa* infections, the clinician should also take the following into account: the patient history, clinical manifestations, physical examinations, serological investigations, radiological examinations, and biochemical investigations, such as oxidase reaction and arginine hydrolysis.

Antibiotic resistance is a matter of primary concern in cases of *P. aeruginosa* infections. Therefore bacterial cultures should be accompanied by sensitivity determination, which can facilitate the right course of treatment, thereby ensuring a good prognosis and outcome.

Treatment

In addition to the broad-spectrum antibiotic resistance seen in the organisms, the high levels of morbidity and mortality associated with *P. aeruginosa* infections makes prevention of infection more effective than the treatment of infections. However, the ubiquitous presence of *P. aeruginosa*, opportunistic nature of the organism, predisposition of certain high-risk patients to infections, and resistance to commonly used disinfectants and antiseptics make prevention of nosocomial and community-acquired infections very challenging in spite of adherence to strict asepsis protocols.

P. aeruginosa infections should be considered a possibility in all high-risk patient groups and should be included in the differential diagnosis. Treatment may require extended courses of IV antibiotics in severe cases or oral antibiotics in mild cases. Some cases that lead to necrotized tissues may indicate the need for surgical debridement. Systemic infections are generally serious and often require intensive care unit (ICU) admissions and supportive care, which involves ventilators, catheters, nasogastric tubes, and endotracheal tubes.

Sensitivity results are empirical in guiding treatment in *P. aeruginosa* infections; however, culture sensitivity tests require time. Timely initiation of treatment is very important to ensure a good prognosis and outcome in *P. aeruginosa* infections. Therefore local antibiograms may prove to be of significant importance in the treatment initiation until sensitivity reports are available. Broad-spectrum antibiotic courses, including antipseudomonal antibiotic groups such as carbapenems, cephalosporins, aminoglycosides, and fluoroquinolones, are indicated for use as first-line therapy in systemic infections presenting with septic shock, respiratory failure requiring mechanical ventilation, a history of multiple ICU admissions, and a history of predisposition to multidrug resistant organisms.[8]

Student Notes

Key take-aways

- *P. aeruginosa* are ubiquitous bacteria found in moist environments, and are responsible for community-acquired or nosocomial opportunistic human infections.
- *P. aeruginosa* infections are associated with a high risk of morbidity and mortality, especially in patients with underlying medical conditions, chronic infections, and an immunocompromised state.
- Resistance to commonly used disinfectants and antiseptics makes it extremely challenging to prevent contamination by *P. aeruginosa*, in spite of adherence to strict asepsis protocols.
- Broad-spectrum antibiotic resistance of *P. aeruginosa* makes effective treatment extremely challenging.
- The inclusion of *P. aeruginosa* infections in a differential diagnosis of high-risk patients facilitates a rapid diagnosis and timely treatment with effective antibiotics, which are very important in ensuring a good prognosis and outcome of the disease.

Problem number 4.3: Acute bronchiolitis caused by respiratory syncytial virus

A 1-month-old infant presented with a 3-day history of a wet but nonproductive cough. He turned blue while coughing continuously, but then regained normal color once the coughing stopped. He had a runny nose, and the parents reported a reduced intake of milk. Since the previous day, he had vomited 4–5 times after feeds; the vomit was nonprojectile and comprised of milk and mucus, but no blood. Reduced activity had been observed since the previous day and his urine had a prominent smell for 2 days. There was no change in bowel movements.

The patient's mother reported an uncomplicated pregnancy, although she had a history of premature rupture of membranes. Antibiotics had been administered to both the mother and baby at birth, which had been through full-term normal vaginal delivery. The baby's birth weight was 3.8 kg and did not have any antenatal or postnatal complications. The patient did not have any known medical conditions and had not required any hospital admissions since birth. There were no known food or drug allergies, and the patient was not on any regular medication. There was no history of recent travel or potential exposure to sick individuals.

The parents are in a nonconsanguineous marriage, and both are healthy. The patient is the couple's first child. The parents denied the presence of any smokers around the child, and there were no pets at home either.

The following are the reports of the investigations conducted:

1. Vitals
   Temperature: 37.8°C
   SpO_2: 97% on room air

2. Physical examination
   Weight: 4.6 kgs
   Well-hydrated
   Capillary refill <2 seconds
   No pain/respiratory distress
   Not cyanosed/pale/jaundiced
   Fontanelles: flat; measuring 3×4 cm
3. Respiratory examination
   Chest wall and lung (EBAE)
   No added sounds
4. Cardiovascular examination
   No murmurs (S1, S2)
5. Abdominal examination
   Soft, lax
   No organomegaly
6. ENT examination
   Congested throat
7. Neurological examination
   Awakened due to examination
   Reflexes intact
8. Radiological Investigations
   Chest X-ray: Hyperinflation
9. Hematological profile

WBC (/L)	11.51×10^9	$5.0–19.5 \times 10^9$
Hb (g/dL)	13.3	10.7–17.1
Platelets (/L)	440×10^9	$150–400 \times 10^9$
Lymphocytes (%)	59.30	25–33
Neutrophils (%)	25.1	20–50
Eosinophils (%)	7.4	1–3

10. Biochemical profile

Sodium (mmol/L)	139	135–145
Potassium (mmol/L)	4.9	3.5–5.3
Chlorine (mmol/L)	107	95–110
Bicarbonate (mmol/L)	26	23–32
Urea (mg/dL)	2.0	3–6.3
Creatinine (μmol/L)	24	26–62
C-reactive protein (μg/mL)	7.22	0.05–57.6

11. Blood culture: Sterile

Diagnosis

The final diagnosis uncovered an acute respiratory syncytial virus (RSV) positive bronchiolitis.

Treatment

1. IV infusion: 500 mL/24 h; ½ NS D5
2. Ventolin Q4 hourly
3. O2: PRN

Question and answer

1. What is the basis for the serotype classification of the RSV?
   The variations in the structural membrane proteins of the virus, such as the attachment protein, is the basis for the serotype classification of the RSV.
2. What is the mode of transmission of the RSV?
   The RSV transmits through direct contact with contaminated respiratory droplets released by infected individuals or indirect contact with contaminated fomites.
3. How does the RSV establish infection in the host?
   The nasopharynx or conjunctival mucosa serve as entry points for the RSV. The virus then rapidly spreads to the respiratory tract, targeting the apical ciliated epithelial cells. The attachment protein of the virus, known as the RSV-G glycoprotein, then binds to the cellular receptors followed by fusion with the host cell membrane using the fusion protein of the virus known as the RSV-F fusion glycoprotein. The established virus then inserts its nucleocapsid into the host cell, hijacking its machinery and initiating its intracellular replication.
4. What is the role of host inflammatory immune responses in exacerbations of RSV infections?
   RSV infections trigger both the humoral and adaptive aspects of the host immune responses. This invariably results in the activation of cytotoxic T-cell responses. The attachment and fusion proteins of the virus also induce IgA and IgM antibodies. The cellular immune response is particularly important for infection clearance. A combination of viral cytotoxicity and the cytotoxic host immune responses results in an uncontrolled secretion of cytokines, precipitating the phenomenon of "cytokine storm." Cytokine storm is known to cause extensive cellular and tissue damage in the host system. Bronchoconstriction is a consequence of inflammatory

responses induced by the antigen – antibody complexes or the secretion of IgE. This demonstrates the role of the host immune response in the exacerbation of RSV infections.

5. What are the histopathological observations in cases of bronchiolitis associated with RSV infections?

   Exacerbated RSV infections are characterized by an intermediate pathology between pneumonia and bronchiolitis. Histopathological observations of bronchiolitis cases, precipitating from RSV infections, include the necrosis of the bronchiolar epithelium, infiltration of ciliated epithelial cells and peri-bronchial leukocytes with edema, and small airway obstruction caused by the formation of cellular and mucus plugs. These histopathological observations are compliant with the demonstration of alveolar obstruction, ciliary dysfunction with impaired mucus clearance, airway edema, and decreased lung compliance in severe cases.

6. Which radiographic findings are considered indicative of RSV infection?

   Radiographic findings, such as hyperinflation, patchy atelectasis, and peri-bronchial thickening, are considered indicative of RSV infections. However, these radiographic findings are generic and similar to findings in bacterial pneumonia. Therefore these findings need to be clinically correlated.

7. What are the treatment and management strategies recommended in moderate and severe cases of respiratory tract infections?

   RSV infections generally affect the upper respiratory tract. Mild infections are self-limiting and do not warrant the need for any medical intervention. However, lower respiratory infections resulting in bronchiolitis of moderate or severe levels may indicate the need for hospitalization and supportive care. While moderate cases can usually be managed by nasal suction and lubrication, antipyretics, and intravenous hydration, severe cases resulting from complications of the infection may need ICU admissions and life support aids, such as ventilators, additional oxygen supply, high-flow nasal cannula, and nasogastric tubes.

8. What is palivizumab?

   Palivizumab is an antibody against the membrane fusion protein of the RSV. The administration of this antibody is an effective means of passive immune prophylaxis for RSV infections. However, the need for monthly administration of this prophylaxis during seasonal peaks of the disease, in addition to its relatively high cost, limits its use in the general population. Palivizumab prophylaxis is therefore recommended for specific high-risk cases, such as infants (less than or equal to 29 weeks gestational age) with underlying predisposition factors like premature birth, congenital heart conditions, chronic pulmonary diseases, and neuromuscular disorders.

9. What are the factors hindering the development of an effective long-term vaccine for RSV infections?

   The possibility of vaccine-induced enhanced disease in individuals vaccinated with attenuated vaccine strains, challenges associated with extrapolation of results from animal studies to humans, and the possibility of frequent reinfections has hampered progress of the development of an effective long-term vaccine for RSV infections.

Learning objectives

1. Determine the logical flow of differential diagnosis based on patient history, clinical manifestations, physical examination, serological investigations, and radiological investigations.
2. Describe the characteristics of the RSV.
3. Summarize the pathogenesis of bronchiolitis caused by RSV.
4. Predict the prognosis of the infection and the outcome based on the clinical manifestations and the extent of pathology presented.
5. Enumerate the methods involved in the diagnosis of acute bronchiolitis.
6. Interpret the correlation between the clinical manifestations shown by the patient and the pathogenicity of the RSV.
7. Determine the appropriate course of treatment and supportive care for the patient based on patient history, the case presentation, and the rapid antigen test reports.

Discussion

Differential diagnosis

Etiology	Rationale
Asthma	Asthma is an allergic condition presenting as breathing difficulty, coughing, wheezing, and shortness of breath. Swelling and the narrowing of the airways and production of extra mucus is also seen in many asthma cases.
Bronchitis	Bronchitis is an acute or chronic inflammation of the lining of the bronchial tubules and may present itself as respiratory distress and productive cough comprised of thickened and discolored mucus.

Continued

Continued

Etiology	Rationale
Croup	Croup is an infection of the respiratory system, usually caused by a virus. The disease may manifest as internal swelling of the trachea interfering with normal breathing processes. A "barking" sound while coughing, stridor, and hoarseness in the voice are the usual clinical presentations of the condition.
Influenza	Influenza is a viral infection usually affecting the upper respiratory tract. Pneumonia is a well-known secondary complication of influenza infection.
Community-acquired pneumonia (CAP)	CAP is a bacterial disease caused usually by *S. pneumoniae*. The disease manifests as mild or severe symptoms related to the lower respiratory tract.
Viral pneumonia	Viral pneumonia manifests symptoms similar to bacterial pneumonia along with some additional generalized symptoms.
Klebsiella pneumonia	Klebsiella pneumonia is a well-known nosocomial infection, although community-acquired forms of the infection are also known. The disease can cause symptoms related to the lower respiratory tract, but has an increased tendency to develop lung abscess, cavitation, empyema, and pleural adhesions.

Background

Human RSV is a common cause of upper respiratory tract infections in infants and children and is also recognized as a significant pathogen in adults, especially in the geriatric population. RSV may occasionally also cause lower respiratory tract infections in young children, accompanied with small airway obstruction. Rarely RSV-associated bronchiolitis may lead to complications, such as pneumonia, respiratory failure, apnea, and even prove fatal.

As the etiological agent in RSV infections is a virus, supportive care and symptomatic relief form the mainstay of treatment for a vast majority of RSV infections.

Etiological agent

Respiratory syncytial virus

Classification and morphology

RSV is a negative-sense single-stranded RNA virus belonging to the *Pneumovirus* genus of the *Paramyxoviridae* family. Human RSV is one of the many RSVs in the *Pneumovirus* genus, which are pathogenic for other animals but do not infect humans. The human RSV serotype is classified into strains A and B based on variations in the structural membrane proteins, especially the attachment protein.[13]

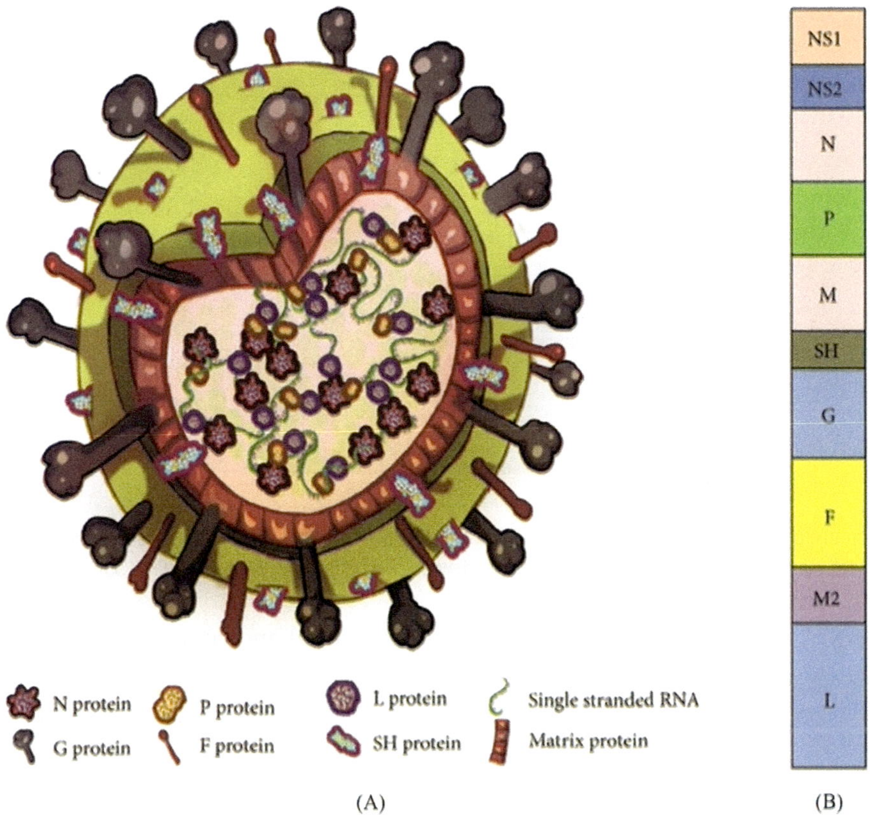

FIGURE 4.13

(A) Structure of respiratory syncytial virus (RSV). (B) Genes comprising the genome of RSV. Source: Swapnil Subhash Bawage, Pooja Munnilal Tiwari, Shreekumar Pillai, Vida Dennis, and Shree Ram Singh/; *https://www.hindawi.com/journals/av/2013/595768/; https://creativecommons.org/licenses/by/3.0/deed.en; https://commons.wikimedia.org/wiki/File:595768.fig.001.jpg.*

RSV comprises of a lipid bilayer envelope surrounding a ribonucleoprotein core comprising of a negative-sense single-strand RNA. The membrane is covered with several membrane proteins, one of which is important for attaching to host cells (attachment glycoprotein—G) while another is indispensable for fusing (fusion protein—F) with host cells (Figs. 4.13 and 4.14).

Epidemiology

RSV is not restricted by any geographical barriers and infects infants, young children, geriatric, and immunocompromised individuals all around the world. Outbreaks generally occur during the winter months in temperate climates, although the peak time may vary among communities[14] and last for

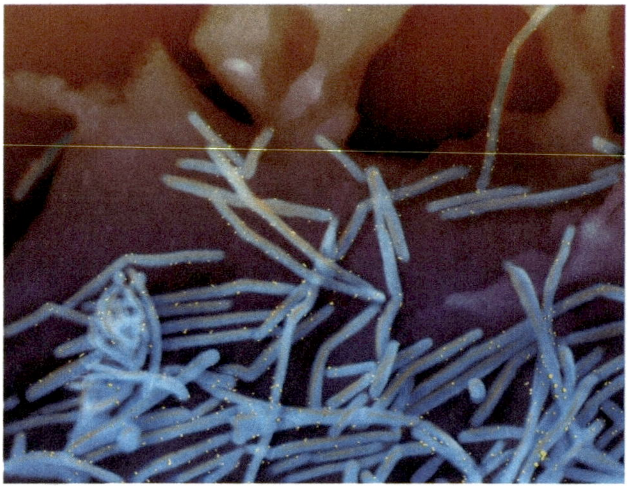

FIGURE 4.14

Scanning electron micrograph of human RSV. Source: National Institute of Allergy and Infectious Diseases, National Institutes of Health (NIH)/; *https://creativecommons.org/licenses/by-nc/2.0/*; *https://www.flickr.com/photos/nihgov/33288028956.*

2–7 months. RSV outbreaks in tropical and equatorial climates, however, are seen as more interseasonal instead of showing pronounced seasonal peaks.

The virus is transmitted via respiratory droplets, and the lack of life-long immunity is responsible for frequent reinfections among the predisposed population, especially during outbreaks. RSV outbreaks usually cause an increased incidence of mild upper respiratory tract infections but also invariably result in an increase in bronchiolitis cases, requiring hospitalizations, especially reported in infants and children. Complications like pneumonia associated with lower respiratory tract infections lead to increased morbidity and mortality, especially in infants between 2 and 6 months of age and geriatric individuals above 65 years of age.[15] The spread of the virus and complications of the infection are directly proportional to increased urbanization and lower socio-economic status.

While community transmission of RSV infections is common, nosocomial transmission may occur in geriatric individuals and patients who are immunocompromised due to underlying respiratory, pulmonary, cardiac, or other chronic medical conditions.

Immune response and pathogenicity

RSV generally establishes itself in the upper respiratory tract, causing mild infections after a mean incubation period of 4–6 days. The incubation period may vary depending on the age of the host and whether the infection is primary or secondary to a previous infection. The virus generally spreads

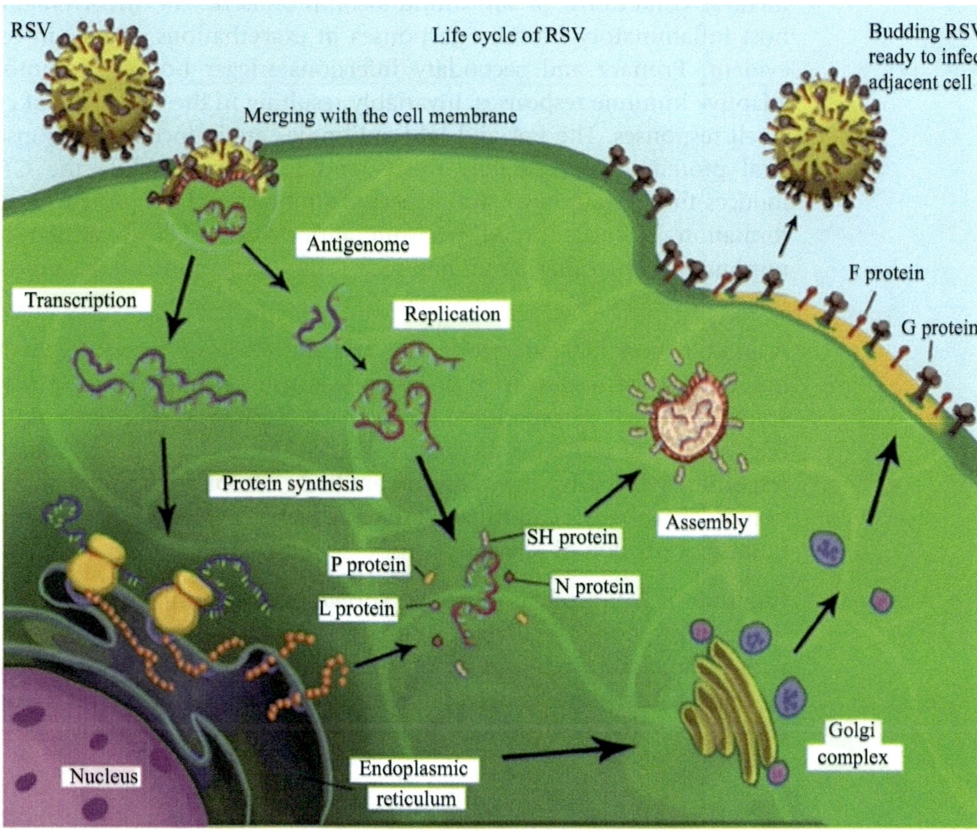

FIGURE 4.15

Schematic representation of the life cycle of RSV. Source: Swapnil Subhash Bawage, Pooja Munnilal Tiwari, Shreekumar Pillai, Vida Dennis, and Shree Ram Singh/; *https://www.hindawi.com/journals/av/2013/595768/*; *https://creativecommons.org/licenses/by/3.0/deed.en*; *https://commons.wikimedia.org/wiki/File:595768.fig.004.jpg*.

through direct contact with contaminated respiratory droplets, and reinfections are common among the predisposed population. The virus primarily infects the nasopharyngeal or conjunctival mucosa from where it rapidly spreads to the respiratory tract, targeting the apical ciliated epithelial cells. The RSV-G glycoprotein then binds to the cellular receptors followed by fusion with the host cell membrane using the RSV-F fusion glycoprotein. The established virus then inserts its nucleocapsid into the host cell, hijacking its machinery and initiating its intracellular replication (Fig. 4.15).

The upper respiratory tract infections may sometimes progress to more serious infections of the lower respiratory tract due to aspiration or cell-to-cell spread in high-risk individuals who are at age extremes, have underlying medical conditions, or are immunocompromised. The involvement of the host inflammatory immune responses in exacerbations of RSV infections is evident. Primary and secondary infections trigger both the humoral and adaptive immune responses, invariably resulting in the activation of cytotoxic T-cell responses. The IgA and IgM antibodies are induced in response to the viral proteins,[16,17] especially the G and F proteins. While the G protein induces the formation of strain-specific antibodies, the F protein induces the formation of more broad-spectrum antibodies.[18,19] The cellular immune response, on the other hand, has been found to be particularly important for infection clearance.[20]

A combination of viral cytotoxicity and the cytotoxic responses of the host immune system results in an uncontrolled secretion of cytokines. The resulting "cytokine storm" is responsible for extensive damage to host cells and tissues[21] (Fig. 4.16). Bronchoconstriction is known to be caused by inflammatory responses induced by the antigen – antibody complexes or through the secretion of IgE, thereby demonstrating the role played by the host immune response in the exacerbation of RSV infections.

Fatalities of RSV bronchiolitis have shown necrosis of the bronchiolar epithelium, infiltration of ciliated epithelial cells and peri-bronchial leukocytes with edema, and small airway obstruction caused by the formation of cellular and mucus plugs.[22] Alveolar obstruction may also occur in some severe cases. Ciliary dysfunction with impaired mucus clearance, airway edema, and

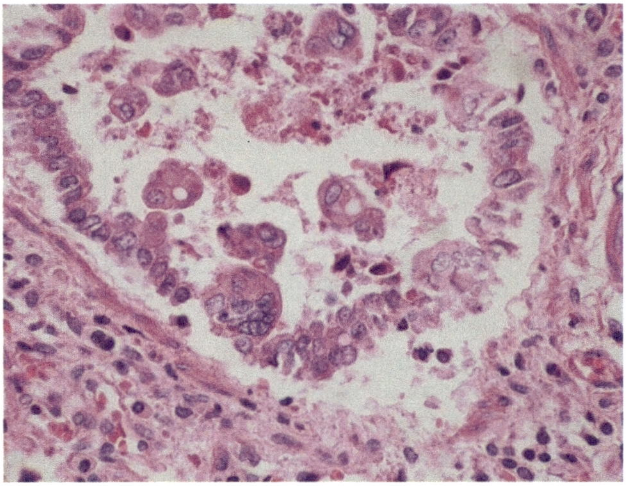

FIGURE 4.16

Image showing the lumen of a bronchiole containing cellular debris and syncytial aggregates of bronchial epithelial cells, although no viral inclusions are seen. Source Yale Rosen/; *https://creativecommons.org/licenses/by-sa/2.0/*; *https://www.flickr.com/photos/pulmonary_pathology/13517060075.*

decreased lung compliance[21] can result in prolonged morbidity and hospitalization. An intermediate pathology between pneumonia and bronchiolitis is a hallmark of exacerbated RSV infections.

Clinical manifestations

RSV infections typically manifest as upper respiratory conditions, although lower respiratory tract involvement is frequently seen in predisposed high-risk individuals. Upper respiratory tract infections caused by RSV generally present with rhinorrhea, nasal congestion, cough, sneezing, general malaise, and are sometimes accompanied with otitis media, fever, and myalgia. Complications associated with lower respiratory tract RSV infections usually manifest 1–3 days after the onset of the primary symptoms in high-risk patients, presenting as classic findings of bronchiolitis, such as rhonchi sounds, tachypnea, accessory muscle use, wheezes, and prolonged expiration. Severe cases may manifest symptoms of viral pneumonia, hypoxia, lethargy, apnea, and acute respiratory failure on rare occasions. RSV infections have also been implicated in a few cases of sudden infant death.[23] Infants and young children rarely harbor asymptomatic RSV infections. A lack of life-long immunity increases the chance of RSV reinfections in children and adults, resulting in exacerbated lower respiratory tract infections and severe disease even in adults.

While serological investigations may not provide much diagnostic value except an indication of raised leukocyte and lymphocyte counts, radiological investigations, such as chest X-rays, may show evidence of air trapping and/or interstitial pneumonitis (Fig. 4.17).

Diagnosis

Clinical manifestation of the disease is dependable for the diagnosis of RSV infections. Presentation of subsequent bronchiolitis establishes the diagnosis of RSV infection and does not require any confirmatory testing or imaging. In fact, specific testing for RSV is recommended only for differentiation from other disorders and can be done either by rapid antigen testing or polymerase chain reaction (PCR)-based testing. Although antigen testing is quick, inexpensive, specific, and can be performed using nasopharyngeal swab samples, PCR testing is preferred due to the higher sensitivity of the test and the ability to detect other pathogens as part of a PCR panel, which expedites the differential diagnosis process.

In addition to antigen-based diagnostic techniques, antibody-based diagnostic techniques targeting the host antibodies toward the viral antigens are also available.

Radiographic findings, such as hyperinflation, patchy atelectasis, and peribronchial thickening in bronchiolitis induced by RSV infections, are generic and similar to findings in bacterial pneumonia; therefore they need to be correlated with the clinical manifestations of the case.

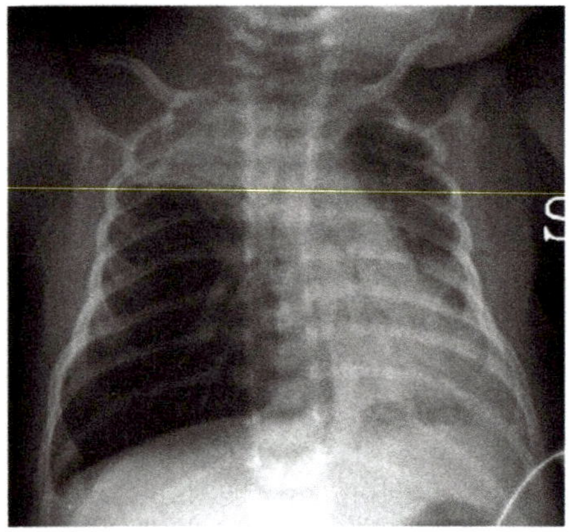

FIGURE 4.17

Chest radiograph of an RSV infection case, demonstrating lung hyperinflation with a flattened diaphragm and bilateral atelectasis in the right apical and left basal regions. Source: Matteo Di Nardo, Daniela Perrotta, Francesca Stoppa, Corrado Cecchetti, Marco Marano and Nicola Pirozzi/ Independent lung ventilation in a newborn with asymmetric acute lung injury due to respiratory syncytial virus: a case report/; *https://creativecommons.org/licenses/by/2.0/deed.en; https://commons.wikimedia.org/wiki/File:Bronchiolitis_chest_X-ray.jpg.*

Treatment and prevention

The viral etiology of RSV infections is the cause for ineffective attempts at the development of prophylactic and therapeutic interventions. Therefore although several means of inhibiting RSV infections exist (Fig. 4.18), treatment of RSV infections mainly relies on supportive care and generic antiviral medication, with immune prophylaxis indicated only in severe cases. Mild upper respiratory tract infections are generally self-limiting and do not require specific medical intervention. Moderate lower respiratory infections, resulting in bronchiolitis, may indicate the need for hospitalization and supportive care, including nasal suction and lubrication, antipyretics, and intravenous hydration, while severe cases resulting from complications of the infection may need ICU admissions and life support aids like ventilators, additional oxygen supply, high-flow nasal cannula, and nasogastric tubes.

Palivizumab, an antibody against the RSV membrane fusion protein, is an effective means of passive immune prophylaxis for RSV infections. However, the need for monthly administration of this prophylaxis during the RSV season, in addition to its relatively high cost, limits its use in the general population. Palivizumab prophylaxis is therefore recommended for specific high-risk cases, such as infants (less than or equal to 29 weeks gestational age) with underlying predisposition factors like premature birth, congenital heart conditions, chronic pulmonary diseases, and neuromuscular disorders. The possibility of vaccine-induced enhanced disease in individuals vaccinated with attenuated vaccine strains, challenges associated with extrapolation of

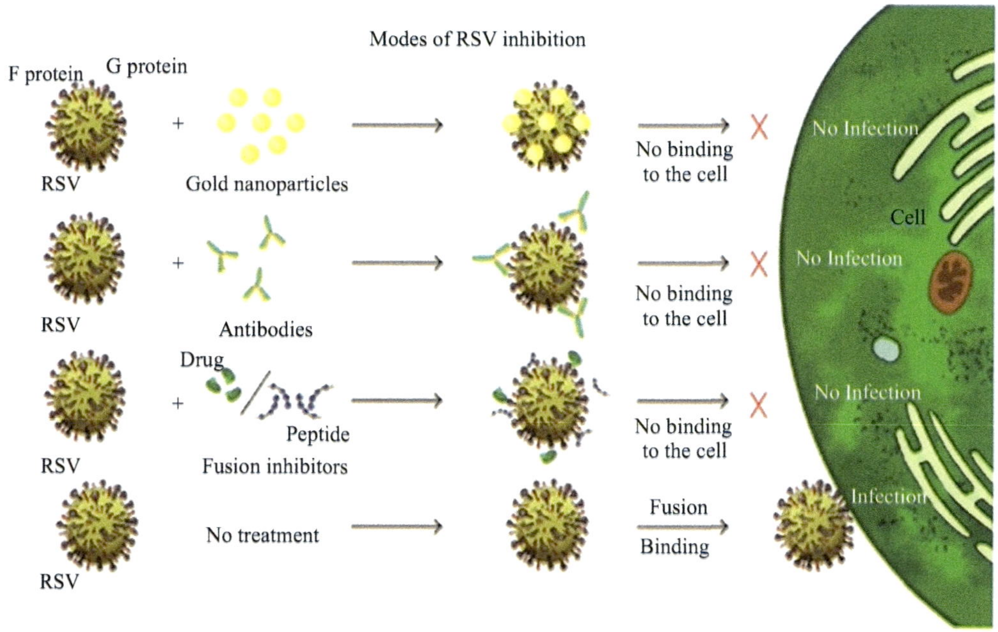

FIGURE 4.18

Schematic representation of various compounds inhibiting the binding of RSV to host cell. Source: Swapnil Subhash Bawage, Pooja Munnilal Tiwari, Shreekumar Pillai, Vida Dennis, and Shree Ram Singh/; *https://www.hindawi.com/journals/av/2013/595768/; https://creativecommons.org/licenses/by/3.0/deed.en; https://commons.wikimedia.org/wiki/File:595768.fig.005.jpg.*

results from animal studies to humans, and the possibility of frequent reinfections, has hampered progress in the development of an effective long-term vaccine for RSV infections.[24]

Ribavirin, a nucleoside analog used as an antiviral in infections caused by several RNA viruses, is the antiviral of choice in RSV infections as well. This antiviral is usually administered in an aerosolized form. However, there are several concerns with regard to the routine use of ribavirin in RSV infections due to its high cost and questionable efficacy, especially in severe cases requiring mechanical ventilation. Therefore the use of ribavirin is limited only to specific cases at risk of serious complications. The use of albuterol, racemic epinephrine, steroids, hypertonic saline, antibiotics, and chest physical therapy are also contraindicated in several instances and may not always be effective, hence they are not recommended for routine cases.[25,26]

Measures, such as strict adherence to asepsis protocols, good hand washing practices, and contact isolation, should be followed to ensure the prevention of nosocomial transmission of RSV infections.

Student Notes

Key take-aways

- RSVs are human pathogens responsible for most of the upper respiratory tract infections seen in infants and young children.
- The general prognosis and outcome of RSV infections is very good, unless underlying risk factors, such age extremities, immunocompromised status, underlying congenital or chronic medical conditions, are involved.
- RSV infections may manifest as lower respiratory tract infections presenting as bronchiolitis in infants and young children.
- Complications of RSV infections, although rare, may manifest in high-risk patients as viral pneumonia, hypoxia, lethargy, apnea, and acute respiratory failure.
- Diagnosis of RSV infections primarily depends on clinical manifestations, and although not required, may be confirmed by rapid antigen tests, PCR, radiological, and serological examinations.
- Supportive care is the mainstay of treatment for RSV infections, although the use of antivirals and passive immunizations may be indicated in some cases.
- Palivizumab prophylaxis is recommended for use only in specific at-risk cases owing to several concerns about its efficacy and cost.
- Nosocomial transmission of RSV infections can prove detrimental for patients with underlying medical conditions and should, therefore, be prevented through adherence to strict asepsis protocols.

Problem number 4.4: Chronic productive cough and low-grade fever caused by *Mycobacterium tuberculosis* infection

A 37-year-old man complained to his doctor about having hemoptysis for 1 day. He mentioned that 3 weeks ago, he had an ineffective course of antibiotics for bronchitis symptoms. The patient had a low-grade fever, breathlessness, loss of appetite, and weight loss of 5 kg over the last few weeks. Social history revealed that the patient was an unemployed, unmarried man, living alone, and a habitually heavy smoker. The patient had also not been immunized with the Bacille Calmette-Guerin (BCG) vaccine, which provides variable immunity against tuberculosis (TB).

The following displays the reports of the investigations carried out:

1. Physical examination
   Temperature: 39.5°C
   Dyspnea and palpable nontender right cervical lymph nodes.

Chest percussion: stony dullness on the right side at the base of the lung, consistent with pleural effusion.
Auscultation (chest): a few crackles on the right apex and right infraclavicular area and decreased vesicular breath sounds on the right base.

2. Chest X-ray
   Evidence of pleural effusion indicative of TB
3. Arterial blood gases

pH	7.44
pCO_2 (mmHg)	51
pO_2 (mmHg)	78

4. ESR: 160 mm in the first hour
5. Gram stain for sputum: Negative
6. Direct sputum stain (Ziehl – Neelsen) for acid-fast bacilli (AFB): Negative (three successive morning specimens).
7. Purified protein (obtained from *M. tuberculosis*) derivative (PPD) skin test (tuberculin skin test) using 5 TU (tuberculin units): Positive (22 mm induration at 72 hours).
8. Morning fasting deep tracheal aspirate sample: Presence of AFB (2–3/field) at high magnification (1000X).
9. Fine needle aspiration cytology of the right cervical lymph node: Granulomatous inflammation with the presence of caseous necrosis consistent with TB.
10. Culture report: Positive for *M. tuberculosis*
11. HIV blood test: Negative

Diagnosis

The final diagnosis was that of TB caused by *M. tuberculosis*.

Treatment

1. Rifampicin
2. Isoniazid
3. Ethambutol
4. Pyrazinamide
5. Pyridoxine
6. Regular liver function test (some prescribed drugs are hepatotoxic).
7. Family members and close contacts of the patient tested and PPD negative individuals vaccinated.

Follow-up and outcome

The patient was discharged after a month of treatment in an afebrile condition, with a 3 kg increase in body weight. As per the recommended TB chemotherapy—direct observation therapy short course (DOTS)—the treatment was to be continued for another month. After the monthly review, a two-drug regimen was prescribed for the next 4 months, thereby completing a total chemotherapy course of 6 months. Weekly follow-ups were recommended for the 6-month chemotherapy period. The recommended chemotherapy course for TB is known as DOTS.

Question and answer

1. What is hemoptysis?
   Hemoptysis is the coughing of blood originating from below the vocal cord, caused by ulcerative lesions in the airways or lungs that lead to erosion of the blood vessels.
2. What are the causes of weight loss in TB?
   Chronic illnesses such as TB result in toxemia, which often causes anorexia, leading to weight loss in TB.
3. What are some of the predisposing factors for chest infections?
   Extremes of age, coexisting chronic diseases, immunocompromised status due to HIV infection, cancer, or steroid use, and a history of smoking, are some of the predisposing factors for chest infections.
4. What is dyspnea?
   Dyspnea is the clinical presentation of difficulty in breathing and can be caused by various pulmonary and cardiac diseases, such as congestive heart failure, respiratory tract infections, acute coronary syndrome, and chronic obstructive pulmonary disease.
5. What is pleural effusion?
   Pleural effusion is the presence of fluid in the pleural cavity, which may be exudate fluid seen in inflammatory conditions, transudate seen in generalized edema, or blood due to hemorrhage. Pleural effusion may interfere with normal ventilation, resulting in compromised respiratory abilities.
6. What is crackle?
   Crackle is a specific sound produced due to the sudden reopening of the collapsed alveoli as air passes through fluid in the alveoli; it is heard as a "crackle" on physical examination of the chest.
7. What are the causes of lymph node enlargement?
   Lymph node enlargement may occur due to the spread of infection, primary lymphomas, secondary metastases, and some other inflammatory conditions.

8. What are the radiological changes observed in a case of TB?
   Radiological changes seen in TB cases include cavitation, fibrosis, enlarged hilar lymph nodes, and multiple homogenous opacities.
9. What is PPD (Mantoux) skin test?
   PPD or Mantoux skin test is an important diagnostic tool for TB. The PPD comprises of an antigen from the *M. tuberculosis* bacteria. The test detects delayed hypersensitivity due to cell-mediated immunity to PPD, thereby indicating exposure to *M. tuberculosis*. A positive PPD test therefore indicates that either the patient has an active TB infection or has a history of TB. Immunization with the BCG vaccine can also result in a positive PPD test.
10. Are there any false negative results for PPD test?
    Yes, false negatives for the PPD test may occur due to decreased cell-mediated immunity caused by lymphomas, viral infections, specific suppressive substances, circulating immune complexes, presence of adherent monocytes, trapping of sensitized T-lymphocytes in lymph nodes, sarcoidosis, and debility due to increased age. Altered-cell mediated immunity can result in skin anergy, causing the tuberculin reaction to decrease or disappear even in the presence of tuberculous infection.
11. What are AFB?
    Bacteria that retain the carbolfuchsin dye after strong decolorization with acid during the Ziehl – Neelsen staining procedure for *Mycobacteria* are known as AFB. Beside *M. tuberculosis*, other atypical *Mycobacteria, Nocardia*, and *Actinomyces* are also acid-fast organisms. Therefore the presence of AFB is not sufficient for a diagnosis of *M. tuberculosis* infection. Cultural tests are thus needed when a positive AFB result is obtained.
12. What tests should be done on a sputum specimen from a suspected TB patient?
    The sputum specimen should be cultured for AFB, which takes 4 – 6 weeks using traditional methods. Presently available PCR techniques and some other new technologies allow for detection within a few days. Antibiotic sensitivity may also be indicated in certain selected clinical situations, such as for cases that have been previously treated for TB or if the treatment is ineffective in resolving the infection.
13. What is granulomatous inflammation?
    Granulomatous inflammation is a specific type of chronic inflammation characterized histologically by the presence of granuloma, and is seen in conditions such as TB, sarcoidosis, bilharziasis, Crohn's disease, and reaction to any foreign body.
14. What is a granuloma?
    A granuloma is a circumscribed collection of epithelioid cells, including altered histiocytes or macrophages, and frequently also containing giant cells, lymphocytes, and fibroblasts at their periphery.

15. What is necrosis?
 Necrosis comprises of the morphological changes that follow cell death in a living body, and may be coagulative, liquefactive, caseous, fatty, fibrinoid, etc.
16. What is caseous necrosis?
 Caseous necrosis is characterized by a cheese-like appearance on gross observation and microscopic observation, demonstrating the presence of eosinophilic amorphous granular material.
17. Why is HIV investigation important in suspected TB cases?
 HIV patients are more susceptible to contracting secondary or reactivation TB due to the immunocompromised state. Hence HIV investigation is important in suspected TB cases.
18. Why was liver function evaluated before initiating treatment?
 The evaluation of liver function is important to set a baseline of the liver function before initiating treatment for TB. Isoniazid (INH), and to a lesser extent rifampicin are hepato-toxic drugs, thereby also necessitating regular evaluation of liver function.
19. Why is pyridoxine included in the TB treatment regimen?
 Isoniazid, used as a part of the TB treatment regimen, is antagonistic to pyridoxine. Pyridoxine deficiency causes peripheral neuritis; therefore it is included in the TB treatment regimen.
20. What are the first-line and second-line (alternative) antituberculous drugs?
 First line: Rifampin, INH, Ethambutol, Pyrazinamide, and Streptomycin.
 Second line (alternative): Kanamycin, Amikacin, Cycloserine, Capreomycin, Quinolones, Clarythromycin (a macrolide).
21. When is the use of the second-line drugs appropriate for treating TB?
 Primarily, first-line drugs are used to treat TB. However, certain conditions such as resistant organism, presence of atypical *Mycobacteria*, and the inability to tolerate first-line drugs are considered as indications for the use of second-line drugs.
22. What tests should be done for the contacts?
 The contacts of the case should be subjected to the PPD test, and if tested positive it should be followed up by a chest radiography.
23. What is the course of action for contacts who are PPD positive and chest X-ray negative (i.e., they do not have active TB)?
 Preventive treatment with INH for 6 months is recommended for contacts who test positive for PPD and initially test negative for PPD, but they are at a high risk of developing active diseases, especially young (5 years old or younger) and HIV-infected close contacts. This is to be followed at least until the repeat skin test done 2–3 months later continues to be negative.
24. What does BCG contain and what type of immunity does the vaccine give?

BCG contains a live-attenuated strain of *Mycobacterium bovis* and gives active immunity to the patient, but its protection is limited to a period of 10 years, and is effective mainly against TB meningitis, miliary, and disseminated forms of TB.

25. How can the microbiological diagnosis of TB be quickened?
    Automated culture method based on the detection of CO_2 production by the growing organisms are replacing the traditional culture methods of TB diagnosis. These methods can also be used for antibiotic sensitivity testing. More recently, PCR is used for the detection of *Mycobacteria* within 24 hours, although false positive results are a major limitation as PCR tests remain positive for a long time after the bacteria are killed by antibiotics.
26. What are the reasons for selecting multiple drugs in treating a patient with TB?
    Combination therapy is used in TB treatment because:
    a. *Mycobacterium* has a tendency to develop drug-resistance.
    b. Intracellular location of the organism prevents some antitubercular drugs from achieving adequate minimum inhibitory concentration (MIC) inside the macrophages.
    c. The organisms are slow growing and are capable of being metabolically inactive, thereby causing some antitubercular drugs to be ineffective on these organisms.
    d. A combination of antitubercular drugs that are bactericidal and bacteriostatic ensure multiple targets/mechanisms for antibacterial action, thereby increasing effectiveness of the treatment.
    e. The duration of treatment for TB is long (6–18 months), and therefore it is important to minimize drug toxicity by using relatively low doses.
27. What is DOTS?
    DOTS is a specific course of TB treatment used for sputum smear positive patients. The therapy comprises of an initial phase of 2 months, during which four drugs, INH, Rifampicin, Pyrazinamide, and Ethambutol, are administered to the patient every day, followed by the continuation phase of 4 months, during which two drugs, INH and Rifampicin, are administered to the patient every week. The treatment regime is to be followed by sputum examination for 3 consecutive days to check for the presence of AFB at the end of 5 and 6 months.

Learning objectives

1. Describe pulmonary infection.
2. Explain the mechanism of crackles.

3. Describe the radiological aspects of intrathoracic TB.
4. Explain the clinical manifestations and patient history of primary, postprimary, miliary, and extra-thoracic tuberculous infections.
5. List the microbiological and pathogenic properties of *M. tuberculosis* and atypical *Mycobacteria*.
6. Explain the association between immunological mechanisms of PPD reaction and delayed type hypersensitivity (DTH).
7. Explain granulomatous inflammation and granuloma formation.
8. Define necrosis and list its types.
9. Outline the pathological changes of TB.
10. Describe the morphology, structure, replication, and cultural characteristics of typical and atypical *Mycobacteria*.
11. Describe the pharmacological properties of antimycobacterial drugs.
12. Summarize the attributes of the BCG vaccine.
13. Describe the epidemiology of TB.
14. Determine the microbiological and clinical criteria for discontinuation of TB treatment.
15. Describe DOTS.

Discussion

Differential diagnosis

Sarcoidosis	Sarcoidosis is characterized by the accumulation of inflammatory cells, forming granulomas, and presents itself with fatigue, swollen lymph nodes, weight loss, joint pain and swelling, wheezing, shortness of breath, chest pain, chronic and persistent dry cough, skin rashes, hyperpigmentation, nodules under the skin, blurred vision, photosensitivity, syncope, palpitations, and edema.
Aspergillosis	Aspergillosis affects the respiratory system and usually manifests with a fever, hemoptysis, respiratory distress, fatigue, unintentional weight loss, wheezing, chest pain, joint pain, chills, and headaches.
Malignancy	Malignancies present with fatigue, sudden and unintended weight changes, breathing problems, persistent cough, long-lasting low-grade fever, difficulty swallowing, hoarseness, and night sweats.
Histoplasmosis	Histoplasmosis, a type of fungal infection of the lungs, presents itself with a fever, chills, headache, chest pain, body aches, and fatigue.
Pneumonia	Pneumonia, an infection that causes inflammation of the lung alveoli, presents with a fever, chills, respiratory distress, productive cough, chest pain, and fatigue.

Continued

Continued

Nontuberculous *Mycobacteria* infection	Nontuberculous *Mycobacteria* infection, caused by atypical *Mycobacteria*, presents itself with a productive chronic cough, fatigue, dyspnea, persistent low-grade fever, night sweats, loss of appetite, unintended weight loss, and excessive mucus production.

Background

TB is a contagious human disease, primarily affecting the respiratory system. Disseminated TB, however, often presents with systemic manifestations. Besides being a pulmonary disease, TB also affects other organ systems such as the gastrointestinal system, the central nervous system, the musculoskeletal system, the reproductive system, the hepatobiliary system, the lymphoreticular system, and the sensory nervous system. The resilience of *M. tuberculosis*, the etiological agent responsible for TB, is evident from its persistence as a human pathogen since ancient times. Even in the face of the advent of modern medicine, *M. tuberculosis* continues to survive as one of the most resilient human pathogens with epidemic and endemic potential. Although concerted global efforts to eradicate TB has resulted in a significant decrease in the global incidence rate, TB continues to account for a formidable global burden of morbidity and mortality, especially among the immunocompromised population. The brunt of the global burden of the disease is borne by the developing countries, owing to the unhygienic and conditions, in addition to overcrowded living conditions.

Etiological agent

Mycobacterium tuberculosis

Classification and morphology

M. tuberculosis, also known as Koch's bacteria, belongs to the genus *Mycobacterium*, which also comprises of some other organisms classified as atypical or nontuberculous mycobacterial organisms. *M. tuberculosis* are slender rod-shaped bacteria (Fig. 4.19) with some branching filamentous forms that resemble fungi, which gives the bacteria their name, meaning "fungi-like" bacteria. These intracellular bacteria are nonspore forming, nonmotile, obligate-aerobic, fastidious, and catalase negative. Weak reactivity to Gram stain, resulting in the "ghost cells" phenomenon, is another characteristic feature of *M. tuberculosis*, contributing to the diagnostic challenges associated with TB. The bacteria resists decolorization by alcohol and acid during Ziehl – Neelsen staining, and is known as alcohol and acid-fast bacillus.

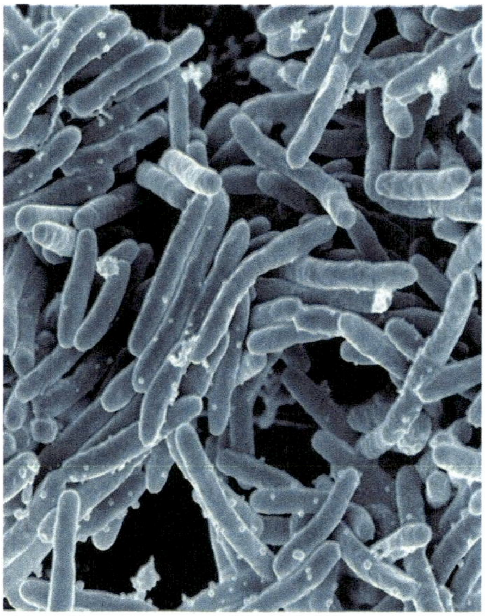

FIGURE 4.19

Scanning electron micrograph of *Mycobacterium tuberculosis* bacteria. Source: NIAID/; *https://www.flickr.com/photos/niaid/16843981465/; https://creativecommons.org/licenses/by/2.0/deed.en; https://commons.wikimedia.org/wiki/File:Mycobacterium_tuberculosis_Bacteria_(16843981465).jpg.*

Epidemiology

TB is a clinically concerning disease worldwide; however, it is especially endemic to developing countries because of its association with socio-economic factors. On the other hand, in developed countries, the burden of the disease is the highest in immunocompromised patients, healthcare workers, and those traveling from TB -endemic regions. *M. tuberculosis* is strictly restricted to being a human pathogen without any animal or environmental reservoir. *M. tuberculosis* is notorious for its ability to develop resistance to the classic antiTB medications, resulting in the emergence of multidrug resistant strains (MDR-TB) and extremely multidrug resistant strains (XDR-TB) of the pathogen.

Pathogenicity and pathogenesis

The group specificity of *Mycobacteria* is determined by the bacterial polysaccharides, while type specificity depends on antigenic proteins. The bacillary protein, tuberculin, is responsible for the delayed hypersensitivity seen in humans infected by *M. tuberculosis*. *M. tuberculosis* bacteria are categorized into two types based on the bacteriocins produced by the rapidly growing bacteria.

Several virulence factors make it challenging for alveolar macrophages to eliminate *M. tuberculosis* bacteria from an infected individual, even in the case of immunocompetent individuals. The high mycolic acid content contained in the outer capsule of the bacteria poses a challenge for the alveolar macrophages to effectively phagocytose the bacteria. Moreover, the cord factor, a bacterial cell wall constituent, is capable of directly damaging the alveolar macrophages. In addition, the ability of the bacteria to prevent the formation of an effective phagolysosome prevents or limits the elimination of the organisms.[27]

The main source of TB infections is through direct or indirect contact with a pulmonary TB patient. TB infections transmit through direct inhalation of aerosolized bacilli present in droplet nuclei of expectorated sputum, released during coughing, sneezing, and speaking. While transmission through direct contact occurs very frequently because of the high infectivity of the released droplets, the possibility of indirect transmission through fomites potentially contaminated by dried bacilli present in dust is much less. As the primary mode of transmission of infection is through infected droplets, forms of TB other than the pulmonary form are considered a lesser threat to public health. Close contacts and household members are at highest risk of contracting the infection due to the presence of large numbers of infectious bacilli in the expectorant sputum of the infected individual. Although rare, infection by other routes such as ingestion may also occur.

The pathology caused by *M. tuberculosis* is not mediated by bacterial toxins. The virulence of these bacteria is attributed to their ability to survive and multiply in macrophages. The tubercle is a characteristic lesion, comprising of an avascular granuloma, formed in tissues infected by *M. tuberculosis*. The tubercle granuloma is made up of a central zone containing giant cells and a peripheral zone containing lymphocytes and fibroblasts. The central zone may or may not show the presence of a firm, and dry mass, resembling the texture of cheese, known as caseation (Fig. 4.20). Exudative tuberculous lesions comprise of accumulated edematous fluid and polymorphonuclear leukocytes resulting from an acute inflammatory reaction followed by an accumulation of lymphocytes and mononuclear cells. Such lesions are typically caused by large numbers of virulent bacilli, eliciting a DTH response in the host rather than protective immunity. Productive tuberculous lesions have a predominantly cellular composition, and are associated primarily with protective immunity than a DTH response.

TB may be primary or secondary depending on the time of infection and the type of host response. Initial infection by *M. tuberculosis* results in primary TB characterized by the engulfment and multiplication of bacilli in alveolar macrophages, giving rise to a subpleural focus of tuberculous pneumonia, known as the "Ghon focus" (Fig. 4.21). The Ghon focus is generally located

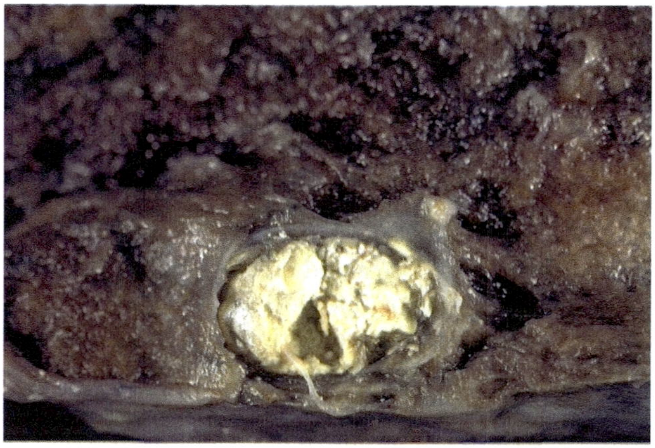

FIGURE 4.20

Caseous necrosis of a tubercle granuloma. Source: Yale Rosen/; *https://www.flickr.com/photos/pulmonary_pathology/6596011395/; https://creativecommons.org/licenses/by-sa/2.0/deed.en; https://commons.wikimedia.org/wiki/File:Tuberculosis_-_Sub-pleural_primary_(Ghon)_focus_(6596011395).jpg.*

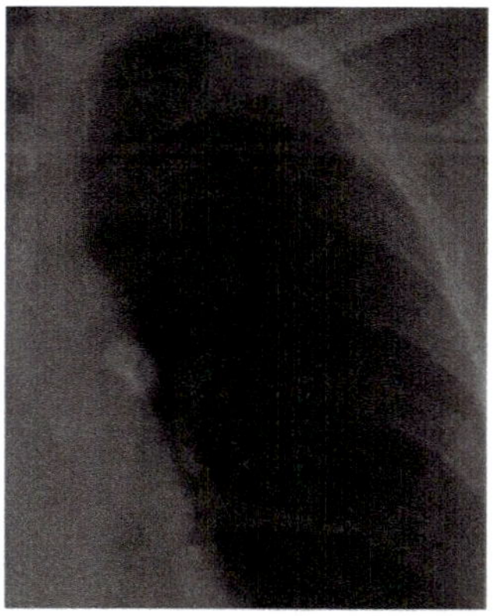

FIGURE 4.21

Chest X-ray showing a Ghon's focus. Source: Basem Abbas Al Ubaidi/; *https://www.clinmedjournals.org/articles/jfmdp/journal-of-family-medicine-and-disease-prevention-jfmdp-4-073.php?jid = jfmdp/; https://creativecommons.org/licenses/by/4.0/deed.en; https://commons.wikimedia.org/wiki/File:Chest_X-ray_of_a_Ghon%E2%80%99s_focus.jpg.*

in the lower lobe or the lower pan of the upper lobe, and along with the enlarged hilar lymph nodes constitutes the primary complex. Hypersensitivity to the bacterial tuberculin protein is responsible for the formation of the primary complex, which occurs about 3–8 weeks from the time of infection. Immunocompetent individuals are capable of suppressing the infection, resulting in the formation of a calcified nodule within 26 months. The presence of a calcified nodule is indicative of a healed primary lesion (Fig. 4.22). Immunocompromised conditions and other high-risk factors may facilitate enlargement of the primary lesion, resulting in disseminated infection. Even in the presence of effective immune responses, healed primary lesions and calcified nodules may shelter surviving bacilli in the latent form, thereby facilitating secondary TB.

Secondary TB can be caused either by endogenous infections triggered by the reactivation of latent infection, or due to exogenous reinfection. There are many differences between primary and secondary TB, but unlike primary infection, secondary TB generally shows its presence in the upper lobes of the lungs, and lymph nodes are usually not involved. The lesions of secondary TB necrotize, resulting in tissue destruction and cavitation. The accumulation and release of the necrotic material into the airways is responsible for the expectoration of bacilli-laden sputum, which facilitates the transmission of the infection. The life-threatening complications of secondary TB commonly seen in immunocompromised individuals is attributable to the lack of cavity formation, which facilitates systemic dissemination of the infectious lesions, commonly known as miliary TB (Figs. 4.23 and 4.24).[28]

Immune response

The natural defenses of the upper respiratory tract in a competent immune system are usually capable of arresting the majority of the inhaled *M. tuberculosis* bacilli. Bacilli that reach the lungs in spite of the primary defenses in the upper

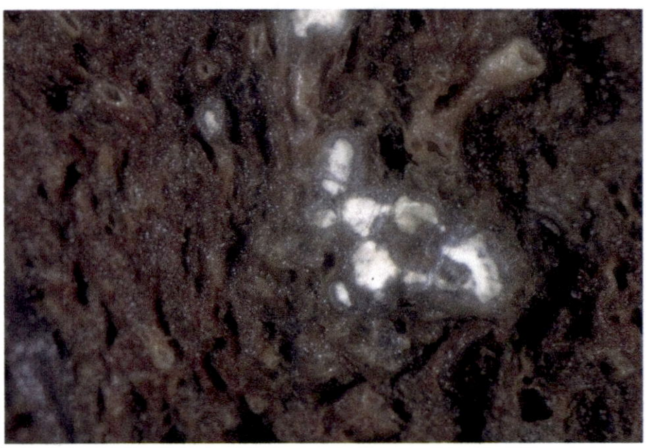

FIGURE 4.22

Heavily calcified and healed focus of primary tuberculosis (TB) infection. Source: Yale Rosen/; *https://www.flickr.com/photos/pulmonary_pathology/6596012049/; https://creativecommons.org/licenses/by-sa/2.0/deed.en; https://commons.wikimedia.org/wiki/File:Tuberculosis_-_Heavily_calcified_healed_focus_of_primary_infection_(Ghon_focus)_(6596012049).jpg.*

FIGURE 4.23

Histopathological view of miliary TB lesions in the lungs. Source: Yale Rosen/; *https://www.flickr.com/photos/pulmonary_pathology/6564687295/; https://creativecommons.org/licenses/by-sa/2.0/deed.en; https://commons.wikimedia.org/wiki/File:Lung_-_Miliary_tuberculosis_(6564687295).jpg.*

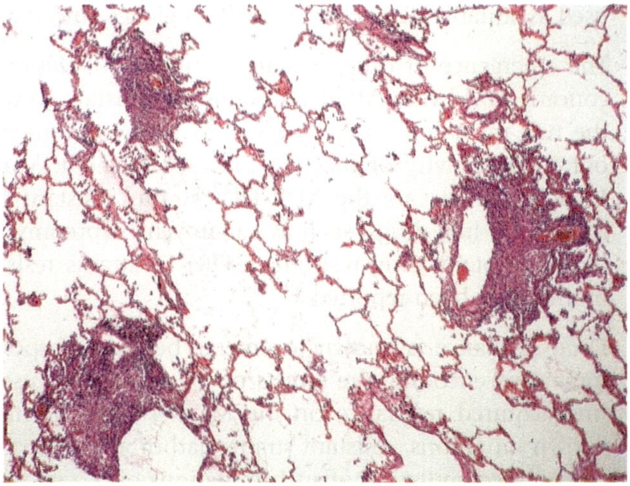

FIGURE 4.24

Microscopic view of miliary TB in the lungs. Source: Yale Rosen/; *https://www.flickr.com/photos/pulmonary_pathology/6564687295/; https://creativecommons.org/licenses/by-sa/2.0/deed.en; https://commons.wikimedia.org/wiki/File:Lung_-_Miliary_tuberculosis_(6564687295).jpg.*

respiratory tract are ingested by the alveolar macrophages. The fact that only about one-tenth of the humans infected by *M. tuberculosis* actually develop active infection is evidence to the ability of a competent immune system to effectively defend the host.[28] The activated CD4 + helper T-cells play an important role in eliciting an effective immune response by releasing cytokines such as interferon (IFN)-γ, interleukin (IL)-1 and 2, and tumor necrosis factor (TNF)-α. While the Th-1 dependent cytokines are responsible for protective immunity and containment of the infection, Th-2 cytokines induce DTH and tissue necrosis, resulting in progressive disease. TB-induced DTH is characterized by the formation of granulomas, comprised of macrophage aggregates in the form of multinucleated giant cells and epithelioid cells, which can effectively phagocytose the bacteria. IFN-γ, in addition to interleukins IL-4, IL-6, and TNF-α, plays an important role in the formation of such granulomas. The unique caseous appearance of TB granulomas on histological examination (Fig. 4.20), and the presence of a central region of eosinophilia seen on microscopic observation by hematoxylin and eosin staining (Fig. 4.24), is attributable to the mycolic acid-rich bacterial cell wall.[27]

Resistance

Mycobacteria are heat-labile organisms that are killed at 60°C within 15–20 minutes. These bacteria are resistant to a wide range of chemical disinfectants up to certain concentrations. However, formaldehyde and glutaraldehyde are effective disinfectants. Exposure to iodine tincture and to 80% ethanol destroys the bacterial cells within 5 minutes of exposure and can therefore be used as suitable disinfectants for skin, gloves, and medical devices.[28]

The emergence of drug-resistant strains of *M. tuberculosis* is of great clinical concern worldwide. MDR-TB strains are resistant to rifampicin and isoniazid, the two most important drugs used in the treatment of TB, thereby ruling out the effectivity of the first line of treatment. More concerning than the MDR-TB strains are the XDR-TB[29] strains resistant to fluroquinolones and the second-line drugs, such as kanamycin, capreomycin, and amikacin. More recently, totally drug resistant (TDR)-TB strains resistant to all known antibiotics have been reported.[30–32]

Drug resistance is known to emerge because of spontaneous bacterial gene mutations encoding the targets of the commonly used antiTB drugs, conferring acquired resistance on the *M. tuberculosis* bacteria. In addition to the known mutations, resistant strains harbor a number of unknown gene mutations.[33] Accumulation of such mutations over a period of time is responsible for the broad-spectrum resistance seen in MDR-TB, XDR-TB, and TDR-TB strains. Noncompliance with treatment regimens owing to the prolonged nature of the treatment regimen has been established as the predominant cause of drug resistance in *M. tuberculosis*.[34] The association between efflux pump mechanisms and the development of drug resistance, or the role of porins in intrinsic drug resistance in *M. tuberculosis*, are important in the analysis of bacterial drug resistance mechanisms.[35] In addition to conferring drug resistance, certain mutations also enhance the transmissibility and virulence of the bacteria. The ability of *M. tuberculosis* to develop resistance to the commonly used antiTB drugs poses a great challenge for effective clinical treatment of the disease and is largely responsible for the high levels of mortality associated with the disease.

Clinical manifestations and complications

Pulmonary TB usually presents itself with a chronic cough, hemoptysis, loss of appetite, unexplained weight loss, persistent low-grade fever, and night sweats (Fig. 4.25). While primary TB usually manifests as symptoms associated with the affected organ system, secondary TB presents itself with more systemic symptoms. TB can remain latent for long periods of time during which there are no clinical manifestations of the disease. However, reactivation or secondary TB manifests with more severe symptoms, owing to the severity of the tissue reaction and hypersensitivity.

Active disease is indicated by the appearance of lesions on a chest X-ray, which may either be contained to the affected organ system, such as the lungs in

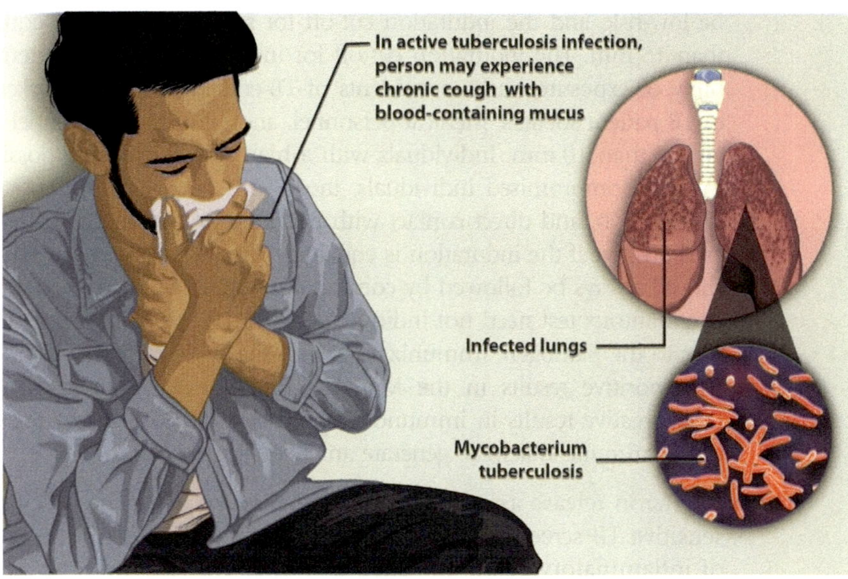

FIGURE 4.25
Depiction of a TB patient. Source: *https://www.myupchar.com/; https://www.myupchar.com/en/disease/tuberculosis-tb/; https://creativecommons.org/licenses/by-sa/4.0/deed.en; https://commons.wikimedia.org/wiki/File:Depiction_of_a_tuberculosis_patient.png.*

pulmonary TB, or may appear systemically in other organ systems, such as the spine, the central nervous system, or the bowel, in case of disseminated TB.

The advent of the effective antiTB drugs has resulted in a good prognosis for a vast majority of TB patients. However, certain high-risk groups, such as extremes of age, extensively disseminated disease, immunosuppression, delayed treatment commencement, infection with multidrug resistant strains, and severe respiratory distress may have some unfavorable outcomes in spite of appropriate treatment. These high-risk group individuals are also prone to suffering some rare complications of the disease, such as acute respiratory distress syndrome, empyema, extensive lung damage, pneumothorax, Horner's syndrome from the damage to the cervical sympathetic ganglia, TB meningitis, and systemic amyloidosis.[27]

Diagnosis

The diagnostic tests for TB are generally divided into screening tests and diagnostic/confirmatory tests. The Mantoux test involves having a PPD skin test during which a dose of the bacterial tuberculin protein is injected subdermally, resulting in the Mantoux reaction, conventionally used to screen individuals potentially exposed to TB. Risk of exposure of the screened individual is a crucial factor in determining the final interpretation of the Mantoux test results. Immunocompetent individuals with minimal possibility of exposure, such as those not residing in a TB-endemic region with no occupational exposure, no travel history, and no direct or indirect contact with TB patients, are considered to be low-risk, and the induration cut-off for such individuals is equal to or more than 15 mm. The induration cut-off for individuals with the intermediate possibility of exposure, such as residents of TB-endemic regions, workers or residents of TB patient shelters, medical personnel, and laboratory personnel, is equal to or greater than 10 mm. Individuals with a high risk of exposure possibility, such as immunocompromised individuals, those with history or evidence of a previous TB infection, and direct contact with potential or known TB patients, are considered positive if the induration is equal to or greater than 5 mm. The Mantoux test should always be followed by confirmatory tests, such as a chest X-ray, as a positive Mantoux test need not indicate an active TB infection; it only indicates exposure to the pathogen. Immunization with the BCG vaccine is also known to give false positive results in the Mantoux test. The Mantoux test may also yield false negative results in immunocompromised patients due to the incompetency of the immune system to generate an adequate reaction to PPD.

Interferon release assays or Quantiferon assays are more specific and equally sensitive TB screening tests as the Mantoux test. This test measures the level of inflammatory cytokines, such as IFN-γ. This assay allows for easier interpretation of results, even in those immunized with the BCG vaccine. However, the high cost of the test and the requirement of skilled professionals are factors hindering the common use of this test.

A chest X-ray is the most common confirmatory test required for a final diagnosis in all cases suspected to be positive based on the screening tests. The acid-fast staining (Ziehl – Neelsen staining) and culture tests are also commonly used for confirming the presence of the pathogen. Nuclear amplification and gene-based tests, such as Genexpert and Dr-MTB, use DNA-based molecular biology techniques to enable identification of the bacteria. These tests enable rapid diagnosis with high precision, thereby facilitating the confirmation of TB within hours rather than days or weeks taken by the conventional culture methods. Molecular biology-based techniques also reduce the pitfalls of false positive and false negative results.[27]

Treatment

Isoniazid is the drug of choice and is prescribed for 6–9 months. This drug has a neuropathic tendency and is therefore usually given in combination with vitamin B6, pyridoxine. Along with isoniazid, WHO also recommends rifapentine weekly for 3 months and rifampicin daily for 3–4 months, followed by rifampicin without isoniazid daily for 3–4 months. Active and confirmed TB cases should always be treated by combination therapy as monotherapy for active TB often proves ineffective and also increases the probability of a relapse.

The combination therapies used for the treatment of TB infections are categorized into first-, second-, and third-line therapies. The first-line medications comprise of group 1 drugs, such as isoniazid, rifampicin, rifabutin, rifapentine pyrazinamide, and ethambutol. The second-line medications comprise of groups 2, 3, and 4. While group 2 drugs include injectable aminoglycosides, such as amikacin, kanamycin, streptomycin, and injectable polypeptides such as capreomycin and viomycin, group 3 drugs include oral and injectable fluoroquinolones, such as levofloxacin, moxifloxacin, ofloxacin, and gatifloxacin. The group 4 antiTB drugs include para-aminosalicylic acid, cycloserine, terizidone, ethionamide, prothionamide, thioacetazone, and linezolid. The third-line medications comprise of group 5 drugs, which are used only as a last resort, especially in cases of TDR-TB due to their variable and unproven efficiency against TB. Clofazimine, linezolid, amoxicillin/clavulanic acid, imipenem/cilastatin, and clarithromycin are some of the third-line medications used in the treatment of TDR-TB and XDR-TB.

MDR-TB cases are also treated using a combination therapy, comprising of first-line and second-line medications at high doses. In addition, bedaquiline is an U.S. Food and Drug Administration (FDA)-approved drug used for the treatment of MDR-TB.

Due to the side effects of these medicines on the nervous system and the hepatobiliary system, regular liver function tests and frequent monitoring for retinopathies are recommended for all patients taking isoniazid and ethambutol treatment.[27]

Student Notes

Key take-aways

- TB is primarily a pulmonary disease, although disseminated TB may present itself with systemic manifestations and affect other organ systems.
- *M. tuberculosis* is the etiological agent responsible for TB and is very resilient and highly adaptable to the advent of modern medicine, thereby continuing to be a human pathogen of significant concern, even today.
- The poor hygiene and sanitation conditions in addition to the overcrowded living conditions prevalent in the developing countries is largely responsible for the huge burden of morbidity and mortality due to TB in these countries.
- TB may manifest as a primary or a secondary infection, which may be a reactivation of a previous dormant infection or a reinfection.
- Immunocompetent individuals are generally capable of suppressing a primary TB infection, forcing the pathogen to go into latency.
- Immunocompromised individuals are especially vulnerable to primary and secondary TB infections.
- Diagnosis of TB comprises of screening tests and confirmatory tests, which are recommended depending on the exposure risk of the individual.
- The Mantoux test, a commonly used screening test for TB, is usually quite effective at screening individuals for exposure to TB, although it has the tendency to yield false positive results in individuals previously immunized with the BCG vaccine and false negative results in immunocompromised individuals.
- The emergence of resistant *M. tuberculosis* strains has posed a challenge for the effective management and treatment of TB and is also responsible for the high levels of mortality associated with the disease.
- Therapy (6–9 months) is recommended for cases that test positive in the screening tests to prevent the adverse effects of the disease and the possibility of reactivation of the pathogen.
- Combination therapy is preferred to monotherapy for high-risk individuals and confirmed TB cases.
- The long duration of TB treatment is often responsible for resistance, occurring due to nonadherence to treatment.

References

1. Cillóniz C, Dominedò C, Garcia-Vidal C, Torres A. Community-acquired pneumonia as an emergency condition. *Curr Opin Crit Care*. 2018;24(6):531–539. Available from: https://doi.org/10.1097/MCC.0000000000000550. PMID: 30239410.

2. Luna CM, Pulido L, Niederman MS, et al. Decreased relative risk of pneumococcal pneumonia during the last decade, a nested case-control study. *Pneumonia (Nathan)*. 2018;10:9. Available from: https://doi.org/10.1186/s41479-018-0053-6. PMID: 30263884; PMCID: PMC6154928.

3. Nunes S, Sá-Leão R, Carriço J, et al. Trends in drug resistance, serotypes, and molecular types of *Streptococcus pneumoniae* colonizing preschool-age children attending day care centers in Lisbon, Portugal: a summary of 4 years of annual surveillance. *J Clin Microbiol*. 2005;43 (3):1285–1293. Available from: https://doi.org/10.1128/JCM.43.3.1285-1293.2005. PMID: 15750097; PMCID: PMC1081291.

4. Dion CF, Ashurst JV. *Streptococcus pneumoniae* [Updated 2020 Aug 10]*StatPearls [Internet]*. Treasure Island, FL: StatPearls Publishing; 2020. Available from: https://www.ncbi.nlm.nih.gov/books/NBK470537/.

5. Cherazard R, Epstein M, Doan TL, Salim T, Bharti S, Smith MA. Antimicrobial resistant *Streptococcus pneumoniae*: prevalence, mechanisms, and clinical implications. *Am J Ther*. 2017;24(3):e361–e369. Available from: https://doi.org/10.1097/MJT.0000000000000551. PMID: 28430673.

6. Lock RA, Paton JC, Hansman D. Purification and immunological characterization of neuraminidase produced by *Streptococcus pneumoniae*. *Microb Pathog*. 1988;4(1):33–43. Available from: https://doi.org/10.1016/0882-4010(88)90046-0. PMID: 3143891.

7. Feldman C, Munro NC, Jeffery PK, et al. Pneumolysin induces the salient histologic features of pneumococcal infection in the rat lung in vivo. *Am J Respir Cell Mol Biol*. 1991;5 (5):416–423. Available from: https://doi.org/10.1165/ajrcmb/5.5.416. PMID: 1834101.

8. Rosenow C, Ryan P, Weiser JN, et al. Contribution of novel choline-binding proteins to adherence, colonization and immunogenicity of *Streptococcus pneumoniae*. *Mol Microbiol*. 1997;25(5):819–829. Available from: https://doi.org/10.1111/j.1365-2958.1997.mmi494.x. PMID: 9364908.

9. Sampson JS, O'Connor SP, Stinson AR, Tharpe JA, Russell H. Cloning and nucleotide sequence analysis of psaA, the *Streptococcus pneumoniae* gene encoding a 37-kilodalton protein homologous to previously reported Streptococcus sp. adhesins. *Infect Immun*. 1994;62 (1):319–324. Available from: https://doi.org/10.1128/IAI.62.1.319-324.1994. PMID: 7505262; PMCID: PMC186105.

10. McDaniel LS, Sheffield JS, Delucchi P, Briles DE. PspA, a surface protein of *Streptococcus pneumoniae*, is capable of eliciting protection against pneumococci of more than one capsular type. *Infect Immun*. 1991;59(1):222–228. Available from: https://doi.org/10.1128/IAI.59.1.222-228.1991. PMID: 1987036; PMCID: PMC257730.

11. Ananthanarayan R, Paniker CK. *Pneumococcus*. 8th ed. *Textbook of Microbiology*. 24. India: Universities Press; 2009:218–223.

12. Corrêa RA, Costa AN, Lundgren F, et al. 2018 recommendations for the management of community acquired pneumonia [published correction appears in *J Bras Pneumol* 2018 November – December;44(6):532] [published correction appears in *J Bras Pneumol* 2019 May 13;45(2):e20180130]*J Bras Pneumol*. 2018;44(5):405–423. Available from: https://doi.org/10.1590/S1806-37562018000000130.

13. Krause CI. The ABCs of RSV. *Nurse Pract*. 2018;43(9):20–26. Available from: https://doi.org/10.1097/01.NPR.0000544277.74514.55. PMID: 30134433.

14. Gilchrist S, Török TJ, Gary Jr HE, Alexander JP, Anderson LJ. National surveillance for respiratory syncytial virus, United States, 1985-1990. *J Infect Dis*. 1994;170(4):986–990. Available from: https://doi.org/10.1093/infdis/170.4.986. PMID: 7930745.

15. Brandt CD, Wha Kim H, Arrobio JO, et al. Epidemiology of respiratory syncytial virus infection in Washington, D.C. III. Composite analysis of eleven consecutive yearly epidemics. *Am J Epidemiol*. 1973;98(5):355–364. Available from: https://doi.org/10.1093/oxfordjournals.aje.a121565.

16. Popow-Kraupp T, Lakits E, Kellner G, Kunz C. Immunoglobulin-class-specific immune response to respiratory syncytial virus structural proteins in infants, children, and adults. *J Med Virol*. 1989;27:215–223. Available from: https://doi.org/10.1002/jmv.1890270307.

17. Erdman DD, Anderson LJ. Monoclonal antibody-based capture enzyme immunoassays for specific serum immunoglobulin G (IgG), IgA, and IgM antibodies to respiratory syncytial virus. *J Clin Microbiology*. 1990;28(12):2744–2749. Available from: https://doi.org/10.1128/jcm.28.12.2744-2749.1990.

18. Cane PA, Thomas HM, Simpson AF, Evans JE, Hart CA, Pringle CR.), Analysis of the human serological immune response to a variable region of the attachment (G) protein of respiratory syncytial virus during primary infection. *J Med Virol*. 1996;48:253–261. 10.1002/(SICI)1096-9071(199603)48:3 < 253::AID-JMV7 > 3.0.CO;2-7.

19. Johnson Jr PR, Olmsted RA, Prince GA, et al. Antigenic relatedness between glycoproteins of human respiratory syncytial virus subgroups A and B: evaluation of the contributions of F and G glycoproteins to immunity. *J Virol*. 1987;61(10):3163–3166. Available from: https://doi.org/10.1128/JVI.61.10.3163-3166.1987. PMID: 3305988; PMCID: PMC255893.

20. Domachowske JB, Rosenberg HF. Respiratory syncytial virus infection: immune response, immunopathogenesis, and treatment. *Clin Microbiol Rev*. 1999;12(2):298–309.

21. Schmidt ME, Varga SM. Cytokines and CD8 T cell immunity during respiratory syncytial virus infection. *Cytokine*. 2020;133:154481. Available from: https://doi.org/10.1016/j.cyto.2018.07.012.

22. Aherne W, Bird T, Court SD, Gardner PS, McQuillin J. Pathological changes in virus infections of the lower respiratory tract in children. *J Clin Pathol*. 1970;23(1):7–18. Available from: https://doi.org/10.1136/jcp.23.1.7. PMID: 4909103; PMCID: PMC474401.

23. Bellini WJ, Rota PA, Anderson LJ. *Paramyxoviruses*. 9th ed. *Topley Wilson's Microbiology Microb Infect*. 23. 2010450–452.

24. Anderson LJ, Dormitzer PR, Nokes DJ, Rappuoli R, Roca A, Graham BS. Strategic priorities for respiratory syncytial virus (RSV) vaccine development. *Vaccine*. 2013;31(2):B209–B215. Available from: https://doi.org/10.1016/j.vaccine.2012.11.106.

25. Perk Y, Özdil M. Respiratory syncytial virüs infections in neonates and infants. *Turk Pediatr Ars*. 2018;53(2):63–70. Available from: https://doi.org/10.5152/TurkPediatriArs.2018.6939. Published June 1, 2018.

26. Foolad F, Aitken SL, Shigle TL, et al. Oral vs aerosolized ribavirin for the treatment of respiratory syncytial virus infections in hematopoietic cell transplant recipients. *Clin Infect Dis*. 2019;68(10):1641–1649. Available from: https://doi.org/10.1093/cid/ciy760.

27. Adigun R, Singh R. *Tuberculosis* [Updated 2020 Oct 27]*StatPearls [Internet]*. Treasure Island, FL: StatPearls Publishing; 2020. Available from: https://www.ncbi.nlm.nih.gov/books/NBK441916/.

28. Ananthanarayan R, Paniker CK. 8th ed. *Textbook of Microbiology*. 39. India: Universities Press; 2009:347–358.

29. Centers for Disease Control and Prevention (CDC). Emergence of *Mycobacterium tuberculosis* with extensive resistance to second-line drugs—worldwide, 2000 – 2004. *MMWR Morb Mortal Wkly Rep*. 2006;55(11):301–305.

30. Udwadia ZF, Amale RA, Ajbani KK, Rodrigues C. Totally drug-resistant tuberculosis in India. *Clin Infect Dis*. 2012;54(4):579–581. Available from: https://doi.org/10.1093/cid/cir889.

31. Migliori GB, Centis R, D'Ambrosio L, et al. Totally drug-resistant and extremely drug-resistant tuberculosis: the same disease? *Clin Infect Dis*. 2012;54(9):1379–1380. Available from: https://doi.org/10.1093/cid/cis128.

32. Velayati AA, Masjedi MR, Farnia P, et al. Emergence of new forms of totally drug-resistant tuberculosis bacilli: super extensively drug-resistant tuberculosis or totally drug-resistant

strains in iran. *Chest.* 2009;136(2):420–425. Available from: https://doi.org/10.1378/chest.08-2427.

33. Zhang H, Li D, Zhao L, et al. Genome sequencing of 161 *Mycobacterium tuberculosis* isolates from China identifies genes and intergenic regions associated with drug resistance. *Nat Genet.* 2013;45(10):1255–1260. Available from: https://doi.org/10.1038/ng.2735.
34. Smith T, Wolff KA, Nguyen L. Molecular biology of drug resistance in *Mycobacterium tuberculosis. Curr Top Microbiol Immunol.* 2013;374:53–80. Available from: https://doi.org/10.1007/82_2012_279.
35. Palomino JC, Martin A. Drug resistance mechanisms in *Mycobacterium tuberculosis. Antibiotics (Basel).* 2014;3(3):317–340. Available from: https://doi.org/10.3390/antibiotics3030317. Published July 2, 2014.

CHAPTER 5

Infections of the cardiovascular system

Problem number 5.1: Rheumatic heart disease caused by *Streptococcus viridans* infection

A 10-year-old boy presented with fever and joint pains of 10-days duration. The joint pain started in the left knee joint 1 week prior to his admission. His left knee was swollen, and he was unable to walk because of the pain. The left knee joint pain and swelling subsided after 6 days, after which, the pain and swelling developed in the right elbow. He had no other complaints. There was no history of prior hospital admissions, and the boy was usually healthy. There was no previous history of joint pains or trauma. He had a history of sore throat with low-grade fever that had lasted for 5 days, about 25 days before the presentation. The sore throat and fever subsided without any medication, and therefore no medical consultation had been taken. The boy did not have any known allergies. His family comprised of seven other siblings and his parents who lived together in a poorly ventilated, two-bedroom house. The doctor suspected acute rheumatic fever (ARF) and recommended hospital admission.

The following are the reports of the investigations carried out:

1. Vitals
   Weight: 28 kg
   Height:130 cm
   Temperature: 39.2°C
   Pulse rate: 135/min (pulse when he was sleeping was 115/min)
   RR: 26/min
2. Physical examination
   a. Skin and mucus membranes: No rashes or subcutaneous nodules noted
   b. Head/ENT: Normal
   c. Neck: No jugular vein distention

CONTENTS

Problem-Based Learning Approach in Microbiology. DOI: https://doi.org/10.1016/B978-0-323-95092-3.00005-6

d. Lung: Normal breath sounds
e. Cardiac:
    i. Active precordium
    ii. Apical pulse in 5th left intercostal space (ICS) at the mid-clavicular line (MCL)
    iii. No thrills were felt
    iv. S1 and S2 were normal
    v. S3 gallop was present
    vi. Auscultation revealed the presence of a grade II/VI pansystolic murmur
    vii. No pericardial friction rubs
f. Abdomen and extremities:
    viii. No organomegaly
    ix. Localized heat, swelling, and tenderness in the right elbow joint
    x. All other joints were normal
g. Neurological:
    xi. Normal, no abnormal movements were noted

3. Complete blood count (CBC)

Test	Value	Reference range
Hb	11.5%	12–16 g/dl
Hematocrit (%)	34	36–46
WBC (/L)	12.4×10^9	$4.5–11 \times 10^9$
Neutrophils (PMN) (%)	65	40–70
Lymphocytes (%)	30	22–44
Monocytes (%)	4	4–11
Eosinophils (%)	1	0–8

4. Erythrocyte sedimentation rate (ESR): 120 mm/1st hour (pre-admission); 70 mm/hour (1 week after hospitalization)
5. C-reactive protein (CRP): Positive (high titer)
6. Antistreptolysin O titer: 500 units (suggestive of ARF)
7. Throat culture: Normal throat bacteria microbiota
   No group A-beta hemolytic streptococci were present
8. Blood culture: Sterile (three sets were collected within 24 hours)
9. Sickling test: Negative
10. G6PD level: Normal
11. Urine analysis: Normal
12. Chest X-ray: No cardiomegaly; lung fields were clear

13. ECG:
    Sinus tachycardia (pre-admission)
    On third day of hospitalization:
    No pericardial fluid collection
    Heart chambers normal in size
    Evidence of mild regurgitation in the mitral valve
    No vegetations were noted in the mitral valve

Diagnosis

The final diagnosis concluded ARF with mild carditis and arthritis caused by Streptococcal infection of the pharynx.

Treatment

1. Benzathine Penicillin G: 1.2 million units (Intramuscular) (single dose)
2. Aspirin: 100 mg/kg/day (every 6 hours)
   Dose adjusted to maintain blood salicylate level at 25 mg/dl (1.8 mmol/L)

Follow-up

The patient was discharged 4 weeks after admission and prescribed aspirin 70 mg/kg/day for 6 weeks. Benzathine Penicillin G injection administration every 4 weeks was prescribed, until advised otherwise.

The patient was advised to remain indoors to prevent exertion until the follow-up visit. The parents were counseled about the medical implications of the disease and the need for rest and regular follow-up visits every 2 weeks after discharge to assess the child's clinical status.

Outcome

Eight years later, the patient was readmitted with history of fever and dyspnea for five days. There was no history of joint pain or swelling in this episode. History revealed nonadherence to the recommended 2 weekly administration of the penicillin injections after the first episode. Three blood cultures during a 12-hour period were ordered.

The following are the reports of the investigation carried out on this visit:

1. Vital signs
   Febrile
   Pulse: 150/min
   RR: 35/min
2. Cardiac:
   Apical impulse in the 6th ICS 2 cm lateral to MCL
   Muffled heart sounds
   Grade III/VI apical pansystolic murmur present
   Grade II/VI mid-diastolic murmur heard best at the apex
3. Chest X-ray:
   Heart enlarged
   Left atrium enlarged
   Residual volume (RV) size increased
   Aorta appeared to be small (due to clockwise rotation of heart)
   Cephalization of pulmonary flow
   Findings suggestive of mitral valve disease
4. ECG:
   Cardiac chamber enlargement
   Presence of an echo-dense mass (vegetation) attached to the anterior leaflet of mitral valve
   Evidence of mitral valve disease
5. Blood culture: Presence of Penicillin-sensitive *S. viridans*

Diagnosis

The final diagnosis revealed rheumatic heart disease (RHD) with mitral valve disease caused by *S. viridans*, resulting from nonadherence to the prophylactic treatment recommended after the previous episode of ARF.

Treatment

Penicillin injection (IV): G2 million units every 4 hours for 4 weeks.

The patient recovered and was advised to attend regular follow-ups for further treatment.

Question and answer

1. What are the probable causes of joint pain and swelling in this case?
   There can be several probable causes of joint pain and swelling in this case:
   a. Infection of the joint.

   b. Infection of a site adjacent to a joint (bone) (osteomyelitis).
   c. Inflammation without infection (rheumatic fever, rheumatoid arthritis, etc.).
   d. Trauma with joint effusion or bleeding into a joint.
   e. Sickle cell disease (vaso-occlusive crisis in bony structures around a joint with secondary joint effusion).
   f. Tumors and other.
2. What is the primary cause of migratory arthritis?
   ARF is the primary cause of migratory arthritis.
3. What are the possible causes of a sore throat and fever?
   The possible causes of sore throat could be infection, for example: pharyngitis or tonsilitis.
4. Which are the pathogens that can cause such a condition?
   The pathogens that can cause such a condition may be viruses, such as the Adenovirus and Epstein – Barr virus, or bacteria like group A ß-hemolytic *Streptococcus* and *Corynebacterium diphtheriae.*
5. What is the usual natural course of streptococcal pharyngitis?
   Streptococcal infections are generally self-limiting. Therefore the usual natural course of streptococcal pharyngitis begins with an acute manifestation of the condition and continues to get resolved after a particular duration.
6. What are the possible complications that may develop after streptococcal infections?
   The complications that may develop after Streptococcal infections may be infectious, such as scarlet fever or noninfectious immune system-mediated complications, like rheumatic fever and acute glomerulonephritis.
7. What are the predisposing factors for ARF?
   Family history, low socioeconomic status, an age of 6–15 years, and a history of streptococcal infections of pharynx are the predisposing factors for ARF.
8. What is the significance of the skin rash and the rash on the mucous membrane?
   Rheumatic fever erythema marginatum and subcutaneous nodules are known to be indications of rheumatic fever and rheumatoid arthritis.
9. What are the possible causes of systolic murmurs?
   Mitral regurgitation, aortic stenosis, ventricular septal defect, tricuspid regurgitation, and pulmonary stenosis are the possible causes of systolic murmurs.
10. What is the significance of the etiology of the previous throat infection?
   The etiology of the previous throat infection is important because of the probable relation between the etiology of the previous throat

infection and the present disease manifestation. CBC, ESR, CRP, and antistreptolysin O titer (ASO) titer are tests that can help to identify the etiology of the previous throat infection.

11. What do the ESR and CRP tests indicate?
    ESR and CRP are acute phase reactants that are nonspecific indicators of inflammation. High values in these tests allude to a systemic inflammatory response.
12. What is the significance of the throat culture and the ASO titer?
    The throat culture and the ASO titer help in detecting any preceding streptococcal infections of the pharynx. An elevated ASO titer is indicative of a recent streptococcal infection.
13. Which are the other tests that can detect previous streptococcal infections?
    Anti-DNAase B, antihyaluronidase, and antistreptokinase are some of the other tests that can detect previous streptococcal infections.
14. Why are the tests for sickle cell disease and G6PD deficiency conducted?
    Sickle cell disease and G6PD deficiency can also manifest as a fever and pain in the extremities. Therefore testing for these conditions is important to rule them out from the differential diagnosis.
15. What is the significance of benzathine penicillin prophylaxis?
    Benzathine penicillin is a sustained-release penicillin drug, which provides long-term protection, and is appropriate for prophylactic measures. Administration of benzathine penicillin at regular intervals prevents the recurrence of streptococcal infections.
16. How can the hemodynamic changes in mitral valve cause endocarditis?
    The pressure gradient across the orifice results in flow changes, which increases the vulnerability to infections, thereby resulting in endocarditis.
17. What are the risks of nonadherence to the monthly prophylactic treatment?
    Nonadherence to the monthly prophylactic treatment can result in the recurrence of the streptococcal infection, complications associated with the infection, and even lead to antibiotic resistance in the pathogenic strains.

Learning objectives

1. List the causes of joint pain.
2. Describe the pathogenesis of ARF.

3. Describe the clinical manifestations that facilitate the diagnosis of ARF.
4. Explain the pathology of rheumatic pancarditis.
5. Describe the hemodynamic changes seen in mitral valve regurgitation.
6. Determine the appropriate differential diagnosis for the heart murmurs based on patient history and clinical manifestations.
7. List the diseases of the pericardium.
8. Explain the pathology of chronic rheumatic valvular diseases.
9. List the investigations that are recommended in suspected streptococcal infections.
10. Explain the significance of ESR and CRP tests.
11. Identify the etiological agent responsible for the particular case of fever and joint pain.
12. Describe the microbiological properties and pathogenicity of *S. viridans*.
13. Describe the pathology of pericardial inflammation and effusion.
14. Explain the significance of echocardiography (ECG) as a diagnostic tool in this case.
15. Determine the appropriate course of treatment and most effective antibiotic regimen for therapy and prophylaxis.
16. Outline the measures that can prevent the occurrence of ARF.
17. Explain the phenomenon of recurrence of ARF.
18. List the causes of bacterial endocarditis.
19. Describe the pathology of endocarditis.
20. Explain the hemodynamics of valvular heart disease.

Discussion

Differential diagnosis

Endocarditis	Endocarditis presents itself with a fever, chills, joint pains, myalgia, dyspnea, night sweats, pale skin, discoloration of the skin, heart murmur, nausea, decreased appetite, unintentional weight loss, bloated feeling in the upper left part of the abdomen, splenomegaly, hematuria, and swelling of legs, feet, and abdomen.
Mitral valve prolapse	Mitral valve prolapse causes fatigue, dizziness, arrythmia, dyspnea, and click-murmur sounds on heart examination by stethoscope.
Myocarditis	Myocarditis can lead to chest pain, dyspnea, arrythmia, fluid retention and swelling in ankles, legs, and feet, fatigue, body aches, joint pain, headache, diarrhea, fever, and sore throat.
Mitral regurgitation	Mitral valve regurgitation manifests with dyspnea, palpitations, swollen feet and ankles, fatigue, and heart murmur.

Background

ARF is a manifestation of an exaggerated autoimmune response to streptococcal pharyngitis, while RHD refers to the manifestations of long-term damage to the heart due to the recurrence of ARF or even one severe episode.[1] ARF and RHD are generally caused by the ß-hemolytic *Streptococcus pyogenes*; however, α-hemolytic *S. viridans* may also be the responsible etiological agent in rare cases. The Viridans streptococci comprise of a large group of α-hemolytic bacteria instead of a single bacterial species, therefore differ from other bacterial species in their nomenclature, being referred to as Viridans streptococci or Viridans group streptococci.[2]

Etiological agent

Streptococcus viridans

Classification and morphology

Viridans streptococci are chain-forming, Gram-positive (Fig. 5.1), facultatively anaerobic, fastidious, and α-hemolytic cocci, which comprise of several streptococcal species, such as *S. anginosus*, *S. mitis*, *S. sanguinis*, *S. salivarius*, and *S. mutans*. Viridans streptococci are resistant to optochin and lack the ability to solubilize in bile, both of which are properties that distinguish this group of α-hemolytic streptococci from *S. pneumoniae*, another species of

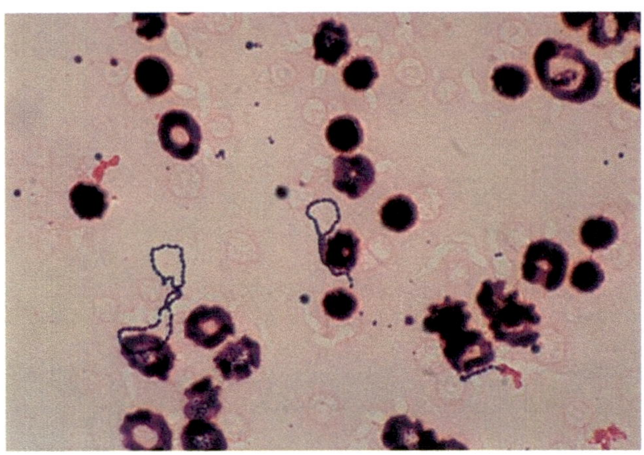

FIGURE 5.1

A photomicrograph of a blood culture specimen, which revealed the presence of Gram-positive, α-hemolytic viridans group streptococci bacteria arranged in long chains resembling strings of pearls. Source*: CDC/Dr. Mike Miller/PHIL#2897; http://phil.cdc.gov/phil/home.asp Public Health Image Library/; https://upload.wikimedia.org/wikipedia/commons/d/d3/Streptococcus_viridans_01.png.*

α-hemolytic streptococcal bacteria. The inability of this group of cocci to grow on sodium chloride containing media distinguishes it from enterococci. Viridans streptococci do not produce catalase or coagulase, another characteristic feature of this group of bacteria.

Epidemiology

Viridans streptococci are an important part of the normal flora of mucous membrane-lined cavities, such as the upper respiratory tract, the gastrointestinal tract, the female genital tract, and the oral cavity of animals and humans. The majority of the flora of the tongue, saliva, dental plaque, and gingival crevices is composed of the viridans streptococci. The presence of this bacterial group as part of the oral flora is essential to prevent colonization by other pathogenic bacteria, such as *Pseudomonas aeruginosa*. The invasion of viridans streptococci into the bloodstream due to the compromised integrity of the mucosal barriers, such as that occurring due to dental extractions and procedures, can result in pathogenic infections, for example, native valve endocarditis, ARF, and RHD.[3]

Pathogenicity and pathogenesis

The pathogenic potential of Viridans streptococci is considered low as this bacterial group forms an important part of the normal human mucosal flora. However, if the bacteria invade the bloodstream, their pathogenicity can be clinically concerning. Viridans streptococci are especially implicated in valvular heart diseases because of their affinity toward colonizing cardiac valves. Although immunocompetent individuals rarely manifest highly pathogenic infections caused by the viridans streptococci bacteria, immunocompromised individuals and those with underlying cardiac conditions or anomalies are vulnerable to infections affecting the cardiac valves caused by this bacterial group. The unique ability of viridans streptococci to use fermentable carbohydrates, like glucose, to synthesize dextrans enhances their adhesiveness, thereby enabling the adherence of these bacteria to heart valves. Fibrin-platelet aggregates may be formed in cardiac valves due to damages caused by underlying cardiac abnormalities or a previous episode of ARF. Such fibrin-platelet aggregates further allow the adherence of these bacteria to the heart valves, thereby increasing the vulnerability of individuals with underlying cardiac abnormalities or conditions to valvular diseases or damage caused by *S. viridans* infection.[4]

Multiple/single severe or untreated episodes of rheumatic fever can result in progressive fibrosis of heart valves that may eventually manifest as rheumatic valvular heart disease. Valvular heart disease is responsible for compromised cardiac function, resulting in impeded blood circulation, and is therefore

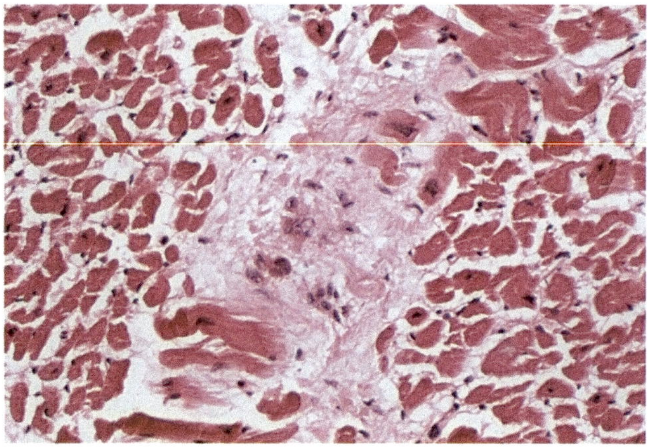

FIGURE 5.2
Aschoff body in rheumatic myocarditis. Source: *Ed Uthman, MD/Reused from https://www.flickr.com/photos/euthman/1859018346/;* Permission for reuse *https://creativecommons.org/licenses/by-sa/2.0/ https://creativecommons.org/licenses/by-sa/2.0/deed.en; https://commons.wikimedia.org/wiki/File:Aschoff_Body_in_Rheumatic_Myocarditis.jpg.*

considered a life-threatening condition. The presence of Aschoff bodies (Fig. 5.2) in heart nodules and MacCallum plaques on the valves and the subendocardium in the left atrium, resulting from inflammation in the heart muscle are characteristic features of RHD.[5]

Clinical manifestations and complications

ARF may follow an episode of pharyngitis or a sore throat as a sequelae of the host immune response to the infection. The condition presents itself with several clinical manifestations, the most common being fever and joint pains. While most clinical manifestations of ARF resolve within weeks, carditis is a rare but severe condition that may result in chronic RHD. Chorea, characterized by involuntary, purposeless, nonrhythmic body movements, restlessness, emotional instability, sudden mood swings, and sudden behavioral outbursts, may also result from ARF. Erythema marginatum (Fig. 5.3), characterized by pink papules spreading outward on the trunk and proximal limbs and subcutaneous nodules (Fig. 5.4) that usually develop on the bony regions, such as the elbows, are some of the skin-related findings seen in ARF. Lethargy, epistaxis, abdominal pain, rapid sleeping pulse rate, and tachycardia disproportionate to fever are some of the other generalized clinical features of ARF.[1]

RHD usually causes a fever and cardiac symptoms depending upon the affected area of the heart. While pericarditis is indicated by pericardial friction rub on auscultation, myocarditis is, for example, indicated by lower extremity edema, short-

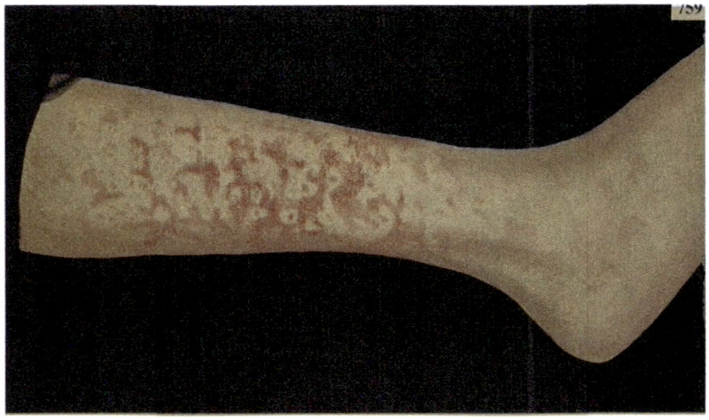

FIGURE 5.3

Representative image of leg with erythema marginatum. Source*: This file comes from Wellcome Images, a website operated by Wellcome Trust, a global charitable foundation based in the United Kingdom. https://wellcomeimages.org/indexplus/obf_images/c1/1a/d35405e8ecf2d2ebc843fa2bf4fa.jpg/; Library reference: St Bartholomew's Hospital Archives & Museum SBHB MU/14/35/1/Photo number: L0061869/; https://creativecommons.org/licenses/by/4.0/deed.en; https://commons.wikimedia.org/wiki/File: Leg_with_erythema_marginatum_Wellcome_L0061869.jpg.*

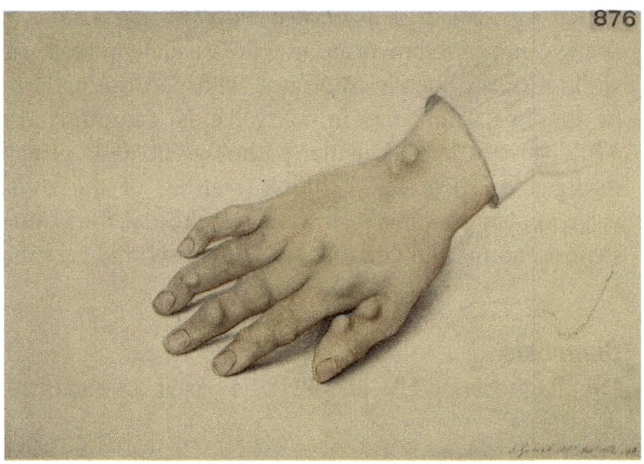

FIGURE 5.4

Subcutaneous rheumatic nodules on the hand. Source*: This file comes from Wellcome Images, a website operated by Wellcome Trust, a global charitable foundation based in the United Kingdom. https://wellcomeimages.org/indexplus/obf_images/86/c8/744291e0dc06314d712d52955d11.jpg/; Library reference: St Bartholomew's Hospital Archives & Museum SBHB MU/14/35/137/2/ Photo number: L0062024/; https://creativecommons.org/licenses/by/4.0/deed.en; https://commons.wikimedia.org/wiki/File: Subcutaneous_rheumatic_nodules_on_the_hand_Wellcome_L0062024.jpg.*

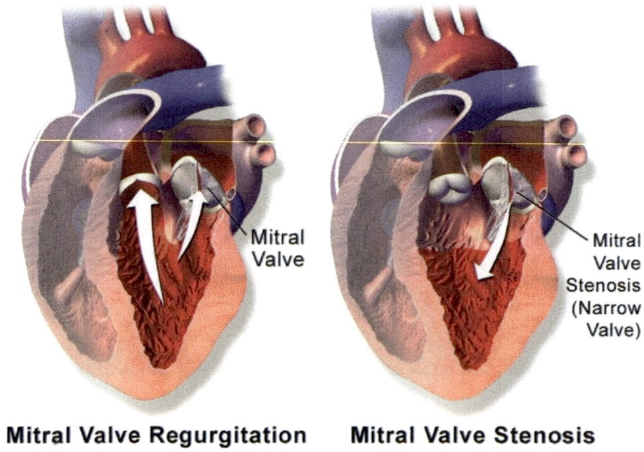

FIGURE 5.5

Mitral valve regurgitation versus mitral valve stenosis. Source: *BlausenBlaus/Own work/ Blausen.com staff (2014). "Medical gallery of Blausen Medical 2014". WikiJournal of Medicine 1 (2). DOI: 10.15347/wjm/2014.010. ISSN 2002–4436/; https://creativecommons.org/licenses/by/3.0/deed.en; https://commons.wikimedia.org/wiki/File:Blausen_0645_MitralValve_RegurgitationvsStenosis.png.*

ness of breath, orthopnea, and abdominal distention. Auscultation to detect an apical systolic or basal diastolic murmur is important to ensure that the suspected myocarditis is associated with valvular disease, thereby implicating RHD as the cause of the myocarditis.[5] ECG usually reveals valvular pathology in RHD, while ARF is generally associated with pan carditis. Involvement of the left-side of the heart and the mitral valve is common, while involvement of the aortic valve is rare.[6] Valvular pathology in RHD generally includes mitral valve regurgitation (Fig. 5.5), chronic scarring of the mitral valve and the mitral valve apparatus, valvular thickening, chordal thickening, mitral prolapse, mitral stenosis, aortic regurgitation, aortic stenosis, and mixed valve disease.[1]

Diagnosis

The diagnosis of ARF and RHD is based on a spectrum of tests and diagnostic criteria.

The Jones criteria (Table 5.1) are the most used basis for the diagnosis of ARF. According to the Jones criteria, the diagnosis of primary ARF requires two major or one major and two minor criteria to be fulfilled, in addition to evidence of preceding streptococcal infection. The Jones criteria also differ depending on the risk level of the population to which the case belongs. The clinical diagnosis of ARF depends on the exclusion of all other probable diagnoses and requires an exhaustive search for other causes of the clinical manifestations.

Table 5.1 Jones criteria for the diagnosis of ARF.[7]

Major criteria	Minor criteria
Arthritis	Fever (≥38.5°F)
Carditis	Sedimentation rate (≥60 mm)
Chorea	CRP (≥3.0 mg/dL)
Erythema marginatum	Prolonged per rectum (PR) interval
Subcutaneous nodules	—

The wide spectrum of cardiac manifestations seen in RHD necessitates the use of various diagnostic and evaluation tests. While an electrocardiogram is required to detect varying degrees of heart block, a chest X-ray can evaluate for cardiomegaly or pulmonary vascular congestion. Auscultation during physical examination must be complemented by a transthoracic echocardiogram to establish the diagnosis of RHD.[6] Echocardiographic diagnosis of RHD is also subject to fulfillment of the age-based criteria given by the World Heart Federation.[5]

As streptococci are generally responsible for manifestations associated with ARF and RHD, streptococcal serology and blood culture are often recommended to determine the etiological agent. The antistreptolysin titer and blood culture reports enable the determination of the appropriate antibiotic regimen, thereby ensuring effective treatment of the condition.

Treatment and management

The treatment of ARF primarily comprises of eradicating the pathogen responsible for pharyngitis, followed by an extended course of secondary prophylaxis with long-lasting intramuscular benzathine penicillin G, unless resistance to penicillin G is indicated. Symptomatic treatment of the arthritic manifestations, management of the symptoms of chorea, and treatment for carditis may also be required depending on the complications and clinical manifestations of the case. The patient and care-givers must be educated and informed about of the importance of adherence to the secondary prophylaxis.

The goal of RHD treatment is to minimize the damage caused to the cardiac muscle and the valves by suppressing the host inflammatory responses. Prevention of recurrent streptococcal infections using extended penicillin G prophylaxis is considered the most effective way to hinder the progression of RHD.[6,8] Severe cases of valvular diseases, however, require surgical interventions, such as valve replacement or repair. Rheumatic mitral valve stenosis without regurgitation, arrythmias, and left atrial thrombus is often dealt with by percutaneous mitral balloon valvuloplasty.[9] angiotensin converting enzyme (ACE) inhibitors, diuretics, and beta-blockers are recommended for patients who develop heart failure due to valve disease.[5]

Student Notes

Key take-aways

- ARF is a manifestation of exaggerated immune responses to streptococcal infections, such as strep throat or pharyngitis.
- The most common streptococcal species implicated in RHD are the group A streptococci, such as *Streptococcus pyogenes*, although the involvement of the viridans group streptococci (VGB) is also seen on rare occasions.
- The VGB are known to have low pathogenicity as they form an important part of the normal mucosal flora of the oral cavity.
- The invasion of the bloodstream by the VGB is known to result in life-threatening conditions like septicemia and carditis.
- Valvular diseases are the most common cardiac manifestations of infections caused by the VGB.
- ARF is the primary manifestation seen in cases of untreated streptococcal infections and is diagnosed based on the Jones criteria and an array of other diagnostic tests.
- The treatment of ARF involves primary eradication of the streptococcal infections followed by an extended secondary prophylaxis with penicillin G.
- Adherence to the secondary prophylaxis regime is vital for preventing chronic RHD caused by recurring damage to the cardiac muscle and valves.
- RHD is a chronic life-threatening cardiac condition resulting from repeated inflammatory assaults of the cardiac muscle and valves.
- Minimization of damage to the cardiac muscle and valves by preventing or reducing the host inflammatory responses is the cornerstone of effective management of RHD.
- As there is no known cure for RHD, the management and treatment of the condition relies on symptomatic management of the clinical manifestations.

Problem number 5.2: Cardiomyopathy caused by *Coxsackie B virus* infection

A 63-year-old woman presented with end-stage renal disease (ESRD). She had undergone a heart transplant about 10 years ago after developing cardiomyopathy following a severe coxsackie B virus infection. About 1 year ago, she was diagnosed with acute kidney injury (AKI) and had been undergoing regular hemodialysis for the past year. Her kidney function, however, had deteriorated further, precipitating the need for a kidney transplant.

Ongoing medications

1. Cyclosporine
2. Prednisone
3. Azathioprine
4. Folic acid

The patient had to undergo an ABO-incompatible kidney transplant from a close relation due to the lack of compatible donors. Due to the high possibility of a severe antigen – antibody reaction, the patient underwent plasmapheresis.

Diagnosis

The final diagnosis concluded ESRD precipitated by AKI, which is the most common complication associated with heart transplantation. The coxsackie B virus infection could have been responsible for the cardiomyopathy, which eventually warranted the need for heart transplantation.

Question and answer

1. What is organ/tissue transplantation? What are the various types of organ transplantations?

   Organ/tissue transplantation is a medical procedure that replaces a damaged or poorly functioning vital organ of the human body with a functional donor organ/tissue. Transplants may be autografted, meaning that it is transplanted from one site to another within the same individual's body, or allografted, which means that it is transplanted from a donor to a recipient belonging to the same species. Allografted transplantations may use living or cadaveric donor organs. Transplantations that take place between genetically identical twins are known as isografts, while transplantations between species are known as xenografts.
2. How is organ rejection after transplantation prevented?

   Organ rejection after transplantation is a major clinical concern and can be life-threatening. Therefore organ transplantations are followed by intensive drug regimens to prevent rejection reactions. The general strategy is to suppress the immune system by targeting multiple immunological cascades. As the risk of organ rejection is the highest immediately after the transplantation, the induction therapy immediately after the transplantation involves relatively high doses of the drugs, followed by tapering off the doses for the maintenance therapy. The multidrug approach ensures that the drugs can be effectively given in relatively low doses, which is very important to avoid exposure to the higher toxic doses of the drugs. However, the unavoidable immunosuppression caused by the drugs results in higher predisposition to various infectious diseases, including a spectrum of opportunistic infections. Long-term use of immunosuppressive therapies can also result in other potentially serious complications, such as nephrotoxicity and malignancies of the skin, the lymph nodes, and the lymphoid organs.

3. Which drugs in the drug regimen are used to prevent organ rejections after transplantations? Describe these drugs with regard to their class, mode of action, and side effects.

The drugs included in antiorgan rejection regimes posttransplantation are categorized according to the class, mode of action, and side effects as follows:

Class	Drugs	Mode of action	Side effects[a]
Cytotoxic drugs	Azathioprine, cyclophosphamide, mycophenolate, and leflunomide.	Inhibit nucleic acid synthesis, thereby preventing clonal expansion of activated lymphocytes.	Nausea, vomiting, skin rash, dizziness, fatigue, anemia, pancreatitis, and skin cancer.
Corticosteroids	Methylprednisolone, prednisone.	Suppresses the migration of polymorphonuclear leukocytes and reverses increased capillary permeability, thereby reducing inflammation and decreasing the activity of the immune system.	Headache, dizziness, slowed wound healing, mood swings, acne, bulging eyes, personality changes, sleep cycle changes, fragile skin, subdermal bruising, menstrual cycle changes, irregular fat distribution, increased hair growth, fatigue, increased sweating, reduced sex drive, and heartburn
Calcineurin inhibitors	Cyclosporine, tacrolimus, and pimecrolimus.	Downregulates T-cell activation by reducing the production of IL-2 and decreasing the expression of IL-2 receptors.	Nephrotoxicity, neurotoxicity, hypertension, lymphoma, hyperkalemia, hyperlipidemia, hyperuricemia, hypomagnesemia, hirsutism, glucose intolerance, gum

Continued

Continued

Class	Drugs	Mode of action	Side effects[a]
			hyperplasia, and hepatotoxicity.
Inhibitors of the mammalian target of rapamycin	Sirolimus, everolimus, and temsirolimus.	Forms a complex with the intracellular protein, FKBP-12, and prevents it from interacting with the raptor-binding protein, thereby interfering with downstream cell signaling.	Stomatitis, rash, anemia, decreased appetite, fatigue, nausea, diarrhea, hyperglycemia, and interstitial lung disease.
Polyclonal antithymoblast/ antilymphocyte globulin and antithymocyte globulin.	Lymphocyte immune globulin, and antithymocyte globulin.	Depletes the circulating population of T-cells and B-cells and also accelerates T-cell apoptosis and complement-mediated lysis.	Nausea, vomiting, diarrhea, skin rash, itching, headache, dizziness, shortness of breath, fatigue, myalgia, swelling of feet, joint pain, tingling/numbness of peripherals, tremors, polyuria, polydipsia, severe abdominal pain, black/bloody stools, personality changes, mood swings, seizures, arrythmias, jaundice, dark urine, anemia, and subdermal bruising.
Monoclonal antibody therapy.	OKT3, daclizumab, and basiliximab.	Antigen-targeted binding to CD25 of the IL-2 T-cell receptor, thereby interrupting T-cell activation and proliferation while preventing unwanted effects on neutrophils, platelets, or T-cells targeted to other antigens.	Fever, chills, hypotension, urticaria, dyspnea, rash, bronchospasm, and emesis.

[a]*Increased susceptibility to infections is a common side effect of all the drugs used to prevent posttransplantation organ rejections due to the immunosuppressive nature of the drugs.*

4. What is ABO-compatibility?

   The blood groups A, B, AB, and O are characterized by the presence of specific antigens on the surface of the red blood cells. Compatibility between recipient and donor blood groups is a very important criteria for successful blood transfusions and organ/tissue transplantations, and is known as ABO-compatibility. ABO-compatibility ensures that transfusions or transplantations do not elicit potentially life-threatening immunological reactions, which can result in the rejection of the transfusion or transplantation.

5. What is human leukocyte antigen (HLA)-compatibility?

   HLAs are specific antigens present on the surface of white blood cells and tissues, also known as the major histocompatibility complex (MHC) molecules. MHC molecules are encoded by a polymorphic set of genes present on the short arm of chromosome 6, known as the MHC complex. The HLA proteins are important regulators of the immune system, and therefore compatibility between the donor and recipient HLA is very important to ensure the success of tissue/organ transplantation. This is known as HLA-compatibility. HLA-incompatibility in tissue/organ transplantations can precipitate potentially life-threatening immunological reactions, resulting in the rejection of the transplanted organ/tissue.

   The MHC antigens are grouped into two classes: MHC class I and class II. While MHC class I comprises of HLA-A, HLA-B, and HLA-C antigens, MHC class II comprises of HLA-Dr, HLA-DP, HLA-DM, HLA-DO, and HLA-DQ antigens. Organ transplantation relies mostly on compatibility in the HLA-A, HLA-B, and HLA-Dr antigens, and HLA typing is therefore highly relevant in ensuring HLA compatibility in organ/tissue transplantation.

While histocompatibility is very important for the success of transplantations, the advent of immunosuppressive therapy has facilitated organ transplants even in cases of HLA incompatibility. Histocompatibility testing is, however, important in the process of screening donors for the best match and also to check for the presence of preformed antibodies in the recipient against the donor antigens to prevent hyperacute graft rejection. Such preformed antibodies may exist due to prior exposure to incompatible HLA antigens during pregnancy, blood transfusion, or previous organ transplant. Histocompatibility testing involves performing a crossmatch between the recipient's serum and the donor lymphocytes.

6. What is the importance of plasmapheresis in cases such as this one?

   ABO-incompatible organ transplants carry a high risk of organ rejection due to hyperacute immunological reactions caused by antibodies circulating in the recipient's system against the donor antigens. Plasmapheresis is a therapy by which the load of the circulating antibodies that could potentially react against the donor antigens are removed from the recipient's system before the organ transplantation, thus decreasing the possibility of organ rejection.

Learning objectives

1. Categorize the drugs included in immunosuppressive drug therapy based on the pharmacokinetics, pharmacodynamics, and drug toxicity aspects.
2. Explain the process of plasmapheresis.
3. Assess the role of plasmapheresis in ABO-incompatible transplants.
4. Summarize the immunological aspects of organ transplantation and rejection.
5. Define tissue typing.
6. Recognize the risks associated with anti-HLA antibodies.
7. Describe the process of HLA cross-matching.

Discussion

Differential diagnosis

Urinary tract infection	Burning sensation during micturition, increased urge to urinate, increased frequency of micturition, inability to empty bladder (passing only small amounts of urine), foamy and turbid urine appearance, dark (red/pink/reddish brown) urine (hematuria), pelvic pain, strong-smelling urine, high-grade fever, chills, loss of appetite, nausea, vomiting, listlessness, and upper back and flank pain.
Atheroembolic disease	Atheroembolic disease can present with discoloration of feet and toes, abdominal pain, diarrhea, purpuras, confusion, weight loss, fever, gradual or acute onset kidney failure, and myalgia.
Tubulointerstitial disease	Tubulointerstitial disease may manifest with fever, pain in the lower back or flank side, and painful urination.
Myeloma kidney	Myeloma kidney presents with a very vague set of symptoms, including loss of appetite, weight loss, bone pain, flank pain, lower back pain, anemia, itching, muscle cramps, nausea, vomiting, swelling in lower extremities, trouble sleeping, and shortness of breath.
Renovascular disease	Renovascular disease presents with high blood pressure, kidney dysfunction, narrowing of systemic arteries, headaches, confusion, nausea, and vision changes.

Background

Coxsackie B viruses usually cause infections of the heart, liver, pancreas, and pleura, resulting in inflammatory conditions, such as pleurodynia, myocarditis, pericarditis, and hepatitis. The pathologies associated with coxsackie B virus infections are mediated through spastic paralysis caused by the degeneration of the neuronal tissue and focal muscle injury.[1]

Cardiomyopathies associated with coxsackie B virus infections often result in fulminant heart failure.[1] While heart transplantations provide long-term solutions to potentially life-threatening cardiac pathologies, the adverse effects associated with the immunosuppressive therapies mandated posttransplantation to prevent organ rejection reactions can be extremely severe. Sustained use of calcineurin inhibitors, such as cyclosporine, is known to be nephrotoxic, often resulting in AKI. AKI leads to the activation of inflammatory cytokines, which in turn causes hyperfiltration. Glomerulosclerosis and tubulointerstitial fibrosis, resulting from the hyperfiltration, eventually lead to ESRD. While the incidence of ESRD in heart transplant patients has been found to be quite rare within 5 years of transplant, the incidence increases between 5 and 10 years of transplant, with the highest incidence occurring after 10 years of transplant.[2]

Etiological agent

Coxsackie B virus

Classification and morphology

Coxsackie viruses are pathogenic, nonenveloped, positive-sense, single-stranded RNA (ssRNA) viruses belonging to the *Picornaviridae* family of the *Enterovirus* genus. The classification into the *Enterovirus* genus is based on the enteric or gastrointestinal route of transmission, and true to their classification in the *Picornaviridae* family, Coxsackie viruses are known for their small icosahedral structure. Coxsackie viruses are divided into group A and group B viruses based on their pathogenicity. While the group A Coxsackie viruses induce flaccid paralysis-based pathologies, the group B Coxsackie viruses induce spastic paralysis-based pathologies. Six serotypes[1–5,10] of Coxsackie B viruses namely CV-B1, CV-B2, CV-B3, CV-B4, CV-B5, and CV-B6 have been recognized.[10]

Epidemiology

Coxsackie B virus infections are pathogenic to humans irrespective of gender and geographical location. Although seasonal changes do not affect the epidemiology of Coxsackie B virus infections, the rate of incidences is generally higher in warm temperatures prevalent during the summer months. Serotypically, Coxsackie B4 viruses are associated with higher mortality rates compared to the other serotypes, while the B1 serotype has been found to be the predominant circulating strain in recent times. Unlike gender, age has been known to play a significant role in the pathogenicity associated with coxsackie B infections.[3] The disease is extremely pathogenic to neonates, while older children and adults are less prone to facing the severe effects of the disease.[4] The Coxsackie B virus, especially the B3 serotype, is known to be the predominant viral etiology for viral myocarditis and cardiomyopathy.[5,6]

Pathogenicity and pathogenesis

The gastrointestinal system and the spleen are the primary sites of replication for the Coxsackie B virus, from where the virus eventually spreads to the heart, its target organ. Coxsackie B viruses are typically known to be cytolytic viruses that mediate their pathogenesis predominantly through certain receptor proteins, coxsackie-adenovirus receptor (CAR) and the decay-accelerating factor (DAF). While DAF is expressed in epithelial and endothelial cells, CAR is present in intercalated disks that facilitate the transmission of signals between myocardial cells. Myocarditis results from the entry of the Coxsackie B virus into the myocardial cells, which is facilitated by the interactions with these receptors.[7] The viral replication induces an influx of leukocytes into the cardiac muscle, resulting in an autoimmune cascade, which eventually injures the infected as well as the healthy cardiac cells. Once triggered, the autoimmune cascade continues to mediate the inflammatory processes, causing catastrophic destruction of the cardiac muscle even after the viral particles have been cleared out of the system. This results in myocarditis and eventual fulminant heart failure. Virus-induced myocarditis has also been implicated in the development of idiopathic dilated cardiomyopathy.[8]

These specific virus – receptor interactions act as the deciding factor, influencing the dissemination of the infection to sites other than the primary one. The spread of coxsackie B virus infections to the central nervous system (CNS) are facilitated either by hematogenous spread or by axonal transport. Damage to the neurons and the tissues of the CNS is attributed to the ability of the Coxsackie B virus to inhibit the production of cellular macromolecules, cell apoptosis induced by viral replication, toxicity of the viral proteins, and possible activation of the autoimmune system.[9]

Coxsackie B virus associated with systemic infections in neonates is usually more severe than in older children and adults.[11] Besides the cardiological and neurological pathologies of the Coxsackie B virus, disseminated infections can also cause exanthema, pleurodynia, pancreatitis, and type I diabetes depending on the organ to which the infection spreads from the primary site of infection.[12]

Immune responses

The pathogenesis of viral myocarditis is a function of the direct cytopathic effects of viral replication as well as the indirect inflammatory effects of the host's autoimmune responses. Myocarditis and dilated cardiomyopathy are the eventualities resulting from the combined effects of T-cell-mediated immunity, antibody-mediated immunity, and innate immunity. While the immune responses in coxsackie B virus infections are important for preventing chronic infection by limiting the spread and eliminating the virus, uncontrolled and excessive immune responses, elicited by the autoimmune cascade, are responsible for the wide-spread destruction of the cardiac tissue, precipitating in cardiomyopathy and fulminant heart failure. The precarious balance between the protective antiviral effects and the damaging immunopathological processes makes the host's immune responses, elicited in coxsackie B virus infections, a double-edged sword.[13]

The translation process involved in the replication of the Coxsackie B virus requires the cell to be in the G1/S phase of the cell cycle, thereby implying that the host's response to the virus is important for the viral replication process.[14] The possibility that the viruses induce the infected cells to enter the cell cycle to facilitate their replication can also be considered. The host's immune responses are initiated through the cross-linking between viruses and the specific cellular receptors that result in signal transduction through the toll-like receptors (TLR), which recognize the ssRNA and double-stranded (dsRNA) viral RNA. TLRs are potent inducers of immune responses that facilitate the proliferation of immune cells and cytokine/chemokine expression.

Although the adaptive immune response is the ultimate facilitator of infection clearance, and is indispensable for preventing chronic infections, the innate immune response plays a very important role in controlling the infection until the adaptive immune responses develop. In addition, the innate immune responses also mediate the development of the virus-specific adaptive immune responses, thus ensuring efficacy of the immune response to the infection.

Clinical manifestations and complications

Coxsackie B virus infections can present with a wide spectrum of symptoms, including gastrointestinal distress, extreme fatigue, nausea, vomiting, headache, fever, chest pain, sore throat, and myalgia. Age and immune status of the individual are very important determinants of the primary site of infection, severity of infection, and clinical manifestations of the disease. Although the incubation period also depends on several variables, 15 days have been found to be the average period during which the spread of the infection is the highest.[15]

Coxsackie B viruses are strongly associated with Bornholm disease or pleurodynia. Chest pain that increases in severity while breathing is a characteristic feature of pleurodynia, but the severity of the pain is such that it is often misdiagnosed as myocardial infarction. The duration of the pain varies between 2 days and 2 weeks during which patients may also present with a fever, tenderness in the muscles of the trunk and the extremities, and abdominal pain.[16]

Neurological manifestations of coxsackie B virus infections include aseptic meningitis, which often presents with varying symptoms depending on the age of the affected individual. Fever is the common presentation, with headaches, nausea, vomiting, and a stiff neck being seen in adults and older children, while irritability is a common characteristic in infants affected by coxsackie B virus infections.[17,18] Severe cases of neurological complications often manifest encephalitis, which presents with confusion and seizures in addition to the spectrum of neurological symptoms seen in cases of aseptic meningitis.

Myopericarditis and myocarditis (Fig. 5.6), resulting in idiopathic dilated cardiomyopathy, are caused by the damage inflicted on the subepicardial myocardium and pericardium due to the aggressive, uncontrolled, and long-lasting inflammatory responses of the host's immune system. Dyspnea, chest pain, and arrythmias are common presentations of the condition. Due to the generic nature of the symptoms, diagnosis poses a challenge. The clinical manifestations can last for up to a few months, thereby undermining the severity of the condition. While the condition may be benign in some cases,

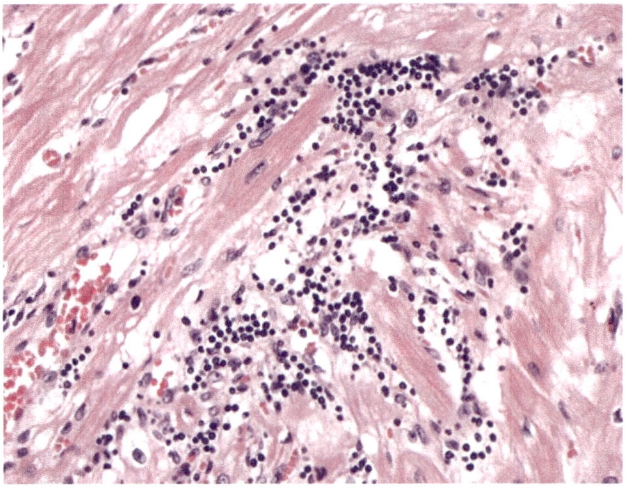

FIGURE 5.6
Histopathological image of viral myocarditis. Source*: KGH/Own work/; https://creativecommons.org/licenses/by-sa/3.0/deed.en; https://commons.wikimedia.org/wiki/File:Viral_myocarditis_(2).JPG.*

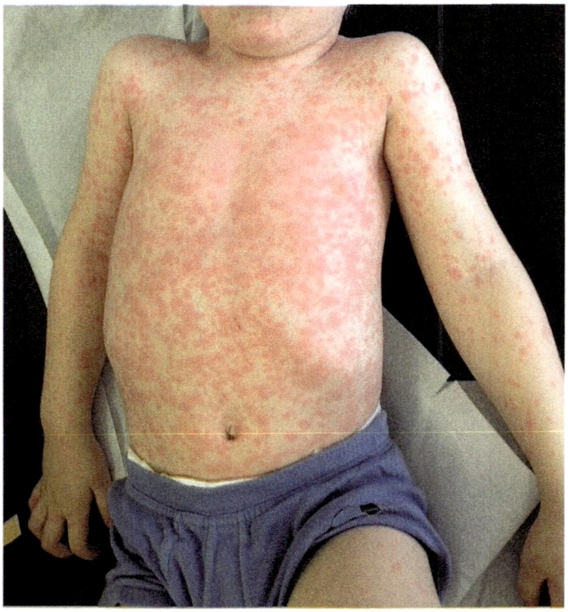

FIGURE 5.7
Representative image of exanthema. Source: *Gzzz/Own work/; https://creativecommons.org/licenses/by-sa/4.0/deed.en https://commons.wikimedia.org/wiki/File:Allergic_exanthema_amoxicillin.jpg.*

severe cases can also lead to dilated cardiomyopathy, fulminant heart failure, and also cardiac arrest.[19]

Some cases of coxsackie B virus infections present themselves with exanthema (Fig. 5.7)—a skin condition that causes maculopapular eruptions with an erythematous halo along its periphery. The lesions first appear on the face and spread all over the body, except the palms and soles, within 24 hours. Petechiae and cervical adenopathy also accompany the spectrum of clinical manifestations seen in coxsackie B virus infections.[16,20]

Pancreatitis (Figs. 5.8 and 5.9) often occurs in cases where pancreas is the primary site of infection or if the infection has spread to the pancreas. Clinical manifestations that resemble those of gastrointestinal distress, such as abdominal pain, nausea, and vomiting, are the most common symptoms of pancreatitis. In addition, laboratory studies indicating elevated pancreatic enzymes, amylase, and lipase imply an inflammation of the pancreas. Severe pancreatitis caused by coxsackie B virus infections often results in pancreatic beta-cell destruction, precipitating the development of autoimmune conditions, such as type I diabetes mellitus in the long run.[12]

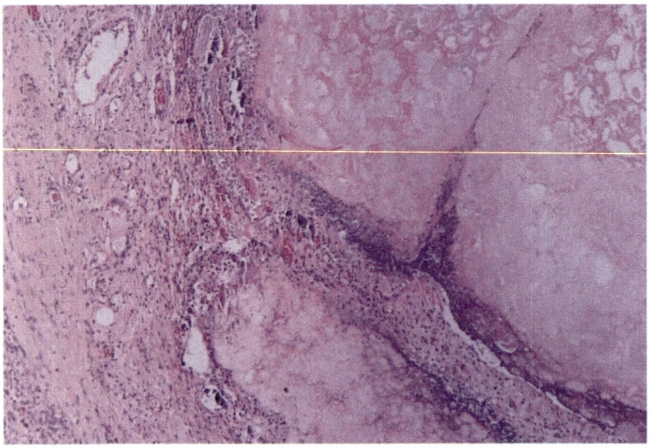

FIGURE 5.8

Tryptic fat tissue necrosis in severe pancreatitis. Source: *Patho/Own work/; https://creativecommons.org/licenses/by-sa/3.0/deed.en; https://commons.wikimedia.org/wiki/File:Tryptic_fat_tissue_necrosis_in_severe_pancreatitis_HE_2.JPG.*

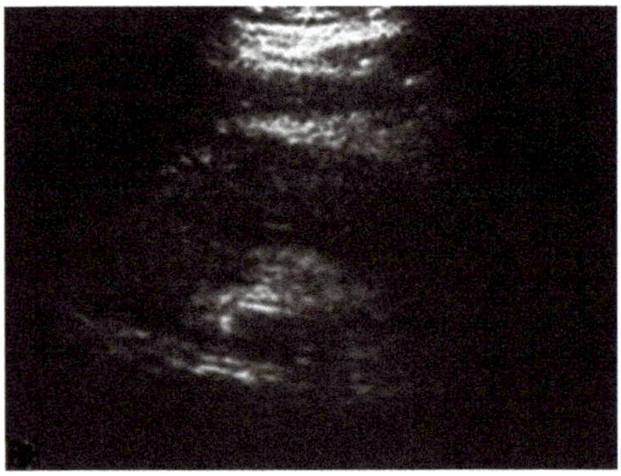

FIGURE 5.9

Ultrasonography image of acute pancreatitis. Source: *Dr. Dileep Kumar Jha, Dr. Mahesh Vadel, Dr. Rahul Sharma/; https://www.ijsr.net/archive/v5i8/ART2016927.pdf/; https://creativecommons.org/licenses/by/4.0/deed.en; https://commons.wikimedia.org/wiki/File:Ultrasonography_of_acute_pancreatitis.jpg.*

Diagnosis

Coxsackie B virus infections are usually self-limiting, and a clinical diagnosis of the infection based on patient history and the clinical manifestations is sufficient. However, severe cases riddled with complications are associated with the cardiovascular and the neurological systems, and cases of systemic neonatal infection warrant identification of the specific etiological organism.

Laboratory investigations that facilitate the diagnostic process include complete blood count, erythrocyte sedimentation rate, lactic acid levels, procalcitonin, and complete metabolic profile. In addition, specific tests, such as chest X-rays, abdominal CTs, head CTs, and lumbar punctures to evaluate the CSF cell count, protein, and glucose levels are recommended for cases suspected of cardiac and neurological complications.

Reverse transcriptase polymerase chain reaction (RT-PCR) is the gold-standard diagnostic method used to detect the specific etiological agent in suspected coxsackie virus infections, taken from urine, serum, and CSF samples due to the time-efficient and sensitive nature of the test, although serotype identification cannot be achieved through RT-PCR.[21,22] Serotype identification warrants the need for diagnosis through viral cultures, although this technique is considered obsolete because of its time-consuming nature.[23] Serological tests are not used in the usual clinical course of diagnosis, but are used during epidemic situations or in cases where the involvement of a particular strain is suspected.[24]

Treatment and prevention

Coxsackie B virus infections are usually self-limiting in immunocompetent individuals, and therefore most cases do not require any specific treatment. Supportive care may be provided depending on the symptoms for symptomatic relief. Neurological complications often warrant the need for antiepileptics to control seizures and sedation to manage delirium.[25] Cases suspected or diagnosed with myocarditis or other cardiac complications may require supplemental oxygen and fluid resuscitation. Analgesics are often prescribed to control the inflammatory damage to the cardiac muscle. Lifestyle and dietary changes aimed at reducing activity levels and decreasing salt intake are recommended for myocarditis cases to prevent additional pressure on cardiac function. The more severe cases indicative of fulminant heart failure and arrythmias may also need specific therapeutic measures. Heart transplants may also be required as the last resort in severe cases of heart failure.

The possibility of severe manifestations and complications associated with coxsackie B virus infections warrant the need for effective preventive measures. As the disease is transmitted via the fecal – oral route, maintaining high standards of sanitation and hygiene, such as good hand washing practices, can help reduce the incidence of coxsackie B virus infection to a large extent.

There are some vaccine candidates, such as PRV-101, that are being investigated for their efficiency in preventing coxsackie B virus infections. These vaccine candidates are being developed with the aim to delay or prevent the onset of autoimmune conditions, such as type 1 diabetes and celiac disease, that may be potentially triggered by coxsackie B virus infections. The principle involved in the development of these vaccines is to prevent autoimmunity by interfering with the initial trigger associated with the viral infection.[26]

Student Notes

Key take-aways

- Coxsackie B viruses are pathogenic enteroviruses transmitted through the fecal – oral route.
- While most infections are self-limiting in immunocompetent individuals, complications related to the cardiac, gastrointestinal, and neurological systems are often associated with coxsackie B virus infections.
- Myocarditis leading to fulminant heart failure is a common complication of coxsackie B virus infections, and often warrants the need for extreme interventions, such as heart transplants.
- Meningitis and encephalitis may also result from the spread of the infection to the neurological system.
- Autoimmune conditions such as type 1 diabetes are known to be triggered by severe pancreatitis caused by coxsackie B virus infections.
- The pathologies of the coxsackie B virus infections are exacerbated by the host's immune responses, which continue to aggressively attack the tissues long after the infection has been cleared, resulting in inflammatory damage and destruction of host cells, eventually causing severe consequences such as organ failure.
- Due to the self-limiting nature of the infection, treatment is usually aimed at symptomatic relief and supportive care.
- Certain complications and severe forms of the disease may require antiinflammatory treatment and interventions, such as supplemental oxygen, and heart transplants in cases of fulminant heart failure.
- Prevention of infection is possible through maintenance of high standards of sanitation and hygiene.

Problem number 5.3: Cardiovascular complications caused by SARS-CoV-2 infection

A 35-year-old male had fever, body pain, and a productive cough for 4 days. On primary evaluation, no other abnormalities were observed, and the patient responded well to medication, which resulted in the patient being discharged. However, 3 days later, the patient returned with a high-grade fever, dry cough, and dyspnea, without any complaints of congestion or rhinorrhea. Chest radiography was performed, and the patient was prescribed amoxicillin and clavulanate (875/125 mg every 12 hours). The patient was then discharged. Two days later, the patient returned complaining of worsened dyspnea, extreme malaise and fatigue, an increased dry cough, and severe sweating, although the fever had subsided. Physical examination revealed decreased bilateral breath sounds and the chest X-ray demonstrated a worsened respiratory condition, evident from the bilateral radiopacity in lung fields, which showed slight improvement on providing oxygen supplementation. The patient was admitted to the hospital for further course of action.

The patient was diabetic but was not on regular treatment. There was no history of travel abroad or contact with individuals with COVID-19-associated symptoms.

The following are the reports of the investigations conducted:

1. Vitals
   Temperature: 37.5°C (day of admission); 38.4°C (next day)
   Blood pressure: 120/70 mmHg (day of admission and next day)
   Heart rate: 92 bpm (day of admission); tachycardiac—105 bpm (next day)
   Glasgow coma score: 15
   85% oxygen desaturation (3 L/min) (next day)
   Tachypneic—85 ipm (next day)
2. Physical examination
   General malaise
   Eupneic in oxygen catheter
   Acyanotic
3. Respiratory examination: Globally decreased vesicular murmur with sparse diffuse snores
4. Cardiovascular examination:
   2-stroke regular heart rhythm
   Tachycardiac heart rhythm with muffled heart sounds
   Chest pain
   Jugular venous distension with signs of diastolic dysfunction
   Vesicular murmur with crackles throughout the right hemithorax
5. Radiological investigations
   Chest X-ray: Bilateral radiopacity pronounced in the lung fields
   Electrocardiography: No signs of ischemia
   Computed tomography of the chest (day after admission):
   Selective intubation of right mainstem bronchus
   Extensive ground-glass opacity in lower lobes of both lungs demonstrating consolidation with air bronchograms suggestive of inflammation/infection
   Myocardial edema with myocardial wall thickening and a slight increase in the cardiac area

6. Hematological profile

Test	Day of admission	Day after admission	Reference values
Hb (g/dL)	15.5	14.4	13.5–17.5
WBC (/L)	16.15×10^9	15.05×10^9	$4.3–10.8 \times 10^9$
Band cells (%)	3.23%	---	< 10%
Platelets (/L)	---	144×10^9	$150–400 \times 10^9$
Hematocrit (%)	---	42	33–40

7. Biochemical profile

Test	Day of admission	Day after admission	Reference values
Glucose (mg/dL)	344	---	200–350
Sodium (mEq/L)	135	135	135–145
Potassium (mmol/L)	4.3	4.2	3.6–5.2
Urea (mmol/L)	2	36	3–6.3
Creatinine (mg/dL)	0.6	0.60	0.9–1.3
C-reactive protein (mg/L)	23.6	25	< 10

8. Quantitative troponin: 0.49 ng/dl
9. D-dimer: 0.1 mg/L
10. Venous lactate: 2.0 mmol/L
11. Rapid HIV test: Negative
12. Rapid COVID-19 test (day after admission): Negative
    COVID-19 PCR (collected post-mortem): Positive
13. Vitals following orotracheal intubation the day after the admission
    Blood pressure: 154×91 mmHg
    HR: 48 bpm
    Sat O_2: 82%
    FiO_2: 100%
    Peep: 12 bpm
14. Cardiovascular condition on the 2nd day after admission
    Supported on mechanical ventilation and sedated
    Tachycardic (hypophonesis of b1 and b3 in outline)
    Tachypneic (Ramsey 6)
    Presence of jugular venous distention
    Edema on the lower limbs (1/4 +)
    Vesicular murmur with diffuse snores

Diagnosis

The diagnostic hypotheses comprised of pneumonia, decompensated diabetes, pulmonary focus sepsis of potential viral etiology, and acute myocarditis.

Treatment

Orotracheal intubation to ensure maintenance of definitive airway.

Broad-spectrum antibiotic therapy.

Measures to prevent acute respiratory failure.

Hemodynamic monitoring.

Vasoactive drugs initiated.

The patient, however, died due to respiratory failure elicited by COVID-19 infection despite all the treatment and management measures.

Question and answer

1. Why are the SARS-CoV-2 viruses known as coronaviruses (CoVs)?
   The spike protein (S) of the SARS-CoV-2 virus is a trimeric form protein embedded in the viral envelope. The family of viruses comprising of the SARS-CoV-2 virus is known as "coronaviruses" due to the characteristic "crown-like" appearance resulting from this arrangement of the S-protein on the viral surface.
2. What is the host receptor for the SARS-CoV-2 virus?
   The SARS-CoV-2 virus has an affinity for the ACE-2 receptor expressed on the surface of host cells of various organs, such as lungs, heart, kidneys, and gastrointestinal tract. The ACE-2 receptor plays a central role in initiating the process of host cell invasion by SARS-CoV-2.
3. What is the role of the two components of S-protein in facilitating the invasion of host cells by the SARS-CoV-2 virus?
   The two components of the S protein (S1 and S2) are crucial for the attachment and entry of the virus into the host cell. The S1 component of the S protein present on the viral surface mediates the attachment of the virus to the ACE-2 receptors on the host cell surface. The attachment of the virus on the host cells is followed by the S2 component of the S protein penetrating it, thereby facilitating

the entry of the virus into the infected host cell through endocytosis or membrane fusion. The S-protein undergoes protease-mediated cleavage to ensure accurate functionality of the two components involved in the fusion of the viral and host cell membranes. The first part of the cleavage process primes the S1/S2 cleavage site and stabilizes the S2 subunit, while the second part elicits conformational changes in the S2 subunit, which eventually results in the fusion of the viral and host cell membranes.

4. What is the phenomenon of the "cytokine storm" that increases the morbidity and mortality associated with COVID-19?
   COVID-19 infected immunocompromised individuals are usually capable of generating adequate and well-controlled immune responses that can effectively contain the pathology of the virus, but the invasion of the type II alveolar epithelial cells by the SARS-CoV-2 virus elicits a deranged immune response. An uncontrolled generation of pro-inflammatory cytokines, such as IFN-α, IFN-γ, IL-1β, IL-6, IL-12, IL-18, IL-33, tumor necrosis factor-alpha (TNF-α), TGFβ, and chemokines CCL2, CCL3, CCL5, CXCL8, CXCL9, CXCL10 can precipitate in the phenomenon of "cytokine storm," which causes extensive systemic inflammation and can potentially result in acute respiratory distress syndrome (ARDS). The cytokines and inflammatory markers act as chemoattractants for neutrophils, CD4 helper T-cells, and CD8 cytotoxic T-cells. Although these cells are crucial for host defense against the viral invasion in the lung tissue, the uncontrolled release of proinflammatory cytokines causes inflammation and damage to the lung tissue. Respiratory distress syndrome is a result of the diffuse alveolar damage caused by persistent injury to both the type 1 and type 2 alveolar epithelial cells.
5. How is respiratory distress syndrome caused by COVID-19 infection associated with myocardial damage?
   Severe pneumonia and respiratory distress syndrome results in persistent hypoxia, which increases the oxygen requirement. The inadequate oxygen supply results in myocardial damage. The resulting acute coronary syndrome may be attributed to acute plaque rupture and the subsequent catecholamine surge. Electrolyte disbalance and hypokalemia have been implicated in arrythmias, and are therefore responsible for cardiac manifestations in COVID-19.
6. What is the characteristic cytokine profile of host immune response to COVID-19 infections?

The characteristic cytokine profile of the host immune response to COVID-19 infections comprises of an increased number of total neutrophils, elevated levels of interluekin-6 (IL-6) and C-reactive protein, increased expression of proinflammatory cytokines and chemokines, such as IP-10, monocyte chemoattractant protein-1 (MCP-1), macrophage inflammatory protein-1α (MIP-1A), and TNF-α.

7. How do humoral immune responses contribute to recovery from COVID-19 infection?

   Cytokines, such as IL-12, induce CD4 + T-cells to activate B-cells, thus eliciting humoral immune responses, resulting in the generation of antigen-specific antibodies, comprising of IgG and IgM. While IgM lasts only for the initial 12 weeks of the infection, IgG persists for a longer duration.

8. What are the characteristic features of high-resolution computed tomography scans of COVID-19-infected individuals?

   The appearance of bilateral multifocal "ground-glass" regions largely concentrated in the lower lobes, associated with consolidation and patchy peripheral distribution, is characteristic of lung tissue damage. The identification of focal opaque patches surrounded by a peripheral consolidation ring results in the formation of a "reversed halo," another characteristic feature of the disease. Computed tomography also facilitates the detection of some other pathological manifestations, such as pleural effusion, cavitation, calcification, and lymphadenopathy.

Learning objectives

1. Determine the appropriate differential diagnosis based on patient history, known underlying medical conditions, and clinical manifestations.
2. Review the epidemiology of SARS-CoV-2 infections.
3. Describe the characteristics of SARS-CoV-2.
4. Summarize the pathogenesis of SARS-CoV-2.
5. Describe the clinical manifestations of COVID-19.
6. Predict the prognosis of a COVID-19 case with cardiac manifestations based on the extent of the pathology presented.
7. Enumerate the methods involved in the diagnosis of COVID-19 infection.
8. Summarize the management and treatment options for a COVID-19 case associated with cardiac manifestations.

Discussion

Differential diagnosis

Cardiogenic edema	Cardiogenic edema presents itself with extreme shortness of breath that results in a feeling of suffocation when lying down, wheezing, gasping for breath, and blood-tinged frothy sputum.
Pulmonary infarctions	Pulmonary infarctions generally present themselves with a fever, severe dyspnea, chest pain that worsens on breathing, hemoptysis, dizziness, weakness, and in some cases, persistent hiccups.
Alveolar hemorrhage	Alveolar hemorrhage presents itself with hypoxemia, dyspnea, and a cough.
Eosinophilic pneumonia	Eosinophilic pneumonia presents itself with dyspnea, a cough, myalgia, fatigue, night sweats, and chest pain that worsens in the supine position.
Drug-induced pneumonitis	Drug-induced pneumonitis presents itself with a rash, acute breathlessness, wheezing, fever, and peripheral eosinophilia.
Radiation pneumonitis	Radiation pneumonitis presents itself with a fever, chest congestion, a cough, chest pain, and dyspnea.
Hypersensitivity pneumonitis	Hypersensitivity pneumonitis presents itself with a fever, chills, myalgia, headaches, a cough, chronic bronchitis, rales, dyspnea, anorexia, weight loss, clubbing of digits, fatigue, and lung fibrosis in extreme cases.
Nonspecific interstitial pneumonia	Nonspecific interstitial pneumonia is a rare lung disorder that presents itself with breathing difficulties, dyspnea, a dry cough, and fatigue.
Adenocarcinoma	Adenocarcinoma presents itself with abdominal pain, nausea, vomiting, unexplained weight loss, anemia, fatigue, jaundiced appearance, and blood-stained or dark-colored stools.
Alveolar proteinosis	Alveolar proteinosis presents itself with chest pain, fever, weight loss, dyspnea, a cough, clubbed growth of finger and toe nails, and low levels of oxygen in blood.

Background

COVID-19 is a novel viral respiratory disease that emerged in 2019. The etiological agent responsible for the disease belongs to the family of CoVs, thereby contributing to the name of the disease, which is essentially an acronym for "coronavirus disease 2019." The homology of the novel CoV, responsible for COVID-19, caused the respiratory pandemic in the last decade, leading scientists to name it "severe acute respiratory syndrome coronavirus 2" (SARS-CoV-2). Although COVID-19 is considered a primarily zoonotic disease, human-to-human transmission has been largely responsible for its community spread globally.

Etiological agent

Severe acute respiratory syndrome coronavirus 2 (SARS-CoV-2)

Classification and morphology

CoVs are the largest group of viruses implicated in respiratory and gastrointestinal infections. Viruses belonging to the *Nidovirales* order in the subfamily of *Coronaviridae* are categorized into four genera, namely *Alphacoronavirus, Betacoronavirus, Gammacoronavirus, and Deltacoronavirus*. SARS-CoV-2 is an enveloped RNA virus species belonging to the *Betacoronavirus* genus of the *Coronaviridae* family.

SARS-CoV-2 viruses are spherical, with a diameter of about 125 nm and contain a nonsegmented, positive-sense, ssRNA at its core. The structural proteins comprise of hemagglutinin esterase (HE), spike (S), small membrane (E), membrane (M), internal (I), and nucleocapsid (N) proteins. The complex of the nucleocapsid protein (N) and the RNA genome forms helically symmetrical capsid structures within the viral envelope. Such helically symmetrical nucleocapsids are commonly seen in negative-sense RNA viruses, but they are unusual among positive-sense RNA viruses. The spike protein (S) in a trimeric form is found to be embedded in the viral envelope, which gives the characteristic "crown-like" appearance,[27] prompting the name "coronavirus" (Fig. 5.10). The two components of the S protein (S1 and S2)

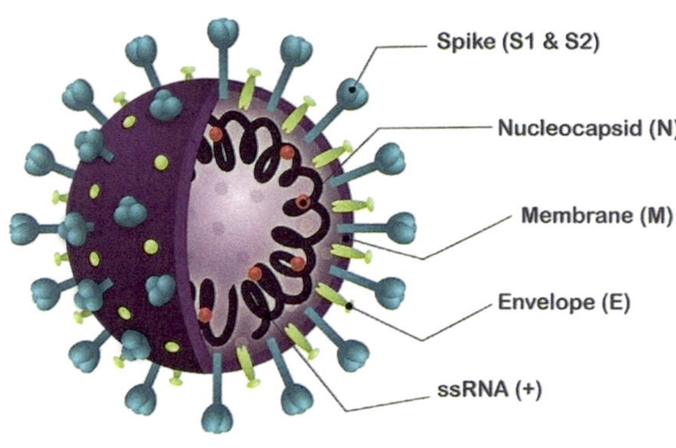

FIGURE 5.10

Schematic structure of SARS-CoV-2. Source*: Igor de Andrade Santos, Victória Riquena Grosche, Fernando Rodrigues Goulart Bergamini, Robinson Sabino-Silva, and Ana Carolina Gomes Jardim/Santos IA, Grosche VR, Bergamini FRG, Sabino-Silva R and Jardim ACG (2020) Antivirals Against Coronaviruses: Candidate Drugs for SARS-CoV-2 Treatment? Front. Microbiol. 11:1818. https://doi.org/10.3389/fmicb.2020.01818/; https://creativecommons.org/licenses/by/4.0/deed.en; https://commons.wikimedia.org/wiki/File:Fmicb-11-01818-g001.jpg.*

are crucial for the attachment and entry of the virus into the host cell.[28] The HE protein forms smaller spikes at the membrane in some CoVs, while the M and E proteins are crucial in the process of virus assembly.[29]

Unlike the small genomic size of RNA viruses, viruses belonging to the *Nidovirales* order have a large genomic size, with the viruses belonging to the *Coronavirinae* subfamily having the largest known RNA genomes.[30,31]

Epidemiology

CoVs are primarily implicated in zoonotic diseases; however, seven CoV species are known to infect humans. While Middle East respiratory syndrome (MERS)-CoV and SARS-CoV cause severe human diseases, the other five viruses generally precipitate mild respiratory manifestations akin to those seen in common cold in immunocompetent individuals. Although a lot is still unknown about the clinical severity of COVID-19, there is ample evidence indicative of the potentially life-threatening prognosis of the disease.[32]

Although CoVs are known for their ability to adapt to a wide range of hosts, including the avian, the feline, the canine, and the swine species, causing respiratory and gastrointestinal ailments, bats are also considered the natural reservoirs due to the highest prevalence of CoVs in bats.[30,33,34] Mutations and genetic variations usually occur in intermediate hosts, such as farm animals, which in turn transmit the infection to humans.[35,36]

Pathogenicity and pathogenesis

Environmental conditions that affect parameters, such as temperature and relative humidity, influence the duration for which SARS-CoV-2 can survive outside a host in the environment. Evidence suggests that the virus can survive for up to 2 weeks in dry conditions, up to 5 days in conditions or temperatures between 22°C and 25°C and relative humidity between 40% and 50% before a gradual decline is demonstrated in the viability of the virus. Higher temperatures and relative humidity levels decrease the viability duration of the virus. Low temperature and humidity conditions are, therefore, known to favor the survival of SARS-CoV-2 in the environment outside the host.[37,38]

The transmission of SARS-CoV-2 among humans generally occurs due to mucous membranes of the eyes, nose, or mouth coming directly or indirectly in contact with infectious macro- or micro-respiratory droplets or aerosols. The risk of transmission is directly proportional with the period and proximity of exposure with infected individuals.[39] Transmission through fomites, such as contaminated linen, medical equipment and devices, utensils, and other items of personal use is also known. SARS-CoV-2 has an affinity for the ACE-2 receptor expressed on the surface of cells of various organs, such as the lungs, heart, kidneys, and gastrointestinal tract.[40] The ACE-2 receptor plays a central role in initiating the process of host cell invasion by SARS-CoV-2[41–43] (Fig. 5.11). The S1 component of the S protein present on the viral surface mediates the attachment of the virus to the ACE-2 receptors on the host cell surface. The attachment of the virus on the host cells is followed by the S2 component of the S protein penetrating it, thereby facilitating the entry of the virus into the infected host cell through endocytosis or membrane fusion. The S protein undergoes protease-mediated cleavage to ensure accurate functionality of the two components involved in the fusion of the viral and host cell membranes. The first part of the cleavage process primes the S1/S2 cleavage site and stabilizes the S2 subunit, while the second part elicits conformational changes in the S2 subunit, which eventually result in the fusion of the viral and host cell membranes.[40,44–46]

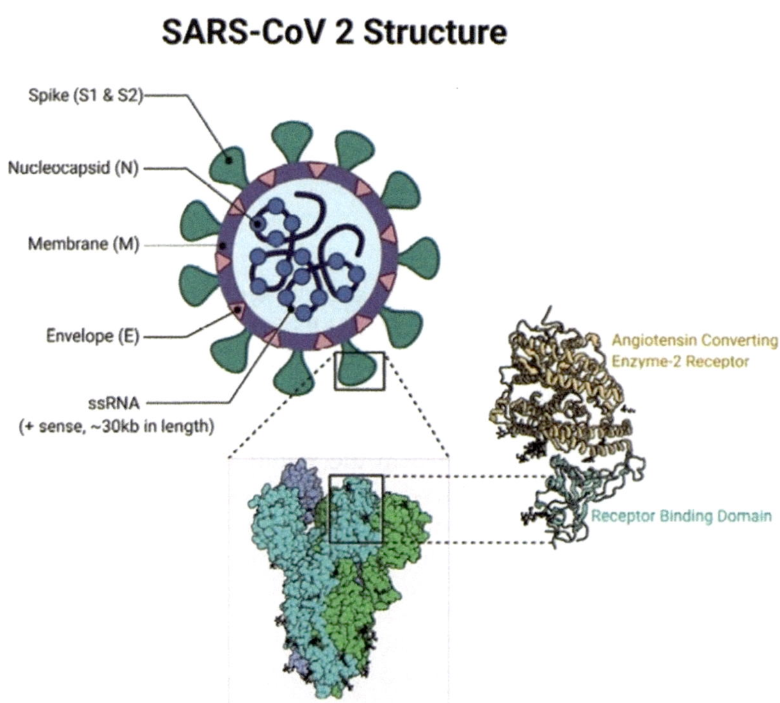

FIGURE 5.11

Molecular architecture of the spike protein and the ACE2-spike protein complex. Source: *Rohan Bir Singh, MD/Marco Cascella; Michael Rajnik; Arturo Cuomo; Scott C. Dulebohn; Raffaela Di Napoli—Features, Evaluation and Treatment Coronavirus (COVID-19)/; https://creativecommons.org/licenses/by/4.0/deed.en; https://commons.wikimedia.org/wiki/File:Struktura_SARS-CoV_2.jpg.*

Subsequent to the entry of the virus into the host cell, the viral RNA and proteins are released into the host cell. The RNA polymerase enzyme transcribes the single-stranded positive sense viral RNA into a negative strand RNA, which subsequently produces new strands of positive RNAs. The process of translation follows in the host cell cytoplasm, resulting in the formation of viral proteins. The viral M protein integrates the viral N protein bound to the viral RNA with the cellular endoplasmic reticulum. The newly formed viral nucleocapsids are subsequently transported to the lumen of the endoplasmic reticulum, and transported farther into the cell membrane by the Golgi vesicles. The process of exocytosis transports the newly formed virions into the extracellular space, where the viral particles undergo a final stage of maturation. The final maturation stage ensures that the newly formed viral particles are infectious and capable of invading and infecting the adjacent epithelial cells. This phase also facilitates human-to-human transmission through respiratory droplets[28] (Fig. 5.12).

The phase of the viral life cycle during which the virus replicates the incubation period of the virus characterized by an asymptomatic phase. The host

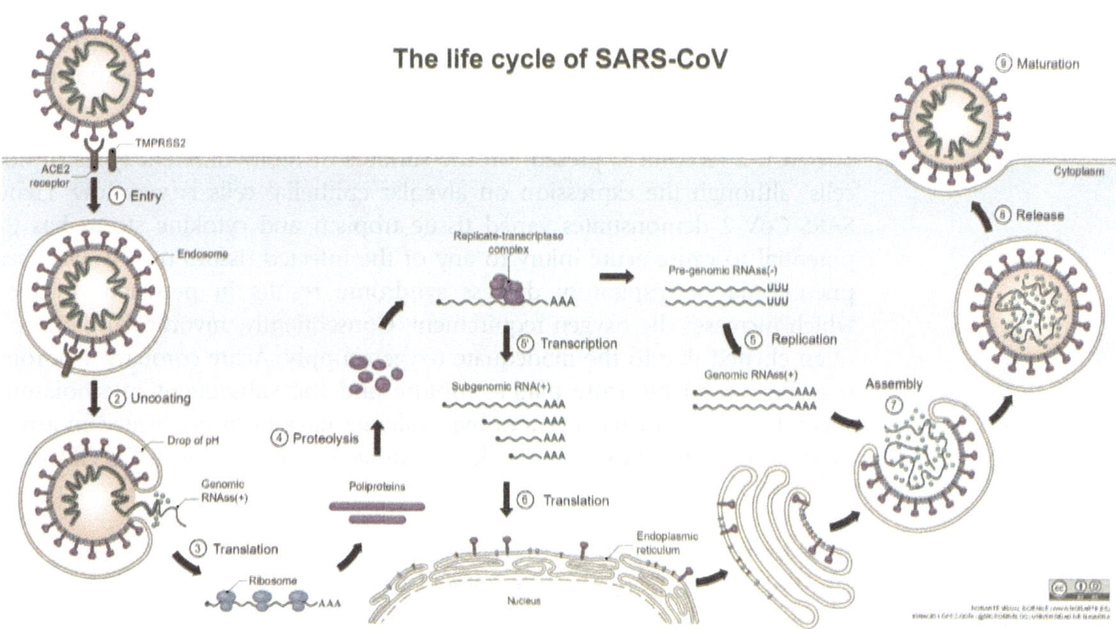

FIGURE 5.12

Schematic illustration of the life cycle of SARS-CoV-2. Source: *Vega Asensio/Own work/; https://creativecommons.org/licenses/by-sa/4.0/deed.en; https://commons.wikimedia.org/wiki/File:The_life_cycle_of_SARS-CoV.svg.*

immune response elicited during this phase is also limited and subtle in its expression. Although the viral load is low during the asymptomatic phase, the virus is detectable in nasal and throat swabs. The person-to-person transmission of infection is the highest during this phase.

The replication phase of the viral life cycle is followed by the migration of the virus from the nasal epithelium to the upper respiratory tract, which results in the initial clinical manifestations of the disease, such as fever, malaise, and a dry cough. The host immune response during this stage involves the release of chemokines and interferons from the virus-infected cells. Generally, immunocompetent individuals generate sufficient immune response during this stage, which effectively contains the spread of the infection and prevents the infected individuals to progress further into the disease.[47] However, some cases may progress further, mediating the invasion of the type 2 alveolar epithelial cells via the ACE-2 receptor. The virus then replicates into these cells, which elicit the release of several cytokines and inflammatory markers, such as interleukins (IL-1, IL-6, IL-8, IL-120 and IL-12), TNF-α, interferons IFN-λ and IFN-β, CXCL-10, MCP-1, and MIP-1α. This process precipitates the phenomenon of "cytokine storm," which acts as a chemoattractant for neutrophils, CD4 helper T-cells, and CD8 cytotoxic T-cells. These cells are subsequently sequestered in the lung tissue and are crucial for host defense against the viral invasion. However, the cytokine storm precipitates subsequent inflammation and damage to the lung tissue. Persistent injury to both the type 1 and type 2 alveolar epithelial cells causes diffuse alveolar damage, ultimately manifesting as ARDS.[28,48]

The ACE-2 receptor is present on the surfaces of many different types of host cells, although the expression on alveolar epithelial cells is very low. Hence SARS-CoV-2 demonstrates varied tissue tropism and cytokine storm has the potential to cause acute injury to any of the infected tissues or organs. Severe pneumonia or respiratory distress syndrome results in persistent hypoxia, which increases the oxygen requirement. Consequently, myocardial damage is often elicited due to the inadequate oxygen supply. Acute coronary syndrome may be caused by acute plaque rupture and the subsequent catecholamine surge. Electrolyte disbalance and hypokalemia have been implicated in arrythmias, and are therefore responsible for cardiac manifestations in COVID-19.[49]

Host immune response

The host immune response mediated by cytokines plays a very important role in the pathogenesis of SARS-CoV-2 infections. Evidence demonstrates an increased number of total neutrophils, elevated levels of interluekin-6 (IL-6) and c-reactive protein, increased expression of proinflammatory cytokines and chemokines, such as IP-10, MCP-1, MIP-1A, and TNF-α.[50,51]

Viral invasion of the host cell triggers the innate immune responses via recognition of the pathogen-associated molecular patterns through the Pattern Recognition Receptors, such as TLRs, non-occlusive disease (NOD)-like receptors (NLRs), and radiologically inserted percutaneous gastrostomy (RIG)-I-like receptors present on the antigen presenting cells, including dendritic cells and macrophages. The triggered immune responses induce cascade signaling that leads to the activation of NF-κB transcription factors and pathogen-activated protein kinases (MAPKs) pathway, which elicits the production of proinflammatory cytokines, such as INF-α and TNF-β. The production of proinflammatory cytokines is the primary response in the first line of defense against virus infection.

The Janus kinase-signal transducer and activator of transcription (JAK-STAT) pathway is activated by the formation of type 1 INF-interferon α/β receptor (IFNAR) complexes through the binding of type I INF to the IFNAR receptors. The JAK-STAT signaling cascade results in the elicitation of IFN-stimulated genes transcription, leading to a suppression of viral replication, thereby hindering further progress of the disease pathology.[52] However, an uncontrolled generation of pro-inflammatory cytokines, such as IFN-α, IFN-γ, IL-1β, IL-6, IL-12, IL-18, IL-33, TNF-α, TGFβ, and chemokines CCL2, CCL3, CCL5, CXCL8, CXCL9, CXCL10 can precipitate in the phenomenon of "cytokine storm," which causes extensive systemic inflammation and can potentially result in ARDS).[52,53]

CD4 + T-helper cells induced by IL-12 stimulates Th1 cell activation. Increased MHC class I expression and NK cell activation also play an important role in the clearance of virus-infected cells through the production of proinflammatory cytokines induced by the activation of the NF-κB signaling pathway. The expression of IL-17 that recruits neutrophils and monocytes to the site of infection and activates several other pro-inflammatory cytokines and chemokines, including IL-1, IL-6, IL-8, IL-21, TNF-β, and MCP-1 also increases significantly in SARS-CoV-2 infections.[54–56] The activated Th1 cells, in turn, stimulate CD8 + T-cells, which have cytotoxic effects on the virus-infected cells.

The CD4 + T-cells also activate the B-cells, thus eliciting humoral immune responses, resulting in the generation of antigen-specific antibodies comprising of IgG and IgM.[54,57] While IgM lasts only for the initial 12 weeks of the infection, IgG persists for a longer duration. The induction of CD4 + and CD8 + cells-associated immune response memory in addition to antibodies has also been documented for CoV infections.[58,59]

Clinical manifestations and complications

The average incubation period for COVID-19 is usually between 2 and 7 days, and during this presymptomatic period, the infected individuals are highly contagious. Clinical manifestations of the disease usually constitute systemic symptoms, such as a high-grade fever, chills, cough, dyspnea, nausea, vomiting, diarrhea, myalgia, fatigue, expectoration, hemoptysis, and in some rare cases, headaches and an altered mental status may also manifest. The similarities between the early signs of COVID-19 and other viral respiratory ailments increases the diagnostic challenges, resulting in a delayed diagnosis and compromised outcome of the disease. Immunocompromised and elderly individuals may not develop a fever due to the inability of their immune system to mount a robust immune response and instead may exhibit an atypical presentation, further increasing the diagnostic challenge. Severe forms of the disease may cause patients to develop pneumonia, cardiac pathologies, and kidney or liver failure. Evidence suggests that the geriatric population, individuals with comorbid medical conditions such as diabetes, obesity, renal pathologies, cardiopathies, and hypertension are especially predisposed to developing the complications associated with COVID-19. Respiratory failure is the most common cause of morbidity and mortality associated with COVID-19. High-risk infected individuals often require supplemental oxygen during the acute clinical phase of the disease, and some cases that progress to ARDS may also require invasive mechanical ventilator support.[60]

Diagnosis

Sputum, nasal and throat swabs, feces, and blood samples are used for various diagnostic evaluations of COVID-19. Although real-time polymerase chain reaction (RT-PCR) is the gold standard of a COVID-19 diagnosis, isolation of SARS-CoV-2 from clinical specimens, immunological tests, such as enzyme-linked immunosorbent assay (ELISA) for the detection of antibodies and antigens, flow-cytometry analysis of CD4 + and CD8 + T-cell counts, chest radiography, computerized tomography, hematological and biochemical profiling are also used. Findings, such as lymphopenia or a normal white blood cell, are considered indicative of poor prognosis. Elevated levels of C-reactive protein, lactate dehydrogenase, d-dimer, neutrophil-to-lymphocyte ratio, creatinine, aspartate aminotransferase, and alanine aminotransferase are suggestive of tissue damage and compromised functionality of organ systems. Abnormalities in coagulation profile parameters, such as international normalized ratio and prothrombin time, may also be observed in some cases. Chest radiography is usually inconclusive in early stages of the disease; however, bilateral multifocal alveolar opacities associated with pleural effusion can be seen in later stages of the disease (Fig. 5.13). High-resolution computed tomography, on the other hand, is extremely useful in diagnosing the disease even in the early stages (Fig. 5.14). The appearance of bilateral multifocal "ground-glass" regions largely concentrated in the lower lobes,

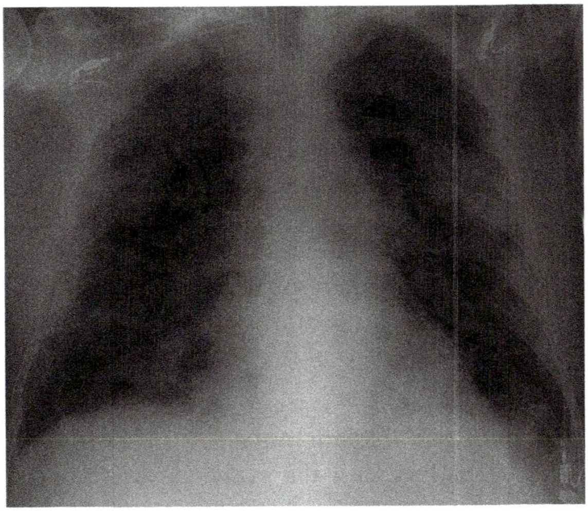

FIGURE 5.13

Chest X-ray demonstrating COVID-19-associated pneumonia. Source: *Hellerhoff/Own work/; https://creativecommons.org/licenses/by-sa/3.0/deed.en; https://commons.wikimedia.org/wiki/File:COVID-19_Pneumonie_-_82m_Roe_Thorax_ap_-_001.jpg.*

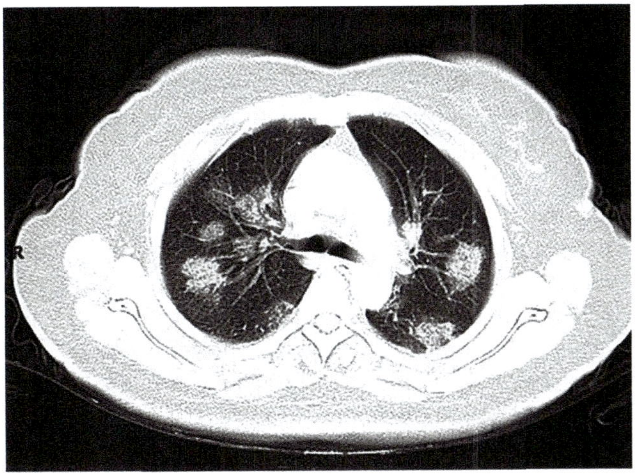

FIGURE 5.14

CT scan demonstrating COVID-19-associated pneumonia. Source: *Jin Y, Cai L, Cheng Z, et al. https://mmrjournal.biomedcentral.com/articles/10.1186/s40779-020-0233-6/; https://creativecommons.org/licenses/by-sa/4.0/deed.en; https://commons.wikimedia.org/wiki/File:COVID19CTPneumonia.jpg.*

associated with consolidation and patchy peripheral distribution, is a characteristic of lung tissue damage. The identification of focal opaque patches surrounded by a peripheral consolidation ring results in the formation of a "reversed halo," another characteristic feature of the disease. Computed tomography also facilitates the detection of some other pathological manifestations, such as pleural effusion, cavitation, calcification, and lymphadenopathy. Transthoracic ECG is recommended for cases manifesting cardiac pathologies, such as arrythmias, ECG changes, cardiomegaly, and heart failure. Cardiac magnetic resonance imaging (MRI) is instrumental in diagnosing myocarditis. A diagnosis of COVID-19 is therefore a function of the clinical manifestations of the disease.

Treatment and management

The viral etiology and the evolving nature of COVID-19 has hindered the development of a specific therapeutic approach, resulting in supportive therapy being the mainstay of COVID-19 treatment. Several repurposed drugs, such as the antiretroviral drugs, lopinavir and ritonavir, remdesivir used in treatment of Ebola infections; interferon-alpha inhalation used for therapy in lung cancer; arbidol used for treatment of influenza infection; chloroquine used as an antimalaria drug; hydroxychloroquine used in treatment of autoimmune conditions such as rheumatoid arthritis and lupus; azithromycin used as an antibacterial drug, and the antiinfluenza and antiSARS-CoV-2 experimental drug favipiravir are thus being used for the treatment of COVID-19 subject to the clinical presentation and medical history of the case. Antiretroviral drugs, such as lopinavir-ritonavir, used in the treatment of HIV infections are being considered for use in combination with ribavirin, another antiretroviral drug for antiviral therapy in the early stages of COVID-19. Nelfinavir, yet another antiHIV drug, has also shown promise as an antiSARS-CoV-2 drug. Corticosteroids, gamma-globulins, interferons, and probiotics are being investigated as supportive therapy to prevent cytokine-storm-mediated tissue damage and maintain good gut health. Antibiotic therapy involving quinolones and higher generation cephalosporins may be required in cases of refractory fever or extremely elevated C-reactive protein levels. Oxygen supplementation, mechanical ventilation, and tracheostomy are required in cases of respiratory failure. Hospital discharge of COVID-19 patients requires two negative RT-PCR reports for samples collected 24 hours apart.[61]

Some novel antiviral therapies, including RNA interference, monoclonal antibody, and synthetic peptides are being developed. Transmission of COVID-19 can be facilitated very easily in the hospital environment and poses a threat to high-risk patients with comorbidities. Therefore strict adherence to infection control guidelines is crucial in high-risk areas such as COVID-19 isolation wards, ICUs, and operating rooms. Invasive and noninvasive procedures required in COVID-19 treatment and management should follow aseptic protocols strictly to reduce complications and the chances of transmission.[60]

Prevention and control

COVID-19 is a highly contagious and communicable disease with proven pandemic potential. Stringent infection control measures and protocols play a very significant role in transmission risk mitigation. Nosocomial COVID-19 infections are commonplace, and the prevention of such transmission is very challenging. Good hand and respiratory hygiene strategies play a pivotal role in reducing the burden of nosocomial and community infections. Sensitization about the disposal of infected tissues and hand hygiene after having direct or indirect contact with respiratory secretions can prove very significant in risk mitigation. Healthcare facilities and commercial spaces should ensure the availability of personal protective equipment, tissues, no-contact receptacles, no-contact alcohol-based hand sanitizer dispensers, and handwashing facilities, such as soap, disposable towels, and a water supply to encourage adherence to good hand and respiratory hygiene practices.

The possibility of abiotic surfaces and materials being contaminated should always be considered, and steps for disinfecting such surfaces and materials should be taken. Use of sodium hypochlorite and ethanol-based disinfectants is recommended. Training and education for housekeeping staff in healthcare facilities and commercial spaces is very important for ensuring effective disinfection of contaminated environments.

Early suspicion of COVID-19, rapid diagnosis, isolation, and stringent infection control measures are instrumental in controlling the spread of this highly contagious disease. Isolation facilities, strict maintenance of hand and respiratory hygiene, contact tracing, and quarantine/isolation of primary contacts play a very important role for containing the nosocomial or community transmission of the disease.

There are several vaccine candidates that are being investigated, while some vaccines have proved efficacy and have been approved for human use after appropriate clinical trials and adherence to regulations, as of April 2021. Pfizer-BioNTech COVID-19 vaccine and Moderna COVID-19 vaccine are RNA vaccines that have been approved for use, while the CVnCoV RNA vaccine from CureVac is in phase III of clinical trials. The Oxford-AstraZeneca COVID-19 vaccine, Russian Sputnik V, Chinese Convidecia, and Johnson & Johnson COVID-19 vaccine are viral vector vaccines, while Chinese CoronaVac, Indian Covaxin, CoviVac, and BBIBP-CorV are approved inactivated virus vaccines. The Valneva COVID-19 vaccine, and subunit vaccines, such as Novavax COVID-19 and RBF-Dimer are all still in clinical trials, while the peptide vaccine, EpiVac Corona, has been approved for use.[62]

Student Notes

Key take-aways

- COVID-19 is a highly contagious viral respiratory infection with a global pandemic implication.
- This disease is caused by the SARS-CoV-2 virus, which shares similarities with the SARS-CoV virus responsible for severe acute respiratory syndrome.
- The SARS-CoV-2 virus has an affinity for the ACE-2 receptor present on the surface of the cells of organs, such as the heart, kidneys, lungs, and gastrointestinal tract.
- Comorbidities, such as hypertension, obesity, diabetes, chronic kidney disorders, cardiac pathologies, and underlying respiratory illnesses predispose an individual to severe COVID-19 disease pathology.
- The "cytokine storm" elicited by the uncontrolled host immune responses to SARS-CoV-2 invasion is responsible for the extensive tissue damage seen in severe COVID-19 cases.
- Respiratory failure is the most common cause of mortality and morbidity in severe COVID-19 cases.
- The similarities between the symptoms exhibited in the early stages of COVID-19 and other viral respiratory illnesses increase the diagnostic challenges associated with the disease.
- Although respiratory manifestations are most commonly associated with COVID-19 infections, cardiac manifestations can also be associated with severe disease in high-risk individuals and can contribute significantly to poor clinical outcome.
- RT-PCRs are considered the gold standard for diagnosis of COVID-19 infections, and certain serological investigations, such as the white blood cell count, c-reactive protein levels, d-dimer levels, and lactate dehydrogenase levels are important prognostic markers.
- Although chest X-rays are not very reliable for a diagnosis in the early stages of the disease, computed tomography scans provide extremely reliable information for an early diagnosis.
- Treatment of COVID-19 infections mainly comprises of supportive therapy and symptom management strategies.
- Repurposed drugs with proved effectiveness in the treatment of other viral diseases, such as influenza, Ebola, and HIV, have been considered for use in the treatment of COVID-19 infections.
- Good respiratory and hand hygiene, effective disinfection of contaminated surfaces, rapid diagnosis, disease reporting, contact tracing, isolation of primary contacts, and adherence to strict infection control protocols are important for restricting the spread of COVID-19.
- A number of vaccines have now been authorized for human use, and many more vaccine candidates are in clinical trials.

Problem number 5.4: Infective endocarditis caused by *Enterococcus faecalis* infection

A 32-year-old female presented with general fatigue, chest pain, dyspnea, and a 3-month history of intermittent fever and night sweats. The woman had undergone a cesarean section for delivery 4 months ago, and the symptoms she complained about at presentation had started to manifest about 3–4 weeks after the delivery. There had also been an episode of lower limb edema and dizziness about a month prior to presentation. She had body pain, and productive cough for 4 days. The patient did not have any history of congenital heart disease as confirmed by ECG during the cesarean delivery, and there was no other significant medical history or any history of travel, exposure to animals, food or drug allergies, or any relevant medical procedures.

The following are the reports of the investigations conducted:

1. Vitals
   Temperature: 37.7°C
   Respiratory rate: 25 breaths/minute
   Systolic blood pressure: 70 mmHg
2. Physical examination
   General fatigue
   Pale conjunctivae
   No petechiae; no Osler nodes/Janeway lesions
3. Respiratory examination:
4. Cardiovascular examination:
   Chest pain
5. Radiological investigations
   Transthoracic ECG: Mobile, 21 × 16 mm vegetation attached to the anterior mitral valve; severe mitral valve regurgitation; worsening symptoms of heart failure
6. Hematological profile

Hb (g/dL)	9.6	12.0–15.5
WBC (/L)	9.1×10^9	$4.3–10.8 \times 10^9$

7. Biochemical profile

Serum creatinine (mg/dL)	0.85	0.6–1.1
C-reactive protein (mg/dL)	22.5	0.8 – 100

8. Mitral valve tissue culture and sensitivity:
   Organism: *E. faecalis*

Susceptibility:
Benzylpenicillin G
Amino-benzyl penicillin
Vancomycin
Gentamicin (MIC—8 mg/L) (low level resistance)

Diagnosis

The final diagnosis concluded as infective endocarditis (IE) based on a positive culture report and the electrocardiogram report, which demonstrated the presence of an oscillating intracardiac mass on the mitral valve.

Treatment

The mitral valve was replaced with a prosthetic valve due to indications of severe mitral valve regurgitation, resulting in worsening symptoms of heart failure.

Antibiotic therapy for 6 weeks:

1. Ampicillin—2 g every 4 hours
2. Gentamicin—1 mg/kg every 8 hours
   Predose monitoring of serum gentamicin levels was performed on the fourth day after treatment initiation—0.4 mg/L

Transthoracic ECG performed after the mitral valve replacement did not show any vegetation. The WBC, serum creatinine, and C-reactive protein were monitored during the 6-week treatment period and the parameters demonstrated improvement over time. The predosing serum concentration of gentamicin was also monitored regularly to ensure early detection of any harmful side-effects of gentamicin. Renal function did not show any deterioration either and was maintained at a creatinine clearance of 100 mL/minute.

The patient was discharged in good health without any complications 6 weeks after admission and was monitored for 20 months after discharge, during which she did not complain of relevant medical problems.

Question and answer

1. What is endocarditis?
   Endocarditis is a condition characterized by an inflammation of the endocardium, the inner lining of the heart, and the valves that separate the four cardiac chambers, usually attributed to a bacterial infection of the cardiac endothelium.

2. Which are the dental conditions that demonstrate the presence of *E. faecalis*?

   E. faecalis has demonstrated a presence in common dental conditions, such as periodontitis, periimplantitis, dental caries, and secondary endodontic infections in previously filled root canals.

3. How does the endocardium get infected?

   An intact and healthy endocardium resists bacterial invasion and protects the cardiac muscle. Blood flow disruption, caused by diseased or defective valves, or by direct tissue injury, inflicted by particulate matter frequently introduced into the system by intravenous drug abuse or the insertion of medical devices such as catheters or electrodes precipitates prodromal endocardial injury. This is often followed by subsequent bacteremia. Such injury creates the perfect environment for the formation of bacterial vegetation invading the cardiac muscle. Injured or compromised endocardium activates the coagulation cascade, resulting in the recruitment of various tissue repair factors, such as platelets, red blood cells, collagen fibers, and fibrin, which facilitates the formation of sterile vegetation. Platelet-fibrin deposits facilitate bacterial evasion of the host immune responses, thereby further enabling the infective process.

4. What is the role of hemodynamics in the pathogenesis of IE?

   The ventricular surface of the aortic valve and the atrial surface of the mitral valve have a high propensity toward the development of bacterial vegetation. Reduced blood flow through blood vessels associated with areas of regurgitant flow has been implicated in mediating the localization of vegetation downstream of such regions. The high-pressure flow associated with high-turbulence lesions, such as a small ventricular septal defect caused by jet lesion or stenotic valves, results in more localized damage compared to large area defects responsible for low-pressure flow. Thus hemodynamics plays a significant role in the pathogenesis of IE.

5. Why are valvular bacterial infections resilient?

   The initial platelet-fibrin deposit precipitated by injury to the endocardium acts as the nucleus for further accumulation of bacteria and tissue repair factors. Mature vegetations, resulting from such gradual accumulation, comprises of a combination of platelets, fibrin, inflammatory cells, and erythrocyte debris. While the initial bacterial colonization is mediated by the platelet-based nucleus, the binding of surface proteins propagates further bacterial aggregation. Bacterial biofilms embedded within these platelet aggregates contribute to the resilience of the valvular infection.

6. Why are the *E. faecalis* endocarditis antigen A (efaA) and cytolysin (cyl) important for the pathogenetic process of IE?

The *E. faecalis* endocarditis antigen A (efaA) facilitates the adhesion of *E. faecalis* to cardiac cells, while cytolysin (cyl), a beta-hemolysin capable of cell lysis, is a potent bacteriocin implicated in the exacerbation of Enterococcal infections in humans.

7. What are the dermatological manifestations of IE?
   Osler nodes, Janeway lesions, and subungual splinter hemorrhages are the dermatological manifestations of IE.
8. Why are splenomegaly, localized peritonitis, and focal motor or sensory deficits presented as clinical manifestations of IE?
   Mesenteric arterial occlusion often results in bowel perforation that can manifest as splenomegaly or localized peritonitis, while focal motor or sensory deficits may be a result of intracerebral embolization affecting specific vascular regions.
9. What are the modified Duke criteria for the diagnosis of IE?
   IE can be diagnosed by the modified Duke criteria, which provide two major criteria, one major and three minor criteria, or five minor criteria. The first major criterion is the confirmation of bacteremia through two separate positive blood cultures of samples drawn 12 hours apart, which demonstrates the presence of the typical culprit pathogens. The second criterion is sonographic evidence, such as echocardiogram, indicative of the presence of a vegetative mass on a cardiac valve, or other nearby structures on a prosthetic device, which is suggestive of endocardial involvement. The five minor criteria are:
   - History of predisposing conditions like underlying valvular disease, structural abnormalities of the heart, or intravenous drug use.
   - Clinical presentation of fever ($>38°C$).
   - Manifestations of vascular pathologies, such as mycotic aneurysms, intracranial hemorrhage, Janeway lesions, major arterial emboli, or septic pulmonary infarcts.
   - Presentation of immunological pathologies, including Osler's nodes, Roth spots, glomerulonephritis, or positive rheumatoid factor.
   - Positive blood cultures that are not exactly adherent to the major criterion or evidence from laboratory work-up that is suggestive of IE.

Learning objectives

1. Determine the appropriate differential diagnosis based on patient history, known underlying medical conditions, and clinical manifestations.
2. Review the epidemiology of IE.
3. Describe the characteristics of *E. faecalis*.
4. Summarize the pathogenesis of *E. faecalis*.

5. Describe the clinical manifestations of IE.
6. Predict the prognosis of IE case based on the extent of the pathology presented.
7. Enumerate the methods involved in the diagnosis of IE.
8. Describe the major and minor parameters of the modified Duke criteria for the diagnosis of IE.
9. Summarize the management and treatment options for IE.

Discussion

Differential diagnosis

Acute coronary syndrome	Acute coronary syndrome presents itself with chest pain, discomfort or tightness in the chest, pain or discomfort in one or both arms, jaw, neck, back, or stomach, dyspnea, lightheadedness, dizziness, nausea, and sweating.
Acute heart failure	Acute heart failure may cause dyspnea even while resting, extreme fatigue, weakness, edema in lower extremities, irregular heartbeat, palpitations, persistent productive cough, wheezing, blood-tinged foamy expectoration, night urination urge, fluid retention causing rapid weight gain, lack of appetite, nausea, altered mental status, and chest pain.
Aortic dissection	Aortic dissection develops a sudden severe shearing and radiating chest, abdominal, or upper back pain, dyspnea, loss of consciousness, loss of vision, one-sided weakness/paralysis, speech difficulty, weak pulse in one arm, difficulty walking, leg pain, or paralysis.
Pulmonary embolism	Pulmonary embolism generally occurs with dyspnea, chest pain that worsens while breathing, anxiety, dizziness, light-headedness, arrythmia, palpitations, blood-tinged expectoration, coughing, sweating, low blood pressure, swelling in the lower extremities, deep-vein thrombosis-associated soreness, redness, or warmth in the leg, and inflamed/discolored skin.
Myopericarditis	Myopericarditis presents itself with chest pain, arrhythmias, dyspnea, swelling in lower extremities due to fluid retention, fatigue, headache, body aches, joint pain, fever, sore throat, and diarrhea.
Pneumonia	Pneumonia causes a fever, chills, sweating, cough, dyspnea, rapid shallow breaths, sharp chest pain that worsens while breathing, loss of appetite, fatigue, low energy, nausea, vomiting, and altered mental status.
Empyema	Empyema presents itself with a fever, night sweats, fatigue, dyspnea, weight loss, chest pain, and cough, often expectorating pus.
Atrial myxoma	Atrial myxoma causes dyspnea when lying in supine position or on one side, difficulty breathing while sleeping, chest pain or feeling of tightness, dizziness, fainting, palpitations, dyspnea during activity, cyanosis, excessive sweating, and deep-vein thrombosis associated with swelling/pain in lower extremities.

Background

IE is a condition characterized by an inflammation of the endocardium, the inner lining of the heart, and the valves separating the four cardiac chambers, usually attributed to a bacterial infection of the cardiac endothelium. The condition is known to have a wide spectrum of clinical manifestations, so early detection is important to prevent intracardiac as well as extracardiac complications. Although the etiological agents responsible for most cases of IE are Gram-positive *Streptococci* and *Staphylococci*, *Enterococci* are also responsible for a considerable number of IE cases. The occurrence of *Enterococcal* endocarditis is attributed to nosocomial causes due to the association of a bacterial species, such as *E. faecalis*, with dental and genitourinary invasive procedures. Unlike *Staphylococcal* endocarditis, valvular heart disease is not a prerequisite for *Enterococcal* endocarditis. *E. faecalis* also forms a part of the normal flora in the human intestine, the female genital tract, and dental cavities. As the patient in this case had not undergone any relevant postpartum dental procedures, opportunistic invasion of the circulation by the intestinal or genital flora during parturition, or through the surgical instrumentation involved in the cesarean section, are the most probable potential sources of infection in this case.

Etiological agent

Enterococcus faecalis

Classification and morphology

E. faecalis is a Gram-positive, facultative anaerobic coccus (Figs. 5.15 and 5.16) known for its resilient nature and ability to persist and survive in hostile conditions within the host and environment. This bacterial species is the predominant species in the *Enterococcus* genus and has been implicated as an opportunistic pathogen, contributing significantly to the burden of nosocomial infections.

Epidemiology

IE is a rare condition, and males have been found to demonstrate a higher propensity toward it.[63] The advent of antibiotics has precipitated a significant evolution of IE. In recent years, *Staphylococcal* species have demonstrated an increased association with IE, rather than *Streptococcal* species. This condition is now seen to occur more frequently in geriatric individuals who are more than 65 years old. However, the potential for nosocomial spread of opportunistic bacteria, such as *E. faecalis*, also implicated in IE, predisposes any individual exposed to dental procedures or other surgical interventions to the disease. Predisposing factors, such as prosthetic valves, indwelling cardiac

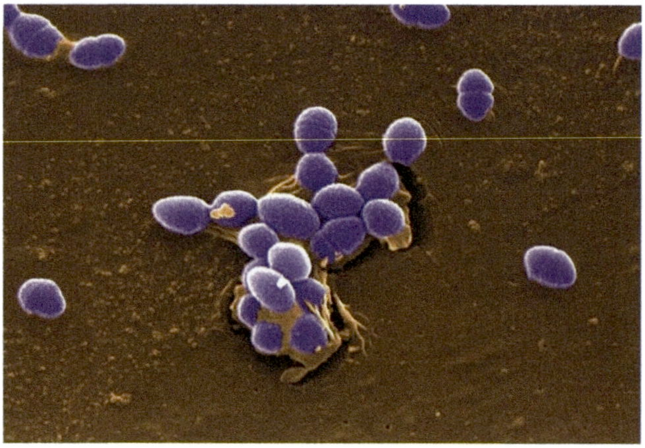

FIGURE 5.15

Gram-positive *Enterococcus. faecalis* bacteria. Source*: Pete Wardell, USCDCP/Own work/; https://creativecommons.org/licenses/publicdomain/; https://pixnio.com/science/microscopy-images/large-numbers-of-gram-positive-enterococcus-faecalis-sp-bacteria.*

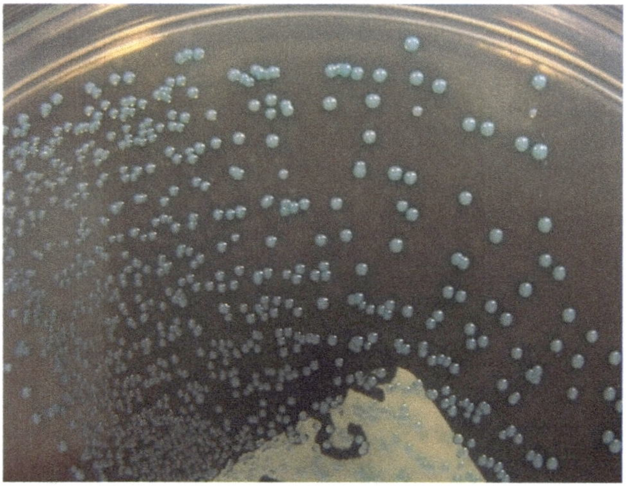

FIGURE 5.16

E. faecalis on chromogenic agar. Source*: Nathan Reading, Halesowen, UK/; https://www.flickr.com/photos/nathanreading/6003039958/; https://creativecommons.org/licenses/by/2.0/deed.en; https://commons.wikimedia.org/wiki/File: Enterococcus_faecalis_on_ChromID_CPS_chromogenic_agar_-_Detail.jpg.*

devices, hemodialysis, history of valvular diseases, intravenous drug abuse, and diabetes mellitus, increase the risk of IE.[64,65] The intracardiac and extracardiac sequelae associated with *Enterococcal* IE are responsible for the rapid deterioration of the disease and high morbidity and mortality rate. Clinical suspicion in relevant cases in early stages and early detection are therefore instrumental in preventing the complications of the disease and reducing the associated morbidity and mortality.

E. faecalis is an important opportunistic pathogen implicated in nosocomial infections. Besides, *E. faecalis* is also a significant part of the normal flora of the human intestine and the female genital tract. Although it does not form a part of the healthy oral flora,[66] *E. faecalis* has demonstrated a presence in common dental conditions like periodontitis, periimplantitis, dental caries, and secondary endodontic infections.[67–69] *E. faecalis* is associated with urinary tract infections, bacteremia, wound infections, meningitis, neonatal infections, and postpartum infections. The association of *E. faecalis* with interventional and prosthetic medical devices is attributed to its fastidious ability to form biofilms.[70] High levels of antibiotic resistance in *E. faecalis* biofilms,[71] prevalence of resistance to commonly used antibiotics,[72,73] and the emergence of vancomycin-resistant enterococci (VRE)[74] are responsible for increased therapeutic challenges. Root canals that have been previously filled provide a conducive environment for *E. faecalis* colonization and biofilm formation, thereby increasing the prevalence of secondary endodontic infections attributable to *E. faecalis*.[75–79]

The possibility of zoonotic transmission of *E. faecalis* has also been considered due to the incidence of this bacterium in livestock.[80,81] Evidence indicates the transmission of porcine-associated *E. faecalis* through food products or raw milk to the human gastrointestinal tract.[82,83]

Pathogenicity and pathogenesis

An intact and healthy endocardium resists bacterial invasion and protects the cardiac muscle. Blood flow disruption caused by diseased or defective valves, or by direct tissue injury, inflicted by particulate matter frequently introduced into the system by intravenous drug abuse[63] or the insertion of medical devices such as catheters or electrodes precipitates prodromal endocardial injury, followed by subsequent bacteremia. Such injury creates the perfect environment for the formation of bacterial vegetation invading the cardiac muscle. The ventricular surface of the aortic valve and the atrial surface of the mitral valve have demonstrated high propensity toward the development of bacterial vegetation, thereby implicating hemodynamics in the pathogenesis. Reduced blood flow through blood vessels associated with areas of regurgitant flow has been implicated in mediating the localization of vegetation downstream of such regions. The high-pressure flow associated with high-turbulence lesions, such as a small ventricular septal defect caused by jet lesion or stenotic valves, results in more localized damage compared to large area defects responsible for low-pressure flow.[84] Injured or compromised endocardium activates the coagulation cascade, resulting in the recruitment of various tissue repair factors, such as platelets, red blood cells, collagen fibers, and fibrin, which facilitates the formation of sterile vegetation. Dental or surgical interventions, compromised asepsis during parturition, and the presence of an established endogenous source of infection can lead to bacterial inoculation, resulting in bacteremia.[85] Platelet-fibrin deposits facilitate bacterial evasion of the host immune responses, which emphasizes the role of bacterial virulence factors capable of binding to and expanding platelet-fibrin deposits in IE pathogenesis[86] (Fig. 5.17).

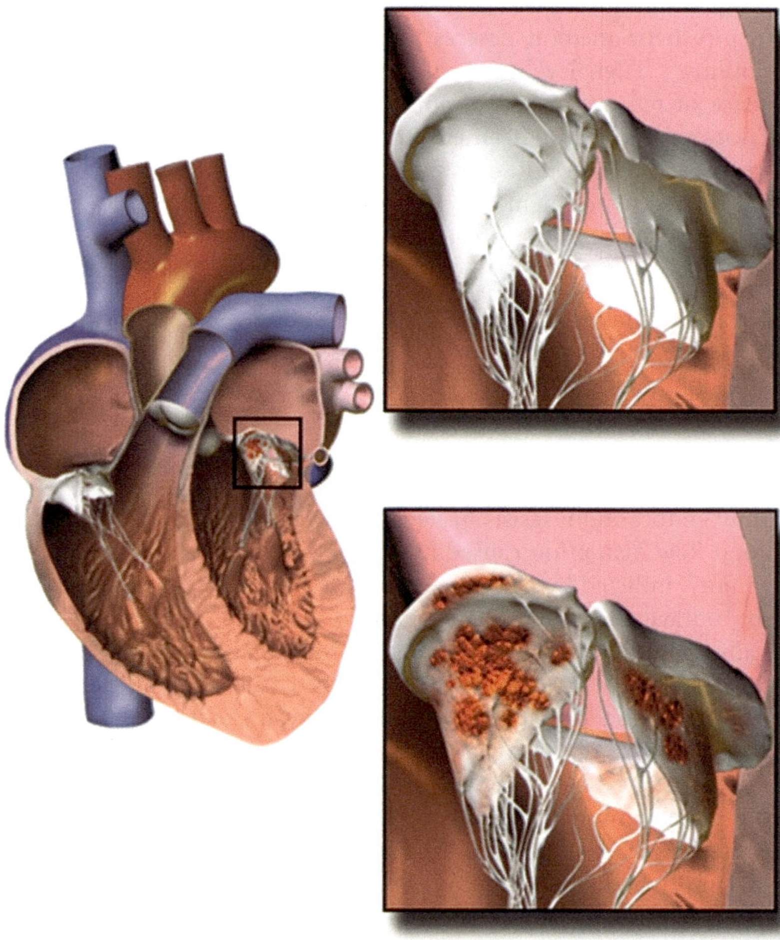

FIGURE 5.17

Illustration of endocarditis. Source: *BruceBlaus/Own work/; https://creativecommons.org/licenses/by-sa/4.0/deed.en; https://commons.wikimedia.org/wiki/File:Endocarditis.png.*

The initial platelet-fibrin deposit acts as the nucleus for further accumulation of bacteria and tissue repair factors. Mature vegetations, resulting from such gradual accumulation, comprises of a combination of platelets, fibrin, inflammatory cells, and erythrocyte debris. Bacterial biofilms embedded within these platelet aggregates contribute to the resilience of the valvular infection. While the initial bacterial colonization is mediated by the platelet-based nucleus, binding of surface proteins propagates further bacterial aggregation.[87] Vegetations may be vascular or avascular depending on whether the condition is gradual or acute. However, even in cases of avascular vegetations, neovascularization, fibroblasts, and fibrosis are initiated once the process of healing begins.[86]

Virulence factors and resistance

Virulence factors associated with *E. faecalis* increases its ability to invade the host system, evade the host immune responses, and contribute to the transmissibility and infectivity of the pathogen. The *Enterococcal* surface protein (esp), aggregation substance, and gelatinase, an extracellular metalloprotease that can hydrolyze gelatin, collagen, and hemoglobin are important for adherence, colonization, and biofilm formation in biotic and abiotic environments.[71,88–95] Hyaluronidase (hyl) is another virulence factor that facilitates *Enterococcal* virulence and enhances its ability to invade host tissue.[90,96] The pathology of endocarditis associated with *E. faecalis* is attributable to the *E. faecalis* endocarditis antigen A (efaA), which facilitates the adhesion of *E. faecalis* to cardiac cells.[97,98] Cytolysin (cyl), a beta-hemolysin capable of cell lysis, is a potent bacteriocin implicated in the exacerbation of *Enterococcal* infections in humans.[99,100] The multiple peptide resistance factor resists immune clearance, slows the process of wound healing, and facilitates the survival of *E. faecalis* in the face of host immune responses.[101] Immunomodulation and resistance to the host immune responses are important contributors in the pathogenesis of *E. faecalis* infections. Capsular polysaccharides in encapsulated *E. faecalis* strains facilitate evasion of innate immune responses, such as complement-mediated opsonophagocytosis, by preventing the detection of the *E. faecalis* surface antigens by the bound C3 component.[102] Evidence has also suggested the active suppression of NF-κß-driven macrophage responses by *E. faecalis*, which is further enhanced in polymicrobial infections.[103]

The emergence of antibiotic resistance in *E. faecalis* has increased the therapeutic challenges posed by *E. faecalis*-associated infections. The proficiency of horizontal gene transfer that mediates the exchange and transfer of the genes conferring antibiotic resistance contributes to the increasing level of antibiotic resistance seen in *E. faecalis*.[71,104] VRE are commonly encountered in healthcare establishments worldwide and pose a formidable threat, which adds to the burden of nosocomial infections. In addition, ß-lactam and aminoglycoside resistant *E. faecalis* strains have also been reported globally.[74]

Clinical manifestations and complications

IE generally presents itself with a multitude of clinical manifestations characterized by a gradual and subtle but persistent onset of fever, chills, malaise, and fatigue.[63] Early clinical suspicion of IE in high-risk patients, such as those with underlying chronic cardiac conditions requiring indwelling prosthetic devices, geriatric individuals, immunosuppressed individuals, those with a history of certain autoimmune conditions like atopic dermatitis, or those on long-term antipyretic, antibiotic, or steroidal use, is instrumental in ensuring a good prognosis and outcome. Some cases may also present with other generic symptoms, such as anorexia, nausea, and headache. Manifestations, such as chest pain, dyspnea, orthopnea, paroxysmal nocturnal dyspnea, and abrupt onset of heart failure are indicative of cardiopulmonary involvement and should raise suspicion of underlying cardiac insufficiency. History of degenerative valve diseases, such as calcific aortic stenosis and mitral valve prolapse, diabetes mellitus, RHD, indwelling catherization, intravenous drug abuse, placement of prosthetic valves or pacemakers, trauma or injury, parturition involving cesarean section, or invasive medical procedures and surgeries, increases predisposition to endocardial injury and infection.[63,65,105]

Valvular insufficiency or systemic infection may manifest as tachypnea and tachycardia in addition to febrile presentations. Systemic infections may result in sepsis or cardiogenic shock due to acute valve perforation, and present as hypotension. Identification of heart murmurs helps to localize the valve involvement. Auscultation demonstrating bilateral pulmonary rales can be indicative of severe mitral or aortic valve regurgitation. Osler nodes (Fig. 5.18), Janeway lesions (Fig. 5.19), and subungual splinter hemorrhages (Fig. 5.20) are rare dermatological manifestations of IE.[65] Bowel perforation, resulting from mesenteric arterial occlusion, can manifest as splenomegaly or localized peritonitis, while intracerebral embolization can cause focal motor or sensory deficits in the affected vascular territories.

Diagnosis

Early clinical suspicion and a rapid diagnosis are instrumental for a good clinical outcome in IE. Electrocardiograms (12-lead) are a rapid and cheap means for the preliminary detection of ischemia, arrythmias, or structural pathologies in cases presenting with cardiac manifestations. Sinus tachycardia (ST) elevations may be suggestive of IE, although they are generally considered as myocardial

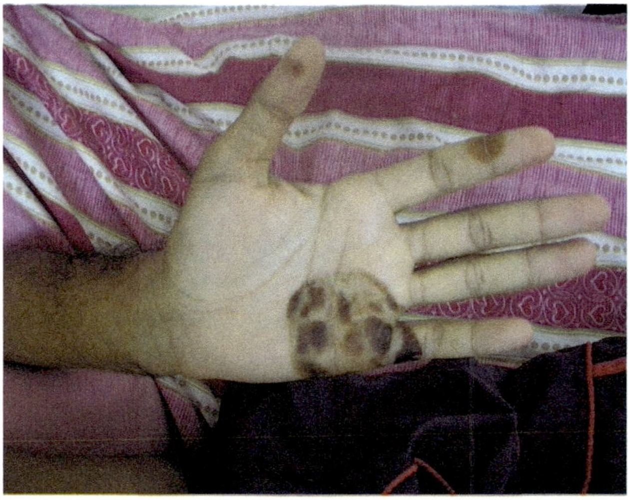

FIGURE 5.18

Osler's nodes. Source: *Roberto J. Galindo/Own work/; https://creativecommons.org/licenses/by-sa/4.0/; https://creativecommons.org/licenses/by-sa/3.0/deed.en; https://creativecommons.org/licenses/by-sa/2.5/deed.en; https://creativecommons.org/licenses/by-sa/2.0/deed.en; https://creativecommons.org/licenses/by-sa/1.0/deed.en; https://commons.wikimedia.org/wiki/File:Osler_Nodules_Hand.jpg.*

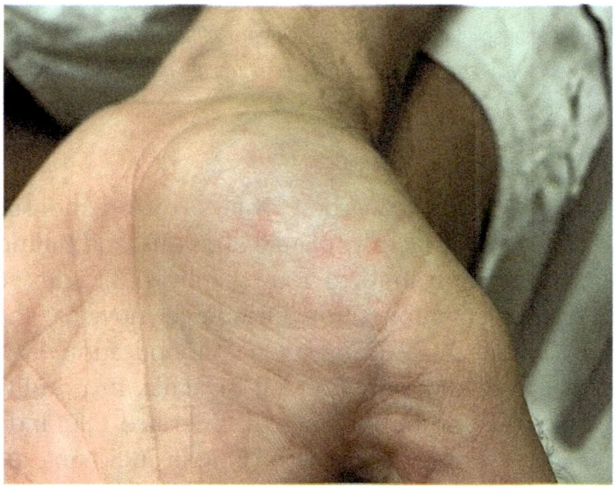

FIGURE 5.19

Janeway lesions. Source: *Warfieldian/Own work/; https://creativecommons.org/licenses/by-sa/4.0/deed.en; https://commons.wikimedia.org/wiki/File:Janeway_lesion.JPG.*

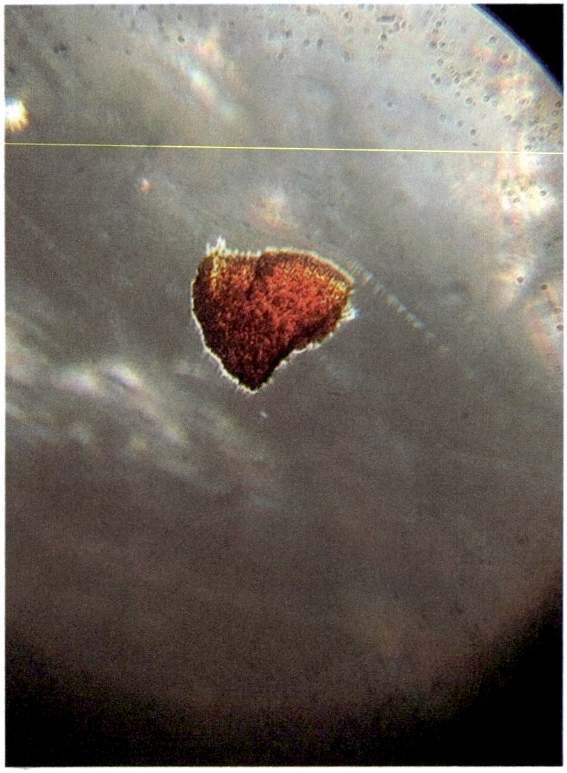

FIGURE 5.20
Microscopically zoomed-in view of subungual splinter hemorrhage. Source: *Dptfhpwlf/Own work/; https://creativecommons.org/licenses/by-sa/4.0/deed.en; https://commons.wikimedia.org/wiki/File:Splinter_hemorrhage_under_the_microscope.jpg.*

infarction. Two-view chest X-rays can demonstrate infiltrates, pulmonary abscesses, cardiomegaly, pleural effusions, and pulmonary vasculature cephalization. Contrast computed tomography (CT) scans or CT angiograms provide more detailed information including empyema, the possibility of pulmonary parenchymal disease, and arterial embolization. Cardiac biomarkers are instrumental in detecting myocardial ischemia, myocarditis, or underlying infarction. Leukocytosis, normocytic anemia, elevated ESR and CRP levels, and electrolyte derangements play a significant role in indicating underlying pathology and set the premise for further detailed evaluations.

IE diagnosis based on the modified Duke criteria is the most reliable diagnostic method. Diagnosis of IE based on the modified Duke criteria requires two major criteria, one major and three minor criteria, or five minor criteria to be satisfied. The confirmation of bacteremia through two separate positive blood cultures of samples drawn 12 hours apart, which demonstrate the presence of the typical culprit pathogens, is the first major criterion.[106] An exception of single positive blood culture or an antiphase 1 IgG antibody titer greater than or equal to 1:800 has been recommended by the American Heart Association for *Coxiella burnetii*.[107] The second criterion is sonographic evidence, such as echocardiogram, indicative of the presence of a vegetative mass on a cardiac valve or other nearby structures, or on a prosthetic device, is suggestive of endocardial involvement (Figs. 5.21 and 5.22). The five minor criteria are: (1) history of predisposing conditions, such as underlying valvular disease, structural abnormalities of the heart, or intravenous drug use; (2) clinical presentation of fever (>38°C); (3) manifestations of vascular pathologies, such as mycotic aneurysms, intracranial hemorrhage, Janeway lesions, major arterial emboli, or septic pulmonary infarcts; (4) Presentation of immunological pathologies, including Osler's nodes, Roth spots, glomerulonephritis, or positive rheumatoid factor; (5) positive blood cultures that are not exactly adherent to the major criterion or evidence from laboratory work-up suggestive of IE.[107]

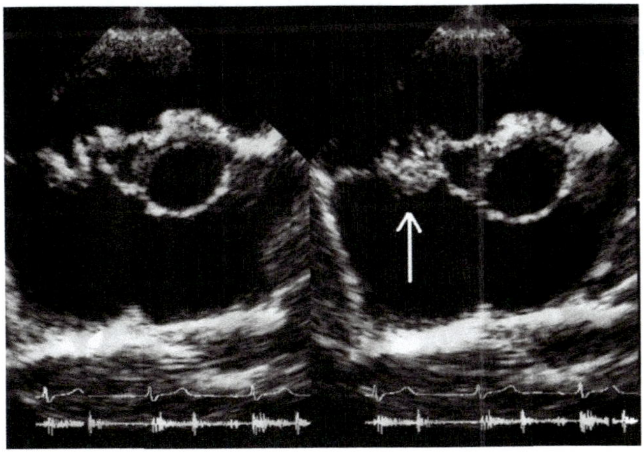

FIGURE 5.21

Electrocardiogram (arrow) demonstrating vegetation on the tricuspid valve. Source*: Daisuke Koya, Kazuyuki Shibuya, Ryuichi Kikkawa and Masakazu Haneda/Daisuke Koya, Kazuyuki Shibuya, Ryuichi Kikkawa and Masakazu Haneda. Successful recovery of infective endocarditis-induced rapidly progressive glomerulonephritis by steroid therapy combined with antibiotics: a case report. BMC Nephrology 2004, 5:18. doi: 10.1186/1471-2369-5-18/; https://creativecommons.org/licenses/by/2.0/deed.en; https://commons.wikimedia.org/wiki/File:Endocarditis_ultrasound.gif.*

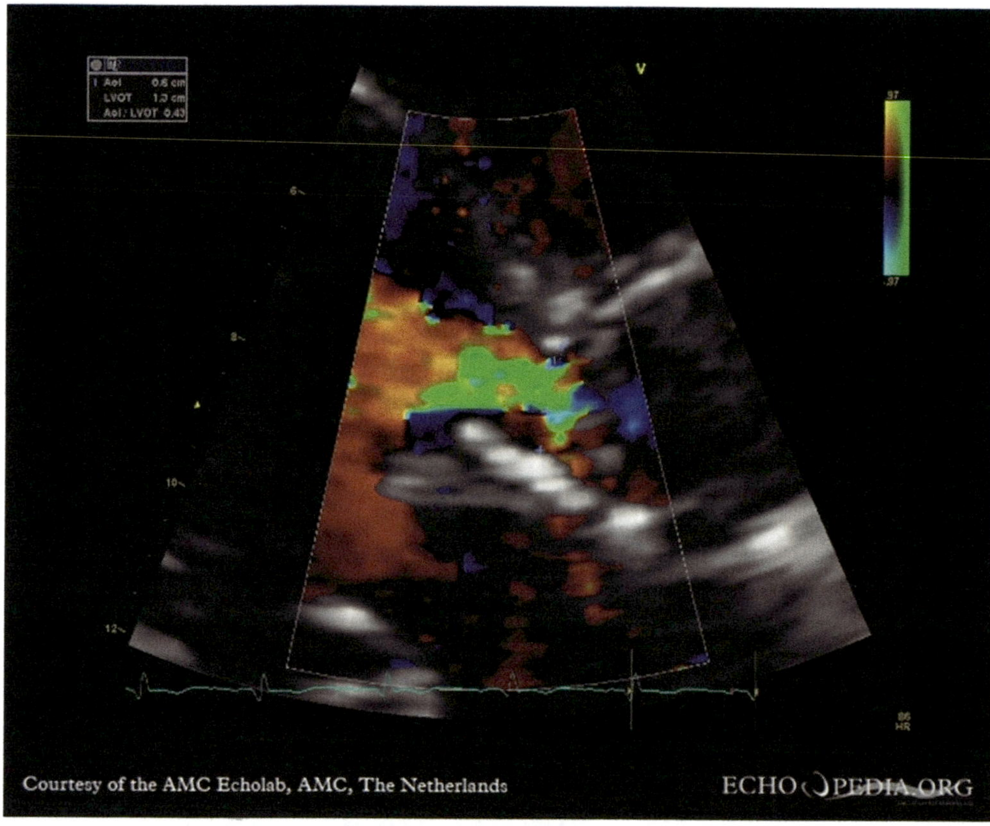

FIGURE 5.22

Vegetation on the aortic valve indicative of endocarditis. Source*: CardioNetworks: Secretariat/CardioNetworks: E00121.jpg. J. Vleugels, AMC, The Netherlands/; https://creativecommons.org/licenses/by-sa/3.0/deed.en; https://commons.wikimedia.org/wiki/File: Aortic_valve_endocarditis_with_vegetation_E00121_(CardioNetworks_ECHOpedia).jpg.*

Treatment and management

The eradication of endocardial vegetation and prevention or reduction of secondary systemic complications are the cornerstones of IE treatment and management. Emergency measures, such as stabilization and resuscitation, are required for cases presenting with acute decompensated heart failure, septic shock, or stroke.

Early cardiothoracic surgical intervention for valve repair or replacement is recommended in cases manifesting acute heart failure, acute valvular compromise, wide-spread infection with localized complications, recurrent arterial embolization, atrioventricular block, paravalvular abscess, presence of destructive infiltrative lesions, recurrent embolic events, or presence of large, mobile native valve vegetations less than 10 mm.[86]

Prolonged bactericidal antibiotic regimes are indispensable in the treatment of IE. The selection of the appropriate antibiotics and the duration of the regime varies depending on several bacterial and host factors, such as the type of valve or cardiac structure affected, and the antibiotic resistance of the pathogenic strain. Prosthetic valve involvement generally requires a longer-duration treatment regime compared to native valve endocarditis. Enterococcal strains like *E. faecalis* are commonly resistant to the bactericidal activity of ß-lactam antibiotic-based monotherapy, and therefore precipitate the need for combination regimens. Ampicillin or penicillin G in combination with an aminoglycoside, such as gentamicin or a dual ß-lactam regimen involving ampicillin in combination with ceftriaxone for a duration of 4 – 6 weeks, is generally recommended for *E. faecalis* infections.[108] Involvement of strains resistant to penicillin warrant the need for vancomycin – gentamicin combination therapy, although effective treatment of infections involving the rapidly emerging strains resistant to penicillin, gentamicin, and vancomycin need linezolid or daptomycin. During the antibiotic regime, it is recommended to draw two blood cultures every 24 – 48 hours to monitor the effectiveness of the regime and guide the ongoing therapy toward complete clearance of the infection. Therefore guidelines for antibiotic treatment need to be routinely reviewed according to the resistance patterns of the commonly circulating pathogenic strain.

Complications associated with post-IE treatment usually occur within the first year after. Therefore 3-monthly vigilant follow-ups are recommended during the first year. Patient counseling to educate and increase awareness regarding the importance of good dental hygiene and the risks involved in intravenous drug use and body piercings/tattoos is also recommended.

Prevention and control

Antibiotic prophylaxis is recommended for high-risk individuals, including those undergoing predisposing surgical procedures; those with prosthetic cardiac valves or devices; a history of IE; cyanotic congenital heart disease; congenital heart disease associated with persistent valvular insufficiency; or cardiac transplants preceding any invasive dental procedures.[109] Amoxicillin or clindamycin are the most recommended antibiotics used for prophylaxis.[110] Strict adherence to aseptic protocols during invasive medical or surgical procedures plays a critical role in preventing nosocomial spread of *E. faecalis* infections.

Student Notes

Key take-aways

- IE is associated with high levels of morbidity and mortality.
- *E. faecalis* is one of the several pathogens implicated in the pathogenesis of IE.
- Early clinical suspicion and rapid diagnosis are important to ensure good prognoses and outcomes in IE.
- The modified Duke criteria are used to help establish a reliable diagnosis of IE.
- Emergency stabilization and resuscitation may be required in cases manifesting acute heart failure.
- Most IE cases require surgical intervention for removing vegetative masses to ensure rapid clearance of the pathogen and prevention of systemic complications.
- Long duration intense antibiotic regimes are indispensable for complete eradication of the pathogen.
- Emergence of antibiotic resistant strains, such as those resistant to ß-lactams and vancomycin, poses a formidable challenge for clinical treatment.
- Antibiotic prophylaxis is recommended for high-risk individuals, such as those with prosthetic heart valves or valve repair, a history of endocarditis, cyanotic congenital heart disease, and underlying autoimmune conditions like atopic dermatitis, preceding invasive dental procedures.

References

1. Sika-Paotonu D, Beaton A, Raghu A, et al. Acute rheumatic fever and rheumatic heart disease, March 10, 2017 [Updated April 3, 2017]In: Ferretti JJ, Stevens DL, Fischetti VA, eds. Streptococcus pyogenes*: Basic Biology to Clinical Manifestations [Internet]*. Oklahoma City, OK: University of Oklahoma Health Sciences Center; 2016. Available from. Available from: https://www.ncbi.nlm.nih.gov/books/NBK425394/.
2. Doern CD, Burnham CA. It's not easy being green: the viridans group streptococci, with a focus on pediatric clinical manifestations. *J Clin Microbiol*. 2010;48(11):3829–3835. Available from: https://doi.org/10.1128/JCM.01563-10.
3. Ayi B. *xPharm: The Comprehensive Pharmacology Reference*. Elsevier; 2007:1–5, Infections caused by Viridans Streptococci. Available from: https://doi.org/10.1016/B978-008055232-3.60855-4.
4. Patterson MJ. Streptococcus. In: Baron S, ed. *Medical Microbiology*. 4th ed. Galveston, TX: University of Texas Medical Branch at Galveston; 1996, Chapter 13. Available from. Available from: https://www.ncbi.nlm.nih.gov/books/NBK7611/.
5. Dass C, Kanmanthareddy A. *Rheumatic heart disease* [Updated November 21, 2020]*StatPearls [Internet]*. Treasure Island (FL): StatPearls Publishing; 2020, January. Available from. Available from: https://www.ncbi.nlm.nih.gov/books/NBK538286/.
6. Reményi B, Wilson N, Steer A, et al. World Heart Federation criteria for echocardiographic diagnosis of rheumatic heart disease—an evidence-based guideline. *Nat Rev Cardiol*. 2012;9 (5):297–309. Available from: https://doi.org/10.1038/nrcardio.2012.7. Published February 28, 2012.

7 Gewitz MH, Baltimore RS, Tani LY, et al. Revision of the Jones Criteria for the diagnosis of acute rheumatic fever in the era of Doppler echocardiography: a scientific statement from the American Heart Association. *Circulation*. 2015;131(20):1806–1818. Available from: https://doi.org/10.1161/CIR.0000000000000205.

8. Watkins DA, Beaton AZ, Carapetis JR, et al. Rheumatic heart disease worldwide: JACC Scientific Expert Panel. *J Am Coll Cardiol*. 2018;72(12):1397–1416. Available from: https://doi.org/10.1016/j.jacc.2018.06.063.

9. Nulu S, Bukhman G, Kwan GF. Rheumatic heart disease: the unfinished global agenda. *Cardiol Clin*. 2017;35(1):165–180. Available from: https://doi.org/10.1016/j.ccl.2016.08.006.

10. Sinner SW, Tunkel AR. 8th ed. *Mandell, Douglas, and Bennett's Principles and Practice of Infectious Diseases*. vol 2. 204. 20152349–2361.e2. Available from: https://doi.org/10.1016/B978-1-4557-4801-3.00204-6.

11. Tariq N, Kyriakopoulos C, Group B. *Coxsackie virus* [Updated December 1, 2020]*StatPearls [Internet]*. Treasure Island, FL: StatPearls Publishing; 2020. Available from: https://www.ncbi.nlm.nih.gov/books/NBK560783/.

12. Roest S, Hesselink DA, Klimczak-Tomaniak D, et al. Incidence of end-stage renal disease after heart transplantation and effect of its treatment on survival. *ESC Heart Fail*. 2020;7(2):533–541. Available from: https://doi.org/10.1002/ehf2.12585.

13. Adams MJ, Lefkowitz EJ, King AM, et al. Ratification vote on taxonomic proposals to the International Committee on Taxonomy of Viruses (2016). *Arch Virol*. 2016;161(10):2921–2949. Available from: https://doi.org/10.1007/s00705-016-2977-6.

14. Johnsson T. Studies on the etiology of Bornholm disease (epidemic pleurodynia). II. Epidemiological observations. *Arch Gesamte Virusforsch*. 1954;5(4):401–412. Available from: https://doi.org/10.1007/BF01243009.

15. Kaplan MH, Klein SW, McPhee J, Harper RG. Group B coxsackievirus infections in infants younger than three months of age: a serious childhood illness. *Rev Infect Dis*. 1983;5(6):1019–1032. Available from: https://doi.org/10.1093/clinids/5.6.1019.

16. Esfandiarei M, McManus BM. Molecular biology and pathogenesis of viral myocarditis. *Annu Rev Pathol*. 2008;3:127–155. Available from: https://doi.org/10.1146/annurev.pathmechdis.3.121806.151534.

17. Blauwet LA, Cooper LT. Myocarditis. *Prog Cardiovasc Dis*. 2010;52(4):274–288. Available from: https://doi.org/10.1016/j.pcad.2009.11.006.

18. Selinka HC, Wolde A, Sauter M, Kandolf R, Klingel K. Virus-receptor interactions of coxsackie B viruses and their putative influence on cardiotropism. *Med Microbiol Immunol*. 2004;193(2–3):127–131. Available from: https://doi.org/10.1007/s00430-003-0193-y.

19. Archard LC, Richardson PJ, Olsen EG, Dubowitz V, Sewry C, Bowles NE. The role of coxsackie B viruses in the pathogenesis of myocarditis, dilated cardiomyopathy and inflammatory muscle disease. *Biochem Soc Symp*. 1987;53:51–62.

20. Berger JR, Chumley W, Pittman T, Given C, Nuovo G. Persistent coxsackie B encephalitis: report of a case and review of the literature. *J Neurovirol*. 2006;12(6):511–516. Available from: https://doi.org/10.1080/13550280601090546.

21. Muehlenbachs A, Bhatnagar J, Zaki SR. Tissue tropism, pathology and pathogenesis of enterovirus infection. *J Pathol*. 2015;235(2):217–228. Available from: https://doi.org/10.1002/path.4438.

22. Sioofy-Khojine AB, Lehtonen J, Nurminen N, et al. Coxsackievirus B1 infections are associated with the initiation of insulin-driven autoimmunity that progresses to type 1 diabetes. *Diabetologia*. 2018;61(5):1193–1202. Available from: https://doi.org/10.1007/s00125-018-4561-y.

23. Zhao L, Fu Z. Roles of host immunity in viral myocarditis and dilated cardiomyopathy. *J Immunol Res*. 2018;2018. Available from: https://doi.org/10.1155/2018/5301548. Article ID 5301548, 12 pages.

24. Feuer R, Mena I, Pagarigan RR, Harkins S, Hassett DE, Whitton JL. Coxsackievirus B3 and the neonatal CNS: the roles of stem cells, developing neurons, and apoptosis in infection, viral dissemination, and disease. *Am J Pathol*. 2003;163(4):1379–1393. Available from: https://doi.org/10.1016/S0002-9440(10)63496-7.

25. Richardson M, Elliman D, Maguire H, Simpson J, Nicoll A. Evidence base of incubation periods, periods of infectiousness and exclusion policies for the control of communicable diseases in schools and preschools [published correction appears in *Pediatr Infect Dis J* 2001 July; 20(7):653]*Pediatr Infect Dis J*. 2001;20(4):380–391. Available from: https://doi.org/10.1097/00006454-200104000-00004.

26. Altman R. Clinical aspects of enterovirus infection. *Postgrad Med*. 1964;35:451–459. Available from: https://doi.org/10.1080/00325481.1964.11695110.

27. Ksiazek TG, Erdman D, Goldsmith CS, et al. A novel coronavirus associated with severe acute respiratory syndrome. *N Engl J Med*. 2003;348(20):1953–1966. Available from: https://doi.org/10.1056/NEJMoa030781.

28. Cascella M, Rajnik M, Cuomo A, Dulebohn SC, Di Napoli R. *Features, evaluation, and treatment of coronavirus (COVID-19)*. *StatPearls*. Treasure Island, FL: StatPearls Publishing; 2021.

29. Weiss SR, Leibowitz JL. Coronavirus pathogenesis. *Adv Virus Res*. 2011;81:85–164. Available from: https://doi.org/10.1016/B978-0-12-385885-6.00009-2. PMID: 22094080; PMCID: PMC7149603.

30. Fehr AR, Perlman S. Coronaviruses: an overview of their replication and pathogenesis. *Methods Mol Biol*. 2015;1282:1–23. Available from: https://doi.org/10.1007/978-1-4939-2438-7_1.

31. Neuman BW, Adair BD, Yoshioka C, et al. Supramolecular architecture of severe acute respiratory syndrome coronavirus revealed by electron cryomicroscopy. *J Virol*. 2006;80(16):7918–7928. Available from: https://doi.org/10.1128/JVI.00645-06.

32. Hopkins, J. Bloomberg School of Public Health. COVID-19. https://www.centerforhealthsecurity.org/resources/COVID-19/index.html. Accessed 30.03.21.

33. Lu R, Zhao X, Li J, et al. Genomic characterisation and epidemiology of 2019 novel coronavirus: implications for virus origins and receptor binding. *Lancet.*. 2020;395(10224):565–574. Available from: https://doi.org/10.1016/S0140-6736(20)30251-8.

34. Chen Y, Liu Q, Guo D. Emerging coronaviruses: genome structure, replication, and pathogenesis [published correction appears in *J Med Virol*. 2020 Oct; 92(10):2249]*J Med Virol*. 2020;92(4):418–423. Available from: https://doi.org/10.1002/jmv.25681.

35. Paules CI, Marston HD, Fauci AS. Coronavirus infections—more than just the common cold. *JAMA*. 2020;323(8):707–708. Available from: https://doi.org/10.1001/jama.2020.0757.

36. de Wit E, van Doremalen N, Falzarano D, Munster VJ. SARS and MERS: recent insights into emerging coronaviruses. *Nat Rev Microbiol*. 2016;14(8):523–534. Available from: https://doi.org/10.1038/nrmicro.2016.81.

37. Chan KH, Peiris JS, Lam SY, Poon LL, Yuen KY, Seto WH. The effects of temperature and relative humidity on the viability of the SARS coronavirus. *Adv Virol*. 2011;2011:734690. Available from: https://doi.org/10.1155/2011/734690.

38. European Centre for Disease Prevention and Control. Interim guidance for environmental cleaning in non-healthcare facilities exposed to SARS-CoV-2. 18 February, 2020. https://www.ecdc.europa.eu/sites/default/files/documents/coronavirus-SARS-CoV-2-guidance-environmental-cleaning-non-healthcare-facilities.pdf. Accessed 31.03.21.

39. Lau AC, Yam LY, So LK. Management of critically ill patients with severe acute respiratory syndrome (SARS. *Int J Med Sci*. 2004;1(1):1–10. Available from: https://doi.org/10.7150/ijms.1.1.

40. Li W, Moore MJ, Vasilieva N, et al. Angiotensin-converting enzyme 2 is a functional receptor for the SARS coronavirus. *Nature*. 2003;426(6965):450–454. Available from: https://doi.org/10.1038/nature02145.

41. Chen Y, Guo Y, Pan Y, Zhao ZJ. Structure analysis of the receptor binding of 2019-nCoV [published online ahead of print, February 17, 2020]*Biochem Biophys Res Commun*. 2020;525(1):135–140. Available from: https://doi.org/10.1016/j.bbrc.2020.02.071.

42. Letko M, Marzi A, Munster V. Functional assessment of cell entry and receptor usage for SARS-CoV-2 and other lineage B betacoronaviruses. *Nat Microbiol*. 2020;5(4):562–569. Available from: https://doi.org/10.1038/s41564-020-0688-y.

43. Walls AC, Park YJ, Tortorici MA, Wall A, McGuire AT, Veesler D. Structure, function, and antigenicity of the SARS-CoV-2 spike glycoprotein [published correction appears in *Cell*. December 10, 2020; 183(6):1735]*Cell*. 2020;181(2):281–292. Available from: https://doi.org/10.1016/j.cell.2020.02.058. e6.

44. Ou X, Liu Y, Lei X, et al. Characterization of spike glycoprotein of SARS-CoV-2 on virus entry and its immune cross-reactivity with SARS-CoV. *Nat Commun*. 2020;11:1620.

45. Belouzard S, Millet JK, Licitra BN, Whittaker GR. Mechanisms of coronavirus cell entry mediated by the viral spike protein. *Viruses*. 2012;4(6):1011–1033. Available from: https://doi.org/10.3390/v4061011.

46. Belouzard S, Chu VC, Whittaker GR. Activation of the SARS coronavirus spike protein via sequential proteolytic cleavage at two distinct sites. *Proc Natl Acad Sci USA*. 2009;106(14):5871–5876. Available from: https://doi.org/10.1073/pnas.0809524106.

47. Tang NL, Chan PK, Wong CK, et al. Early enhanced expression of interferon-inducible protein-10 (CXCL-10) and other chemokines predicts adverse outcome in severe acute respiratory syndrome. *Clin Chem*. 2005;51(12):2333–2340. Available from: https://doi.org/10.1373/clinchem.2005.054460.

48. Xu Z, Shi L, Wang Y, et al. Pathological findings of COVID-19 associated with acute respiratory distress syndrome [published correction appears in *Lancet Respir Med*. February 25, 2020]*Lancet Respir Med*. 2020;8(4):420–422. Available from: https://doi.org/10.1016/S2213-2600(20)30076-X.

49. Xiong TY, Redwood S, Prendergast B, Chen M. Coronaviruses and the cardiovascular system: acute and long-term implications. *Eur Heart J*. 2020;41(19):1798–1800. Available from: https://doi.org/10.1093/eurheartj/ehaa231.

50. Chen N, Zhou M, Dong X, et al. Epidemiological and clinical characteristics of 99 cases of 2019 novel coronavirus pneumonia in Wuhan, China: a descriptive study. *Lancet*. 2020;395(10223):507–513. Available from: https://doi.org/10.1016/S0140-6736(20)30211-7.

51. Huang C, Wang Y, Li X, et al. Clinical features of patients infected with 2019 novel coronavirus in Wuhan, China [published correction appears in *Lancet*. January 30, 2020]*Lancet*. 2020;395(10223):497–506. Available from: https://doi.org/10.1016/S0140-6736(20)30183-5.

52. Prompetchara E, Ketloy C, Palaga T. Immune responses in COVID-19 and potential vaccines: lessons learned from SARS and MERS epidemic. *Asian Pac J Allergy Immunol*. 2020;38(1):1–9. Available from: https://doi.org/10.12932/AP-200220-0772.

53. Chen J, Lau YF, Lamirande EW, et al. Cellular immune responses to severe acute respiratory syndrome coronavirus (SARS-CoV) infection in senescent BALB/c mice: CD4 + T cells are important in control of SARS-CoV infection. *J Virol*. 2010;84(3):1289–1301. Available from: https://doi.org/10.1128/JVI.01281-09.

54. Rabi FA, Al Zoubi MS, Kasasbeh GA, Salameh DM, Al-Nasser AD. SARS-CoV-2 and coronavirus disease 2019: what we know so far. *Pathogens*. 2020;9(3):231. Available from: https://doi.org/10.3390/pathogens9030231. Published March 20, 2020.

55. Li G, Fan Y, Lai Y, et al. Coronavirus infections and immune responses. *J Med Virol.* 2020;92 (4):424–432. Available from: https://doi.org/10.1002/jmv.25685.

56. Zumla A, Hui DS, Azhar EI, Memish ZA, Maeurer M. Reducing mortality from 2019-nCoV: host-directed therapies should be an option. *Lancet.*. 2020;395(10224):e35–e36. Available from: https://doi.org/10.1016/S0140-6736(20)30305-6.

57. Li X, Geng M, Peng Y, Meng L, Lu S. Molecular immune pathogenesis and diagnosis of COVID-19. *J Pharm Anal.* 2020;10(2):102–108. Available from: https://doi.org/10.1016/j.jpha.2020.03.001. Epub March 5, 2020. PMID: 32282863; PMCID: PMC7104082.

58. Fan YY, Huang ZT, Li L, et al. Characterization of SARS-CoV-specific memory T cells from recovered individuals 4 years after infection. *Arch Virol.* 2009;154(7):1093–1099. Available from: https://doi.org/10.1007/s00705-009-0409-6.

59. Tang F, Quan Y, Xin ZT, et al. Lack of peripheral memory B cell responses in recovered patients with severe acute respiratory syndrome: a six-year follow-up study. *J Immunol.* 2011;186(12):7264–7268. Available from: https://doi.org/10.4049/jimmunol.0903490.

60. Pal M, Berhanu G, Desalegn C, Kandi V. Severe acute respiratory syndrome coronavirus-2 (SARS-CoV-2): an update. *Cureus.* 2020;12(3):e7423. Available from: https://doi.org/10.7759/cureus.7423. Published March 26, 2020.

61. Xu XW, Wu XX, Jiang XG, et al. Clinical findings in a group of patients infected with the 2019 novel coronavirus (SARS-Cov-2) outside of Wuhan, China: retrospective case series [published correction appears in *BMJ.* February 27, 2020;368: m792]*BMJ.* 2020;368:m606. Available from: https://doi.org/10.1136/bmj.m606. Published February 19, 2020.

62. Centers for Disease Control and Prevention. Different COVID-19 vaccines. Updated March 4, 2021. https://www.cdc.gov/coronavirus/2019-ncov/vaccines/different-vaccines.html. Accessed 31.03.21.

63. Cahill TJ, Prendergast BD. Infective endocarditis. *Lancet.*. 2016;387(10021):882–893. Available from: https://doi.org/10.1016/S0140-6736(15)00067-7.

64. Correa de Sa DD, Tleyjeh IM, Anavekar NS, et al. Epidemiological trends of infective endocarditis: a population-based study in Olmsted County, Minnesota [published correction appears in *Mayo Clin Proc.* August 2010; 85(8):772]*Mayo Clin Proc.* 2010;85(5):422–426. Available from: https://doi.org/10.4065/mcp.2009.0585.

65. Murdoch DR, Corey GR, Hoen B, et al. Clinical presentation, etiology, and outcome of infective endocarditis in the 21st century: The International Collaboration on Endocarditis-Prospective Cohort Study. *Arch Intern Med.* 2009;169(5):463–473. Available from: https://doi.org/10.1001/archinternmed.2008.603.

66. Aas JA, Paster BJ, Stokes LN, Olsen I, Dewhirst FE. Defining the normal bacterial flora of the oral cavity. *J Clin Microbiol.* 2005;43(11):5721–5732. Available from: https://doi.org/10.1128/JCM.43.11.5721-5732.2005.

67. Kouidhi B, Zmantar T, Mahdouani K, Hentati H, Bakhrouf A. Antibiotic resistance and adhesion properties of oral Enterococci associated to dental caries. *BMC Microbiol.* 2011;11:155. Available from: https://doi.org/10.1186/1471-2180-11-155. Published June 29, 2011.

68. Dahlén G, Blomqvist S, Almståhl A, Carlén A. Virulence factors and antibiotic susceptibility in enterococci isolated from oral mucosal and deep infections. *J Oral Microbiol.* 2012;4. Available from: https://doi.org/10.3402/jom.v4i0.10855.

69. Rams TE, Degener JE, van Winkelhoff AJ. Prevalence of β-lactamase-producing bacteria in human periodontitis. *J Periodontal Res.* 2013;48(4):493–499. Available from: https://doi.org/10.1111/jre.12031.

70. Sandoe JAT, Witherden IR, Cove JH, Heritage J, Wilcox MH. Correlation between enterococcal biofilm formation in vitro and medical-device-related infection potential in vivo. *J Med Microbiol.* 2003;52(Pt 7):547–550. Available from: https://doi.org/10.1099/jmm.0.05201-0.

71. Paganelli FL, Willems RJ, Leavis HL. Optimizing future treatment of enterococcal infections: attacking the biofilm? *Trends Microbiol.* 2012;20(1):40–49. Available from: https://doi.org/10.1016/j.tim.2011.11.001.

72. Arias CA, Murray BE. Emergence and management of drug-resistant enterococcal infections. *Expert Rev Anti Infect Ther.* 2008;6(5):637–655. Available from: https://doi.org/10.1586/14787210.6.5.637.

73. Gilmore MS, Lebreton F, van Schaik W. Genomic transition of enterococci from gut commensals to leading causes of multidrug-resistant hospital infection in the antibiotic era. *Curr Opin Microbiol.* 2013;16(1):10–16. Available from: https://doi.org/10.1016/j.mib.2013.01.006.

74. Cetinkaya Y, Falk P, Mayhall CG. Vancomycin-resistant enterococci. *Clin Microbiol Rev.* 2000;13(4):686–707. Available from: https://doi.org/10.1128/cmr.13.4.686-707.2000.

75. Siqueira Jr JF, Rôças IN. Diversity of endodontic microbiota revisited. *J Dent Res.* 2009;88(11):969–981. Available from: https://doi.org/10.1177/0022034509346549.

76. Vidana R, Sullivan A, Billström H, Ahlquist M, Lund B. *Enterococcus faecalis* infection in root canals - host-derived or exogenous source? *Lett Appl Microbiol.* 2011;52(2):109–115. Available from: https://doi.org/10.1111/j.1472-765X.2010.02972.x.

77. Al-Ahmad A, Ameen H, Pelz K, et al. Antibiotic resistance and capacity for biofilm formation of different bacteria isolated from endodontic infections associated with root-filled teeth. *J Endod.* 2014;40(2):223–230. Available from: https://doi.org/10.1016/j.joen.2013.07.023.

78. Ran S, Liu B, Jiang W, Sun Z, Liang J. Transcriptome analysis of *Enterococcus faecalis* in response to alkaline stress. *Front Microbiol.* 2015;6:795. Available from: https://doi.org/10.3389/fmicb.2015.00795. Published August 7, 2015.

79. Anderson AC, Al-Ahmad A, Elamin F, et al. Comparison of the bacterial composition and structure in symptomatic and asymptomatic endodontic infections associated with root-filled teeth using pyrosequencing. *PLoS One.* 2013;8(12):e84960. Available from: https://doi.org/10.1371/journal.pone.0084960. Published December 30, 2013.

80. Larsen J, Schønheyder HC, Singh KV, et al. Porcine and human community reservoirs of *Enterococcus faecalis*, Denmark. *Emerg Infect Dis.* 2011;17(12):2395–2397. Available from: https://doi.org/10.3201/eid1712.101584.

81. Hammerum AM. Enterococci of animal origin and their significance for public health. *Clin Microbiol Infect.* 2012;18(7):619–625. Available from: https://doi.org/10.1111/j.1469-0691.2012.03829.x.

82. Larsen J, Schønheyder HC, Lester CH, et al. Porcine-origin gentamicin-resistant *Enterococcus faecalis* in humans, Denmark. *Emerg Infect Dis.* 2010;16(4):682–684. Available from: https://doi.org/10.3201/eid1604.090500.

83. Gelsomino R, Vancanneyt M, Cogan TM, Condon S, Swings J. Source of enterococci in a farmhouse raw-milk cheese. *Appl Environ Microbiol.* 2002;68(7):3560–3565. Available from: https://doi.org/10.1128/aem.68.7.3560-3565.2002.

84. Rodbard S. Hemodynamic considerations in endocarditis. *Med Sci.* 1964;15:43–49.

85. Veloso TR, Chaouch A, Roger T, et al. Use of a human-like low-grade bacteremia model of experimental endocarditis to study the role of *Staphylococcus aureus* adhesins and platelet aggregation in early endocarditis. *Infect Immun.* 2013;81(3):697–703. Available from: https://doi.org/10.1128/IAI.01030-12.

86. Yallowitz AW, Decker LC. *Infectious endocarditis* [Updated February 8, 2021]*StatPearls [Internet]*. Treasure Island, FL: StatPearls Publishing; 2021, January. Available from. Available from: https://www.ncbi.nlm.nih.gov/books/NBK557641/.

87. Jung CJ, Yeh CY, Shun CT, et al. Platelets enhance biofilm formation and resistance of endocarditis-inducing streptococci on the injured heart valve. *J Infect Dis.* 2012;205 (7):1066–1075. Available from: https://doi.org/10.1093/infdis/jis021.

88. Toledo-Arana A, Valle J, Solano C, et al. The enterococcal surface protein, Esp, is involved in *Enterococcus faecalis* biofilm formation. *Appl Environ Microbiol.* 2001;67(10):4538–4545. Available from: https://doi.org/10.1128/aem.67.10.4538-4545.2001.

89. Franz CM, Stiles ME, Schleifer KH, Holzapfel WH. Enterococci in foods—a conundrum for food safety. *Int J Food Microbiol.* 2003;88(2-3):105–122. Available from: https://doi.org/10.1016/s0168-1605(03)00174-0.

90. Kayaoglu G, Ørstavik D. Virulence factors of *Enterococcus faecalis*: relationship to endodontic disease. *Crit Rev Oral Biol Med.* 2004;15(5):308–320. Available from: https://doi.org/10.1177/154411130401500506. Published September 1, 2004.

91. Kreft B, Marre R, Schramm U, Wirth R. Aggregation substance of *Enterococcus faecalis* mediates adhesion to cultured renal tubular cells. *Infect Immun.* 1992;60(1):25–30. Available from: https://doi.org/10.1128/IAI.60.1.25-30.1992.

92. Olmsted SB, Dunny GM, Erlandsen SL, Wells CL. A plasmid-encoded surface protein on *Enterococcus faecalis* augments its internalization by cultured intestinal epithelial cells. *J Infect Dis.* 1994;170(6):1549–1556. Available from: https://doi.org/10.1093/infdis/170.6.1549.

93. Süssmuth SD, Muscholl-Silberhorn A, Wirth R, Susa M, Marre R, Rozdzinski E. Aggregation substance promotes adherence, phagocytosis, and intracellular survival of *Enterococcus faecalis* within human macrophages and suppresses respiratory burst. *Infect Immun.* 2000;68 (9):4900–4906. Available from: https://doi.org/10.1128/iai.68.9.4900-4906.2000.

94. Chuang-Smith ON, Wells CL, Henry-Stanley MJ, Dunny GM. Acceleration of *Enterococcus faecalis* biofilm formation by aggregation substance expression in an ex vivo model of cardiac valve colonization. *PLoS One.* 2010;5(12):e15798. Available from: https://doi.org/10.1371/journal.pone.0015798. Published 2010 Dec 30.

95. Anderson AC, Jonas D, Huber I, et al. *Enterococcus faecalis* from food, clinical specimens, and oral sites: prevalence of virulence factors in association with biofilm formation. *Front Microbiol.* 2016;6:1534. Available from: https://doi.org/10.3389/fmicb.2015.01534. Published January 11, 2016.

96. Fisher K, Phillips C. The ecology, epidemiology and virulence of Enterococcus. *Microbiology (Read).* 2009;155(Pt 6):1749–1757. Available from: https://doi.org/10.1099/mic.0.026385-0.

97. Singh KV, Coque TM, Weinstock GM, Murray BE. In vivo testing of an *Enterococcus faecalis* efaA mutant and use of efaA homologs for species identification. *FEMS Immunol Med Microbiol.* 1998;21(4):323–331. Available from: https://doi.org/10.1111/j.1574-695X.1998.tb01180.x.

98. Reynaudaf Geijersstam A, Culak R, Molenaar L, et al. Comparative analysis of virulence determinants and mass spectral profiles of Finnish and Lithuanian endodontic *Enterococcus faecalis* isolates. *Oral Microbiol Immunol.* 2007;22(2):87–94. Available from: https://doi.org/10.1111/j.1399-302X.2007.00327.x.

99. Van Tyne D, Martin MJ, Gilmore MS. Structure, function, and biology of the *Enterococcus faecalis* cytolysin. *Toxins (Basel).* 2013;5(5):895–911. Available from: https://doi.org/10.3390/toxins5050895. Published April 29, 2013.

100. Van Tyne D, Gilmore MS. Friend turned foe: evolution of enterococcal virulence and antibiotic resistance. *Annu Rev Microbiol.* 2014;68:337–356. Available from: https://doi.org/10.1146/annurev-micro-091213-113003.

101. Chong KKL, Tay WH, Janela B, et al. *Enterococcus faecalis* modulates immune activation and slows healing during wound infection. *J Infect Dis.* 2017;216(12):1644–1654. Available from: https://doi.org/10.1093/infdis/jix541.

102. Thurlow LR, Thomas VC, Fleming SD, Hancock LE. *Enterococcus faecalis* capsular polysaccharide serotypes C and D and their contributions to host innate immune evasion. *Infect Immun.* 2009;77(12):5551–5557. Available from: https://doi.org/10.1128/IAI.00576-09.

103. Tien BYQ, Goh HMS, Chong KKL, et al. *Enterococcus faecalis* promotes innate immune suppression and polymicrobial catheter-associated urinary tract infection. *Infect Immun.* 2017;85(12). Available from: https://doi.org/10.1128/IAI.00378-17. e00378-17. Published November 17, 2017.

104. Manson JM, Hancock LE, Gilmore MS. Mechanism of chromosomal transfer of *Enterococcus faecalis* pathogenicity island, capsule, antimicrobial resistance, and other traits. *Proc Natl Acad Sci USA.* 2010;107(27):12269–12274. Available from: https://doi.org/10.1073/pnas.1000139107. Epub June 21, 2010. PMID: 20566881; PMCID: PMC2901427.

105. Nakagawa T, Wada H, Sakakura K, et al. Clinical features of infective endocarditis: comparison between the 1990s and 2000s. *J Cardiol.* 2014;63(2):145–148. Available from: https://doi.org/10.1016/j.jjcc.2013.06.007.

106. Li JS, Sexton DJ, Mick N, et al. Proposed modifications to the Duke criteria for the diagnosis of infective endocarditis. *Clin Infect Dis.* 2000;30(4):633–638. Available from: https://doi.org/10.1086/313753.

107. Baddour LM, Wilson WR, Bayer AS, et al. Infective endocarditis in adults: diagnosis, antimicrobial therapy, and management of complications: a scientific statement for healthcare professionals from the American Heart Association [published correction appears in *Circulation.* October 27, 2015; 132(17):e215] [published correction appears in *Circulation.* August 23, 2016; 134(8):e113] [published correction appears in *Circulation.* July 31, 2018; 138(5):e78-e79]*Circulation.* 2015;132(15):1435–1486. Available from: https://doi.org/10.1161/CIR.0000000000000296.

108. Gavaldà J, Len O, Miró JM, et al. Brief communication: treatment of *Enterococcus faecalis* endocarditis with ampicillin plus ceftriaxone. *Ann Intern Med.* 2007;146(8):574–579. Available from: https://doi.org/10.7326/0003-4819-146-8-200704170-00008.

109. Nishimura RA, Otto CM, Bonow RO, et al. 2017 AHA/ACC focused update of the 2014 AHA/ACC guideline for the management of patients with valvular heart disease: a report of the American College of Cardiology/American Heart Association Task Force on Clinical Practice Guidelines. *Circulation.* 2017;135(25):e1159–e1195. Available from: https://doi.org/10.1161/CIR.0000000000000503.

110. Hoen B, Duval X. Clinical practice. Infective endocarditis [published correction appears in *N Engl J Med.* June 27, 2013; 368(26):2536]*N Engl J Med.* 2013;368(15):1425–1433. Available from: https://doi.org/10.1056/NEJMcp1206782.

CHAPTER 6

Infections of the urogenital tract

CONTENTS

Problem number 6.1: Urinary tract infection caused by *Escherichia coli* (extended-spectrum ß-lactamase)

A 3-month-old girl, with a history of allergy to cow's milk, presented with fever without any complaints of runny nose, cough, flu-like symptoms, rashes, or urinary problems. Although her stool appeared to be greenish, there were no complaints of diarrhea or constipation. Her feed intake had reduced and her activity levels were reported to be decreased. She did not have any known food/drug allergies, and her vaccinations were up-to-date. She had been delivered through normal vaginal delivery at full term; however, her mother had urinary tract infection (UTI) and gestational diabetes during pregnancy. There were no pre-or postnatal complications. She showed normal development and had achieved all the milestones on the appropriate timeline. She did not have any history of other known medical illnesses or surgeries and was not taking any medications. She had been admitted to the hospital once at the age of 24 days for 12 days for UTI and respiratory syncytial virus infection. Antibiotic treatment was given to resolve the infections.

1. Vitals
   Temperature: 38.4°C
   Blood pressure: 110/70 mmHg
2. Physical examination
   Comfortably asleep
   No muscle wasting/deformities/missing limbs
   No swollen lymph nodes
   No visible rashes, bruises, scars, lesions, or ulcers
   No evidence of pallor or jaundice
3. Respiratory examination: Equal air entry bilaterally; no added sounds
4. Cardiovascular examination: Normal S1, S2; no added sounds
5. Abdominal examination: Soft; lax; non-tender
6. Ear and throat examination: No congestion
7. Radiological investigations

Problem-Based Learning Approach in Microbiology. DOI: https://doi.org/10.1016/B978-0-323-95092-3.00004-4

X-ray: Normal

8. C-reactive protein: 14.5
9. Urine examination (Culture + Sensitivity) (obtained using a straight catheter)
   Microorganism: *E. coli* (ESBL); 10,000–100,000 cfu/mL

Antibiotic	Sensitivity
Amoxicillin/Clavulanate	Resistant
Cephalothin	Resistant
Cefuroxime	Resistant
Ceftriaxone	Resistant
Meropenem	Susceptible
Gentamycin	Susceptible
Nitrofurantoin	Susceptible
Trimethoprim/Sulphamet	Susceptible
Ciprofloxacin	Susceptible
Fosfomycin/Trometamol	Susceptible

10. Blood culture: Sterile
11. Cerebrospinal fluid (CSF) examination
    White blood cells: 2 cells/μL
    Red blood cells: 6 cells/mL
    CSF culture: Sterile
    Gram stain: No organism seen
12. Biochemical profile

Glucose (mmol/L)	3.4	4.4–7.8
Lactic acid (mmol/L)	2.9	0.5–2.2
Protein (g/dL)	3.58	4.4–6.6

Diagnosis

The final diagnosis was that of UTI caused by extended-spectrum ß-lactamase (ESBL)-producing *E. coli* infection.

Treatment

Admission to the hospital for 10 days

Antibiotics administered:

1. Ampicillin 200 mg/day Q 6 hours

2. Claforan 200 mg/day Q 8 hours
3. Meropenem 150 mg Q 8 hours

Question and answer

1. What causes ESBL-producing *E. coli* infection in neonates and infants?
   ESBL-producing *E. coli* does not cause vertical transmission of infection from an infected to an unborn fetus. However, neonates and infants may be infected by ESBL-producing *E. coli* during the process of parturition as a result of transmission by contact in the vaginal delivery of mothers with active infection.
2. What distinguishes *E. coli* from other coliforms?
   The ability to ferment lactose at temperatures higher than that favorable for other coliforms distinguishes *E. coli* from other coliforms.
3. Which risk factors predispose individuals to recurrent *E. coli* infections?
   Risk factors such as diabetes mellitus, pregnancy, obesity, vaginal infections, age above 60 years, and congenital defects of the renal system such as vesicoureteral reflux predispose individuals to recurrent UTIs caused by *E. coli*.
4. What is the virulence attributes essential for the colonization and survival of uropathogenic *E. coli* (UPEC) in the urinary tract?
   Adhesion factors, known as adhesins, allow the UPEC bacteria to adhere to the receptors of the uroepithelial cells and are instrumental in facilitating the invasion of the urinary tract. Invasion, formation of colonies, and proliferation of the bacteria enable the establishment of infection in the host at a location other than the original niche of the microorganism. UPEC uses the ability to produce biofilms as one of the strategies to invade the urinary tract, while effectively evading the host immune system. Iron acquisition systems consisting of heme and siderophores are also essential for the virulence of UPEC in the urinary tract.
5. How are resistance genes such as blaCTX-M transferred by ESBL-producing UPEC?
   Biofilms in ESBL-producing UPEC use conjugation to facilitate the transfer of plasmid-encoded resistance genes such as blaCTX-M.
6. What does an alkaline pH above 8.5 of a urine sample indicate?
   Normal urine has a pH ranging between 4.5 and 8.0, and an alkaline pH of more than 8.5 is indicative of either bacterial infections of the urinary tract or struvite stones formed as a result of the infection caused by urease-producing pathogens.
7. What is leukocyte esterase and what does its presence in urine samples indicate?

Leukocyte esterase is an enzyme produced by leukocytes and released in urine in response to bacterial infection. Therefore the presence of leukocyte esterase in urine samples is a good indicator of an active UTI.

Learning objectives

1. Determine the appropriate differential diagnosis based on patient history, known underlying medical conditions, and clinical manifestations.
2. Differentiate the clinical manifestations of UTIs from other causes based on urinalysis (UA) reports.
3. Describe the characteristics of *E. coli* in the context of UPEC.
4. Identify ESBL-producing *E. coli* strains based on urine culture reports.
5. Summarize the pathogenesis of UPEC.
6. Predict the prognosis of the case based on the clinical manifestations and the extent of pathology presented.
7. Enumerate the methods involved in the diagnosis of UPEC infection.
8. Determine the appropriate antibiotic regimen based on the culture sensitivity reports to ensure rapid and effective resolution of the infection.

Discussion

Differential diagnosis

Pyelonephritis	Pyelonephritis is an inflammatory condition of the kidney caused because of a bacterial infection. The condition manifests with a fever, chills, pain and burning during urination and frequent urge for urination.
Kidney stones	Kidney stones may present themselves with abdominal, flank, or back pain, chills, fever, nausea, and vomiting.
Vaginitis	Vaginitis is an inflammation of the vagina and can cause discharge, unpleasant odors, pain, and itching.
Pelvic inflammatory disease	Pelvic inflammatory disease results from an infection of the female reproductive organs and can manifest as a fever, chills, mild to severe pain, abnormal or heavy vaginal discharge with an unpleasant odor, abnormal uterine bleeding, pain during intercourse, painful and frequent urination.
Genital herpes simplex virus 1 infection	Genital herpes is usually caused by HSV 2 in sexually active adults but may be caused by HSV 1 in infants and small children. The infection manifests with painful blisters, which break to form tender sores resulting in pain, burning, and itching.

Background

ESBL-producing *E. coli* are resistant to most of the ß-lactam antibiotics, and are, therefore, associated with complicated treatment regimes. Multidrug resistance results in a poor outcome for infections associated with these pathogens and facilitate the effective spread of the pathogen. While ESBL-producing *E. coli* have been implicated in a large number of infections, complicated UTIs is the most common pathological manifestation of ESBL-producing *E. coli*. UTIs caused by ESBL-producing *E. coli* are generally difficult to treat and result in an increased burden of morbidity requiring hospital admissions, intensive care, extended hospital stays, and a reduction in favorable patient response.

Although vertical transmission of ESBL-producing *E. coli* has not been reported, neonates and infants may be infected by ESBL-producing *E. coli* as a result of transmission by contact during vaginal delivery.[1] Therefore a history of UTIs during pregnancy should be considered a potential risk factor for the neonate and should indicate the need for postpartum screening of the neonate for potential infections caused by *E. coli*.

Etiological agent

Extended-spectrum ß-lactamase producing E. coli

Classification and morphology

ESBL-producing *E. coli* are a multidrug resistant variant of the *E. coli* bacteria, which are a subset of the group of coliform bacteria. *E. coli* are Gram-negative, rod-shaped (Fig. 6.1), nonspore-forming, and motile bacteria belonging to the genus *Escherichia*. Some of the *E. coli* strains may be capsulated. Although *E. coli* are preferably aerobic organisms, they can survive facultatively in anaerobic conditions. The ability to ferment lactose at temperatures higher than that favorable for other coliforms distinguishes *E. coli* from other coliforms.

Epidemiology

E. coli are parasites strictly restricted to human and animal intestines, where these bacteria are present as part of the normal flora without causing infection and disease in the host. Certain risk factors such as diabetes mellitus, pregnancy, obesity, vaginal infections, age above 60 years, and congenital defects of the renal system such as vesicoureteral reflux predispose individuals to recurrent UTIs caused by *E. coli*.[2]

ESBL-producing *E. coli* were considered to be primarily nosocomial, but have now become increasingly prevalent in the community worldwide. This radical change in the epidemiology of ESBL-producing *E. coli* poses a serious public health threat. Community-acquired UTIs caused by ESBL-producing

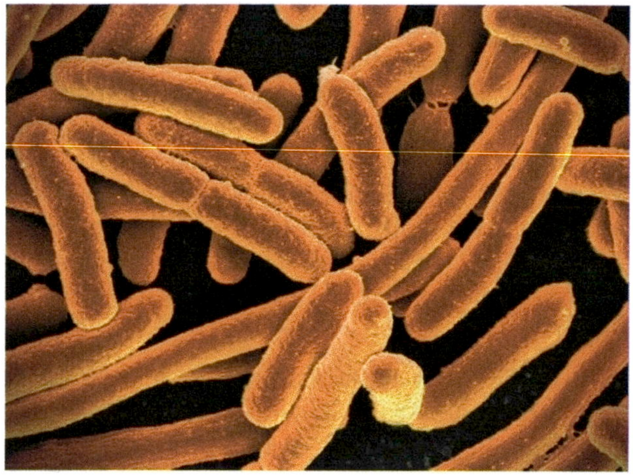

FIGURE 6.1
Colorized scanning electron micrograph of *Escherichia coli.* Source: *National Institute of Allergy and Infectious Diseases, National Institutes of Health/; https://creativecommons.org/licenses/by-nc/2.0/; https://www.flickr.com/photos/nihgov/24661308922.*

E. coli are prone to developing complications resulting in acute bloodstream infections reported to have high levels of mortality. The widespread prevalence of ESBL-producing *E. coli* resistant to most antibiotics in communities globally has led to an increased need for reevaluation of the epidemiological characteristics of these multidrug resistant pathogens.

Virulence and pathogenicity

E. coli strains capable of causing UTIs, also known as UPEC, acquire certain specific virulence attributes that are essential for the colonization and survival of the bacteria in the urinary tract. Adhesion factors, known as adhesins that allow the bacteria to adhere to the receptors of the uroepithelial cells, are instrumental in conferring invasive abilities to the bacteria.[3,4] Invasion, formation of colonies, and proliferation of bacteria are the primary steps required for establishing infection in the host at a location other than the original niche of the microorganism. Biofilm production is one of the strategies used by UPEC to invade the urinary tract, while effectively evading the host immune system.[5] Biofilm synthesis enabled by the *pgaABCD* gene locus depends on various factors intrinsic to the pathogen and derives its adhesive abilities from the polysaccharides comprising the biofilm.[6] Iron acquisition systems consisting of heme and siderophores are also essential for the virulence of *E. coli* in the urinary tract.[7] The bacteria also release various toxins such as hemolysins, cytotoxic necrotizing factor-1, and the members of the auto-transporter family[8] (Fig. 6.3).

Broad-spectrum resistance to antibiotics in *E. coli* has led to the emergence of ESBL-producing *E. coli,* which are implicated in a large number of community-associated complicated UTIs. CTX-M enzymes, a group of ESBLs, are widely disseminated in ESBL-producing UPEC worldwide,[9,10] and are responsible for the resistance of the bacteria to a broad spectrum of antibiotics. Biofilms in ESBL-producing UPEC use conjugation to facilitate the transfer of resistance genes such as $bla_{CTX\text{-}M}$ encoded by plasmids.[11,12] ESBLs confer resistance to most of the ß-lactam antibiotics including penicillin, cephalosporin, norfloxacin, cefotaxime, and cotrimoxazole,[13] thereby increasing the challenge associated with effective treatment of ESBL-producing UPEC infections.

Clinical manifestations

UPEC infections present with pain during urination (dysuria), frequent urination urge, inability to start the urine stream, sudden urge to urinate immediately, cloudy and foul-smelling urine, and blood in the urine (hematuria). Uncomplicated UTIs usually do not manifest signs of kidney involvement of upper tract disease such as pyelonephritis. Therefore fever, chills, nausea, vomiting, and back pain may not present in all cases of UTIs. The ambiguity associated with differentiating the diagnosis of uncomplicated UTIs from kidney infections emphasizes the need for aggressive treatment for upper UTIs in cases where a diagnosis is ambiguous.

Patients with indwelling catheters and those with congenital or acquired renal conditions are at high risk of developing complicated ESBL-producing UPEC infections owing to repeated exposures to various antibiotics for recurrent UTIs. Such patients also tend to have vague symptoms such as an elevated white blood cells (WBC) count, fatigue, and fever. Pyuria resulting from elevated bacterial colony counts in the urine indicate UTI. Pyonephrosis and sepsis leading to septic shock are rare complications of untreated UTIs and are associated with high levels of morbidity and mortality. Urinary tract obstruction presenting as a complication of infections can lead to further stagnation of urine, worsening causes a rapid deterioration in the patient's clinical conditions.

Laboratory diagnosis

UA is considered to be the gold standard for diagnosing an UTI, and for determining culture sensitivity. A clean-catch specimen collected mid-stream is vital for ensuring an accurate UA. The collected urine sample should be sent to the lab immediately or stored under refrigerated conditions to avoid rapid bacterial growth that can occur at room temperature, and thus compromise the accuracy of the test.

Although pyuria may be detected by visual examination of the urine, the diagnosis of UTI should be confirmed through accurate UA tests. Dipstick testing can prove to be rapid and effective with pH, leukocyte esterase, presence of blood, and nitrites acting as useful indicators for diagnosis. The dipstick test done using the first-morning urine has more than 90% specificity as the presence of bacteria that causes uncomplicated UTIs results in the conversion of nitrates present in the urine to nitrites over 6 hours.

Normal urine has a pH ranging between 4.5 and 8.0, and an alkaline pH of more than 8.5 is indicative of either bacterial infections of the urinary tract or struvite stones formed as a result of the infection caused by urease-producing pathogens. Leukocyte esterase is released by WBCs in urine in response to bacterial infection, and is, therefore, a good indicator of UTIs, although the presence of WBCs may also indicate inflammatory conditions associated with the bladder. Traces of blood in urine differentiate UTIs from vaginitis and urethritis as hematuria is generally indicative of damage to the bladder lining caused by bacterial infections[14] (Figs. 6.2 and 6.3).

Microscopic evaluation of the urine sample indicated by the results of the UA facilitates detection of WBCs and bacterial cells in the urine sample, thereby confirming an infection of the urinary tract. Urine culture analyses are not required in uncomplicated UTIs; however, risk factors such as immunosuppression, diabetes mellitus, pregnancy, and congenital or acquired renal conditions may indicate the need for urine culture depending on the results of the microscopic evaluation of the kidney. A urine culture may also be combined with antibiotic sensitivity assays if resistance to antibiotics is suspected.

Treatment and prevention

The advent of antibiotics has been directly responsible for improved prognosis and outcome in severe infections. However, the widespread and generalized use of antibiotics has resulted in the formation of the vicious circle of the emergence of resistant bacteria necessitating the use of more effective and aggressive

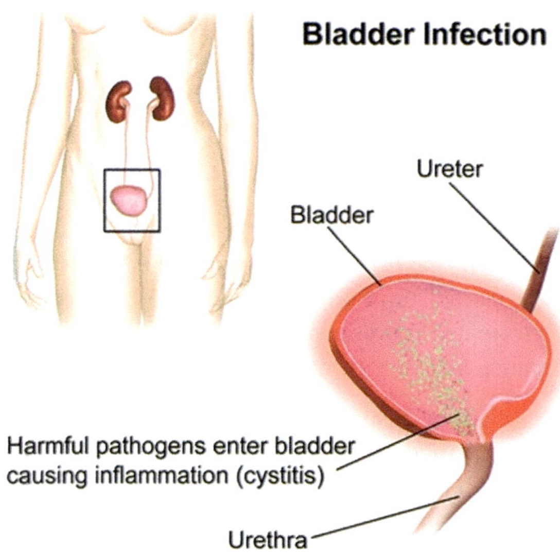

FIGURE 6.2
Bladder infection. Source: *BruceBlaus/Own work/; https://creativecommons.org/licenses/by-sa/4.0/deed.en; https://commons.wikimedia.org/wiki/File:Bladder_Infection.png.*

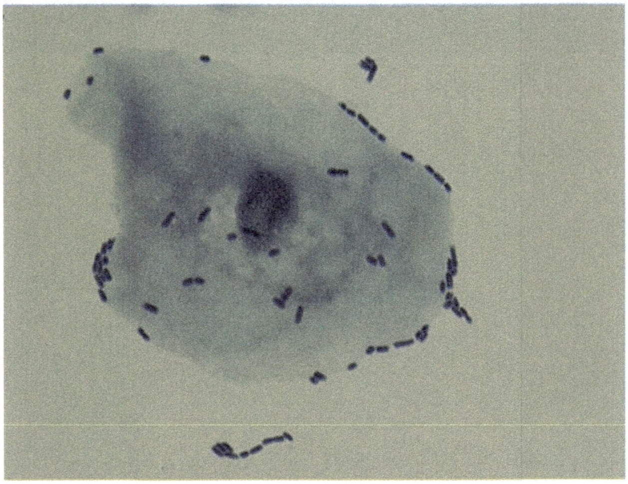

FIGURE 6.3
Uropathogenic *Escherichia coli* (UPEC) adhering to bladder epithelial cell. Source*: Stefan Walkowski/Own work/; https://creativecommons.org/licenses/by-sa/4.0/deed.en; https://commons.wikimedia.org/wiki/File:UPEC_adhered_to_BEC.jpg.*

antibiotics. As *E. coli* is the most common pathogen responsible for UTIs, the prevalence of resistance to the commonly used antibiotics has been observed among *E. coli*. The challenges posed by the widespread prevalence of the ESBL-producing *E. coli* strains are indicative of the catastrophic effect of the emergence of multidrug resistance among human pathogens. ESBL-producing strains are resistant to all ß-lactam-based antibiotics, which include all the penicillins and the majority of the cephalosporins. The resistance is not only to the first- and second-generation antibiotics but also to the third and fourth generation ones, in addition to cross-resistance to some other antibiotics.

The limitations associated with the diagnosis and identification of ESBL infections result in a delay in treatment initiation, which further complicates the prognosis and tilts the scale toward a bad outcome. Therefore early identification of ESBL-producing UPEC is instrumental in ensuring effective treatment and infection control. An appropriate antibiotic regimen based on culture sensitivity reports ensures judicial and effective use of antibiotics, thereby ensuring good prognosis and outcome. Antibiotic regimens including carbapenems may need to be considered for ESBL-producing *E. coli* strains resistant to penicillins and cephalosporins.

The timely and effective treatment of UTIs caused by ESBL-producing *E. coli* not only ensures a good prognosis but also prevents the spread of the resistant pathogen. This is especially important in hospitals and healthcare facilities, where the threat of nosocomial infections caused by such multidrug resistant pathogens can prove to be catastrophic.

Student Notes

Key take-aways

- *E. coli*, a part of the normal human intestinal flora, is the most common bacteria responsible for UTIs in humans.
- UPEC acquires specific attributes that allow them to invade and survive in the human urinary tract.
- The widespread use of antibiotics has led to the emergence of certain multidrug resistant strains such as the ESBL-producing *E. coli* strains that are resistant to all ß-lactam producing antibiotics such as penicillins and some cephalosporins.
- While UTIs are usually uncomplicated and can be resolved by simple regimens of commonly used antibiotics without resulting in any long-term damage to the renal or the urinary system, UTIs are caused by multidrug-resistant pathogens, such as ESBL-producing UPECs, can be very complicated, and can prove to be a challenge for effective treatment.
- Complicated UTIs caused by ESBL-producing *E. coli* can result in delayed and ineffective treatment because of the lack of effective antibiotics, thereby resulting in long-term damage to the renal and urinary system contributing to poor prognosis and outcome.
- Old age, pregnancy, diabetes mellitus, immunosuppression, and congenital or acquired renal conditions are risk factors associated with recurrent UTIs, especially those caused by multidrug resistant pathogens.
- The differentiation of UTIs from other causes of the clinical manifestations is important and should be based on UA, instead of just visual analysis of the urine.
- A microscopic evaluation may be indicated for confirmation of the diagnosis based on the results of UA, and urine culture may be required if antibiotic resistance is indicated.
- Effective and timely antibiotic treatment based on culture sensitivity reports can ensure a good prognosis and prevent the further spread of the multidrug resistant pathogen.

Problem number 6.2: Vulvovaginal candidiasis caused by *Candida albicans* infection

A 31-year-old woman presented with complaints of vaginal discharge, pruritus vulvae, and painful micturition for a 3-days duration. She mentioned that she had a regular menstrual cycle lasting for 6 days every month, with her last menstrual period being 2 weeks ago. The patient was diabetic but was on diet control only without the need for any medications.

The patient was married for 13 years and had four children. Her youngest child was 1 year old, and after the delivery of the last child, the couple had been using appropriate birth control measures.

The following are the reports of the investigations carried out:

1. Vitals
   BMI: 32 kg/m^2
2. Physical examination
   a. Chest and cardiac: Normal
   b. Abdomen: Soft; no tenderness, rigidity, or palpable masses
   c. Vaginal:
      Vulva, perineum, and inner thighs: Red, inflamed, scratching-induced excoriated skin
      Pain on gentle vaginal and speculum examinations
      Generalized vaginal erythema with curdy thick odorless whitish discharge
      No vesicles were observed on the vulva or the vaginal wall
   d. Pelvic:
      Uterus and adnexae—Normal
      Cervix—Healthy
3. High vaginal and endocervical swab cultures for direct microscopic examination, Gram staining, culture, and sensitivity: *C. albicans*
4. Cervical smear for exfoliative cytology: Low-grade squamous intraepithelial lesion

Diagnosis

The final diagnosis was that of candidiasis caused by *C. albicans* infection.

Treatment

1. Miconazole vaginal pessaries (200 mg): at bedtime (3 days)
2. Miconazole vaginal cream: 7 nights

The patient was also educated about the importance of personal hygiene, weight reduction through controlled diet, physical exercise, and regular blood sugar monitoring.

Follow-up and outcome

The patient reported back for a follow-up after 2 weeks as advised. There were no complaints of any continuing discomfort or symptoms.

A repeat cervical smear taken after 6 months was negative with no abnormal cells.

Question and answer

1. What are the potential reasons responsible for vaginal discharge?
   Vaginal discharge, also known as leukorrhea, could be normal during midcycle, pre- or postmenses, intercourse, and during pregnancy. However, abnormal discharge may also be due to infection caused by *Candida, Trichomonas, Neisseria gonorrhoeae, and Chlamydia* or due to atrophic vaginitis caused by chemical irritation.
2. What does the association of vaginal discharge with pruritus vulvae indicate?
   The association of vaginal discharge with pruritis vulvae indicates vulval irritation caused by an infection resulting in ulcers or lesions.
3. What is the significance of the 2-days duration of the onset of the complaint?
   The sudden onset of the complaint for 2 days indicates an acute condition rather than a chronic one.
4. What is the significance of diabetes in the case?
   Diabetes, especially uncontrolled blood sugar levels, is one of the precipitating factors for infections, especially fungal infections such as candidiasis.
5. What is the cause of the redness and the inflammation seen in the vulva and the perineum?
   The infection results in irritation and scratching of the vulva and the perineum, thereby resulting in inflammation and redness.
6. What are the causes of vulval ulcerations?
   Vulval ulcerations can be caused by bacterial, viral, helminthic, or protozoal infections. Vulval ulcers can also be a sign of malignancy or trauma.
7. What caused the vaginal and speculum examinations to be painful in this case?
   The inflammation of the vagina and the vulva resulting in vaginitis and vulvitis caused the vaginal and speculum examinations to be painful.
8. What microbiological agents may be associated with the etiology of the case?
   Trichomonas vaginalis, C. albicans, Gardnerella vaginalis, and bacterial vaginosis caused by an overgrowth of anaerobes may be associated with the etiology of the case.
9. Which pathogens are known to be involved in the genesis of vesicles?
   Herpes simplex virus type 2 is known to cause genital herpes, which manifests as vesicles similar to the oral ulcers seen in herpes simplex type 1 infection.
10. Name some sexually transmitted diseases (STDs).

Gonorrhea, chlamydia, syphilis, human papilloma virus infection, human immunodeficiency virus infection, and hepatitis B infection are some of the known STDs.

11. What is the importance of cervical examination?
    A cervical examination is important to rule out chronic cervicitis, cervical ectropion, and malignancy.
12. What are the various criteria associated with vaginal discharge used to differentiate among the different causative agents of abnormal vaginal discharge?

Criteria	Observation	Cause
pH	< 5	Candidiasis
	> 5	Bacterial vaginosis, trichomoniasis
Odor	Bad	Bacterial vaginosis/trichomoniasis
Color	Greyish	Bacterial vaginosis
	Greenish-yellow	Gonorrhea, trichomoniasis
Consistency	Curd-like	Candidiasis
	Liquid	Bacterial vaginosis
	Frothy	Trichomoniasis
Pruritus vulvae		Candidiasis

13. What is the swab sample expected to show under the microscope?
    The swab sample was expected to show epithelial cells, clue cells that are covered with a lot of bacteria, polymorphonuclear leukocytes (PMNs) > 10/high power field 40X, yeast such as *Candida*, and protozoa such as *T. vaginalis*.
14. What is the normal microbiota of the vagina?
    The normal microbiota of the vagina during the reproductive years (between menarche to menopause) consists of an abundance of lactobacilli, and are important for maintaining a low physiological pH (3.5–4.5) and also for competing with invading pathogens, thereby having a protective role. Some other microorganisms such as those belonging to the *Streptococcus* and *Staphylococcus* genus also form an important part of the normal vaginal microbiota.
15. What is the significance of the culture and sensitivity test?
    The culture and sensitivity tests enable the determination of the organism responsible for the infection and the sensitivity of the causative agent to antimicrobials, which is important for ensuring effective treatment.
16. What pathogens primarily affect the cervix?
    Trichomonas, Herpes virus, Papilloma virus, *Chlamydia trachomatis*, and *Neisseria gonorrhea* are the pathogens that primarily affect the cervix.

17. What is exfoliative cytology or cervical smear?
    Exfoliative cytology or cervical smear are diagnostic tests used as means of screening the cervix of the female uterus for cells indicative of premalignant conditions and viruses such as the human papilloma virus, *C. albicans*, and Trichomonas.
18. How was the diagnosis of candidiasis made?
    The curd-like vaginal discharge associated with pruritis vulvae is indicative of candidiasis infection.
19. What could be the precipitating factor of vaginal candidiasis?
    Diabetes and multiple pregnancies could have been the precipitating factors for vaginal candidiasis. Immunosuppression, abuse of broad-spectrum antibiotics, and combination pills also increase the vulnerability of an individual to acquiring a vaginal candidiasis infection.
20. Why was the patient instructed about personal hygiene?
    Personal hygiene is very important for the prevention of reinfection, especially because the patient is diabetic, which predisposes her to develop the infection.

Learning objectives

1. List the causes of vaginal discharge.
2. Enumerate the precipitating factors for candidiasis.
3. Determine the differential diagnosis for vaginal discharge.
4. List the microbiological agents implicated in vaginitis and cervicitis.
5. Describe the normal microbiota of the vagina.
6. Explain the laboratory diagnosis of the etiology of vaginal discharge.
7. Describe the clinical manifestations of vaginal candidiasis.
8. Explain the microbiology of candidiasis.
9. Explain the importance of educating candidiasis patients about personal hygiene.
10. Describe the management of vaginal infections such as those caused by fungal, Trichomonas, and Gardnerella.

Discussion

Differential diagnosis

Allergic reaction	Allergic reactions can present themselves with skin irritation, watery eyes, and/or sneezing. Severe cases may even manifest as anaphylactic shock.

Continued

Continued

Sexual abuse	Sexual abuse manifests as a wide spectrum of emotional and physical manifestations. The most prominent and commonly encountered signs of sexual abuse include pain in the abdominal or pelvic region, bruises, scars, chafing, vaginal discharge or bleeding, rectal bleeding, or anal tears.
Psoriasis	Psoriasis can cause red scaly skin plaques, discolored, pitted, brittle, and detached nails, and crusty areas of skin on the scalp.
Atopic dermatitis	Atopic dermatitis is characterized by dry and itchy skin, increased itching at night, red, brownish, or gray patches, tiny raised bumps that leak fluid and forms crusts, thick skin that is cracked and scaly, and tender, sensitive, swollen skin due to scratching.
Physiologic leukorrhea	Physiologic leukorrhea is a natural phenomenon that occurs as a part of the menstrual cycle; however, the possibility of pathological causes cannot be ruled out. Physiologic leukorrhea is characterized by a white/straw-colored copious discharge that is odorless and does not cause any discomfort or irritation. However, if the leukorrhea is a result of an infection or other pathological causes, the discharge generally has a foul smell and results in itching, burning, irritation, and discomfort.
Lichen sclerosis	Lichen sclerosis is identified by severe itching, smooth white skin patches, redness, pain, discomfort, bleeding, skin tears, blotchy and wrinkled skin patches, blisters, ulcerated sores, and painful intercourse.
Paget's disease of the vulva	Paget's disease of the vulva, a rare type of skin cancer arising from glandular cells, presents itself with red or pink velvety areas having white islands of tissue and moist, pus-oozing, and bleeding ulcers.
Lichen simplex chronicus	Lichen simplex chronicus is characterized by a chronic scratch-itch cycle resulting from repetitive itching leading to gradual skin thickening. The condition generally manifests on the inner sides of the wrist, forearms, elbows, sides and back of neck, upper thighs, knees, shins, ankles, feet, scalp, vulva, scrotum, and anus.
Neoplasm	Neoplasms present themselves with a wide spectrum of symptoms depending on the affected area. Some of the general manifestations include abdominal pain, fatigue, fever, chills, diarrhea, lesions, ulcers, skin masses, shortness of breath, abnormal bleeding, and discharge.
Vulvodynia	Vulvodynia manifests as chronic vulval pain, burning, pricking, itching, stinging, tenderness, and sores.

Background

Candidiasis is usually seen as an opportunistic infection in immunocompromised individuals, although the occurrence of the disease in immunocompetent individuals belonging to age extremes and in those with underlying chronic conditions such as diabetes and obesity are also known. The most common etiological agent of candidiasis is a fungus known as *C. albicans* that forms an abundant part of the normal oral, gastrointestinal, and urogenital flora in humans. Some other nonalbicans *Candida* species such as *C. lusitaniae, C. parapsilosis, C. glabrata, C. kefyr, C. famata, C. africana,* and *C. orthopsilosis* may also be responsible for certain candidiasis infections. The majority of the cases of vulvovaginal candidiasis are caused by *C. albicans,* while a few cases may also be caused by *C. glabrata.* Immunocompetent and healthy individuals are usually capable of preventing infections caused by *C. albicans,* and hence the organism acts as a commensal. Healthy normal microbiota is also an important deterrent for *C. albicans* infections. Therefore besides immunocompromised and chronic conditions, antibiotic abuse is also known to cause *C. albicans* infection. While candidiasis can affect any part of the body, the oral cavity and the genitals are the most commonly affected. *C. albicans* infection is usually superficial and rarely results in invasive disease, although severely immunocompromised individuals may suffer from disseminated forms of infection, which are potentially life-threatening.

Etiological agent

Candida albicans

Classification and morphology

C. albicans is a dimorphic pathogenic fungus belonging to the *Saccharomycetaceae* family and *Candida* genus, which can grow as unicellular yeast cells as well as filamentous cells (Fig. 6.4). *C. albicans* grows partly as a chain of single round cells resembling yeast cells and partly as elongated cells that appear like hyphae. This

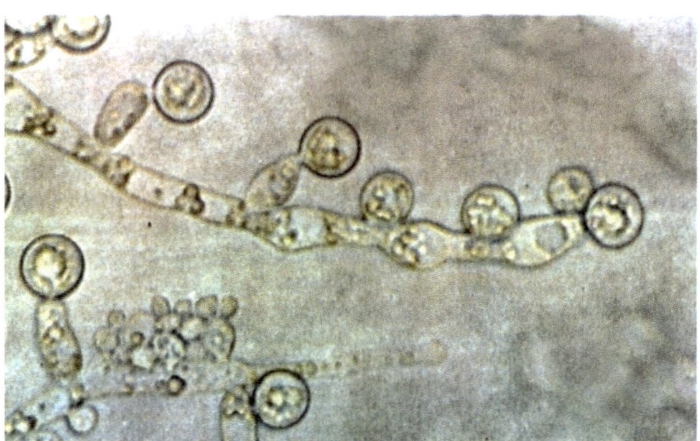

FIGURE 6.4

Candida albicans under the microscope. Source*: GrahamColm/Own work/; https://creativecommons.org/licenses/by-sa/3.0/deed.en; https://commons.wikimedia.org/wiki/File:Candida_albicans_2.jpg.*

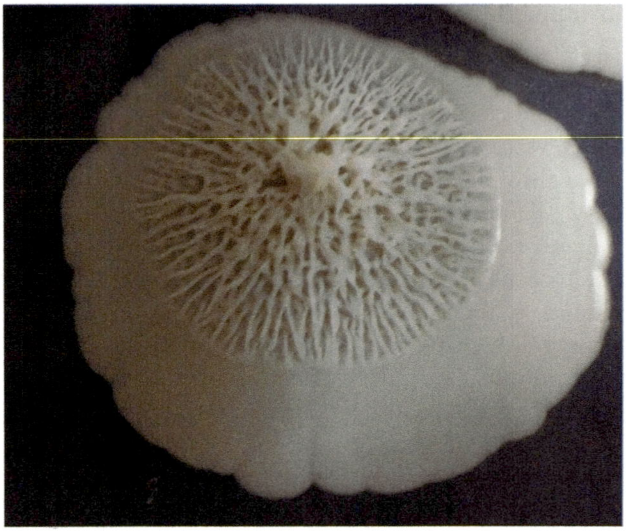

FIGURE 6.5

Candida albicans colony exhibiting both smooth and wrinkled forms. Source*: Garnhami/Own work/; https://creativecommons.org/licenses/by-sa/4.0/deed.en; https://commons.wikimedia.org/wiki/File:Candida_albicans_mixed_colony.jpg.*

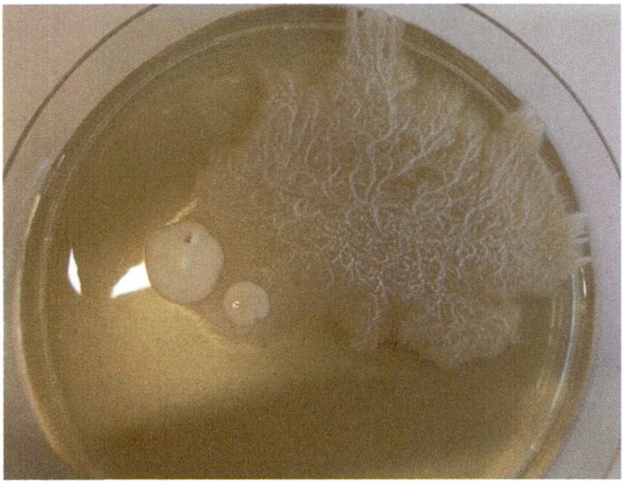

FIGURE 6.6

Candida albicans growing both as yeast cells and filamentous cells. Source*: Garnhami/Own work/; https://creativecommons.org/licenses/by-sa/4.0/deed.en; https://commons.wikimedia.org/wiki/File:Candida_albicans_growing_as_yeast_cells_and_filamentous_(hypha)_cells.jpg.*

fungus is also well-known for its ability to exhibit various morphological phenotypes (Figs. 6.5 and 6.6), owing to the phenotypic switching and transition between the bud and hypha forms.[15] This pleomorphic ability enables the

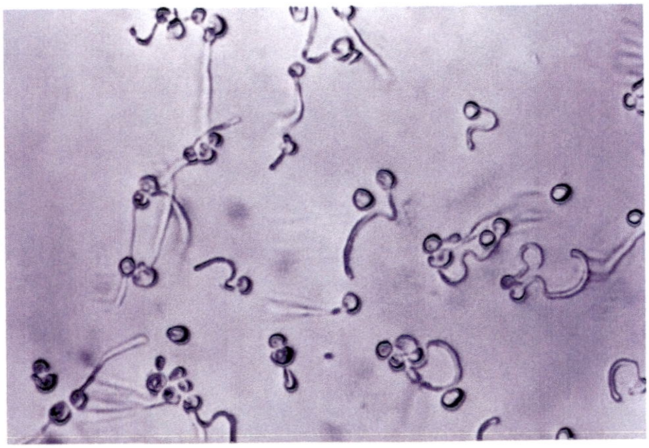

FIGURE 6.7
Photomicrograph of *Candida albicans* spores (chlamydospores). Source*: https://www.rawpixel.com/image/2288324/free-photo-image-bacteria-biology-lab/ https://creativecommons.org/publicdomain/zero/1.0/; Centers for Disease Control and Prevention.*

organism to grow as a commensal as well as a pathogen under various environmental conditions.[16] *C. albicans* forms spores known as chlamydospores (Fig. 6.7), which gives it the ability to survive for long periods even outside the body of the host in hostile environmental conditions.[17] The Gastrointestinally indUced Transition phenotype of *C. albicans* is a specialized adaptable form of the organism that facilitates survival in the hostile conditions of the digestive tract, and the transition to this commensal cell type is triggered by the unfavorable environment of the digestive tract.[18,19]

Epidemiology

The prevalence of candidiasis is found to be extremely high among infants and those of advanced age, individuals on immunosuppressive therapy due to organ transplants or cancer treatment, individuals with congenital or acquired immunodeficiencies, malnourished populations, pregnant women, and those with underlying chronic medical conditions such as diabetes and obesity. Increased estrogen exposure due to endogenous elevation of estrogen levels in pregnancy and obesity or exogenous use of estrogen in the form of contraceptives and topical ointments have been recognized as predisposing factors for vulvovaginal candidiasis. Thrush (Fig. 6.8) is the most common manifestation of *C. albicans* infection. Candidiasis occurs in both males and females, with increased incidences in populations prone to malnourishment. Although candidiasis is usually a superficial infection, incidences of invasive and disseminative forms of the disease, especially in vulnerable populations and high-risk individuals, have increased in recent times owing to antibiotic abuse.[20]

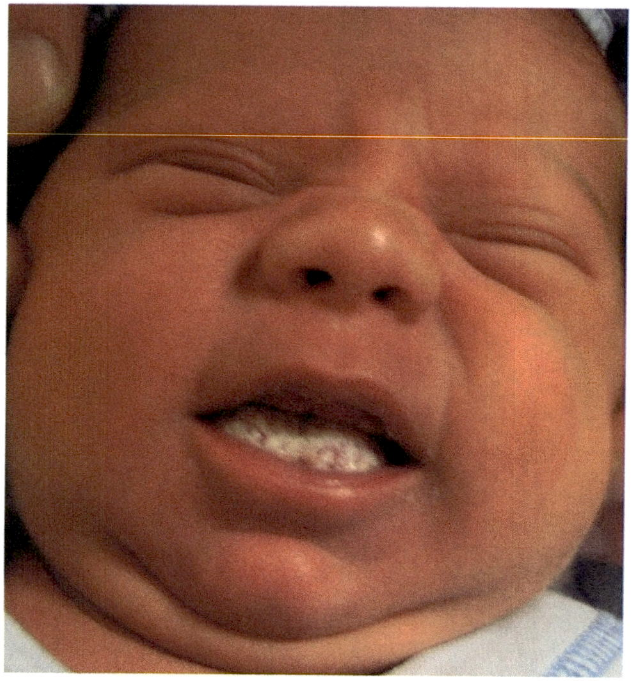

FIGURE 6.8

Oral candidiasis (thrush). Source*: James Heilman, MD/Own work/; https://creativecommons.org/licenses/by-sa/3.0/deed.en; https://commons.wikimedia.org/wiki/File:Thrush2010.JPG.*

Pathogenicity and pathogenesis

Incompetency of the immune system allows *C. albicans* to proliferate resulting in an overgrowth of the organism on mucosal surfaces. Such overgrowth causes desquamation of epithelial cells, thereby compromising the integrity of the epithelial layer of protection and increasing vulnerability to infection. This causes an accumulation of keratin, bacteria, and necrotic tissue, all of which form a layer close to the infected mucosal layer. Edema, ulceration, and necrosis of underlying mucosa are rare in candidiasis due to the formation of this layer close to the infected mucosal layer. Vulvovaginal candidiasis is a result of the superficial penetration of the vaginal mucosa by the organism. The inflammatory response elicited by the mucosal invasion results in an accumulation of polymorphonuclear cells and macrophages and is responsible for the clinical manifestations of vaginal burning, itching, redness, soreness, and swelling.

The ability of *C. albicans* to produce biofilms[21] on the surfaces of medical intervention devices such as indwelling catheters, ventilators, stents, shunts, prostheses, implants, endotracheal tubes, and pacemakers is clinically concerning.[22]

While superficial *C. albicans* infection is common, invasive *C. albicans* is relatively rare and occurs only if the organism reaches the bloodstream. Invasion of the bloodstream may occur either through the gastrointestinal tract mucosal barrier due to immune-incompetency, interventional medical devices, and rarely through localized infections such as an ascending *C. albicans* UTI in individuals with an intrinsic obstruction or extensive compression.[23]

Normal microbiota plays a very important role in maintaining the commensalism of *C. albicans,* and therefore any change in the composition of the normal flora facilitates the proliferation of *C. albicans* precipitating an infection in the process. Antibiotic abuse is, therefore, the most common precipitating factor for *C. albicans* infection in an immunocompetent host. Vulvovaginal candidiasis is often a result of reduced or changed normal vaginal flora and pH balance caused by misuse of topical or oral antibiotics, lack of personal hygiene, improper use of vaginal lubricants containing glycerin, uncontrolled diabetes, pregnancy, hormonal imbalance, stress, inadequate sleep, and a poor diet.

Clinical manifestations and complications

The clinical manifestations of candidiasis differ based on the site of the infection.

Vulvovaginitis generally presents with severe itching and irritation of the vulva, pain and soreness of the vagina, burning sensation during micturition similar to that experienced in UTIs, vaginal and vulval dryness, erythematous rash, causing redness of the vulva, excoriations, painful intercourse, and thick curd-like odorless adherent discharge (Fig. 6.9).

An oral *C. albicans* infection, commonly known as thrush (Fig. 6.10), manifests as a white or straw-colored rash that cannot be brushed or scraped off on the tongue and the mucous membranes of the mouth, often accompanied by redness, soreness, and cracking at the corners of the mouth. Pain and difficulty in swallowing may be experienced in some extreme cases where the infection may have extended to the oral pharynx.

Laryngeal candidiasis (Fig. 6.11) is rare, and predominant in females. This form of candidiasis is generally associated with long-term usage of steroid inhalers. Leukoplastic lesions are usually present on the glottis, which affects the normal functioning of the glottis and the larynx. Dysphoria is a common manifestation of laryngeal candidiasis.[24]

C. albicans infection of the intestines may occur as a sequel to oral antibiotic therapy and presents with diarrhea, abdominal cramps, fever, fatigue, loss of appetite, unexplained weight loss, and bloody stools.

While the pulmonary system is a very uncommon site for candidiasis infection, underlying pulmonary conditions can result in a rare complication of candidiasis known as bronchopulmonary candidiasis.

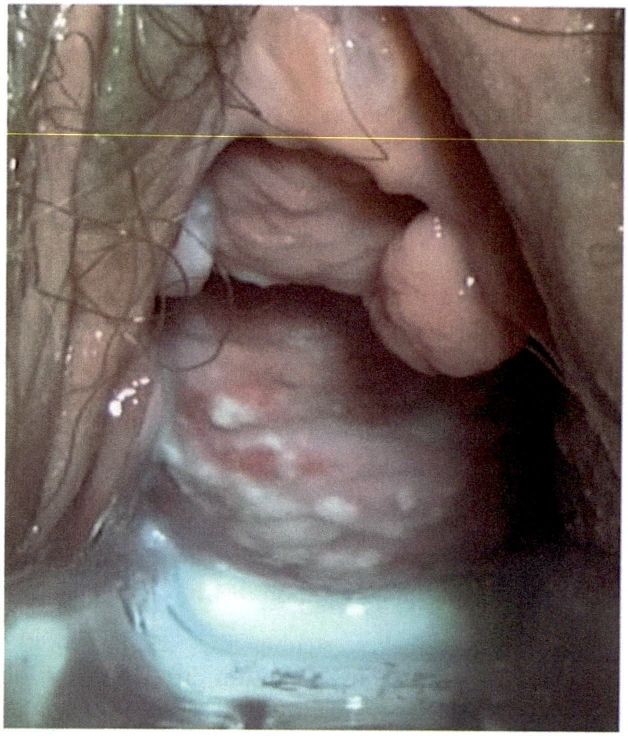

FIGURE 6.9

Speculum exam in vulvovaginitis, showing the presence of thick, curd-like adherent discharge and erythema. Source: *Mikael Häggström/ Own work/; https://creativecommons.org/publicdomain/zero/1.0/deed.en; https://commons.wikimedia.org/wiki/File: Speculum_exam_in_candidal_vulvovaginitis.jpg.*

Pregnancy results in certain changes in the pH and the normal flora of the vagina, which acts as a precipitating factor for the proliferation of *C. albicans* in the vagina.[25] Vaginal candidiasis during pregnancy could result in low birth weight and premature delivery of the neonate. Neonates may also acquire the infection during the process of parturition and develop systemic infections owing to the low immunity at birth.

Inflammatory dermatosis conditions such as intertrigo (Fig. 6.12) affects the areas between skin folds such as the armpits, genital area, abdomen, and beneath the breasts. Intertrigo is commonly seen in vulnerable individuals such as the obese, diabetics, paraplegics, immunocompromised, and the aged. The condition presents with redness, pain, and itching and is often associated with high levels of morbidity and compromised quality of life.

While the superficial localized forms of candidiasis are the most common, rarely untreated infections precipitate invasive disseminated infections

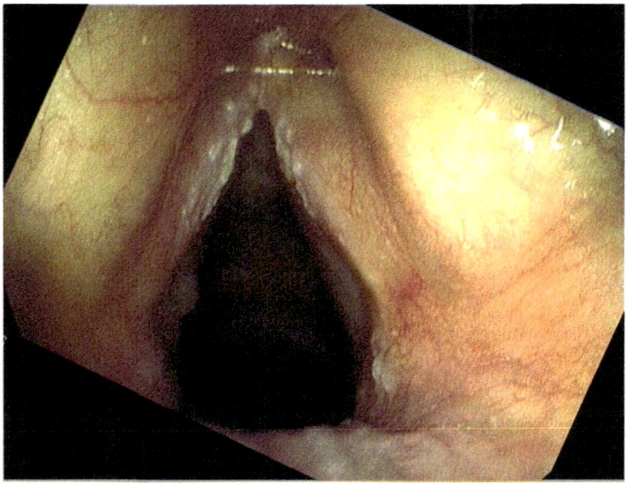

FIGURE 6.10
Laryngitis caused due to fungal infection. Source: *James P. Thomas, MD/Own work/; https://creativecommons.org/licenses/by-nc-sa/3.0/; https://www.voicedoctor.net/media/structural-injury-inflammation-fungal/candida/fungal-laryngitis.*

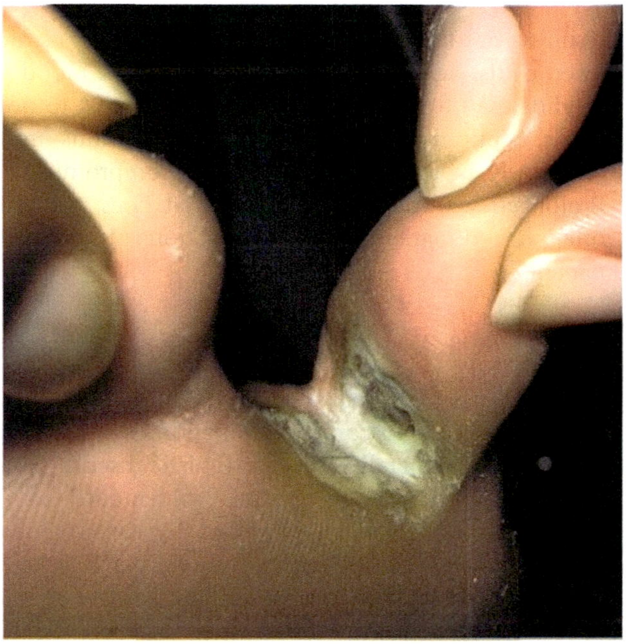

FIGURE 6.11
Interdigital intertrigo caused by *Candida albicans* infection. Source: *Grook Da Oger/Own work/; https://creativecommons.org/licenses/by-sa/3.0/deed.en; https://commons.wikimedia.org/wiki/File:Intertrigo_interdigital_%C3%A0_CA.jpg.*

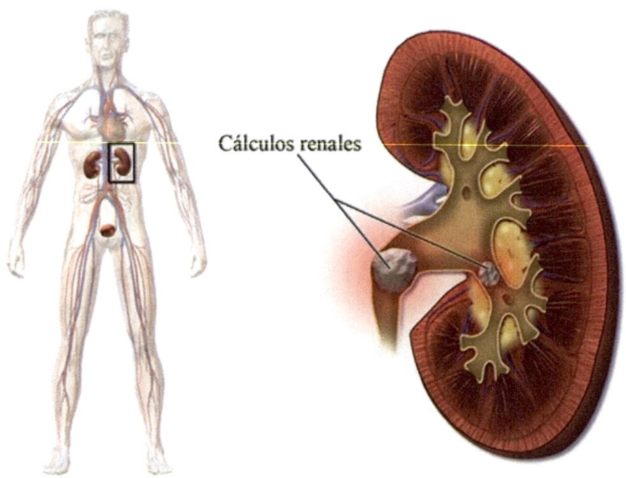

FIGURE 6.12

Renal stones. Source: *De la traducción Ortisa/; http://Blausen.com staff (2014). "Medical gallery of Blausen Medical 2014". WikiJournal of Medicine 1(2). DOI: 10.15347/wjm/2014.010. ISSN 2002–4436/; https://creativecommons.org/licenses/by-sa/3.0/deed.en; https://commons.wikimedia.org/wiki/File:Blausen_0595_KidneyStones-es.png.*

resulting in candidemia, which presents with systemic manifestations such as fever, chills, delirium, and a sudden drop in blood pressure levels. Systemic candidemia often carries the risk of advancing to potentially life-threatening sepsis. Septicemia, meningitis, and endocarditis are rare complications seen exclusively in terminally ill, immunocompromised, and leukemic patients. Prolonged neutropenia and fever that does not respond to antibiotics are often the hallmark clinical manifestations of systemic candidiasis in leukemia patients.

Diagnosis

The diagnosis of vaginal candidiasis is a function of patient history, physical examination, microscopic examination, and culture sensitivity reports. As candidiasis is generally an opportunistic disease, patient history is important for the analysis of the patient's vulnerability to acquiring the infection. Vaginal candidiasis is characterized by evident inflammation of the vulva and vagina and noninflamed cervix on pelvic examination. Pelvic examination is also important for ensuring the absence of masses or irregular bumps in the vulva, vagina, or cervix, thus enabling the clinician to rule out any neoplastic causes for the presenting condition. A vaginal pH test demonstrating a pH of 5 or less is usually indicative of vulvovaginal candidiasis.

Microscopic analysis of a vaginal discharge sample or a vaginal swab sample from a case of vulvovaginal candidiasis demonstrates an abundance of lactobacilli and inflammatory cells, along with the presence of budding yeast and hyphae. A Whiff test involving the application of potassium hydroxide to the vaginal discharge or swab sample gives a distinct fishy odor in case of vulvovaginal candidiasis.[26,27]

It is also important to determine whether the candidiasis is a primary infection or secondary to other infections such as bacterial vaginosis or trichomoniasis. While the presence of clue cells in the vaginal discharge or swab samples is indicative of bacterial vaginosis, trichomonads indicate trichomoniasis infection.

Routine microscopic cultures are usually discouraged in asymptomatic women since *C. albicans* forms a significant part of the normal vaginal flora. Women presenting with symptoms of vulvovaginal candidiasis, however, should be evaluated for the presence of *C. albicans* infection through microscopic analysis. While a microscopic culture of vaginal discharge or swab may be required in some symptomatic patients, the demonstration of budding yeast cells in microscopic analysis (Fig. 6.4) is often sufficient for diagnosing vulvovaginal candidiasis. However, culturing a vaginal discharge or swab sample is recommended in women with a history of recurrent *C. albicans* infection to determine the possibility of resistance of the infecting strain to the drugs being used to treat the infection.[23]

Treatment and management

Antifungal agents are generally the primary line of treatment for cases of acute vulvovaginal candidiasis. *C. albicans* is known to be sensitive to azole-based antifungal agents. Antifungal agents such as fluconazole are oral drugs, while antifungal ointments and intravaginal pessaries are meant for topical application on the affected area. Immunocompetent patients without any history of recurrent *C. albicans* infections or pregnancy usually benefit from either therapy taken for a short term of about 7 days. Nonresponsiveness to azole-based therapy warrants the need for culture sensitivity tests to detect strains that may be resistant to fluconazole. Pregnancy is a contraindication for oral antifungal therapy although it is safe for use during breastfeeding, and therefore pregnant women can only be given a 7-day course of intravaginal therapy.

Immunocompromised patients or those with underlying conditions that make them vulnerable to *C. albicans* infection may need more extensive therapy. Patients presenting with a history of recurrent vulvovaginal candidiasis may require therapy for as long as up to 6 months to ensure its efficacy.[23,28]

Student Notes

Key take-aways

- Candidiasis is an opportunistic infection, often seen in immunocompromised individuals or immunocompetent individuals belonging to age extremes and in those with underlying chronic conditions such as diabetes and obesity.
- Although candidiasis can be a result of an infection caused by several fungal species belonging to the *Candida* genus, *C. albicans* is the most commonly known etiological agent for candidiasis.
- *C. albicans* is a commensal and forms a significant part of the normal flora of the oral cavity, the mucosal layer of the digestive tract, and the mucosal layer of the genital tract.
- Vulvovaginal candidiasis is often seen in women of reproductive age and is generally caused by *C. albicans*, although a few cases may also be caused by *C. glabrata*.
- Antibiotic abuse, prolonged use of steroids, and elevated estrogen levels due to hormonal imbalance or the use of estrogen-based creams have also been implicated in increased vulnerability to *C. albicans* infection.
- Candidiasis is usually a superficial infection and is therefore not associated with high levels of morbidity and mortality.
- Although candidiasis in immunocompetent individuals is not a life-threatening condition, an acute form of infection can have social implications such as awkwardness, embarrassment, and avoidance of sexual activities.
- Invasive disease is rare but severely immunocompromised individuals are often prone to disseminated forms of infection leading to potentially life-threatening conditions such as septicemia, meningitis, and endocarditis.
- While diagnosing vulvovaginal candidiasis, differential diagnoses, including trauma, sexual abuse, malignancy, autoimmune diseases, and STDs should be considered, due to the similarities in the clinical manifestations.
- The diagnosis of candidiasis should be based on patient history, physical examination, and microscopic analysis of samples.
- Treatment of vulvovaginal candidiasis using oral and intravaginal antifungal agents for 7 days is usually effective in immunocompetent patients, and those without a history of recurrent candidiasis infections.
- High-risk patients may need more extensive treatment going up to 6 months for certain recurrent cases.
- Pregnancy is a contraindication for oral antifungal agents, although these agents can be safely prescribed to breast-feeding women.
- Recurrent infections should also be evaluated with cultural sensitivity to determine the possibility of strains resistant to azole-based antifungal agents, which are the most commonly used form of therapy in candidiasis infections.

Problem number 6.3: Urinary tract infection caused by *Proteus mirabilis* infection

A 73-year-old man presented with fever, lower abdominal pain, and inability to pass urine. While the fever and abdominal pain had been there for 3 days, he had been unable to pass urine only since the past 24 hours. The patient mentioned a history of five recurrent episodes with similar symptoms during the last 8 months, except that the inability to pass urine had occurred only this time.

The following are the reports of the investigations carried out:

1. Temperature: 38.4°C
2. Physical examination
   Abdomen: Tender swelling in the supra-pubic region indicative of a distended bladder, also giving rise to the possibility of an inflammatory state of the bladder known as cystitis
   Digital rectal examination: Smooth nodular enlarged mass felt at the position of the prostate gland, which could be indicative of inflammation of the prostate, hyperplasia, or a benign/malignant growth
3. Urine analysis
   Sample obtained through bladder drainage by insertion of sterile catheter
   Appearance: Turbid
   WBCs present in urine
   Leukocyte esterase: + +
   Culture: *P. mirabilis*
4. Prostate specific antigen: 4.4 ng/mL (Normal $<$ 4 ng/mL)
5. Radiology
   Urinary tract ultrasonography: Enlarged prostate; mild unilateral dilatation of ureter and pelvis; no other abnormalities of the urinary tract
   X-ray of the kidney, ureter, bladder: No stones were detected
6. Complete blood count (CBC): Leukocytosis
7. Serum creatinine: Normal
8. Serum electrolytes: Normal
9. Urea: Normal

Diagnosis

The final diagnosis was that of UTI caused by the *P. mirabilis* bacteria, with the histopathology report of the prostate biopsy sample pending.

Treatment

1. Bladder drained by insertion of sterile catheter to relieve obstruction and pain
2. Ciprofloxacin
3. Trans-urethral resection of the prostate was performed and a biopsy sample was collected for histopathological analysis

Follow-up and outcome

The patient recovered remarkably well and did not have any complications related to the infection or the surgery.

Question and answer

1. What causes the urinary bladder to get infected in a UTI?
   Certain pathogens responsible for UTIs can ascend up the urinary tract reaching the bladder, the ureters, and the kidneys, establishing infections in these organs.
   The bladder infection established by the pathogens results in an inflammatory condition of the bladder known as cystitis.
2. What does the presence of WBCs in urine indicate?
   WBCs are normally absent in urine. As WBCs are responsible for protection against infections, their presence in urine is indicative of infection and/or inflammation of the bladder or the kidneys, usually of a bacterial etiology.
3. What does the fever and the leukocytosis in the CBC test indicate?
   The fever and the leukocytosis in the CBC test indicate an infection in the body—usually an acute bacterial infection.
4. Why is ciprofloxacin often prescribed in UTIs?
   UTIs and lower respiratory tract infections are generally caused by Gram-negative bacteria. Ciprofloxacin is a quinolone derived from nalidixic acid that inhibits DNA gyrase, an enzyme essential for bacterial DNA synthesis. This antibiotic is known to be very effective against Gram-negative bacteria.
5. What is *P. mirabilis*?
   P. mirabilis is an anaerobic, Gram-negative rod-shaped bacteria that is commonly found as a commensal in the human intestinal tract. The facultative anaerobic nature of the organism supports its survival in the urinary tract in case it gains entry into the urinary tract.
6. Which other commensals of the human gastrointestinal tract are also implicated in UTIs?

E. coli, *Klebsiella pneumoniae*, *Staphylococcus saprophyticus*, and *Enterococcus faecalis* are some commensals of the human gastrointestinal tract that are also implicated in UTIs.

7. What are the predisposing factors for UTIs?

   The female anatomy is responsible for the higher incidences of UTIs among females due to the shorter urethral length. Sexual activity, lack of personal hygiene, metabolic disorders such as diabetes, changes in the pH of the genitourinary tract, changes in the normal flora of the genitourinary tract, menarche, menopause, long-term use of birth control, anatomical anomalies, prostate enlargement, pregnancy, and holding urine for long periods of time are all predisposing factors for UTIs.

8. What do recurrent UTIs indicate?

   Recurrent UTIs usually indicate anatomical anomalies, benign or malignant growths, ineffectively resolved UTIs, lack of personal hygiene, uncontrolled blood sugar levels, or resistant and persistent pathogens.

Learning objectives

1. List the causes of UTI.
2. Summarize the complications associated with UTIs.
3. Describe the clinical features of UTI.
4. Enumerate the laboratory and radiological investigations recommended for UTI.
5. Interpret the results of the radiological investigations and the urine analysis investigations.
6. Recognize the risks associated with UTIs.
7. Explain the recommended process for the management of UTIs.
8. List the organisms responsible for UTIs.
9. Enumerate the predisposing factors for UTI.
10. Describe the laboratory diagnosis of UTI.
11. Summarize the pathological changes that may occur in the kidneys due to UTI.

Discussion

Differential diagnosis

Etiology	Rationale
Pyelonephritis	Pyelonephritis presents with high-grade fever, flank pain, abdomen, or back, pain and/or burning during urination, turbid urine, hematuria, frequent urination, fishy-smelling urine, chills, nausea, vomiting, malaise, fatigue, neurological symptoms, and moist skin.

Continued

Continued

Etiology	Rationale
Renal stone	Renal stones often manifest with dark-colored urine, turbid or frothy urine, frequent urge for urination, inability to empty the bladder, nausea, vomiting, fever, chills, and flank pain, abdomen, or back.
Perinephric abscess	Perinephric abscess presents with a gradual and subtle onset of fever and pain in the abdomen, flanks, groin, legs, and back.
Urethral syndrome	Urethral syndrome presents with lower abdominal pain, feeling of pressure in the abdomen, frequent urination, extreme urge to urinate, pain during urination, pain during intercourse, and hematuria.
Chronic prostatitis	Chronic prostatitis mainly manifests with pelvic or genital pain resembling a gnawing ache or an intense sharp pain, pain in the region between the scrotum and the rectum, pain in the lower back, rectum, penis, or testicles, difficulty and pain during ejaculation, pain during urination, extreme urge to urinate, and frequent urination including increased frequency at night. The involvement of an infection is indicated by fever, chills, nausea, and vomiting.
Renal tuberculosis	Renal tuberculosis usually presents with back, flank, and suprapubic pain, increased urination frequency, nocturia, and hematuria.

Background

P. mirabilis is commonly found in the human gastrointestinal tract either as a commensal or as a pathogen. The involvement of *P. mirabilis* in UTIs may result in symptomatic infections manifesting as cystitis and pyelonephritis or can also result in asymptomatic bacteriuria in individuals such as diabetics, those with anomalies of the urinary tract, and the elderly who are at high risk for UTIs.[29,30] UTIs caused by *P. mirabilis* may be transmitted from person-to-person, especially as a nosocomial infection. However, most cases result from ascension of the bacteria from the gastrointestinal tract to the urinary tract.[31] While most UTIs can be easily resolved, some may develop life-threatening complications such as bacteremia and urosepsis. Concurrent UTIs have been linked to a higher risk of developing *P. mirabilis* associated Gram-negative bacteremia, which has a notoriously high mortality rate. Besides UTIs, *P. mirabilis* has also been implicated in empyema, meningoencephalitis, and osteomyelitis in neonates, urolithiasis, and in infections of the respiratory tract, eye, ear, nose, throat, and skin. Potentially life-threatening infections of burns and wounds have also been attributed to *P. mirabilis* infections.[31–33]

Etiological agent

Proteus mirabilis

Classification and morphology

P. mirabilis are rod-shaped, motile bacteria belonging to the *Enterobacteriaceae* family. This bacterial species is flagellated, and is especially known for its robust swarming-type motility. In addition to motility, evidence suggests that the flagella enable colonization and are responsible for the biofilm-forming ability of the bacteria. The ability of the bacteria to evade the host immune responses and resistance to certain antibiotics has also been attributed to the bacterial flagella.[34] *P. mirabilis* is a Gram-negative, facultative anaerobe capable of producing urease and hydrogen sulfide. While this bacterial species can ferment maltose, it is unable to ferment lactose. In addition to the hallmark features of swarming motility and urease production, *P. mirabilis* is also known for its ability to self-elongate. The attachment and motility of the bacteria on various surfaces, including medical equipment, can be attributed to its ability to secrete a polysaccharide when it comes in contact with solid surfaces.

Epidemiology

Bacterial species belonging to the *Proteus* genus are found abundantly in soil and water, in addition to being an important part of normal flora of the human gastrointestinal tract. *P. mirabilis* has been found to be responsible for most of the community-acquired *Proteus* infections. Bacterial migration into the urinary tract through in-dwelling catheters is a common source of *P. mirabilis* infection in morbid patients. Abnormalities of the urinary tract and interventions in the urinary tract are predisposing factors for nosocomial *P. mirabilis* infections of the urinary tract. While most of the UTIs in immunocompetent healthy individuals are caused by *E. coli*, most nosocomial UTIs and UTIs resulting from anatomical or interventional complications are found to be caused by *P. mirabilis*. The incidence of *P. mirabilis*-associated UTIs is the highest in the geriatric population with long-term in-dwelling catheters. Although the incidence of *P. mirabilis*-associated UTIs does not vary based on geographical locations, the predisposition of the female gender to UTIs makes females more prone to *P. mirabilis*-associated UTIs.[33,35]

Pathogenicity and pathogenesis

The ability of *P. mirabilis* to colonize the urinary tract either acquired from an external source or through ascension of the urinary tract depends on a constellation of highly connected viral and host factors. The extra-cytoplasmic membrane of the bacteria comprises of several elements that act as determinants for bacterial virulence. *P. mirabilis* employ a wide array of virulence factors such as urease production, stone formation, fimbriae, adhesins, biofilm formation, iron and zinc acquisition, proteases, and endotoxins to exert its pathogenic effects in the human urinary tract. In addition to the viral and host factors, the infecting inoculum size also has a strong influence on the virulence of the infection.

The infection process begins with the attachment of the bacteria to the host cells, a process facilitated by the fimbriae projecting externally from the bacterial surface. These fimbriae secrete a polysaccharide on contact with any solid surface, which allows the bacteria to adhere to the surfaces of medical equipment and form biofilms, in addition to facilitating their attachment to the host cells of the urinary tract. Biofilm formation is responsible for the increased propensity of patients with in-dwelling catheters toward *P. mirabilis*-associated UTIs.[36]

Crystalline biofilms notorious for blocking urine flow through catheters is a hallmark feature of *P. mirabilis* invasion. The urease production ability of *P. mirabilis* is responsible for the increased vulnerability to urolithiasis, pyelonephritis, and upper UTIs attributed to this bacterial species. Urease hydrolyzes urea into ammonia and carbon dioxide, of which ammonia acts as the preferred nitrogen source for bacteria. The generation of ammonia increases the pH of the urinary tract resulting in the precipitation of mineral ions leading to the formation of urinary stones such as struvite (magnesium ammonium phosphate) and apatite (calcium phosphate). While in the urinary tract these stones can block urine flow and damage tissue,[37] in the form of crystalline biofilms adhering to the surface and the lumen of the catheter they can result in catheter obstruction.[38]

Motility in the form of swimming and swarming[39] in *P. mirabilis* is facilitated by flagella. The flagella also contribute significantly to the virulence of this bacterial species. On contact with solid nutrient surfaces, *P. mirabilis* has been found to differentiate into swarmer cells, which are phenotypically distinct from swimmer or vegetative bacterial cells. Swarmer cells are elongated hyperflagellated cells formed through self-elongation of bacterial cells, which move together forming rafts of parallel cells.[40] Certain polysaccharides associated with swarmer cells have been implicated in enabling the motility of the bacteria across various surfaces.[41,42] The process of swarming induces certain virulence factors such as urease, protease, and hemolysin, which may be responsible for priming the bacteria swarming up catheters to invade the urinary tract, although the role of swarming in UTI virulence is debatable.[43,44] However, the role of swarming in facilitating the invasion of the urinary tract in catharized individuals is undebatable.

A calcium-independent hemolysin and the autotransporter *Proteus* toxic agglutinin (*Pta*) are important virulence factors that mediate cytopathic effect, lysing nucleated cells as well as red blood cells.[45,46] Proteases such as *ZapA* encoded by *P. mirabilis* enables its survival in the urinary tract by degrading numerous host antibodies and antimicrobial peptides.[47,48] In addition, iron and zinc acquisition systems also play very important roles in the infection process in *P. mirabilis*-associated UTIs. While iron plays a significant role in the survival and the infectivity of the bacteria in the urinary tract,[49] zinc is essential for other virulence factors such as flagella[50] and *ZapA*.[51] The limited quantities of zinc in the urinary tract gives a competitive advantage to organisms capable of zinc uptake. Lack of a functional zinc uptake system can, therefore, result in an improper regulation of flagellin and *ZapA*, weakening the ability of the bacteria to colonize the urinary tract.

Immune responses

The attachment of *P. mirabilis* to host cells triggers a cascade of host immune responses such as the secretion of cytokines and interleukins, IL-6 and IL-8. Flagellin is recognized by the toll-like receptors and nucleotide-binding oligomerization domain (NOD)-like receptors, which induces the expression of pro-inflammatory chemokines. The elicitation of inflammation by flagellin results in the recruitment of leukocytes to the bladder epithelium.[52] In addition, certain modifications in lipopolysaccharides and proteases, such as *ZapA*[51] and some bacterial virulence factors, such as the *Pta* result in resistance to the host antibodies and various antimicrobial peptides such as protegrin, polymyxin B, defensin, and LL-37.[53,54] This favors the survival of the bacteria and enhances its ability to establish an infection.[47,48] The fight that ensues between the defense mechanisms of the pathogen and the host results in cell apoptosis and wide-spread desquamation of the epithelial cell layer, resulting in organ damage. *P. mirabilis* also has a tendency to get encased in urinary stones and in the biofilm crystals, thereby enabling the bacteria to evade immune responses and therapeutic measures. In addition, endotoxins released by the cell walls of Gram-negative bacteria such as *P. mirabilis* precipitate bloodstream invasion, which in turn trigger additional host immune responses that increases the risk of potentially life-threatening conditions such as systemic inflammatory response syndrome and sepsis.[35]

Resistance

The treatment of *P. mirabilis* infections has become very complicated in recent years due to the advent of strains resistant to several classes of antibiotics. While the resistance to sulfamethoxazole/trimethoprim (SXT) is widespread, more recently strains resistant to ß-lactam-based antibiotics such as penicillin and cephalosporins have also emerged. Resistance to tetracycline is especially common among the *P. mirabilis* bacteria,[31] although the organism has also developed resistance to nitrofurantoin, fluroquinolones, fosfomycin, sulfonamides, and aminoglycosides.[55,56]

Clinical manifestations and complications

Clinical manifestations of *P. mirabilis* infections differ depending on the affected organ in the urinary system. Urethritis manifests in cases of urethral infection, bladder infections manifest as cystitis, cases involving inflammation of the prostate manifest as prostatitis (Fig. 6.13), while pyelonephritis (Fig. 6.14) occurs in cases where a wide-spread inflammation of the nephrons is involved. Due to the involvement of *P. mirabilis* in urolithiasis, recurrent manifestations of renal stones are considered to be indicative of an underlying *P. mirabilis* infection.

Although broadly the symptoms of UTIs are similar irrespective of the organ involved, certain symptoms help in clinical distinction of the organs involved in the infection. While urethritis presenting with dysuria, pyuria (Fig. 6.15A and B),

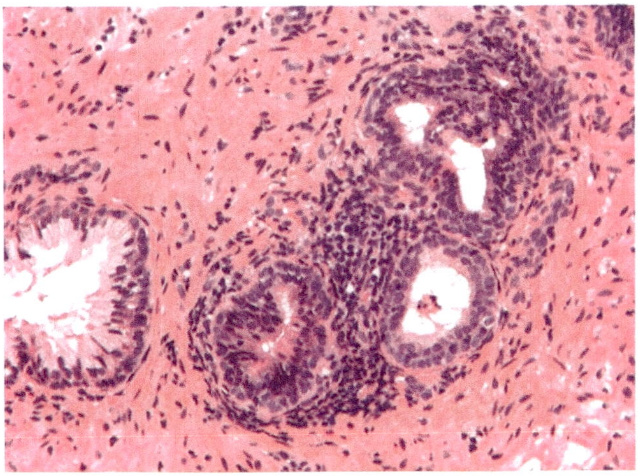

FIGURE 6.13
Micrograph of prostate gland showing chronic inflammatory infiltrate (Hematoxylin and Eosin (H&E), a Standard Tissue Stain) indicating prostatitis. Source*: Nephron/Own work/; https://creativecommons.org/licenses/by-sa/3.0/deed.en; https://commons.wikimedia.org/wiki/File:Inflammation_of_prostate.jpg.*

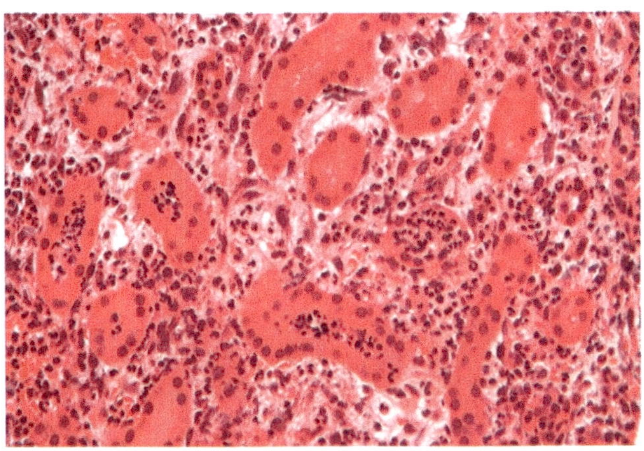

FIGURE 6.14
Micrograph of acute pyelonephritis in a kidney biopsy (H&E stain). Source*: Nephron/Own work/; https://creativecommons.org/licenses/by-sa/3.0/deed.en; https://commons.wikimedia.org/wiki/File:Acute_pyelonephritis_-_2_-_very_high_mag.jpg.*

urethral discharge, and increased urinary frequency often manifests as mild disease that is typically ignored, cystitis is an acute condition presenting with dysuria, increased frequency of micturition, urgency to urinate, pain in the suprapubic region, back, or flanks (Fig. 6.16), small volume of urine, and hematuria (Fig. 6.17). Pyelonephritis is a more severe condition that may occur as a complication of either urethritis, cystitis, or prostatitis. Pyelonephritis is often suspected

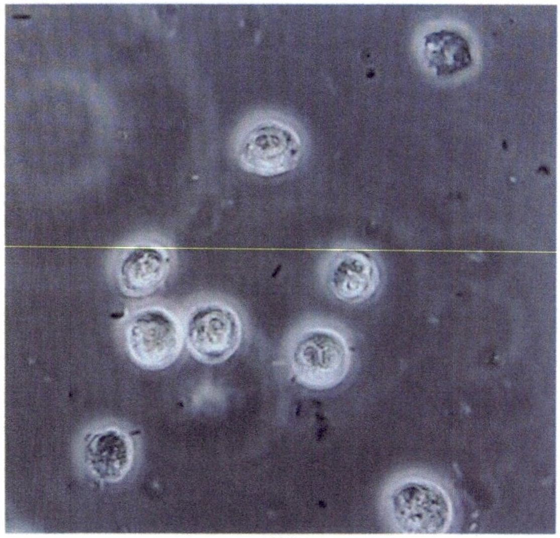

(A)

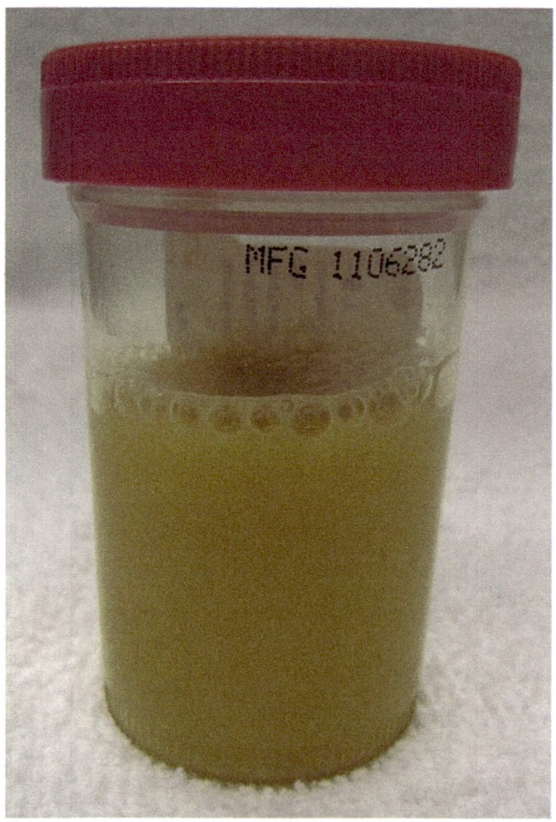

(B)

FIGURE 6.15

(A) Bacteriuria and pyuria demonstrated in microscopic analysis of urine. (B) Pyuria indicated by turbid urine in a case of urosepsis. Source*: Steven Fruitsmaak/Own work/; https://creativecommons.org/licenses/by/3.0/deed.en; https://commons.wikimedia.org/wiki/File: Bacteriuria_pyuria.jpg; James Heilman MD/Own work/; https://creativecommons.org/licenses/by-sa/3.0/deed.en; https://commons. wikimedia.org/wiki/File:Pyuria2011.JPG.*

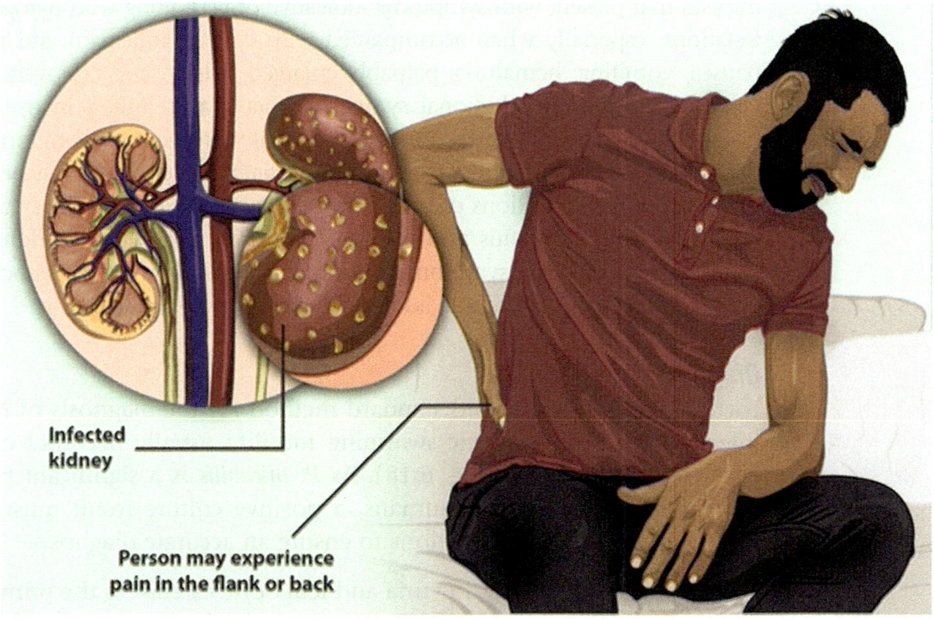

FIGURE 6.16

Depiction of a man suffering from a kidney infection indicating pain in the flank region along with a view of an infected kidney. Source*: https://www.myupchar.com/en/ https://www.myupchar.com/en/disease/kidney-infection-pyelonephritis/; https://creativecommons.org/licenses/by-sa/4.0/deed.en; https://commons.wikimedia.org/wiki/File:Depiction_of_a_man_suffering_from_a_Kidney_Infection.png.*

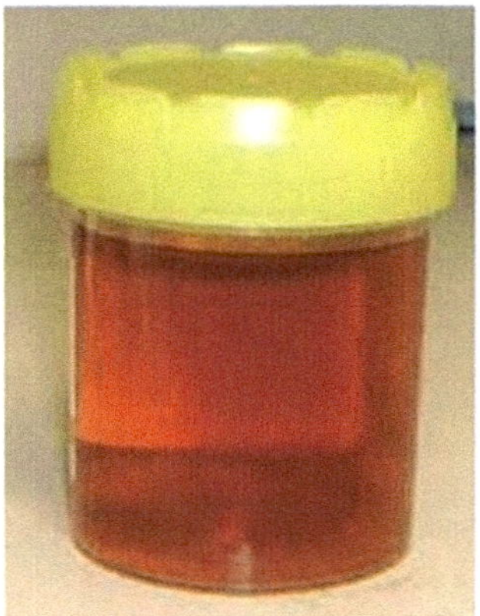

FIGURE 6.17

Hematuria in urine sample. Source*: omicsonline.org/; https://www.omicsonline.org/mexico/hematuria-peer-reviewed-pdf-ppt-articles/; https://creativecommons.org/licenses/by/4.0/deed.en; https://commons.wikimedia.org/wiki/File:Urinary_sample_with_hematuria.jpg.*

in cases that present with symptoms indicative of UTI along with pyrogenic manifestations, especially when accompanied with chills. Flank pain and tenderness, nausea, vomiting, hematuria, palpably enlarged kidney, and costovertebral angle tenderness are some additional symptoms that may manifest in pyelonephritic cases. Bacteremia and sepsis are potentially life-threatening conditions resulting from the invasion of the bloodstream by *P. mirabilis* bacteria. Men generally show more severe manifestations of prostatitis than cystitis, although the symptoms are often the same. Prostatitis also presents with fever and chills in addition to the usual constellation of symptoms. Obstruction of urinary tract associated with prostatitis can also result in an additional complain of perianal pain.[35]

Diagnosis

Bacterial culture is the gold standard method for the diagnosis of *P. mirabilis* infection. The characteristic swarming motility usually becomes evident on bacterial culture plates (Fig. 6.18). As *P. mirabilis* is a significant part of the normal intestinal flora in humans, a positive culture result must be correlated with clinical manifestations to ensure an accurate diagnosis.

Urine analysis to check for pyuria and leukocyte esterase is the primary evaluation in cases indicative of UTI. Pyuria is a characteristic feature of UTIs caused by bacterial etiologies, and therefore the lack of pyuria is considered to be suggestive of the need for a different approach. Gram staining is also

FIGURE 6.18

Proteus mirabilis cultivation on blood agar showing the characteristic swarming motility. Source*: Microrao/Own work/; https://creativecommons.org/licenses/by-sa/4.0/deed.en; https://commons.wikimedia.org/wiki/File:Proteus_swarming_on_blood_agar.jpg.*

useful for detecting bacteriuria, although the possibility of false negatives and inconclusive reports is very high (Fig. 6.19). Leukocyte esterase dipstick is a more rapid albeit less sensitive alternative to microscopic analysis. The pH of the urine sample is also a very important indicator of problems related to the urinary tract. A chronically alkaline pH of urine combined with a positive *P. mirabilis* culture is strongly suggestive of renal stones.

Complications such as perinephric abscess, renal stones, or involvement of a resistant pathogenic strain must be considered in cases that do not resolve with antibiotic therapy alone.

Treatment and management

Uncomplicated UTIs caused by *P. mirabilis* infection are empirically treated with a 3-day SXT course unless resistance is indicated. In regions where SXT-resistant strains are endemic, alternative antibiotics such as fluroquinolones, fosfomycin, and nitrofurantoin may be prescribed, although strains resistant to these therapies have also been reported.[55–57]

Although pyelonephritis is an acute condition, an uncomplicated case can be treated on an outpatient basis using a 7–14 days regimen comprising of fluroquinolones or cephalosporin, sometimes used in combination with a one-time dose of ceftriaxone or gentamycin. Severe cases presenting with high-grade fever usually require intravenous treatment with ceftriaxone, fluoroquinolone,

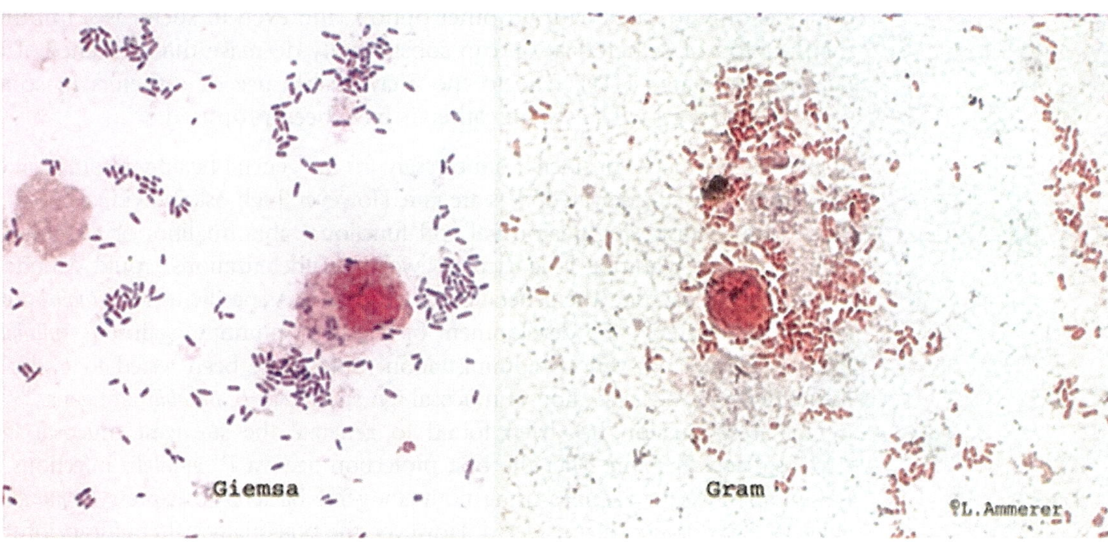

FIGURE 6.19

Proteus mirabilis (Giemsa and Gram stain). Source*: Kreisleriana10/Own work/; https://creativecommons.org/licenses/by-sa/4.0/deed.en; https://commons.wikimedia.org/wiki/File:Proteus_mir_stain.jpg.*

gentamycin, gentamycin with ampicillin, ceftriaxone, gentamycin with ampicillin, or aztreonam until fever resolves, after which the therapy is switched to the 7–14 days oral therapy comprising of cephalosporin, SXT, or fluroquinolone.[35]

Complicated and catheter-associated UTIs are often treated based on the severity of the clinical manifestations. The presence of renal stones in cases of UTIs associated with *P. mirabilis* infections warrants a more careful treatment approach owing to the bacterial tendency of being encased inside the stone. Techniques such as shockwave lithotripsy or interventions for removal of the renal stones need to be applied before antibiotic therapy is administered. Such cases also need monthly urine culture follow-ups for up to 3 months after completion of the antibiotic therapy to ensure complete resolution of the infection.[58]

Prevention

The prevention of *P. mirabilis* infections is possible with appropriate sanitation and hygiene practices and effective sterilization of medical equipment and surfaces. Anatomical and functional abnormalities of the urinary tract significantly increase the risk of contracting *P. mirabilis*-associated UTIs. Therefore interventions aimed at resolving or managing such abnormalities should be considered whenever possible. Long-term in-dwelling catheterization is another factor that significantly increases the risk of *P. mirabilis* associated UTIs. The use of intermittent clean catheterization is, therefore, preferred over in-dwelling catheters as far as possible.[59] Limiting the use of catheters only to cases with no other option, and even in such cases limiting the duration of catheterization can substantially decrease the incidence of *P. mirabilis*-associated UTIs. Due to the unavoidable use of catheters in some cases, bacterial growth-resistant catheters have been proposed.

P. mirabilis vaccine is considered unnecessary for the general healthy population as cases of *P. mirabilis*-associated UTIs are rare. However, high-risk individuals such as those with predisposing anatomical and functional abnormalities of the urinary tract and those requiring long-term in-dwelling catheterizations would definitely benefit from a vaccine. The generation of mucosal IgA specific to *P. mirabilis* antigens is important to the development of effective immunity against *P. mirabilis* infections. Therefore various administration routes have been tested to evaluate their efficacy in the generation of mucosal IgA specific to *P. mirabilis* antigens.[60–62] Intranasal vaccination has been found to generate the strongest mucosal IgA response, and therefore offer the best protection against *P. mirabilis* infections.[63] Recombinant-vector vaccines using nonpathogenic bacteria to express *P. mirabilis* antigens have also been considered. However, the potentially better efficacy of the recombinant *MrpA* vaccine contends the benefits of the recombinant-vector vaccines.[64,65] Vaccine candidates targeting the fimbrial and nonfimbrial proteins including iron acquisition genes are also being contemplated.

Student Notes

Key take-aways

- *P. mirabilis* is an important part of the normal human intestinal flora, but can behave as a pathogen causing UTIs.
- Flagellar motility in the form of swarming by holding together self-elongated cells in a raft formation and urease production are characteristic of *P. mirabilis*.
- Urease production ability is responsible for the involvement of this organism in the formation of renal stones and crystalline biofilms.
- UTIs caused by *P. mirabilis* infection can manifest as cystitis, prostatitis, urethritis, and pyelonephritis.
- Structural and functional abnormalities of the urinary tract and long-term in-dwelling catheterizations are predisposing factors responsible for increased incidences of *P. mirabilis*-associated UTIs.
- The tendency of this organism to get encased in the renal stones and in the crystals of the biofilm makes it a challenge for clinical treatment of UTIs.
- Resistance to antibiotic therapies such as STX, tetracyclines, and ß-lactam-based antibiotics is a matter of increasing concern for treatment of UTIs caused by *P. mirabilis*.
- Vaccinating high-risk individuals against *P. mirabilis* is a promising option for drastically reducing the incidence of *P. mirabilis*-associated UTIs.
- The development of bacterial-growth resistant catheters and vaccines against *P. mirabilis* has gained impetus in recent times due to the promising benefits for high-risk individuals.

Problem number 6.4: Postinfectious glomerulonephritis secondary to Epstein – Barr virus infection

A 12-year-old boy complained of puffiness around the eyes every morning for the past 4 days and a general feeling of malaise for the last 2 weeks with occasional bouts of headaches and vomiting, although there was no history of convulsions. The patient also mentioned difficulty in breathing in the supine position and passing dark-colored urine. There was a history of a sore throat about 3 weeks ago, which had been treated with antibiotics, although no laboratory investigation had been done to identify the etiological agent responsible for the sore throat.

The following are the reports of the investigations carried out:

1. Vitals:
   Temperature: 37°C
   Blood pressure: 145/92 mm Hg
   Pulse: 112/min
   Weight: 37 kgs (Weight was 32 kgs few weeks ago)

Height: 137 cm

2. Physical examination

   Eyes: Swelling around eyes; normal appearance of optic fundi
   Skin: No rash/infection
   Limbs: Mild pitting edema in both legs
   Abdomen: Liver felt 3 cm below costal margin; tender on palpation; no ascites on percussion
   Chest: Wheezes and basal crackles (both sides)

3. Urine analysis

   Appearance: Cloudy/Turbid
   Specific gravity: 1.030
   Microscopy
   RBC: + + +
   WBC: + +
   Protein: + +
   RBC Casts: +
   Protein/24 hours: 900 mg/24 hours (Normal $<$ 150 mg/24 hours)
   Urine volume/24 hours: 350 mL (Normal $>$ 600 mL)
   Creatinine: 18.05 mmol/L (Normal: 44–177 mmol/L)
   Culture: No significant growth

4. CBC

Test	Values	Reference range
Hemoglobin (g/dL)	11.2	11–13
Hematocrit	0.31	0.33–0.40
Neutrophils (%)	57	54–62
Lymphocytes (%)	45	25–33
Monocytes (%)	6	3–7
Eosinophils (%)	2	1–3

5. Radiology

   Chest X-ray: Cardiomegaly; prominent pulmonary blood vessels; bilateral basal congestion
   Renal ultrasound: Normal size of kidneys
   ECG: Normal sinus rhythm; left axis deviation; left ventricular hypertrophy (R in V5 and S in V1$>$ 40 mm)

6. Throat swab culture: Normal throat microbiota
7. Biochemistry

Test	Values	Reference range
Urea (mmol/L)	13	3–6.3
Sodium (mmol/L)	131	135–145

Continued

Continued

Test	Values	Reference range
Potassium (mmol/L)	4.4	3.5–5.3
Chlorine (mmol/L)	102	95–110
Bicarbonate (mmol/L)	16	23–32
Creatinine (mmol/L)	98	44–88
Creatinine clearance (ml/min/1.73 m^2)	47	90–139
Total protein (g/L)	65	68–80
Albumin (g/L)	32	35–55
Globulin (g/L)	38	15–30
Serum C3 (mg/L)	25	70–158
Serum C4 (mg/dL)	19	20–40
ASO titer	1/150	<1/200
Anti-EBV VCA IgG (AU/mL)	78	<20
Anti-EBV VCA IgM (AU/mL)	107	<20
Anti-EBV EA IgG (AU/mL)	68	<20

8. Kidney biopsy: Although a kidney biopsy was considered, it was not performed because a kidney biopsy is suggested only in cases of irretractable rapidly declining renal function, especially when the etiology responsible for the pathological changes cannot be determined. However, in this case, the increased levels of anti-EBV viral capsid antigen (VCA) IgM and IgG antibodies clearly indicate an active Epstein – Barr virus (EBV) infection, and the clinical manifestations of the case in combination with laboratory evidence of immune-mediated processes, such as low levels of complement factors (C3 and C4), are suggestive of acute glomerulonephritis precipitated by an immune complex disease.

Diagnosis

The final diagnosis concluded acute glomerulonephritis, probably secondary to EBV infection [infectious mononucleosis (IM)]. The clinical manifestations of hypertension, hypervolemia, cardiac involvement, dyspnea, edema, a history of recent throat infection combined with the laboratory evidence of immune-mediated processes, such as low levels of complement factors (C3 and C4), were considered as the basis for the final diagnosis.

Treatment

The patient was hospitalized, monitored, and treated as follows:

1. Dexamethasone—10 mg/day (Day 1)
2. Methylprednisolone—24 mg/day tapered by 4 mg/day
3. Diuretics
4. Bed rest
5. Daily body weight measurement
6. Measurement of urine output
7. Daily blood pressure monitoring
8. Diet modifications
9. Restricted fluid intake
10. Chest X-ray repeated after 2 days of treatment—Decreased lung infiltrate and reduction in cardiomegaly

Follow-up and outcome

The chest X-ray taken 2 days after treatment showed decreased lung infiltrate and reduced cardiomegaly. A complete resolution of proteinuria was seen 3 days after the completion of the treatment. The patient was discharged from the hospital following this positive outcome and was recommended regular follow-ups with a pediatric nephrologist for a period of 1 year.

Question and answer

1. What is the significance of the sore throat history in this case?
   Although most cases of a sore throat resolve without any complications, sore throats caused due to streptococcal or EBV infections can result in nephrotic complications in some rare cases. Therefore the history of the sore throat may be significant in this case.
2. Which are the most common pathogens responsible for throat infections?
   Group A ß-hemolytic *Streptococcus pyogenes*, and less commonly, Group C *Streptococcus* are the most common pathogens responsible for throat infections. EBV infections are relatively rare causes of throat infections. Besides, other viral etiologies, such as influenza, rhinoviruses, coronaviruses, and herpes viruses, can also cause throat infections.
3. What is the body's response to infection?
   An infection usually triggers a cascade of defensive events in the body, including inflammation, phagocytosis, antibody production, and cell-mediated immune reaction.
4. What can cause dark-colored urine?
   The presence of blood, bilirubin, certain drugs, or high concentrations of various cells, protein, and salts in urine can result in its dark color. While some causes of dark-colored urine, such as the expulsion of

certain drugs, may not be a cause for concern, most of the other causes, like the presence of blood or protein, indicate a serious underlying pathology affecting the urinary system. The presence of bilirubin in the urine is suggestive of a pathology associated with the liver or the hepatobiliary system instead of the urinary system.

5. What causes the episodes of headaches and vomiting in this case?

   Hypertension caused by the renin-angiotensin aldosterone system, triggered by renal ischemia, and the retention of fluids due to the renal disease, is responsible for the episodes of headache and vomiting in this case.

6. Does the tenderness and enlargement of the liver indicate an infection in this case?

   While the tenderness and enlargement of the liver may indicate an infection, in this case, it is caused by the congestion of the liver precipitated by cardiac disease.

7. How is the rapid increase in body weight associated with the pathology indicated in this case?

   The rapid increase in body weight is caused by fluid retention resulting from the impaired renal functioning.

8. How do the different parameters of routine urine analysis help to establish a diagnosis for this case?

   Typically, urine is a clear, straw-colored liquid, and any deviation is considered a primary indication of pathological changes in the urinary tract. The pH of urine is usually acidic, and any change in the pH, such as an increased pH (alkaline), suggests certain bacterial infections, such as *P. mirabilis*. Similarly, the presence of casts or crystals in urine also implies certain specific bacterial infection. Reduced or increased urine output is suggestive of clinically significant changes in renal function. Changes in specific gravity of urine indicates a change in its composition, which alludes to pathological changes in the urinary system. The presence of RBCs, WBCs, or protein in urine indicates impaired filtration processes of the kidney.

9. What is the significance of measuring blood urea, creatinine, and electrolytes?

   Blood urea, creatinine, and electrolytes are tests that enable the analysis of kidney function.

10. What is the expected finding in throat culture swabs?

    Throat culture swabs are usually expected to show the presence of bacterial pathogens responsible for the pathology; however, viral etiologies do not show in throat swab cultures. Therefore although the majority of throat infections are caused by viral etiologies, throat culture swabs cannot detect them.

11. What is the significance of measuring the C3 and C4 levels in this case?
    C3 and C4 levels are indicative of the functioning of the complement system, an important part of the human immune system. Immune complex diseases, such as glomerulonephritis and leukocytoclastic vasculitis associated with certain bacterial and viral infections, can show a decrease in the C3 and C4 levels. The alternative and the classical pathways of the complement system can be activated by the microbial polysaccharides and the antigen – antibody complexes, respectively.
12. How is the renal pathology associated with the manifestations of congestive heart failure in this case?
    The hypertension and hypervolemia resulting from the impaired renal function is responsible for congestive heart failure in this case.
13. What is the significance of the daily body weight measurement of the patient?
    The body weight measurement is important for monitoring the changes in the levels of fluid retention.
14. Why is it important to measure the daily urine output?
    Urine output is important for monitoring the kidney function and fluid needs of the patient.
15. What is the role of diuretics?
    Diuretics increase the urine output, thereby reducing hypervolemia and relieving edema and congestion.
16. How are dietary changes important in the resolution of renal pathologies?
    Dietary changes, such as limited salt intake, help to prevent the impairment of kidney function by reducing the fluid retention in the body. Hypervolemia caused by fluid retention is known to cause several other pathologies like hypertension and congestive heart failure.
17. Why is proteinuria measured based on a 24-hour collected urine sample?
    Proteinuria may be intermittent, and therefore a 24-hour collected urine sample is a better representation, which also enables the calculation of the glomerular filtration rate.
18. How are the serum urea and creatinine levels associated with the glomerular filtration rate?
    The serum levels of urea and creatinine reflect the glomerular filtration rate because both urea and creatinine are waste products in the blood, and increased levels of these two parameters is indicative of impaired glomerular filtration.

19. How are certain bacterial and viral infections associated with glomerulonephritis?

    Certain bacterial and viral infections, such as streptococcal and EBV infection, result in the production of antibodies and the formation and deposition of immune complexes in the glomeruli of kidneys precipitating an inflammation of the cells, which manifests as glomerulonephritis.

20. What is the interpretation for the antistreptolysin O (ASO) titer, anti-EBV VCA, and anti-EBV early antigen (EA) tests?

    The ASO titer measures the levels of antibodies against the streptococcal antigen (streptolysin O), which indicates the presence or absence of streptococcal infection. As the Streptococcal bacteria are commensals in the human body, therefore only antibody levels higher than a certain threshold (> 200) signal streptococcal infections. Anti-EBV VCA and anti-EBV EA tests measure the levels of antibodies against specific EBV antigens in the blood, and the anti-EBV VCA test checks for IgG and IgM antibodies. While the increased levels of IgG antibodies denote to a past EBV infection, increased IgM antibody levels signify active EBV infections. An increased level of anti-EBV EA antibodies is indicative of the acute phase of the disease.

21. How is complement fixation and the formation of immune complexes associated with pathological changes in the kidney?

    Immune complex diseases are characterized by the presence of granular deposits of antigen – antibody complexes in the affected organ. The usual sites of detection of these deposits are the bloodstream and the glomeruli of the kidney due to the hemodynamics of the process of formation and deposition of immune complexes. The process by which antigens are detected and bound by specific antibodies, known as complement fixation, is responsible for generating these immune complexes. Peripheral mononuclear cells recruited from the C5a and C3a components of the complement pathway form immune complexes which are phagocytosed and deposited in the glomeruli of the kidney. This is followed by the secretion of proteolytic enzymes by the PMNs, resulting in destruction of the glomerular basement membrane.

22. What results in the alleviation of symptoms in cases of immune complex disease?

    The clearance of excess antibodies generated against antigens and accumulated antigen – antibody complexes due to the absence of the stimulatory antigen leads to symptom relief in cases of immune complex disease.

Learning objectives

1. Describe the characteristics of EBV.
2. Summarize the pathogenesis of EBV infections.
3. Summarize the biochemical and pathological changes seen in glomerulonephritis.
4. Explain the pathology of glomerulonephritis secondary to EBV infections.
5. Summarize the pathogenesis of immune complex disease.
6. Describe the pathologies associated with the glomeruli in immune complex disease.
7. Compare the pathologies of immune complex mediated glomerulonephritis and rheumatic heart disease.
8. Enumerate the methods involved in the diagnosis of EBV infections.
9. Summarize the principles of urine sample collection and culture.
10. Describe the laboratory diagnosis of immune complex disease.
11. Determine the prognosis of EBV-associated glomerulonephritis.

Discussion

Differential diagnosis

Poststreptococcal glomerulonephritis	Poststreptococcal glomerulonephritis is a rare condition that follows a streptococcal throat infection. The condition generally presents itself with dark urine, edema of the face and limbs, reduced urine output, and fatigue.
Interstitial nephritis	Interstitial nephritis is characterized by a fever, decreased urine output, hematuria, nausea, vomiting, edema, sudden unexplained weight gain due to fluid retention, and changes in mental status.
Crescentic glomerulonephritis	Crescentic glomerulonephritis manifests in various forms depending on the etiology responsible for the condition. It usually presents itself with oliguria, uremia, fluid retention, hematuria, anemia, abdominal or flank pain, hemoptysis, hypertension, pallor, skin rash, fatigue, nausea, and vomiting. Arthritis, neuritis, encephalopathy, episcleritis, encephalopathy, malaise, myalgia, and otitis are also caused in cases of multisystem involvement.
Diffuse proliferative glomerulonephritis	Diffuse proliferative glomerulonephritis develops pedal edema, decreased urine output, hypertension, fluid retention, proteinuria, hematuria, uremia, fatigue, nausea, and vomiting.
Focal segmental glomerulosclerosis	Focal segmental glomerulosclerosis usually causes fluid retention, peripheral and facial edema, rapid and unexplained weight gain, high cholesterol levels, and low serum protein levels. Critical and advanced cases often

Continued

Continued

	develop a loss of appetite, fatigue, headache, itchy skin, nausea, vomiting, and dyspnea.
Membranoproliferative glomerulonephritis	Membranoproliferative glomerulonephritis presents itself with dark-colored urine or hematuria, turbid urine, reduced urine output, edema, decreased alertness, decreased concentration, and behavioral changes.
Goodpasture syndrome	Goodpasture syndrome is an autoimmune condition that usually occurs with edema, hypertension, uremia, hemoptysis, and dyspnea.
Acute kidney injury	Acute kidney injury presents itself with a feeling of malaise, dehydration, diarrhea, reduced urine output, back or flank pain, edema, confusion, nausea, arrythmias, dyspnea, and drowsiness.
Lupus nephritis	Lupus nephritis causes hematuria, hypertension, peripheral edema, high serum creatinine levels, and foamy urine.
Chronic kidney disease	Chronic kidney disease usually presents itself with fatigue, weight loss, poor appetite, insomnia, itchy skin, muscle cramps, headaches, increased night-time urge to urinate, peripheral edema, dyspnea, hematuria, malaise, and erectile dysfunction in men.
Azotemia	Azotemia causes signs of acute kidney injury or renal failure, loss of appetite, lethargy, fatigue, fluid retention, dark-colored urine or hematuria, nausea, and vomiting.

Background

Postinfectious glomerulonephritis (PIGN) is usually associated with a preceding group A ß-hemolytic streptococcal infection. Rarely, EBV infections are also responsible for PIGN.[66–68] Although EBV-associated renal impairment typically manifests as tubulointerstitial nephritis or acute tubular necrosis, the possibility of immune-complex-associated glomerulonephritis as a rare postinfection associated with EBV infections has also been considered.[66,69,70]

IM is the most common manifestation of childhood EBV infections, which are usually self-limiting and typically benign. Although subclinical renal involvement is known, critical renal impairment leading to kidney failure is rare.

Etiological agent

Epstein – Barr virus

Classification and morphology

The EBV is an enveloped, double-stranded DNA virus belonging to the gammaherpesviridae subfamily within the *Herpesviridae* family. The classification is based on the similarities of the nucleic acid structure and composition of EBV and other gammaherpesviruses. The lymphotropism of EBV is also an important determinant for the classification of the virus.

The structure of EBV is typical of herpesviruses, with the double-stranded DNA being encased in a protein core. The protein shell or nucleocapsid is icosahedral and is composed of 162 capsomeres. The virus has an outer envelope derived from host cell membranes punctuated with glycoprotein spike proteins encoded by the viral genome (Fig. 6.20). The glycoprotein spike proteins are important for attachment and entry into the host cells.

There are two types of EBV (types A and B) that are known to exist, although the types are not considered clinically significant as they are not associated with any specific disease types. The two types of the virus share high levels of genomic homology, except for some differences in the regions that encode the EBV nuclear antigens (EBNAs) and the small viral RNAs. Type A is the globally predominant type and is more efficient at immortalizing cells. Type B, on the other hand, is commonly seen in African communities and in immunocompromised individuals.[71]

Epidemiology

EBV infections are prevalent in almost 95% of the world's population, with higher prevalence among adolescents and children. While gender and race differences show minimal effects on the epidemiology of EBV infections, poor sanitation, over-crowded living conditions, and a poor literacy rate have been associated with a higher prevalence of EBV infections in certain

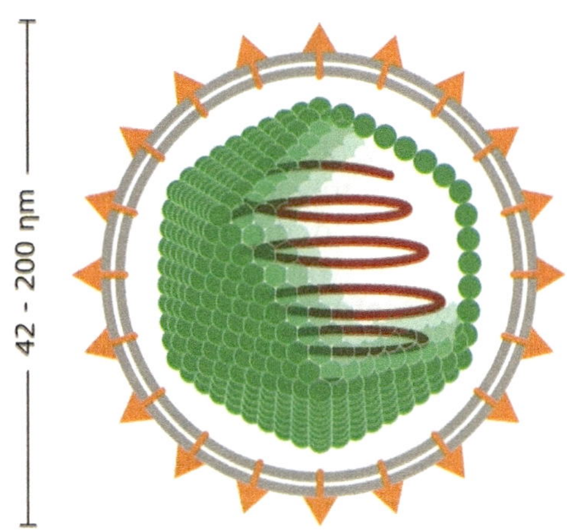

FIGURE 6.20

Representative structure of an enveloped icosahedral virus. Source: *Nossedotti (Anderson Britto)/Own work/; https://creativecommons.org/licenses/by-sa/3.0/deed.en; https://commons.wikimedia.org/wiki/File:Enveloped_icosahedral_virus.svg.*

populations. The virus is typically known for establishing life-long persistence akin to all herpesviruses. Humans are the natural host for this virus, although experimental establishment of infections in certain nonhuman primates is also known.[71] EBV is transmitted mainly through saliva, and while young children are infected early in life indirectly through contact with contaminated objects, sexually active adults are infected through direct salivary transfer during kissing. Rare cases of transmission through contaminated blood and organs during transfusions and transplants are also known.[72,73] Children in developing countries are usually infected within the first 5 years of their life due to the poor standards of living, while developed countries show a lower prevalence of EBV infections during early life.

Primary infection early in life is usually asymptomatic or mildly symptomatic, and triggers the host immune system to generate antibodies and specific cell-mediated immune responses against the viral antigens. These immune responses are instrumental in curbing the EBV infection and keeping it under control for the rest of the individual's life. Delayed primary infections, however, manifest in the form of IM.

Pathogenicity and pathogenesis

The pathogenicity of EBV and pathogenesis of infections caused by the virus are a direct function of the precarious virus – host balance, which maintains the latency of the virus in the body after a primary infection. The virus targets small resting B-cells and enters through a ligand – receptor interaction between the viral glycoprotein gp340/220 and CD21 expressed on the surface of mature circulating B-lymphocytes.[71] This binding of the virus to the host cell induces initial activation signals, which facilitate the fusion of the viral envelope with the cell membrane, mediating the release of the viral nucleocapsid into the host cell cytoplasm. The viral DNA then enters the host cell nucleus and integrates with the host DNA or circularizes to exist in the form of an episome. The characteristic ability of EBV to maintain life-long persistency in the host through latency is attributed to the formation of episomes.[74] The infected B-cells are induced to differentiate into memory B-cells that circulate through the system. The EBV-specific cytotoxic T-cell responses elicited by the formation of memory B-cells are responsible for keeping the EBV infection under control. Some of the infected B-lymphocytes may, however, enter the lytic phase, thereby triggering the replication of the viral DNA and the formation of many virions. This results in occasional infections in the buccal cavity, the pharynx, and the urogenital tract, during which the virus-shedding occurs in the saliva and genital secretions, thereby increasing the chances of transmission of the infection.

The immunological reactions in children are not as exuberant as in adults, which is perhaps the reason why acquiring EBV infections in early childhood causes a benign course of the disease, eventually leading to latency of the virus. The mode of transmission among children is usually through indirect contact with contaminated objects, which ensures a small infective dose compared to adults where the mode of transmission is direct exposure to contaminated saliva during mouth-to-mouth kissing. Therefore the infective dose may also be partly responsible for the reduced exuberance of the immune responsiveness in childhood EBV infections. The large number of activated CD8 + T-cells in circulation or accumulated in the lymph nodes, tonsils, spleen, liver, and the lymphoid tissues of the buccal cavity and the pharynx due to stimulation by the B-cells is responsible for the characteristic exaggerated pathogenesis of IM. The cytotoxic action of these activated T-cells results in the destruction of the EBV-infected host cells through the secretion of various lymphokines. The immunopathological mechanisms responsible for the clinical manifestations, such as a fever, sore throat, tonsillitis, lymphadenopathy, hepatosplenomegaly, and malaise are attributable to the effects of these secreted lymphokines.

Latent EBV kept in check by the immune system can be triggered into reactivation by factors that disrupt the precarious virus – host balance. The reactivation of a latent EBV infection, reinfection, or rare secondary complications are often implicated in some serious pathologies, such as autoimmune manifestations and lymphoproliferative diseases. Certain moieties encoded by EBV interact with or mimic some antiapoptotic molecules, signal transducers, and cytokines, which play a significant role in the processes of infection, immortalization, and oncogenesis in host cells.[75] Autoimmune manifestations, on the other hand, are precipitated due to a cascade triggered by specific T-cell responses that elicit inflammatory responses in healthy host cells. Acute tubulointerstitial nephritis, acute glomerulonephritis (Fig. 6.21), or minimal change disease (Fig. 6.22) are clinical implications of the infiltration of these cells into the renal interstitium. The induction of inflammatory processes in these cells causes them to secrete cytokines, thereby promoting the formation of immune complexes. The deposition of these immune complexes plays an important role in mediating inflammatory damage within the kidneys.[67,76]

Viral antigens and host immune responses

The establishment of latency requires EBV to express only a restricted range of viral genes, which are crucial for continual cell survival and division while avoiding viral replication. Certain triggers act as stimuli that induce the activation and differentiation of the infected B-cells, allowing viral replication to be initiated in the cells. This facilitates the switch to lytic infection, enabling the production of a new viral progeny, eventually resulting in cell death. The set of viral genes expressed in latency is unique to EBV, and is

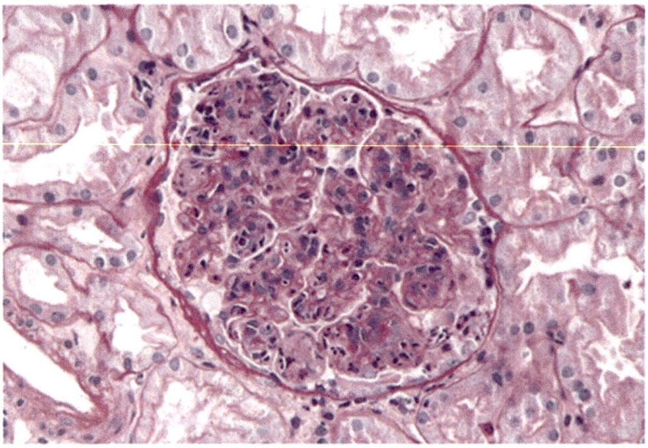

FIGURE 6.21
Postinfectious glomerulonephritis. Source*: Nephron/Own work/; https://creativecommons.org/licenses/by-sa/3.0/deed.en; https://commons.wikimedia.org/wiki/File:Post-infectious_glomerulonephritis_-_very_high_mag.jpg.*

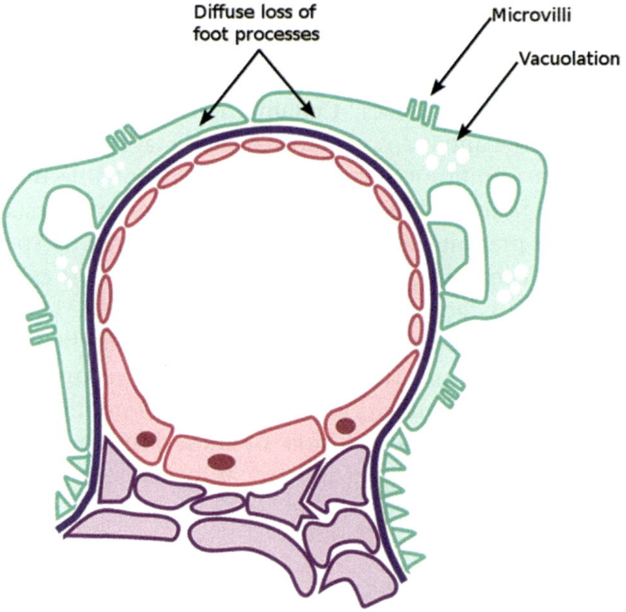

FIGURE 6.22
Schematic representation of the pathological changes seen under the electron microscope in minimal change disease. Source*: M. Komorniczak/; Renal_corpuscle.svg/retouched picture – modified by Huckfinne/; https://creativecommons.org/licenses/by-sa/3.0/deed.en; https://commons.wikimedia.org/wiki/File:Minimal_Change_Disease_Pathology_Diagram.svg.*

pretranscribed in infected B-cells, thereby influencing subsequent lytic gene expression.[77]

The viral genes expressed during latency encode six EBNAs, latent membrane protein (LMP), terminal protein 1 (TP1), terminal protein 2 (TP2), and two EBV-encoded small RNAs (EBERs). Evidence suggests that EBNAs 2, 3, 6, and LMP are essential for the immortalization ability of the virus. EBNA1 synchronizes the replication of the viral genome with host cell division and is also essential to the maintenance of the viral episome during latency. Although EBNA5 is not required for the immortalization of cells, studies suggest its role in controlling cellular proliferation. The most abundant of the viral transcripts found in infected B-cells during latency is the LMP mRNA, which translates into a significant viral oncoprotein with an important role in B-cell immortalization. LMP has been implicated in the induction of NF-κß and in the upregulation of the oncogene *bcl-2* that encodes an inhibitor of apoptosis. It has also been suggested that LMP is associated with the signal transduction pathways used by TNF. TP1 and 2 are instrumental in preventing the activation of the viral lytic cycle by inhibiting calcium mobilization during latency in activated B-lymphocytes that are EBV infected. Protein kinase R (PKR) mediates antiviral interferon activity in host cells. EBER1 and 2 interact with PKR and regulate its activity, thereby facilitating immune evasion by inhibiting the antiviral and antiproliferative effects of interferons.[71]

EBV early antigen is expressed early in the infection process before viral DNA synthesis, whereas viral capsid antigen is expressed later in the process after viral DNA synthesis. This is also confirmed by the early appearance of anti-VCA IgM and IgG, and anti-EA IgG serum antibodies during the clinical course of the disease, whereas anti-EBNA1 antibodies are generally only detectable during recovery. Although anti-VCA IgM and anti-EA IgG antibodies cease to exist after recovery, anti-VCA and anti-EBNA1 IgG antibodies persist for life. The presence of IgG anti-VCA and anti-EBNA1 antibodies is therefore useful in retrospective diagnosis, and the presence of IgM anti-VCA and IgG anti-EA antibodies in the absence of IgG anti-VCA and anti-EBNA1 is considered useful in indicating a primary infection. In contrast, the presence of anti-VCA IgM and anti-EA IgG antibodies in the presence of anti-VCA and anti-EBNA1 IgG antibodies indicates seroconversion or reactivation of latent EBV infection.

The T-cell responses to EBV-infected B-cells during a primary IM infection are excessive and ineffective. However, as IM is usually a self-limiting, lymphoproliferative disease, it resolves within a few days without any complications.[78] The establishment of effective CD4+ and CD8+ responses after the resolution of a self-limiting course of primary EBV infection plays a pivotal role in achieving the virus–host homeostasis, essential for keeping the life-long persistence of EBV under control.[79] T-cell surveillance is vital for preventing seroconversion. Therefore immunocompromised individuals are vulnerable to EBV reactivations. Individuals receiving immunosuppressive therapy often require infusions of EBV-specific T-cells to treat potentially fatal lymphoproliferative lesions comprising of EBV-transformed B-cells.[80,81]

Complications of a primary EBV infection manifested as IM are rare but have been documented. Renal involvement in EBV infections is usually in the form of interstitial nephritis, although some glomerulopathies have also been associated with the deposition of the immune complexes formed because of exaggerated immune responses to EBV infections.[66,69,70] The damage resulting in the renal pathologies may be mediated by the immune complexes generated in the humoral immune response or by direct, toxic cytopathic damage precipitated by the cell-mediated immune response. The deposition of many such immune complexes results in inflammatory conditions further enhanced by the hyperactivity of the host immune system, causing a self-reactive state that can persist for long periods of time, sometimes also leading to the establishment of an autoimmune condition. While such autoimmune conditions are implicated in chronic conditions and cause gradual damage, the deposition of immune complexes and cytopathic toxicity can also result in acute conditions.

Clinical manifestations and complications

The clinical presentation of EBV infections is vague and nonspecific, and may also be asymptomatic, especially in children. The symptoms, such as headache, malaise, fever, splenomegaly, lymphadenopathy, and sore throat are generic and associated with a wide-spectrum of diseases. However, the persistence of such symptoms in addition to unexplained fatigue often leads to consideration of EBV etiology. Laboratory investigations during the symptomatic phase of the disease may reveal lymphocytosis, atypical lymphocytes, and abnormally elevated liver enzymes.

Although IM is usually a self-limiting disease, the lymphoproliferative pathology of the disease can sometimes result in potentially life-threatening complications, such as splenomegaly (Fig. 6.23), which may lead to splenic rupture. Airway obstruction caused by edema of the pharyngeal tissues is another acute and critical complication of IM, occurring most commonly in children. Acute acalculous cholecystitis, an inflammatory disease of the

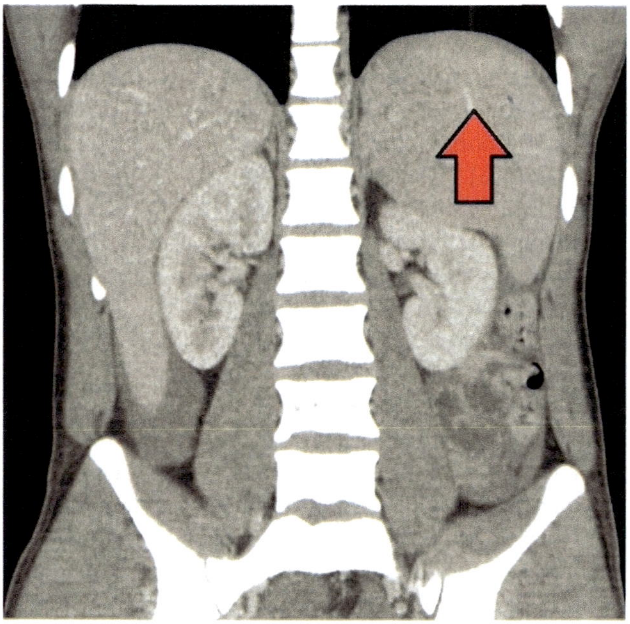

FIGURE 6.23
Splenomegaly caused by infectious mononucleosis, resulting in a subcapsular hematoma. Source: *James Heilman, MD/Own work/; https://creativecommons.org/licenses/by-sa/4.0/deed.en; https://commons.wikimedia.org/wiki/File:SplenomegalyandsubcaphematomaCorMark.png.*

gallbladder without any evident cause, such as gallstones or cystic duct obstruction, is another complication of IM, dreaded for its unbearable abdominal pain.

EBV may also result in secondary complications like myocarditis, autoimmune hemolytic anemia, encephalitis, hemophagocytic lymphohistiocytosis, pancreatitis, lymphomas, and nasopharyngeal cancers.[72,82–85]

Immune complex-mediated nephritis is a rarely occurring secondary complication of IM that presents itself with symptoms of renal, cardiac, and respiratory distress, such as back or flank pain, hematuria or dark urine, turbid or frothy urine, fever, chills, fatigue, dyspnea, edema, hypervolemia, hypertension, and signs of congestive heart failure.

Diagnosis

The establishment of EBV as the etiology for certain presentations of clinical manifestations require an amalgamation of patient history, physical examination, and serological testing. A microscopic examination of blood smears is the most preliminary test that may show atypical lymphocytosis, which is suggestive of EBV infection.[72] Antibody tests are highly recommended in cases suspected of EBV infection as they are rapid, inexpensive, considerably sensitive (63–84%) and have high specificity (84–100%).[86,87] However, the lack of antibodies due to the subtlety of the immune reactions may result in false negative antibody tests in children. Specific antibody tests, such as the anti-VCA IgM and IgG are used to confirm the diagnosis, and to differentiate between acute and chronic or past infections, respectively.[72,88]

Treatment and management

Primary EBV infections are generally self-limiting, especially in early life. Therefore supportive care and symptomatic relief is usually indicated. Evidence has indicated that the use of certain antivirals decreases the amount of virus-shedding in the saliva, thereby reducing the rate of transmission. However, antivirals have not been seen to offer any relief to the prevailing clinical symptoms of the disease.[86] Complicated cases involving extreme inflammatory conditions or autoimmune diseases may be treated using corticosteroids, which may not have therapeutic benefits, but they effectively bring the activity of the immune system under control and reduce the damage caused by inflammation.[72]

Prevention

The unavailability of an effective vaccine against EBV and the potential complications associated with the infection warrant the need for effective preventive practices. As the primary mode of transmission is through the saliva, avoidance of sharing utensils, food, beverages, face masks, handkerchiefs, and mouth-to-mouth kissing with infected or symptomatic individuals drastically reduces the incidence of the disease.[72]

Individuals with a predisposition to developing complications associated with EBV infections should be aware of the warning signs and should get tested to ensure an early diagnosis and better prognosis. Complicated cases, such as those with splenomegaly, should take necessary precautions to avoid irreversible outcomes, like splenic rupture.[86] A history of symptoms associated with IM succeeded by signs of any inflammatory condition must generate a clinical suspicion of EBV infection and necessary diagnostic measures must be undertaken immediately.

Student Notes

Key take-aways

- Acute glomerulonephritis is an inflammatory condition that damages the glomeruli of the nephrons, thereby affecting renal function.
- Poststreptococcal glomerulonephritis is well-documented; however, post-PIGN succeeding a primary EBV infection is rare.
- EBV infections are usually associated with interstitial nephritis caused by the deposition of immune complexes in the renal interstitium.
- EBV is primarily transmitted through contaminated saliva, although rare events of transmission during blood transfusions and organ transplants are also known.
- The incidence of childhood EBV infections is higher in developing countries due to lower standards of living, overcrowding, lack of adequate sanitation, and lack of awareness and literacy.
- IM is the manifestation of a primary EBV infection that generally occurs in early childhood, and is caused mainly due to indirect contact with contaminated saliva on surfaces of shared toys or food.
- In sexually active adults, IM usually occurs through the direct transfer of saliva during mouth-to-mouth kissing.
- The lower viral load involved in indirect transmission occurring in early childhood combined with the subtlety of the immune responses in children results in mild or asymptomatic disease.
- EBV is known for its unique ability to remain latent in infected B-lymphocytes and establish life-long persistence in the host.
- EBV is potent at the immortalization of mammalian cells, and is therefore implicated in several lymphoproliferative and malignant diseases.
- IM is generally a self-limiting disease and has a good prognosis; however, potentially life-threatening complications of the disease may result in a bad outcome.
- Diagnosis of EBV infections must be based on patient history, physical examinations, and serological testing, including specific antibody tests.
- Antibody tests must aim at confirming the diagnosis and differentiating between an acute and chronic infection.
- Treatment of EBV infections and complications associated with it mainly rely on supportive care and symptomatic relief, although corticosteroids may provide some benefits by controlling the immune responses and reducing the inflammatory damages.
- Although there is no vaccine available for EBV, infections can be prevented by good hygienic practices, avoiding mouth-to-mouth kissing, and not sharing food and beverages with an infected or symptomatic individual.

References

1. Tschudin-Sutter S, Frei R, Battegay M, Hoesli I, Widmer AF. Extended spectrum ß-lactamase-producing *Escherichia coli* in neonatal care unit. *Emerg Infect Dis.* 2010;16(11):1758–1760. Available from: https://doi.org/10.3201/eid1611.100366.
2. Flores-Mireles AL, Walker JN, Caparon M, Hultgren SJ. Urinary tract infections: epidemiology, mechanisms of infection and treatment options. *Nat Rev Microbiol.* 2015;13:269–284.
3. Oliveira FA, Paludo KS, Arend LNVS, Farah SMSS, Pedrosa FO, Souza EM. Virulence characteristics and antimicrobial susceptibility of uropathogenic *Escherichia coli* strains. *Genet Mol Res.* 2011;10(4):4114–4125.
4. Kaper JB, Nataro JP, Mobley HLT. Pathogenic *Escherichia coli*. *Nat Rev Microbiol.* 2004;2 (2):123–140.
5. Donlan RM. Biofilms: microbial life on surfaces. *Emerg Infect Dis.* 2002;8:881–890.
6. Wang X, Iii JFP, Romeo T. The pgaABCD locus of *Escherichia coli* promotes the synthesis polysaccharide adhesin required for biofilm formation. *J Bacterio ASM.* 2004;186:2724–2734.
7. Lüthje P, Brauner A. Chapter Seven—Virulence factors of uropathogenic *E. coli* and their interaction with the host. In: Poole RK, ed. *Advances in Microbial Physiology.* Volume 65. Academic Press; 2014:337–372. ISSN 0065-2911. ISBN 9780128001424. Available from: https://doi.org/10.1016/bs.ampbs.2014.08.006.
8. Welch RA. Uropathogenic *Escherichia coli*-associated exotoxins. *Microbiol Spectr.* 2016;4(3). Available from: https://doi.org/10.1128/microbiolspec.UTI-0011-2012.
9. Ghafourian S, Sadeghifard N, Soheili S, Sekawi Z. Extended spectrum beta-lactamases: definition, classification and epidemiology. *Curr Issues Mol Biol.* 2015;17(1):11–22.
10. Rodríguez-baño J, Navarro MD, Martínez-martínez L, et al. Epidemiology and clinical features of infections caused by extended-spectrum beta-lactamase-producing *Escherichia coli* in nonhospitalized patients. *J Clin Microbiol.* 2004;42:1089–1094.
11. Wragg R, Harris A, Patel M, Robb A, Chandran H, McCarthy L. Extended spectrum beta lactamase (ESBL) producing bacteria urinary tract infections and complex pediatric urology. *J Pediatr Surg.* 2017;52:286–288.
12. El Bouamri MC, Arsalane L, Zerouali K, Katfy K, El Kamouni Y, Zouhair S. Molecular characterization of extended spectrum β-lactamase-producing *Escherichia coli* in a university hospital in Morocco, North Africa. *Afr J Urol.* 2015;21:161–166.
13. Shrestha R, Khanal S, Poudel P, et al. Extended spectrum β-lactamase producing uropathogenic *Escherichia coli* and the correlation of biofilm with antibiotics resistance in Nepal. *Ann Clin Microbiol Antimicrob.* 2019;18:42. Available from: https://doi.org/10.1186/s12941-019-0340-y.
14. Bono MJ, Reygaert WC. *Urinary tract infection* [Updated July 2, 2020] *StatPearls [Internet].* Treasure Island, FL: StatPearls Publishing; 2020. Available from: https://www.ncbi.nlm.nih.gov/books/NBK470195/.
15. Basso V, d'Enfert C, Znaidi S, Bachellier-Bassi S. From genes to networks: the regulatory circuitry controlling *Candida albicans* morphogenesis. *Curr Top Microbiol Immunol.* 2019;422:61–99. Available from: https://doi.org/10.1007/82_2018_144.
16. Tao L, Du H, Guan G, et al. Discovery of a "white-gray-opaque" tristable phenotypic switching system in *Candida albicans*: roles of non-genetic diversity in host adaptation. *PLoS Biol.* 2014;12(4):e1001830. Available from: https://doi.org/10.1371/journal.pbio.1001830. Published April 1, 2014.
17. Staib P, Morschhäuser J. Chlamydospore formation in *Candida albicans* and *Candida dubliniensis*—an enigmatic developmental programme. *Mycoses.* 2007;50(1):1–12. Available from: https://doi.org/10.1111/j.1439-0507.2006.01308.x.

18. Pande K, Chen C, Noble SM. Passage through the mammalian gut triggers a phenotypic switch that promotes *Candida albicans* commensalism. *Nat Genet.* 2013;45(9):1088–1091. Available from: https://doi.org/10.1038/ng.2710.

19. Noble SM, Gianetti BA, Witchley JN. *Candida albicans* cell-type switching and functional plasticity in the mammalian host. *Nat Rev Microbiol.* 2017;15(2):96–108. Available from: https://doi.org/10.1038/nrmicro.2016.157.

20. Dubey AK, Singla RK. Current trends in anti-Candida drug development. *Curr Top Med Chem.* 2019;19(28):2525–2526. Available from: https://doi.org/10.2174/156802661928191206162925.

21. Ramage G, Saville SP, Thomas DP, López-Ribot JL. Candida biofilms: an update. *Eukaryot Cell.* 2005;4(4):633–638. Available from: https://doi.org/10.1128/EC.4.4.633-638.2005.

22. Kojic EM, Darouiche RO. Candida infections of medical devices. *Clin Microbiol Rev.* 2004;17 (2):255–267. Available from: https://doi.org/10.1128/cmr.17.2.255-267.2004.

23. An R, Rafiq NB. *Candidiasis* [Updated November 20, 2020]*StatPearls [Internet]*. Treasure Island, FL: StatPearls Publishing; 2020. Available from: https://www.ncbi.nlm.nih.gov/books/NBK560624/.

24. Valente P, Ferreira J, Pinto I, et al. Management of laryngeal candidiasis: an evidence-based approach for the otolaryngologist. *Eur Arch Otorhinolaryngol.* 2020;277(5):1267–1272. Available from: https://doi.org/10.1007/s00405-020-05865-4.

25. Ghaddar N, El Roz A, Ghssein G, Ibrahim JN. Emergence of vulvovaginal candidiasis among Lebanese pregnant women: prevalence, risk factors, and species distribution. *Infect Dis Obstet Gynecol.* 2019;2019:5016810. Available from: https://doi.org/10.1155/2019/5016810. Published July 28, 2019.

26. Donders GGG, Ravel J, Vitali B, Netea MG, Salumets A, Unemo M. Role of molecular biology in diagnosis and characterization of vulvo-vaginitis in clinical practice. *Gynecol Obstet Invest.* 2017;82(6):607–616. Available from: https://doi.org/10.1159/000478982.

27. van Schalkwyk J, Yudin MH, Infectious Disease Committee. Vulvovaginitis: screening for and management of trichomoniasis, vulvovaginal candidiasis, and bacterial vaginosis. *J Obstet Gynaecol Can.* 2015;37(3):266–274. Available from: https://doi.org/10.1016/S1701-2163 (15)30316-9.

28. Crouss T, Sobel JD, Smith K, Nyirjesy P. Long-term outcomes of women with recurrent vulvovaginal candidiasis after a course of maintenance antifungal therapy. *J Low Genit Tract Dis.* 2018;22(4):382–386. Available from: https://doi.org/10.1097/LGT.0000000000000413.

29. Matthews SJ, Lancaster JW. Urinary tract infections in the elderly population. *Am J Geriatr Pharmacother.* 2011;9(5):286–309. Available from: https://doi.org/10.1016/j.amjopharm.2011.07.002.

30. Papazafiropoulou A, Daniil I, Sotiropoulos A, et al. Prevalence of asymptomatic bacteriuria in type 2 diabetic subjects with and without microalbuminuria. *BMC Res Notes.* 2010;3:169. Available from: https://doi.org/10.1186/1756-0500-3-169. Published June 17, 2010.

31. O'Hara CM, Brenner FW, Miller JM. Classification, identification, and clinical significance of Proteus, Providencia, and Morganella. *Clin Microbiol Rev.* 2000;13(4):534–546. Available from: https://doi.org/10.1128/cmr.13.4.534-546.2000.

32. Jacobsen SM, Stickler DJ, Mobley HL, Shirtliff ME. Complicated catheter-associated urinary tract infections due to *Escherichia coli* and *Proteus mirabilis. Clin Microbiol Rev.* 2008;21 (1):26–59. Available from: https://doi.org/10.1128/CMR.00019-07.

33. Schaffer JN, Pearson MM. *Proteus mirabilis* and urinary tract infections. *Microbiol Spectr.* 2015;3(5). Available from: https://doi.org/10.1128/microbiolspec.UTI-0017-2013.

34. Majeed HT, Aljanaby AAJ. Antibiotic susceptibility patterns and prevalence of some extended spectrum beta-lactamases genes in Gram-negative bacteria isolated from patients infected

with urinary tract infections in Al-Najaf City, Iraq. *Avicenna J Med Biotechnol.* 2019;11 (2):192–201.

35. Jamil RT, Foris LA, Snowden J. Proteus mirabilis *infections* [Updated November 30, 2020] *StatPearls [Internet]*. Treasure Island, FL: StatPearls Publishing; 2020. Available from: https://www.ncbi.nlm.nih.gov/books/NBK442017/.
36. Stickler DJ. Bacterial biofilms in patients with indwelling urinary catheters. *Nat Clin Pract Urol.* 2008;5(11):598–608. Available from: https://doi.org/10.1038/ncpuro1231.
37. Chew R, Thomas S, Mantha ML, Killen JP, Cho Y, Baer RA. Large urate cystolith associated with Proteus urinary tract infection. *Kidney Int.* 2012;81(8):802. Available from: https://doi.org/10.1038/ki.2011.506.
38. Mobley HL, Warren JW. Urease-positive bacteriuria and obstruction of long-term urinary catheters. *J Clin Microbiol.* 1987;25(11):2216–2217. Available from: https://doi.org/10.1128/JCM.25.11.2216-2217.1987.
39. Williams FD, Schwarzhoff RH. Nature of the swarming phenomenon in Proteus. *Annu Rev Microbiol.* 1978;32:101–122. Available from: https://doi.org/10.1146/annurev.mi.32.100178.000533.
40. Jones BV, Young R, Mahenthiralingam E, Stickler DJ. Ultrastructure of *Proteus mirabilis* swarmer cell rafts and role of swarming in catheter-associated urinary tract infection. *Infect Immun.* 2004;72(7):3941–3950. Available from: https://doi.org/10.1128/IAI.72.7.3941-3950.2004.
41. Stahl SJ, Stewart KR, Williams FD. Extracellular slime associated with *Proteus mirabilis* during swarming. *J Bacteriol.* 1983;154(2):930–937. Available from: https://doi.org/10.1128/JB.154.2.930-937.1983.
42. Rahman MM, Guard-Petter J, Asokan K, Hughes C, Carlson RW. The structure of the colony migration factor from pathogenic *Proteus mirabilis*. A capsular polysaccharide that facilitates swarming. *J Biol Chem.* 1999;274(33):22993–22998. Available from: https://doi.org/10.1074/jbc.274.33.22993.
43. Allison C, Emödy L, Coleman N, Hughes C. The role of swarm cell differentiation and multicellular migration in the uropathogenicity of *Proteus mirabilis*. *J Infect Dis.* 1994;169 (5):1155–1158. Available from: https://doi.org/10.1093/infdis/169.5.1155.
44. Jansen AM, Lockatell CV, Johnson DE, Mobley HL. Visualization of *Proteus mirabilis* morphotypes in the urinary tract: the elongated swarmer cell is rarely observed in ascending urinary tract infection. *Infect Immun.* 2003;71(6):3607–3613. Available from: https://doi.org/10.1128/iai.71.6.3607-3613.2003.
45. Alamuri P, Mobley HL. A novel autotransporter of uropathogenic *Proteus mirabilis* is both a cytotoxin and an agglutinin. *Mol Microbiol.* 2008;68(4):997–1017. Available from: https://doi.org/10.1111/j.1365-2958.2008.06199.x.
46. Swihart KG, Welch RA. Cytotoxic activity of the Proteus hemolysin HpmA. *Infect Immun.* 1990;58(6):1861–1869. Available from: https://doi.org/10.1128/IAI.58.6.1861-1869.1990.
47. Senior BW, Loomes LM, Kerr MA. The production and activity in vivo of *Proteus mirabilis* IgA protease in infections of the urinary tract. *J Med Microbiol.* 1991;35(4):203–207. Available from: https://doi.org/10.1099/00222615-35-4-203.
48. Pearson MM, Sebaihia M, Churcher C, et al. Complete genome sequence of uropathogenic *Proteus mirabilis*, a master of both adherence and motility. *J Bacteriol.* 2008;190 (11):4027–4037. Available from: https://doi.org/10.1128/JB.01981-07.
49. Piccini CD, Barbé FM, Legnani-Fajardo CL. Identification of iron-regulated outer membrane proteins in uropathogenic *Proteus mirabilis* and its relationship with heme uptake. *FEMS Microbiol Lett.* 1998;166(2):243–248. Available from: https://doi.org/10.1111/j.1574-6968.1998.tb13897.x.

50. Nielubowicz GR, Smith SN, Mobley HL. Zinc uptake contributes to motility and provides a competitive advantage to *Proteus mirabilis* during experimental urinary tract infection. *Infect Immun*. 2010;78(6):2823–2833. Available from: https://doi.org/10.1128/IAI.01220-09.

51. Belas R, Manos J, Suvanasuthi R. *Proteus mirabilis* ZapA metalloprotease degrades a broad spectrum of substrates, including antimicrobial peptides. *Infect Immun*. 2004;72 (9):5159–5167. Available from: https://doi.org/10.1128/IAI.72.9.5159-5167.2004.

52. Umpiérrez A, Scavone P, Romanin D, et al. Innate immune responses to *Proteus mirabilis* flagellin in the urinary tract. *Microbes Infect*. 2013;15(10–11):688–696. Available from: https://doi.org/10.1016/j.micinf.2013.06.007.

53. McCoy AJ, Liu H, Falla TJ, Gunn JS. Identification of *Proteus mirabilis* mutants with increased sensitivity to antimicrobial peptides. *Antimicrob Agents Chemother*. 2001;45(7):2030–2037. Available from: https://doi.org/10.1128/AAC.45.7.2030-2037.2001.

54. Kaca W, Radziejewska-Lebrecht J, Bhat UR. Effect of polymyxins on the lipopolysaccharide-defective mutants of *Proteus mirabilis*. *Microbios*. 1990;61(246):23–32.

55. Ma KL, Wang CX. Analysis of the spectrum and antibiotic resistance of uropathogens in vitro: results based on a retrospective study from a tertiary hospital. *Am J Infect Control*. 2013;41(7):601–606. Available from: https://doi.org/10.1016/j.ajic.2012.09.015.

56. Schito GC, Naber KG, Botto H, et al. The ARESC study: an international survey on the antimicrobial resistance of pathogens involved in uncomplicated urinary tract infections. *Int J Antimicrob Agents*. 2009;34(5):407–413. Available from: https://doi.org/10.1016/j.ijantimicag.2009.04.012.

57. Armbruster CE, Mobley HLT, Pearson MM. Pathogenesis of *Proteus mirabilis* infection. *EcoSal Plus*. 2018;8(1). Available from: https://doi.org/10.1128/ecosalplus.ESP-0009-2017.

58. Bichler KH, Eipper E, Naber K, Braun V, Zimmermann R, Lahme S. Urinary infection stones. *Int J Antimicrob Agents*. 2002;19(6):488–498. Available from: https://doi.org/10.1016/s0924-8579(02)00088-2.

59. Nicolle LE. Catheter-related urinary tract infection. *Drugs Aging*. 2005;22(8):627–639. Available from: https://doi.org/10.2165/00002512-200522080-00001.

60. Jones RJ. Oral vaccination against *Proteus mirabilis*. *Br J Exp Pathol*. 1976;57(4):395–399.

61. Moayeri N, Collins CM, O'Hanley P. Efficacy of a *Proteus mirabilis* outer membrane protein vaccine in preventing experimental Proteus pyelonephritis in a BALB/c mouse model. *Infect Immun*. 1991;59(10):3778–3786. Available from: https://doi.org/10.1128/IAI.59.10.3778-3786.1991.

62. Pellegrino R, Galvalisi U, Scavone P, Sosa V, Zunino P. Evaluation of *Proteus mirabilis* structural fimbrial proteins as antigens against urinary tract infections. *FEMS Immunol Med Microbiol*. 2003;36(1–2):103–110. Available from: https://doi.org/10.1016/S0928-8244(03)00103-2.

63. Li X, Lockatell CV, Johnson DE, Lane MC, Warren JW, Mobley HL. Development of an intranasal vaccine to prevent urinary tract infection by *Proteus mirabilis*. *Infect Immun*. 2004;72 (1):66–75. Available from: https://doi.org/10.1128/iai.72.1.66-75.2004.

64. Scavone P, Miyoshi A, Rial A, et al. Intranasal immunisation with recombinant *Lactococcus lactis* displaying either anchored or secreted forms of *Proteus mirabilis* MrpA fimbrial protein confers specific immune response and induces a significant reduction of kidney bacterial colonisation in mice. *Microbes Infect*. 2007;9(7):821–828. Available from: https://doi.org/10.1016/j.micinf.2007.02.023.

65. Scavone P, Umpiérrez A, Maskell DJ, Zunino P. Nasal immunization with attenuated *Salmonella yyphimurium* expressing an MrpA-TetC fusion protein significantly reduces *Proteus mirabilis* colonization in the mouse urinary tract. *J Med Microbiol*. 2011;60(Pt 7):899–904. Available from: https://doi.org/10.1099/jmm.0.030460-0.

66. Mikhalkova D, Khanna S, Vaidya R, Sethi S, Hogan MC. Epstein – Barr virus-associated nephrotic syndrome. *Clin Kidney J.* 2012;5(1):50–52. Available from: https://doi.org/10.1093/ndtplus/sfr119.

67. Subat-Dezulović M, Dessardo NS, Dezulović M, Dessardo S, Hadzisejdić I, Brncic N. Postinfectious glomerulonephritis and Epstein – Barr virus co-infection. *Coll Antropol.* 2010;34(Suppl. 2):229–232.

68. Wenderfer SE. Viral-associated glomerulopathies in children. *Pediatr Nephrol.* 2015;30(11):1929–1938. Available from: https://doi.org/10.1007/s00467-015-3057-y.

69. Karamadoukis L, Toth T, Tomson C. Membranoproliferative glomerulonephritis associated with an Epstein – Barr virus infection. *NDT Plus.* 2008;1(5):319–321. Available from: https://doi.org/10.1093/ndtplus/sfn098.

70. Kano K, Yamada Y, Sato Y, Arisaka O, Ono Y, Ueda Y. Glomerulonephritis in a patient with chronic active Epstein – Barr virus infection. *Pediatr Nephrol.* 2005;20(1):89–92. Available from: https://doi.org/10.1007/s00467-004-1645-3.

71. Epstein MA, Crawford DH. *Gammaherpesviruses: Epstein – Barr Virus. Topley and Wilson's Microbiology and Microbial Infections.* 9th ed. vol. 1. Wiley; 2010:351–366.

72. Odumade OA, Hogquist KA, Balfour Jr. HH. Progress and problems in understanding and managing primary Epstein – Barr virus infections. *Clin Microbiol Rev.* 2011;24(1):193–209. Available from: https://doi.org/10.1128/CMR.00044-10.

73. Dunmire SK, Verghese PS, Balfour Jr. HH. Primary Epstein – Barr virus infection. *J Clin Virol.* 2018;102:84–92. Available from: https://doi.org/10.1016/j.jcv.2018.03.001.

74. Lieberman PM. Chromatin structure of Epstein – Barr virus latent episomes. *Curr Top Microbiol Immunol.* 2015;390(Pt 1):71–102. Available from: https://doi.org/10.1007/978-3-319-22822-8_5.

75. Carbone A, Gloghini A, Dotti G. EBV-associated lymphoproliferative disorders: classification and treatment. *Oncologist.* 2008;13(5):577–585. Available from: https://doi.org/10.1634/theoncologist.2008-0036.

76. Okada H, Ikeda N, Kobayashi T, et al. An atypical pattern of Epstein – Barr virus infection in a case with idiopathic tubulointerstitial nephritis. *Nephron.* 2002;92(2):440–444. Available from: https://doi.org/10.1159/000063322.

77. Rowe M, Kelly GL, Bell AI, Rickinson AB. Burkitt's lymphoma: the Rosetta Stone deciphering Epstein – Barr virus biology. *Semin Cancer Biol.* 2009;19(6):377–388. Available from: https://doi.org/10.1016/j.semcancer.2009.07.004.

78. Callan MF, Steven N, Krausa P, et al. Large clonal expansions of CD8 + T cells in acute infectious mononucleosis. *Nat Med.* 1996;2(8):906–911. Available from: https://doi.org/10.1038/nm0896-906.

79. Rowe M, Zuo J. Immune responses to Epstein – Barr virus: molecular interactions in the virus evasion of CD8 + T cell immunity. *Microbes Infect.* 2010;12(3):173–181. Available from: https://doi.org/10.1016/j.micinf.2009.12.001.

80. Haque T, Wilkie GM, Taylor C, et al. Treatment of Epstein – Barr-virus-positive post-transplantation lymphoproliferative disease with partly HLA-matched allogeneic cytotoxic T cells. *Lancet.* 2002;360(9331):436–442. Available from: https://doi.org/10.1016/S0140-6736(02)09672-1.

81. Rooney CM, Smith CA, Ng CY, et al. Use of gene-modified virus-specific T lymphocytes to control Epstein – Barr-virus-related lymphoproliferation. *Lancet.* 1995;345(8941):9–13. Available from: https://doi.org/10.1016/s0140-6736(95)91150-2.

82. Watanabe M, Panetta GL, Piccirillo F, et al. Acute Epstein – Barr related myocarditis: an unusual but life-threatening disease in an immunocompetent patient. *J Cardiol Cases.*

2019;21(4):137–140. Available from: https://doi.org/10.1016/j.jccase.2019.12.001. Published December 26, 2019.

83. Akkoc G, Kadayifci EK, Karaaslan A, et al. Epstein – Barr virus encephalitis in an immunocompetent child: a case report and management of Epstein – Barr virus encephalitis. *Case Rep Infect Dis*. 2016;2016:7549252. Available from: https://doi.org/10.1155/2016/7549252.

84. Mărginean MO, Molnar E, Chinceşan MI. Epstein – Barr virus-associated hemophagocytic lymphohistiocytosis in a small child: a case report. *Med (Baltim)*. 2020;99(3):e18759. Available from: https://doi.org/10.1097/MD.0000000000018759.

85. Hoover K, Higginbotham K. *Epstein – Barr virus* [Updated November 20, 2020] *StatPearls [Internet]*. Treasure Island, FL: StatPearls Publishing; 2020. Available from: https://www.ncbi.nlm.nih.gov/books/NBK559285/#.

86. Womack J, Jimenez M. Common questions about infectious mononucleosis. *Am Fam Physician*. 2015;91(6):372–376.

87. Fugl A, Andersen CL. Epstein – Barr virus and its association with disease—a review of relevance to general practice. *BMC Fam Pract*. 2019;20(1):62. Available from: https://doi.org/10.1186/s12875-019-0954-3. Published May 14, 2019.

88. Rea TD, Russo JE, Katon W, Ashley RL, Buchwald DS. Prospective study of the natural history of infectious mononucleosis caused by Epstein – Barr virus. *J Am Board Fam Pract*. 2001;14(4):234–242.

CHAPTER 7

Infections of the blood, skin, and musculoskeletal system

Problem 7.1: Meningococcemia caused by *Neisseria meningitidis* infection

CONTENTS

A 49-year-old male presented with a 4-day fever, chest pain radiating to the left arm, and orthopnea that developed 48 hours before admission. He had a generalized rash, which had been there for 2 days, and it had extended to the mucus membranes on the day he arrived at the hospital. The patient also complained of vomiting gastric contents several times. He had visited a private medical center upon the onset of symptoms and had been prescribed analgesics, assured of his well-being, and discharged.

The condition of the patient deteriorated rapidly during the initial examination itself, precipitating a decreased level of consciousness. He was admitted to the high-dependency unit for further assessment and management.

The following are the reports of the investigations conducted:

1. Vitals
   Temperature: 39.4°C
   Blood pressure: 105/60 mmHg
   Heart rate: 100/min
   Oxygen saturation: 92% on 6 L FM
   Respiratory rate: 22/min
2. Physical examination
   Conscious, oriented to time, place, and person on initial presentation to the emergency department
   Appeared generally sick
   Wide-spread, nonblanching, petechial purpuric rash, mainly concentrated on the upper and lower limbs
   Petechial lesions on the mouth mucus membranes with surrounding erythema
   Bilateral basal inspiratory coarse crepitations

Problem-Based Learning Approach in Microbiology. DOI: https://doi.org/10.1016/B978-0-323-95092-3.00002-0

3. Neurological examination
   Equally reactive pupils
   Ability to move all limbs according to command
   Normal communication abilities
   All limb reflexes were 2 +
   Down-going plantars
   No neck stiffness
   Kernig's and Brudzinski signs: Negative
4. Complete blood count (CBC): 12% bandemia
5. Arterial blood gas
   Pending-type respiratory failure and respiratory alkalosis

pH	7.48	7.35–7.45
PaO_2 (mmHg)	63.9	75–100
$PaCO_2$ (mmHg)	29.5	38–42
HCO_3 (mmol/L)	21.4	22–26

6. Blood culture: Primary cultures showed the presence of Gram-negative diplococci, highly suggestive of *N. meningitidis* infection
7. Radiological investigations

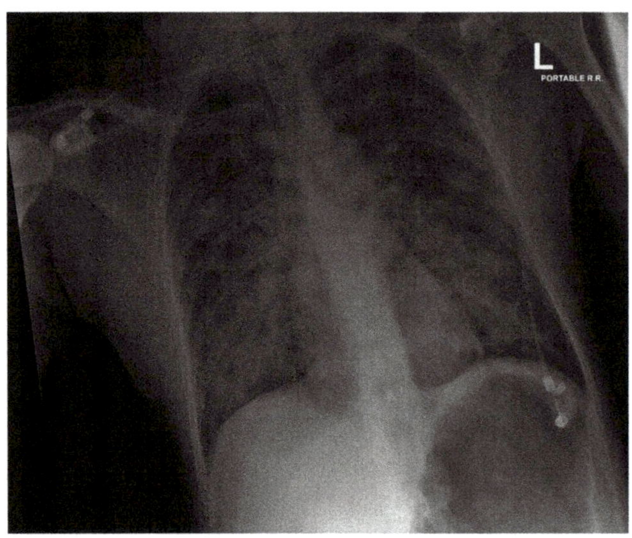

FIGURE 7.1
Chest X-ray showing bilateral reticulonodular infiltrates.

Chest X-ray: Bilateral reticulonodular infiltrates (Fig. 7.1)
Echocardiogram: Ejection fraction—45%; no regional wall motion abnormality; no evidence of vegetations; no pericardial effusion
Abdominal ultrasound: Perinephric fluid collection; renal parenchymal changes; no other abnormality
Computerized tomography (CT) of abdomen and brain: No remarkable changes

8. Lumbar puncture and cerebrospinal fluid (CSF) analysis
   Gram stain: Negative

Appearance	Blood-stained	Clear
Red blood cell (RBC)	6,70,000	<1/μL
White blood cells (WBC)	520	<5/μL
Neutrophil	77	0/μL
Lymphocyte	16	<5/μL

9. Coagulation profile (after bleeding into the airway during intubation)
   Prolonged value, suggesting disseminated intravascular coagulation
   Drop in platelet count to 33

Diagnosis

The final diagnosis concluded meningococcemia based on the clinical manifestations and the presence of Gram-negative diplococci, highly suggestive of *N. meningitidis* infection.

Treatment

1. Ceftriaxone—2 g twice per day
2. Vancomycin—1 g twice per day
3. The patient was electively intubated after around 8 hours of presentation as he became drowsier and was gasping for breath. The patient started to bleed inside the airway during the intubation procedure, from the oral cavity and the nasogastric tube.
4. Fresh frozen plasma and packed RBC transfused

The patient continued to show a drop in blood pressure despite boluses of fluids and inotropic support. Ten hours after his initial presentation, the patient became asystole. Cardiopulmonary resuscitation was initiated according to Basic Life Support protocol; however, the process was unsuccessful, and the patient was declared dead.

Question and answer

1. What is the role of asymptomatic carriers in the spread of *N. meningitidis*?

   N. meningitidis infection is transmitted through close contact with infected nasopharyngeal droplets and contaminated fomites. Asymptomatic nasopharyngeal carriers do not manifest the clinical signs of the disease caused by *N. meningitidis* and are very instrumental in spreading the infection to unsuspecting contacts.
2. Which are the factors that contribute to the virulence of *N. meningitidis*?

   Capsular polysaccharides, surface adhesive proteins, iron sequestration mechanisms, and endotoxins play an important role in facilitating virulence of *N. meningitidis*.
3. Why is the uncontrolled multiplication of the *N. meningitidis* bacteria facilitated in the CSF?

   The local deficiency of complement factors and immunoglobulins in the CSF facilitates uncontrollable multiplication of bacteria, although the local inflammatory response induces an influx of polymorphonuclear leukocytes.
4. Why is the clinical management of meningococcaemia particularly challenging?

   The clinical stage of meningococcemia is characterized by rapid evolvement from minor symptoms to hypotension, septic shock, and disseminated intravascular coagulation. The time taken for diagnosis is a crucial determinant of the prognosis and outcome of the disease. The fulminant nature of the disease's progression makes the clinical management of meningococcemia particularly challenging.
5. What are the effects of the inflammatory cascade triggered in meningococcemia infections?

   The humoral and cell-mediated immune responses trigger major inflammatory cascade systems, which cause a systemic upregulation of cytokines and nitric oxide levels. The increased nitric oxide levels result in widespread vasodilation, leading to a sudden spike in blood pressure levels. The upregulation of cytokines, on the other hand, results in a "cytokine storm," which causes systemic cellular damage.

The precipitating hemorrhage and profound shock eventually result in death.

6. What causes false negative results in blood cultures of *N. meningitidis*?
   Blood culture is the most used method for the diagnosis of *N. meningitidis* infection. However, broad-spectrum antibiotics used in the empiric treatment of the disease often impede bacterial growth on blood culture, thereby resulting in false negative results.
7. Why does the capsular polysaccharide of the B serotype of *N. meningitidis* have low immunogenicity?
   The capsular polysaccharide of the B serotype of *N. meningitidis* shares several similarities with human neural adhesion molecules. This causes this capsular polysaccharide to have low immunogenicity, and therefore it is not a good candidate for a vaccine antigen.

Learning objectives

1. Determine the logical flow of differential diagnosis based on patient history, clinical manifestations, physical examination, serological investigations, and radiological investigations.
2. Describe the characteristics of *N. meningitidis*, the etiological agent of meningococcal disease.
3. Predict the prognosis of the disease based on the clinical manifestations and the extent of the pathology presented.
4. Interpret the correlation between clinical manifestations and the pathogenicity of *N. meningitidis*.
5. Enumerate the methods involved in the diagnosis of *N. meningitidis* infections.
6. Determine the appropriate antibiotic-based treatment regimen for the patient based on a history of allergies, underlying medical conditions, and culture sensitivity reports.

Discussion

Differential diagnosis

Etiology	Rationale
Rocky mountain spotted fever	Rocky mountain spotted fever is a bacterial disease transmitted through ticks. The disease usually presents itself with a fever, headache, rash, nausea, vomiting, abdominal pain, myalgia, and loss of appetite.

Continued

Continued

Etiology	Rationale
Hemophilus influenza type B	Hemophilus influenza B is a bacterial disease that develops with a fever, chills, dyspnea, sweating, chest pain, headache, myalgia, and fatigue.
Ehrlichiosis	Ehrlichiosis is a tick-borne bacterial infection characterized by a rash, severe headache, fever, chills, muscle aches, nausea, vomiting, abdominal cramps, diarrhea, loss of appetite, and an altered mental status.
Streptococcal pneumoniae	Streptococcal pneumoniae manifests with a fever, chills, cough, dyspnea, chest pain, stiff neck, confusion, increased sensitivity to light, joint pain, ear pain, sleeplessness, and irritability.
Staphylococcus aureus infection	*Staphylococcus aureus* infections of the bloodstream are characterized by a high-grade fever, chills, altered mental status, petechial rash, myalgia, nausea, and vomiting.
Gonococcemia	Gonococcemia presents itself with a fever, chills, malaise, joint pains, swollen joints, pain associated with the tendons of the wrists or heels, and petechial or blistering rash.
Henoch – Schonlein purpura	Henoch – Schonlein purpura is characterized by a hemorrhagic rash, arthritis, internal hemorrhage, abdominal pain, bleeding associated with the digestive tract, subdermal swelling, swollen kidneys, testicular inflammation, and neurological pathologies.
Typhus	Typhus presents itself with a fever, chills, myalgia, nausea, vomiting, loss of appetite, abdominal pains, cough, and rash.
Collagen vascular disease	Collagen vascular disease causes a fever, fatigue, body aches, joint pain, muscle weakness, and skin rash.
Idiopathic thrombocytopenic purpura	Idiopathic thrombocytopenic purpura is characterized by purple subdermal bruising caused by hemorrhage, petechiae, nosebleeds, heavy menstruation, blood-stained vomit and/or stool, hematuria, bleeding in the buccal cavity, bleeding gums, and intracranial hemorrhage.

Background

Meningococcal disease, also known as meningococcal meningitis, is a bacterial infection commonly seen in the developing countries of Asia and Africa.[1] The disease can present as a spectrum of transient fever and bacteremia, fulminating to severe disease and even death within hours of the first set of symptoms. The presentation of the disease can be in one of three forms: (1) meningitis, (2) meningitis with meningococcemia, or (3) meningococcemia without clinical evidence of meningitis.

Meningococcal disease is known to have high mortality, particularly among children and young adolescents. The close contacts of infected patients are 300–1000 times more prone to getting the infection compared to the rest of the population.[2] Patients with immunodeficiencies caused by underlying medical conditions, such as human immunodeficiency virus (HIV), diabetes, asplenia, complement system defects, and elderly patients, are more prone to rapid deterioration during the clinical course of the disease.

Etiological agent

Neisseria meningitidis

Classification and morphology

N. meningitidis belongs to the genus *Neisseria*, which consists of Gram-negative (Fig. 7.2), nonsporulating, nonmotile, catalase and oxidase-positive cocci. These cocci are oval or spherical, and typically occur in pairs, precipitating the diplococci reference (Fig. 7.3). These bacteria are strictly aerobic organisms growing optimally between a pH range of 7.4 and 7.6 and at temperatures between 35°C and 36°C, while temperatures below 30°C prevent growth. The genus comprises of two important

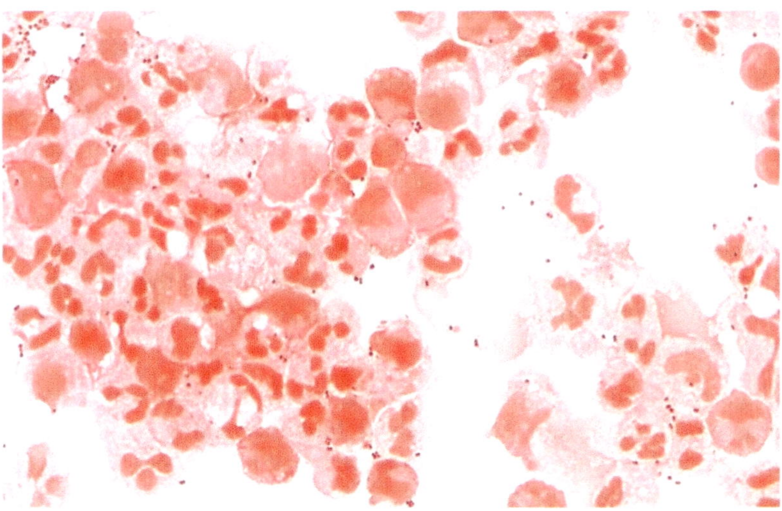

FIGURE 7.2

Gram-stained image of the Gram-negative diplococci form of *Neisseria meningitidis* in spinal fluid. Source*: Microman 12345/Own work/; https://creativecommons.org/licenses/by-sa/4.0/deed.en; https://commons.wikimedia.org/wiki/File:Neisseria_meningitidis_CSF_Gram_1000.jpg.*

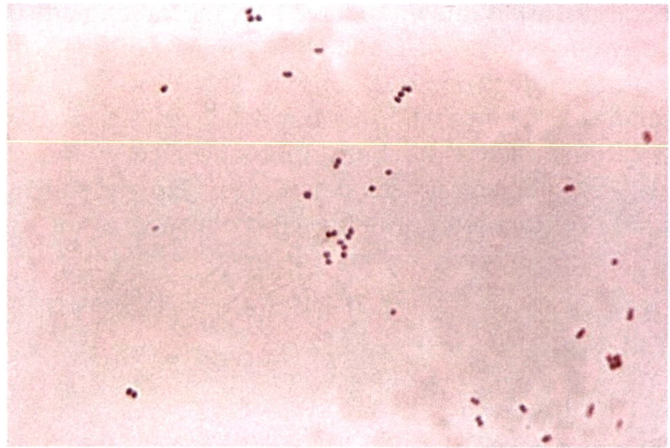

FIGURE 7.3

Micrograph of *N. meningitidis*. Source: שועל*/Own Work/; https://creativecommons.org/licenses/by-sa/4.0/deed.en; https://commons.wikimedia.org/wiki/File:Neisseria_meningitidis_(Sharpened).jpg.*

pathogens, *N. meningitidis* and *N. gonorrhoeae*, along with many other bacterial species that are commensals living in the mouth and the upper respiratory tract of humans.

While meningococci are capsulated, the other bacterial species belonging to the *Neisseria* genus are noncapsulated. The polysaccharide antigens on the capsules of meningococci have resulted in the classification of the species into 13 serogroups. Serotypes and subtypes are formed based on outer membrane proteins (OMPs) and polysaccharides among the serogroups. Epidemics are usually attributed to group A, and group C is mostly associated with localized sporadic outbreaks, while group B is known to cause both epidemics and sporadic outbreaks. While any of the serogroups may colonize the human nasopharynx, groups A, B, C, W-135, X, and Y usually cause life-threatening diseases.[3]

Meningococci cultures require very specific media enriched with components, such as blood and serum. High levels of humidity and 5–10% CO_2 content facilitate growth in culture. Old cultures are prone to autolysis, which results in considerable variation in size, shape, and staining properties. On the other hand, fresh isolates are generally found to be capsulated. Lesion smears typically show the cocci as being intracellular. The oxidase positivity attribute, resulting in *N. meningitidis* colonies turning deep purple on addition of an oxidase reagent, is the biochemical identification for *N. meningitidis* in mixed cultures.

Epidemiology

N. meningitidis is an obligate human parasite, and the human nasopharynx is the only known reservoir for the bacterium. Asymptomatic nasopharyngeal carriers, who rarely experience the clinical manifestations, are instrumental in spreading the infection to unsuspecting contacts. The disease is transmitted through close contact with infected nasopharyngeal droplets and also by fomites in rare cases. The occurrence of the disease may be sporadic in the form of localized outbreaks or may occur as epidemics. An epidemic is marked by an increase in the percentage of carriers from 5% in interepidemic periods to 90% during epidemics. Crowded living conditions have been known to cause epidemics of the meningococcal disease.

Resistance

Meningococci have very low resistance to changes in environmental factors, such as temperature, pH, moisture content, and the presence of chemicals. These bacteria are, therefore, highly susceptible to heat, pH changes, desiccation, and chemicals in disinfectants. Although the organism is usually sensitive to first-, second-, and third-line antibiotics, emergence of resistant strains has been reported in recent times, owing to the widespread use of antibiotics. Strains resistant to both penicillin and ciprofloxacin have been reported.[4] This has substantially increased the challenge involved in clinical treatment of the life-threatening diseases, such as cerebrospinal meningitis and meningococcal septicemia or meningococcemia caused by this organism.

Virulence

Multiple factors, such as capsular polysaccharides, surface adhesive proteins, iron sequestration mechanisms, and endotoxins contribute to *N. meningitidis* virulence. The organism is also known to evolve genetic mechanisms responsible for antigenic variations, allowing the organism to successfully adapt at mucosal surfaces and invade the host.[5] *N. meningitidis* is known for its immune evasive abilities through capsule switching and molecular mimicry, which is partly how bacteria adapt through genetic transformation.[6]

Pathogenicity

Meningococcal disease manifests either as cerebrospinal meningitis or as meningococcal septicemia, also referred to as meningococcemia. Asymptomatic carriage of *N. meningitidis* is very common. Local inflammation presenting as rhinitis and pharyngitis following infection is seen in some cases.

Extremities of age, widespread skin lesions, the acute onset of the infection, the absence of signs of meningitis, elevated inflammatory markers, and the presence of shock are the poor prognostic indicators in infected patients.[7] Disseminated intravascular coagulopathy is a common complication of meningococcal infection.

Although the blood – brain barrier (BBB) usually protects the central nervous system (CNS) effectively against invading pathogens, some extracellular bacteria can cross the BBB and invade the meninges. *N. meningitidis* is especially effective at invading the meninges once it invades the bloodstream. The bacteria may spread from the nasopharynx to the meninges directly along the perineural sheath of the olfactory nerve, through the cribriform plate to the subarachnoid space, or through the bloodstream. The local deficiency of complement factors and immunoglobulins results in uncontrollable multiplication of bacteria in the CSF, despite the influx of polymorphonuclear leukocytes induced by the local inflammatory response.[8] Analysis of infected spinal fluid, therefore, invariably reveals the presence of the bacteria, both free and within the leukocytes.

The bacteria are also capable of multiplying rapidly in the bloodstream and are responsible for a severe form of septic shock, resulting from extensive thrombosis, vascular leakage, and cardiovascular failure, a constellation of clinical symptoms known as the purpura fulminans disorder. Inflammatory response is observed in the CSF usually after the occurrence of purpura fulminans at an early stage of the bloodstream infection.[8] Therefore it is difficult to prevent the onset of purpura fulminans, which once manifested cascades into acute pathology rapidly, thus resulting in high mortality rates associated with bloodstream infection.

Immune response and clinical manifestations

The initial manifestations of the disease are nonspecific and can be confused with other transient viral illnesses. The disease presents itself with a fever, loss of concentration, nausea, vomiting, and myalgia. The first classic sign of the disease is a rash that may progress from a nonspecific rash to petechiae (Fig. 7.4), which hemorrhage within few hours.[9]

The condition is characterized by rapid evolvement from minor symptoms to hypotension, septic shock, and disseminated intravascular coagulation. This makes clinical management of the disease especially challenging, with time taken for diagnosis being the crucial determinant for the prognosis and outcome of the disease.

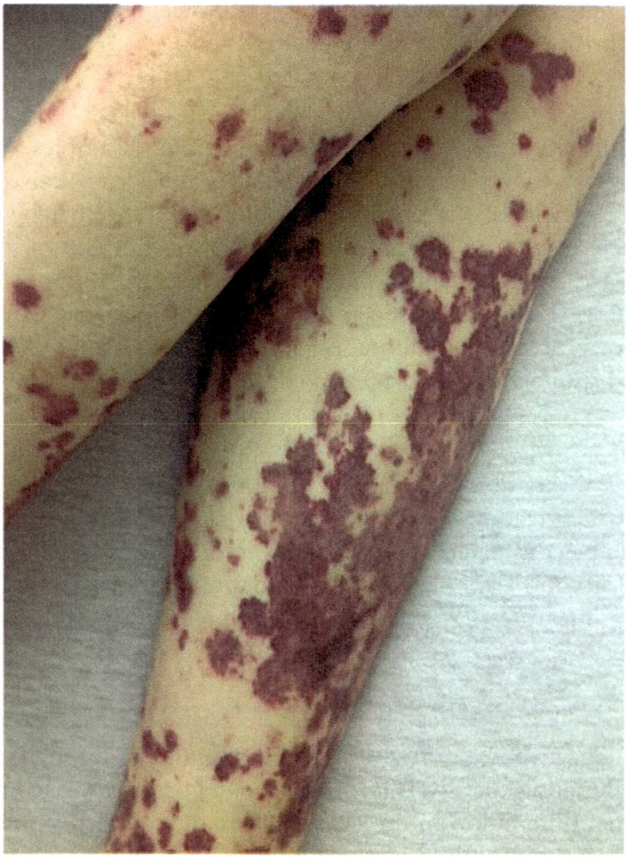

FIGURE 7.4
Petechial rash. Source: *DrFO. Jr. Tn/Own work/; https://creativecommons.org/licenses/by-sa/3.0/deed.en; https://commons.wikimedia.org/wiki/File:Petechial_rash.JPG.*

Meningococcemia can develop into a fatal condition, identified by shock, disseminated intravascular coagulation, and multisystem failure. Chronic meningococcemia may occur in rare cases.

The humoral and cell-mediated immune responses trigger all the major inflammatory cascade systems, resulting in a systemic upregulation of cytokines and nitric oxide levels. The increased nitric oxide levels result in widespread vasodilation, leading to a sudden spike in blood pressure levels. The upregulation of cytokines, on the other hand, causes a "cytokine storm," which leads to systemic cellular damage. The precipitating hemorrhage and profound shock eventually result in death.

Diagnosis

The organism can be detected by Gram staining blood samples, cerebrospinal samples, or skin biopsy samples. Widespread antibiotic use can, on occasion, impede the growth of the organism on blood culture, thereby resulting in false negative results, which can interfere with effective clinical management of the disease. Some diagnostic methods, such as skin biopsy, polymerase chain reaction (PCR), and antigen detection are not affected by using antibiotics. However, blood culture is the more frequently used diagnostic method in hospitals as it is easier and cheaper.[10]

Treatment

Penicillin is the antibiotic of choice for treating meningococcal infections. However, widespread resistance to penicillin requires third-generation cephalosporins to be considered as the drug of choice, especially ceftriaxone, which is known for good CNS penetration. Chloramphenicol is another antibiotic of choice in cases known to be hypersensitive to penicillin or cephalosporin.[11]

Prevention

The most effective way to prevent the disease is through vaccination. However, the antigenic diversity of the meningococcal surface proteins has been the main limitation in the design of broadly protective meningococcal vaccines.[12] The most common serogroups responsible for clinical infection are A, B, C, W135, X, and Y. The quadrivalent polysaccharide vaccine available can prevent infections caused by serogroups A, C, W135, and Y,[13] as the capsular polysaccharide represents the major antigen, which is the basis for the conjugate vaccine. Serogroup B capsular polysaccharide, however, has low immunogenicity because of its similarity to human neural adhesion molecules and is, therefore, not a good candidate for a vaccine antigen.[14] The MenB vaccine containing the outer membrane vesicles (OMVs) has been shown to be efficient against the OMV-derived strains and is used as the vaccine against the serogroup B strains.[15]

High-risk groups, such as healthcare workers and primary contacts of infected patients, should be recommended chemoprophylaxis using antibiotics, including ciprofloxacin or rifampicin. Ceftriaxone should be used as the prophylactic antibiotic for pregnant women. However, it is important to understand that prophylaxis only prevents nasopharyngeal carriage of the infection but does not prevent the disease if the colonization has already caused an infection.[16]

Student Notes

Key take-aways

- Meningococcal disease is a serious condition that progresses rapidly and is notorious for a high mortality rate.
- *N. meningitidis* is one of the bacterial species responsible for meningococcal diseases in the form of either meningitis, meningococcemia, or both.
- *N. meningitidis* is known for its ability to breach the BBB once it has invaded the blood stream.
- Inclusion of the disease in the differential diagnosis of cases that demonstrate implicating clinical manifestations from the beginning of the clinical presentation can be instrumental in improving the prognosis and outcome of the disease.
- The fast-paced nature of the disease's progression, especially in cases of meningococcemia, poses a challenge in ensuring effective clinical treatment.
- Rapid and aggressive treatment regimen comprising of fluid and appropriate antibiotics is the cornerstone of effectively treating the infection.
- Antibiotic resistant strains of *N. meningitidis* have increased the challenge in effectively treating meningococcal diseases.
- The disease can be potentially life-threatening in high-risk individuals, such as the immunocompromised or those at extremes of the age spectrum.
- Prophylactic antibiotic regimes are strongly recommended to ensure prevention of the disease, especially in the high-risk groups, such as healthcare personnel and the primary contacts of infected individuals.
- Vaccination is the most effective means of avoiding the risks associated with meningococcal diseases; however, the development of broadly protective meningococcal vaccines is riddled with challenges.

Problem number 7.2: Malaria

A 34-year-old male had a fever and fatigue for 7 days and vomiting for 3 days. He denied having a history of any known previous medical illness and claimed to be in perfect health until 7 days ago when the fever developed. The patient mentioned that the onset of the fever was sudden and is not constant (waxing and waning). The fever was especially apparent at night. His febrile condition was not relieved by paracetamol and there was no known aggravating factor or trigger. His fever was accompanied by sweating, chills, nausea, vomiting, and abdominal pain. There were, however, no other associated symptoms, such as diarrhea, myalgia, cough, and sore throat. The onset of the vomiting was abrupt, around 3 days prior to the presentation, and was accompanied with a loss of appetite, although no weight loss was documented. The vomitus from the first episode of vomiting contained food material with no blood, but on the day of presentation, the vomitus was bloody and scanty in volume. The vomiting occurred twice per day but was not related to food or any other factors. The patient also complained about an epigastric pain (score of 4 out of 10) that was dull, intermittent in character, and did not radiate.

Patient history revealed no major medical illness or genetic diseases since childhood, and no history of hypertension, diabetes, dyslipidemia, cancer, liver, or heart disease. In addition, there was no history of any injuries, trauma, surgeries, blood transfusions, or hospitalization since childhood. His vaccination profile is complete, and he had not traveled to foreign destinations in the last 5 months. The patient did not report any known food or drug allergies and is not on any medications or vitamin supplements.

The patient's parents and siblings are all healthy, and there is no known history of family illnesses. The patient denied smoking or abusing drugs or alcohol. His occupation is that of a cleaner and his lifestyle lacked the consumption of healthy and nutritious food.

The following are the reports of the investigations conducted:

1. Vitals
   Temperature: 39.6°C
   Pulse: 118 beats/minute, regular and with normal volume, normal in character, symmetrical and vessel wall not palpable, no signs of collapsing pulse.
   No radio-radial/radio-femoral delay
   Blood pressure: 109/69 mmHg while lying
   Respiratory rate: 20 breaths/minute
2. Physical examination
   Thin, middle-aged male, age: 34 years
   Jaundiced appearance, but no pallor
   Spider naevi on the face
   Scleral icterus of the eye noted
   Pupils: 3 mm equally round and reactive to light
   Comfortable in supine position
   Conscious, alert, and oriented
   Scleral icterus present
   No muscle wasting/deformities/missing limbs
   No signs of dehydration or cyanosis
   No swollen lymph nodes
   Some loss of appetite, but no weight loss
   Fatigue
   No sleeping problem
   No rash or itchy skin

No lumps/swelling
No recent trauma

3. Cardiovascular examination
   Tachycardia
   No chest pains
   No palpitations
   No dizzy spells
   No swollen ankles
   No dyspnea
   No orthopnea
   Normal exercising capacity
4. Respiratory examination
   No dyspnea
   No wheezing
   No hemoptysis
   No cough
5. Gastrointestinal examination
   Abdominal pain
   Nausea/vomiting (twice/day)
   Hematemesis
   Some loss of appetite
   No heartburn
   No constipation
   No melena
   No diarrhea
   No dysphagia
6. Genitourinary examination
   No difficulty/pain during urination
   No history of hematuria
   No urgency or incontinence
   No history of sexual transmitted diseases
   No prostate disease
7. Neurological examination
   Memory and speech are good
   No syncope
   No dizziness
   No headache
   No seizures
   No altered sensation or weakness
   No difficulty in walking
   No visual disturbances
   No hearing loss
   No memory/concentration changes

8. Locomotor examination
   No joint pain/back pain
   No stiffness/joint swelling
   No diurnal variation
   No functional deficit
   Normal spine structure and function
   Normal movement of limbs
9. Endocrinological examination
   No polyuria
   No polydipsia
   No excessive sweating
   No heat or cold intolerance
   No irritability
   No dry skin or hair changes
10. Psychiatric examination
   Feels healthy
   No history of psychiatric disorders
   No mental health medications
11. Hematological profile

WBC (/L)	5.67×10^9	$4.3–10.8 \times 10^9$
RBC (/L)	5.66×10^{12}	$4.2–5.9 \times 10^{12}$
Hb (g/dL)	13.9	13.5–17.5
Hematocrit (%)	39.20	33–40
Mean corpuscular volume (MCV) (fL)	85.3	80–100
Mean corpuscular Hb (MCH) (pg)	34.2	27–32
Mean corpuscular Hb concentration (MCHC) (g/dL)	37.2	33.4–35.5
RDW (%)	13.2	11.8–14.5
Platelets (/L)	102×10^9	$150–400 \times 10^9$
Mean platelet volume (MPV) (fL)	8.9	7.5–12.0
Plateletcrit (%)	0.1	0.20–0.36
Platelet size distribution (%)	51.8	8.3–56.6
Neutrophils (%)	79.3	54–62
Lymphocytes (%)	15.80	25–33
Monocyte (%)	5.2	3–7
Eosinophils (%)	1.2	1–3
Basophils (%)	0.4	0.5–1
Neutrophils absolute count (/L)	5.0×10^9	$1.7–7.0 \times 10^9$
Lymphocyte absolute count (/L)	0.63×10^9	$0.90–2.90 \times 10^9$
Monocytes absolute count (/L)	0.25×10^9	$0.30–0.90 \times 10^9$
Eosinophils absolute count (/L)	0.07×10^9	$0.05–0.50 \times 10^9$
Basophils absolute count (/L)	0.0×10^9	$0.00–0.30 \times 10^9$
Nucleated RBC (%)	0	0

12. Malarial parasites in blood smear: Seen (*Plasmodium vivax* schizonts)
13. Coagulation profile

Prothrombin time (seconds)	14.60	11–13.5
INR	1.21	≤1.1
Activated partial thromboplastin time (aPTT) (seconds)	24.60	30–40
aPTT ratio	0.94	1.8–3.3 seconds
Fibrinogen (mg/dL)	436.00	200–400
D-dimer	1.33 mg/L	<0.50

14. Biochemical profile

Creatinine (μmol/L)	84	61.9–114.9
Amylase (U/L)	68.8	30–110
Troponin-1 (ng/mL)	0.0	<0.04
Glucose (mmol/L)	5.3	4.4–7.8
Urea (mmol/L)	5.6	3–6.3
Calcium (mmol/L)	2.14	2.2–2.7
Phosphorus (mmol/L)	0.3	1.12–1.45
Magnesium (mmol/L)	0.60	1.7–2.2
Potassium (mmol/L)	3.9	3.5–5.3
Sodium (mmol/L)	138	135–145
Chlorine (mmol/L)	106	95–110
Bicarbonate (mmol/L)	22	–32

15. Liver function test

Protein (total) (g/L)	73	68–80
Albumin (g/L)	38	35–55
Globulin (g/L)	35	15–30
Total bilirubin (μmol/L)	38	1.71–20.5
Bilirubin (Direct) (μmol/L)	14	1.7–5.1
Bilirubin (Indirect) (μmol/L)	24	3.4–12.0
Alkaline phosphatase (U/L)	90	44–147

16. Cardiac enzymes

CK–MB (IU/L)	22	5–25
Creatine kinase (total) (U/L)	64	55–170
Lactic dehydrogenase (U/L)	245	140–280

17. Urine analysis

pH	6
Protein	Negative
Glucose	Normal
Ketones	Negative
Blood hemoglobin	Negative
Leukocyte esterase	Negative
Nitrite	Negative
Bilirubin	Negative
Bilirubinogen	Normal
RBC	0–2 cells\hpf
WBC	0–2 cells\hpf
Crystals	Not seen
Casts	Not seen
Epithelial cells	Not seen
Yeast cells	Not seen
Parasite	Not seen

Diagnosis

The final diagnosis concluded malaria caused by *P. vivax* infection.

Treatment

1. IV hydration
2. Antimalarial medication:
   a. Chloroquine
   b. Primaquine
3. Omeprazole

Question and answer

1. What distinguishes malarial infections caused by *P. vivax* from infections caused by other *Plasmodium* species?

   Malarial infections caused by *P. vivax* can be distinguished from infections caused by other *Plasmodium* species by the presence of certain unique morphological features, such as small dark granules in the reticulocyte cytoplasm, known as Schüffner's dots, which represent caveola–vesicle complexes of parasite origin exported into the cytoplasm of infected reticulocytes.

2. Microscopy is an important aspect of a malaria diagnosis. What are considered diagnostic markers during microscopic examination of blood smears?

   The similarity in the round shape of *P. vivax* parasites in the sexual and asexual stages, and the presence of round shaped *P. vivax* gametocytes in blood smears, are good diagnostic markers for malaria.
3. Why is *P. falciparum* malaria more prevalent in the African continent, although *P. vivax* malaria is more prevalent globally?

   The African population has a low expression of the Duffy antigen on the surface of red blood cells. The presence of the Duffy antigen on the surface of red blood cells is important for the establishment of *P. vivax* infection. This causes the rate of *P. vivax* infections to be lower in the African continent compared to *P. falciparum* infection. The pathogenicity of *P. falciparum* infections is more compared to *P. vivax* infections, thereby resulting in a higher mortality rate associated with malaria in the African region.
4. Why is malaria known as "tertian fever"?

   The peculiar pattern of intermittent symptomatic and asymptomatic periods gives *P. vivax* malaria the name, "tertian fever." *P. vivax* sporozoite cells localized in the salivary gland of the female *Anopheles* mosquito are injected into humans along with the anticoagulant-containing saliva of the mosquito. The sporozoites travel rapidly from the bloodstream to the liver and invade the hepatocytes within 30 minutes of entering the bloodstream. The sporozoites may become dormant, forming hypnozoites or may replicate forming schizonts. Numerous merozoites are formed from the schizonts and are released from the liver into the blood stream where they infect erythrocytes. This process occurs as a cycle at intervals of 48 hours. The release of the *P. vivax* merozoites from the erythrocytes triggers the clinical manifestations of malaria. The 48-hour cycle of the formation of merozoites from schizonts and their release from the erythrocytes causes the clinical manifestations of malaria to demonstrate an alteration of asymptomatic periods and symptomatic periods.
5. How are sporozoites formed in the female *Anopheles* mosquito?

   The merozoites that do not infect erythrocytes are known as gametocytes. These gametocytes are ingested when an *Anopheles* mosquito bites an infected person. These gametocytes mature and form gametes within the mosquito. The fusion of two gametes results in the formation of a zygote, which then migrates to the mosquito's intestine and enlarges there, forming numerous sporozoites.

6. What are the characteristics of immune response to *P. vivax* malaria infection?
   Immune response to *P. vivax* malaria infection is characterized by toxin-induced interferon (IFN)-γ and tumor necrosis factor-α (TNF$-\alpha$) secretion. While the innate immune response is dominated by monocyte and macrophage-associated phagocytosis within the spleen, adaptive immunity is identified by IFN-γ and TNF-α-induced increase in CD4 + lymphocytes.
7. What causes a relapse of a malaria infection and how can this be prevented?
   The relapse of a malaria infection and the appearance of clinical symptoms of the disease are attributed to the release of dormant hypnozoites into the bloodstream. The addition of primaquine to treatment regimens for malaria eliminates the dormant hypnozoites, thereby preventing the relapse of the disease.
8. Which are the most important diagnostic modalities for malaria?
   Light microscopy, rapid diagnostic tests (RDTs), and PCR are the three most important diagnostic modalities for malaria.
9. Why is the use of primaquine contraindicated in individuals with G6PD deficiency and in pregnant women?
   Primaquine is known to cause severe hemolytic anemia in individuals with G6PD deficiency and fetal hemolysis in pregnant women. Therefore primaquine is contraindicated in individuals with G6PD deficiency and in pregnant women.

Learning objectives

1. Determine the logical flow of differential diagnosis based on clinical manifestations, physical examination, and serological investigations.
2. Describe the characteristics of *P. vivax*, the *Plasmodium* species associated with most of the malaria cases.
3. Describe the life cycle of *P. vivax*.
4. Summarize the pathogenesis of *P. vivax* malarial infection.
5. Predict the prognosis of the disease based on the clinical manifestations and the extent of pathology presented.
6. Enumerate the methods involved in the diagnosis of malaria.
7. Interpret the correlation between the clinical manifestations shown by the patient and the pathogenicity of *P. vivax*.
8. Determine the appropriate treatment regimen for the patient based on the case presentation and the drug-sensitivity reports.

Discussion

Differential diagnosis

Etiology	Rationale
Dengue fever	Dengue is a viral vector-borne disease spread by mosquito bites. This disease is characterized by muscle and joint pain, fever, headaches, generalized rash, nausea, and vomiting.
Chikungunya	Chikungunya is a viral vector-borne disease spread by mosquito bites. This disease is identified by extreme debilitating joint pain and swelling, fever, muscle pain, headaches, and a generalized rash.
Typhoid	Typhoid is a bacterial infection of the digestive system caused by an intermittent fever, indigestion, heartburn, abdominal pain, loss of appetite, diarrhea, nausea, and vomiting.
Sepsis from bacteremia	Bacteremia is the presence of bacteria in circulation, resulting in sepsis, which is manifested as systemic responses, including fever, weakness, rapid heart rate, palpitations, rapid breathing, and increased white blood cell count.
Viral hemorrhagic fever	Viral hemorrhagic fevers are severe diseases characterized by multiorgan involvement, acute damage to blood vessels, and dysregulated homeostasis.
Leptospirosis	Leptospirosis is an infection of the blood caused by the *Leptospira* bacteria. Clinical manifestations of this disease are broad-spectrum, ranging from mild symptoms, such as headaches, muscle pains, and fevers to more severe symptoms like meningitis and hemorrhage in the lungs.
Rickettsial diseases	Rickettsial diseases are vector-borne characterized by fever, maculopapular rash, or an eschar.
Viral hepatitis	Viral hepatitis is manifested as liver inflammation and damage due to viral infections. The clinical manifestations include jaundice, loss of appetite, vomiting, nausea, itchy skin, abdominal pain, heartburn, and fever.
Brucellosis	Brucellosis is a zoonotic disease caused by the ingestion of contaminated milk or meat from infected animals. The clinical manifestations include intermittent fever with chills, headaches, abdominal pain, and general malaise.

Background

Malaria, derived from the words "bad air" in Italian, is a common and widespread parasitic infection in humans. The disease poses a significant global health problem and is responsible for a substantial disease burden worldwide. Malaria is a vector-borne disease caused by a protozoal infection, spread via the female mosquitoes belonging to the *Anopheles* genus. Although several effective therapeutic options are available, malaria continues to be one of the most important infectious human diseases. The African continent is the worst affected by the morbidity and mortality caused by malaria.

The elusive pattern of infection, characteristic of the malarial parasite, has resulted in significant challenges in the development of effective diagnostic, therapeutic, and especially prophylactic options. Understanding the disease, having good knowledge of the typical and atypical symptoms, getting a rapid and accurate diagnosis, and starting effective and timely treatment are the mainstays of ensuring a good prognosis and outcome in malaria cases. As diagnosis of malaria may sometimes be misleading, accurate interpretation of the clinical manifestations by the clinician becomes vital for ensuring timely and effective treatment.

Etiological agent

Plasmodium vivax

Classification and morphology

The malarial pathogen is a single-celled parasite belonging to the *Plasmodium* genus. There are five species of parasites in the *Plasmodium* genus, known to be of pathogenic significance to humans, namely: *P. falciparum*, *P. vivax*, *P. ovale*, *P. malariae*, and *P. knowlesi*. Of these, *P. falciparum* and *P. vivax* account for the majority of the malaria cases worldwide. While *P. falciparum* is more fatal, *P. vivax* is the most common parasite among the *Plasmodium* species affecting humans, and can cause severe infections that can result in fatalities if untreated, thereby being responsible for high rates of morbidity and mortality.

P. vivax infections can be distinguished from the other *Plasmodium* species by the presence of the unique morphological features, such as the small dark granules in the reticulocyte cytoplasm, known as Schüffner's dots (Figs. 7.5 and 7.6), which represent caveola–vesicle complexes of parasite origin that are exported into the cytoplasm of infected reticulocytes.[17,18] The similarity in the round shape of parasites in the sexual and asexual stages is another characteristic of the *P. vivax* parasite. All stages of parasitic development occurring in the bloodstream can be observed microscopically in peripheral blood samples. Round-shaped *P. vivax* gametocytes are also known to be good diagnostic markers because of their early appearance in blood smears during the infection (Fig. 7.7).

Epidemiology

Humans are the mammalian reservoir, while female *Anopheles* mosquitoes are the vector for the malarial protozoa. The life cycle of this parasite occurs partly in the human reservoir and partly in the female *Anopheles* mosquito vector, which then enables the transmission of the parasite from person to person. The *Anopheles* mosquitoes thrive in warmer climates, especially wet, low-lying areas. Consequently, malaria occurs predominantly in the tropics and subtropics. *P. vivax*, however, can survive better in colder climates compared to the other *Plasmodium* species, owing to its dormant liver phase. While *P. vivax* is the more widespread cause of malaria

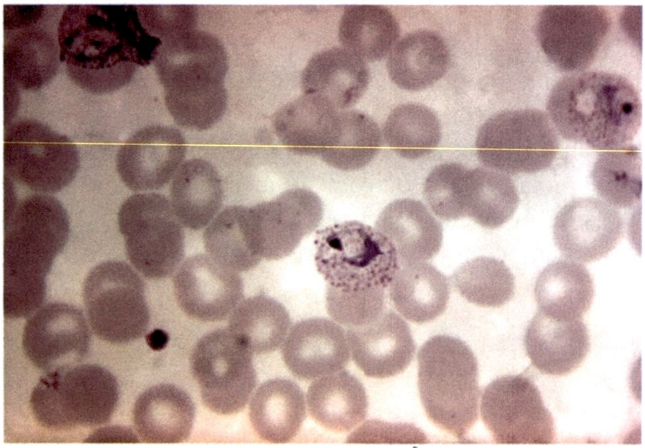

FIGURE 7.5

Microscopic view of blood smear showing distorted RBCs and Schüffner's dots characteristic of *P. vivax* infection. Source*: USCDCP/Own work/; https://creativecommons.org/licenses/publicdomain/; https://pixnio.com/science/microscopy-images/malaria-plasmodium/rbcs-are-enlarged-by-15x-to-2x-and-may-be-distorted-in-shape-containing-fine-schuffners-dots.*

worldwide, the African continent has a higher prevalence of *P. falciparum*. The African population has a low expression of the Duffy antigen on the surface of red blood cells, which is required by *P. vivax* to infect the cells, thereby resulting in lower rates of *P. vivax* infection compared to *P. falciparum* infection.[19] The higher mortality rate associated with malaria in the African region is, therefore, attributable to the higher level of pathogenicity of *P. falciparum* infections.

Pathogenicity and life cycle of P. vivax

The life cycle of the malarial parasite requires both the human host and the female *Anopheles* mosquito to complete (Fig. 7.8). Small, elongated plasmodial sporozoite cells produced in the female *Anopheles* mosquito are localized in the salivary gland of the insect and are injected into humans along with the anticoagulant-containing saliva of the mosquito. Once in the bloodstream, the sporozoites rapidly travel to the liver, invading the hepatocytes within 30 minutes of entering the bloodstream where they may become dormant, forming hypnozoites or may replicate and form schizonts. The schizonts then divide into numerous merozoites, which are released from the liver into the blood stream where they infect red blood cells (erythrocytes). The cycle between the schizont and the merozoite stages proceeds in the infected erythrocytes at synchronized intervals of 48 hours. The release of the *P. vivax* merozoites from the erythrocytes triggers the clinical manifestations of malaria, characterized by chills followed by fever of up to 40°C (104°F). The intermittent nature of the clinical manifestations comprising of

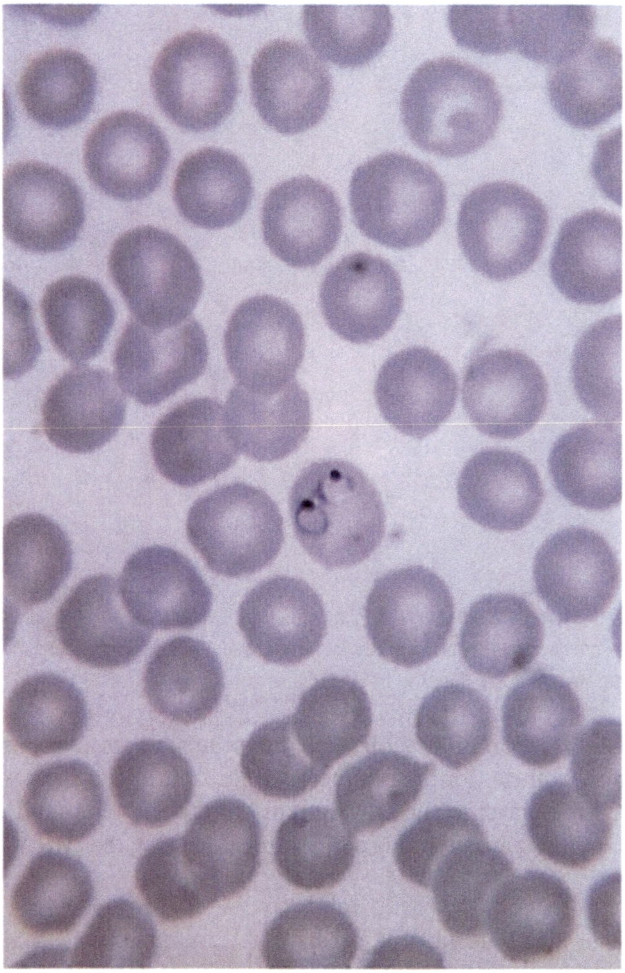

FIGURE 7.6

Micrograph of a blood smear showing an RBCs with two ring-form *P. vivax* that contains a large quantity of cytoplasm, a large chromatin dot, and occasional pseudopods. Although the red blood cells are normal, fine Schüffner's dots, often containing multiple parasites, are seen. Source: *(Originally uploaded by Merlin Magicien at French Wikipedia) CDC/Dr Mae Melvin/Icke G, Davis R, McConnell W. Teaching health workers malaria diagnosis.* PLoS Med. *2005;2(2):e11/; https://creativecommons.org/licenses/by/2.0/deed.en; https://commons.wikimedia.org/wiki/File:P_vivax_rings.jpg.*

asymptomatic periods alternating with symptomatic periods, peculiar to malaria, is a result of this 48-hour cycle.

The merozoites that do not infect erythrocytes are called gametocytes, which are ingested when another *Anopheles* mosquito bites an infected person. These gametocytes mature and form gametes within the mosquito. The fusion of two gametes

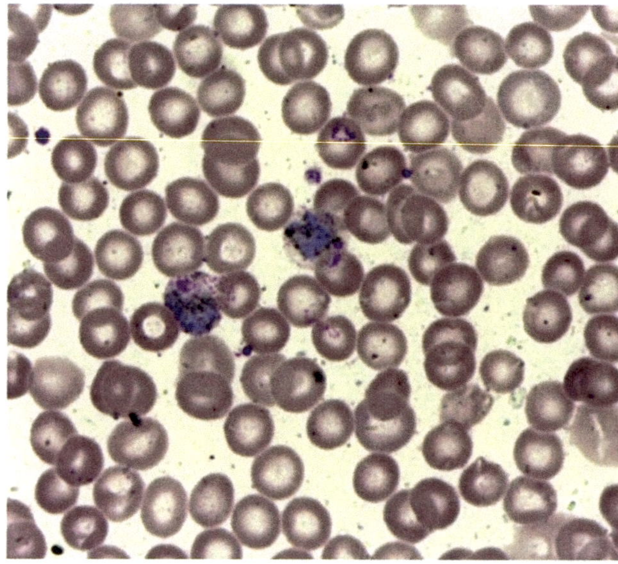

FIGURE 7.7

Peripheral smear showing schizont stage of *P. vivax*. Source*: Department of Pathology/Calicut Medical College/; https://creativecommons.org/licenses/by-sa/4.0/deed.en; https://commons.wikimedia.org/wiki/File:Plasmodium_vivax_malaria_1.jpg.*

results in the formation of a zygote, which then migrates to the mosquito's intestine and enlarges there, forming numerous sporozoites. The sporozoites released from the intestine travel to the salivary gland of the mosquito and can be introduced into another human during a blood meal, causing the cycle to begin again.

Resistance

The diagnosis and detection of resistant *P. vivax* strains is challenging and therefore hinders the development of strategies for detection and tracking of drug-resistant *P. vivax*. In addition, the emergence and spread of chloroquine, sulfadoxine-pyrimethamine, and mefloquine-resistant *P. vivax* strains is now well-known.[20–22] Therefore the development of effective techniques for the assessment of drug resistance and sensitivity is essential for ensuring the establishment of novel and effective treatment strategies.

Immune response

Immune response to *P. vivax* malaria is characterized by toxin-induced IFN-γ and TNF-α secretion.[23] Monocyte and macrophage-associated phagocytosis within the spleen dominates the innate immune response whereas adaptive immunity is characterized by IFN-γ and TNF-α-induced increase in CD4 + lymphocytes.[23] Parasitemia can range between 0.002% in unexposed mild cases and 5% in cases showing severe disease.

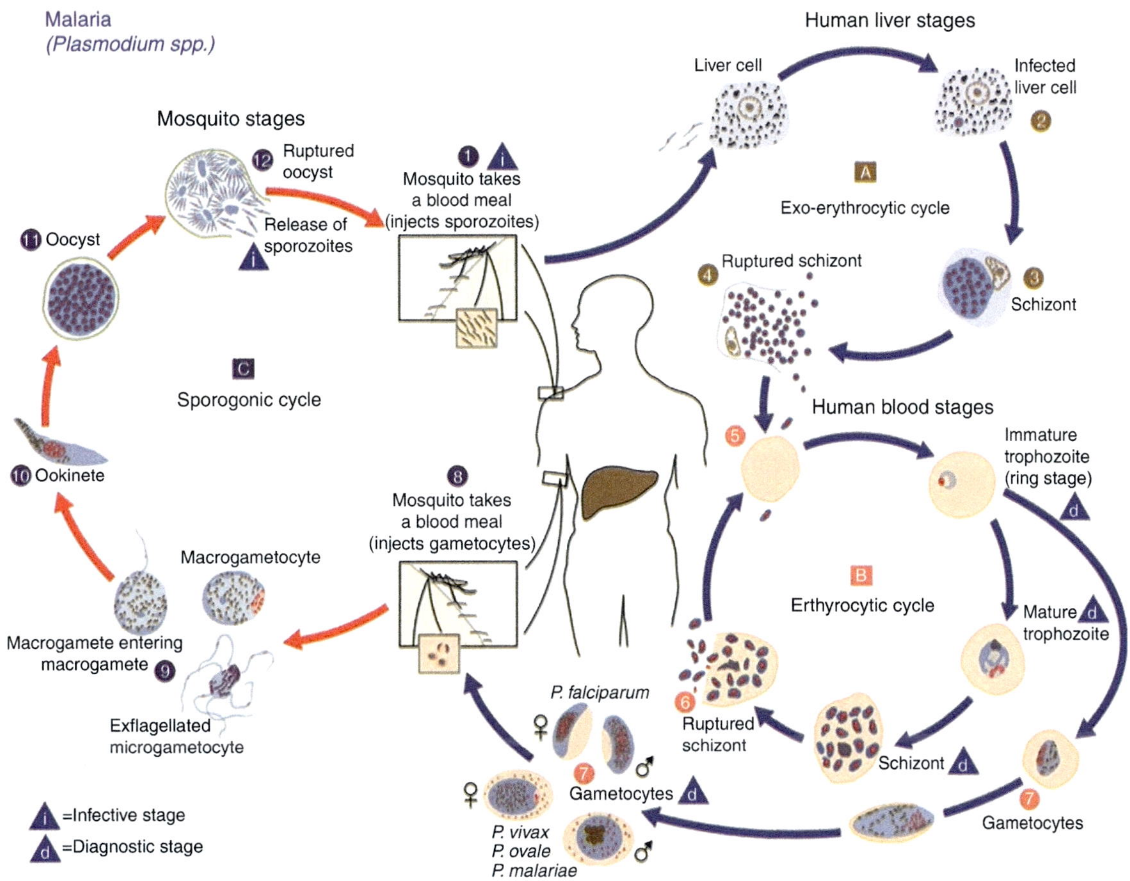

FIGURE 7.8

Life cycle of *P. vivax*. Source*: CDC/Alexander J. da Silva, PhD/Melanie Moser/Centers for Disease Control and Prevention's Public Health Image Library/; https://creativecommons.org/licenses/by-sa/3.0/; https://commons.wikimedia.org/wiki/File: Plasmodium_lifecycle_PHIL_3405_lores.jpg.*

Clinical manifestations and complications

Malaria, caused by *P. vivax*, typically has an incubation period of 12–17 days; however, rare cases relapse up to 2 years later and are seen due to dormant hypnozoites being released into the blood stream.[24] Generally, the clinical manifestations of malaria include fever, headache, nausea, vomiting, body aches, and jaundice. Anemia and splenomegaly may also result from the loss of red blood cells. The nonspecificity of the clinical manifestations overlapping with those of any febrile disease increases the challenge associated with an accurate clinical diagnosis of malaria. However, the peculiar pattern of intermittent symptomatic and asymptomatic periods, resulting from the life cycle of the malarial parasite,[25] which gives *P. vivax* malaria the name, "tertian fever," is considered the tell-tale sign for considering *P. vivax* malaria in differential diagnosis. Cases from *P. vivax* endemic areas have been found to invariably present with anemia. Although severe cases of *P. vivax*, causing cerebral malaria, renal failure, splenic rupture, acute respiratory distress, and shock, have been reported, the chances of multiorgan failure due to sequestration are found to be low compared to *P. falciparum* malaria.[24–27]

High-risk groups, such as pregnant women, are at increased risk of adverse outcomes like higher parasitic load, increased risk of premature delivery, miscarriage, low birth weight, neonatal, or maternal death from all forms of malaria, including that caused by *P. vivax* infection.

Laboratory diagnosis

The diagnosis of malaria in the clinical setting depends on three main diagnostic modalities:

1. Light microscopy
2. RDTs
3. PCR

1. Light microscopy

   This method of diagnosis considered as the gold standard for the diagnosis of *P. vivax malaria* involves the direct visualization of the *P. vivax* schizonts in a Giemsa-stained blood smear by light microscopy, facilitating evaluation of the presence of parasitemia, determination of the species, and calculation of the parasite density. While good sensitivity and low cost of the technique promote its frequent use in clinical settings, the need for skilled and experienced laboratory personnel and the availability of microscope may act as a hinderance.[28]

   The lower parasite density and possibility of the dormant hypnozoite form in *P. vivax* malaria make it more challenging to diagnose compared to *P. falciparum* malaria.

2. RDTs

   RDTs can be extremely useful in diagnosing malaria, especially in the resource-limited setting.[29] RDTs can detect various types of *Plasmodium* antigens in the sample and are not dependent on the skill-level of the technician, while also enabling a rapid diagnosis. However, the sensitivity of RDTs increases with a growing parasite load, thereby boosting the chances of false negative results in cases with lower parasite load.[30] Therefore the sensitivity of RDTs in the detection of *P. vivax* malaria is significantly lower than that for the detection of *P. falciparum* malaria.[31–33] As RDTs may yield positive results up to 1 month after successful treatment of malaria,[34] light microscopy may be recommended for the confirmation of RDT reports.

3. PCR

 PCR is used more frequently in the research setting, where highly skilled laboratory personnel and high-throughput PCR devices are available. The malarial species involved in a particular infection may also be determined using PCR in a clinical setting.

Treatment

The primary drug used for the treatment of malaria caused by any of the *Plasmodium* species has been chloroquine unless resistance to chloroquine is indicated.[26] Chloroquine-sensitive, uncomplicated *P. vivax* malaria is usually treated using chloroquine phosphate or hydroxychloroquine along with primaquine. Chloroquine-resistant *P. vivax* malaria can be treated using atovaquone-proguanil and quinine, in addition to tetracycline, doxycycline, or mefloquine. The addition of primaquine to the treatment regimens is necessitated due to the possibility of the dormant liver hypnozoite phase. However, the use of primaquine is contraindicated in cases with G6PD deficiency due to the risk of hemolytic anemia and in pregnant women due to the risk for fetal hemolysis.

In patients with G6PD deficiency, primaquine may cause severe hemolytic anemia. Decisions for its use in this patient population must occur on a case-by-case basis.[35] In pregnant women, primaquine is contraindicated because of the risk for fetal hemolysis.[36]

Severe malaria may benefit from the use of intravenous artesunate;[37] however, the lack of approval from U.S. Food and Drug Administration (FDA) has restricted its use. Intravenous quinidine and tetracycline, or doxycycline, or clindamycin is the more frequently used mode of treatment for severe malaria.

Prevention

The challenge associated with the development of an effective vaccine for malaria has resulted in a multifaceted approach being involved in the prevention of malaria. The diligent use of appropriate prophylactic medications is considered the key for malaria prevention among travelers. Clothing that ensures maximum body coverage, in addition to the use of diethyltoluamide (DEET) and/or permethrin and mosquito nets, has been shown to substantially reduce the risk of malaria.[38]

Elimination of the *Anopheles* mosquito, resulting in an interruption in the life cycle of the malarial parasite, is considered the best way to prevent malaria infections since antimalarial drug treatment is an expensive and short-term solution, even when drug-resistant strains are not involved.

Student Notes

Key take-aways

- Malaria is a parasitic disease in humans responsible for huge social impact globally.
- *P. vivax*, one of the *Plasmodium* protozoal species responsible for malaria in humans, is the most widespread cause of malaria and is usually responsible for a milder form of disease compared to malaria caused by *P. falciparum*, although severe cases have also been reported.
- The generalized nature of the clinical manifestations of malaria, overlapping with other febrile diseases, makes the diagnosis of malaria challenging for clinicians. The "tertian fever," characteristic of malarial infections, however, is a hallmark sign and should lead to the inclusion of malaria in the differential diagnosis process.
- The dormant phase of the *P. vivax* malarial parasite in the form of hypnozoites manifested in the human liver can result in delayed onset or relapse of symptoms after months or sometimes even years.
- The elusive life cycle of the malarial parasite is largely responsible for the difficulties involved in the development of effective prophylactic and therapeutic strategies against malaria.
- Elimination of the mosquito vector and prevention of mosquito bites is considered the mainstay of malaria prevention.
- The advent of drug-resistant strains of *P. vivax* has substantially increased the challenge involved in the treatment of malarial infections.

Problem 7.3: Posttraumatic foot infection caused by *Staphylococcus aureus*

A 14-year-old boy presented with a fever and the inability to move his left leg due to a painful wound on the left foot. The boy had been febrile for the past 4 days. Examination of the left leg revealed the presence of a skin ulcer, measuring 4×3 cm on the plantar surface of the left foot. The ulcer oozed pus with the edges and surrounding tissue showing all the cardinal signs of inflammation (redness, swelling, warmth, and painful to touch). Inspection of the groin area disclosed no evidence of lymphadenitis secondary to the lesion on his foot.

The patient history revealed that his current condition started with a small cut inflicted by a broken piece of glass while playing football bare-footed. The bleeding stopped after applying pressure, but the following day the site of injury became swollen, red, and painful. The patient has been vaccinated against tetanus.

A swab sample of the aspirated pus was collected for culture and sensitivity tests and a blood sample was gathered for a complete blood count.

The following are the reports of the investigations carried out:

1. Vitals
   Temperature: 39.0°C
   Pulse: 125/min
2. Pus culture and sensitivity test
   Type of specimen: Swab sample of the aspirated pus from the wound
   Bacterial isolate: *S. aureus* + +
   Culture sensitivity

Ampicillin	Resistant
Ciprofloxacin	Sensitive
Cotrimoxazole	Resistant
Erythromycin	Sensitive
Fucidin	Resistant
Penicillin G	Resistant
Tetracycline	Resistant

3. Complete Blood Count

Hemoglobin (g/dl)	12.3	11.5–15.5
Total WBC (/L)	23.7×10^9	$4–11 \times 10^9$
Differential WBC		
Neutrophil (%)	72	45–60
Lymphocyte (%)	25	30–50
Monocyte (%)	5	2–8
Eosinophil (%)	2	1–6

Diagnosis

The final diagnosis concluded a posttraumatic foot infection caused by *S. aureus* infection.

Treatment

The wound was cleaned with antiseptic and dressed, and the following medications were prescribed:

1. Analgesic
2. Erythromycin

The patient was asked to attend a follow-up as an outpatient 3 days later.

The patient presented as afebrile on the follow-up 3 days later, and the pain in his left leg had subsided. A prescription for a continuation of the medications for 4 more days was made and a follow-up appointment after 2–3 weeks was recommended. During the next meeting 3 weeks later, the patient could walk normally without any complaints, and the wound had healed completely, leaving behind a clean scar without any complications.

Question and answer

1. What is the mechanism of a fever?
   A fever is caused by the effect of pyrogens in bacterial toxins, products of tissue damage, and mediators, such as prostaglandins and cytokines, released during inflammation on the temperature-regulating ability of the hypothalamus. The mechanisms responsible for the conservation of heat and increased heat production mechanisms come into effect, which in turn causes the optimum body temperature to rise.
2. What is the process of inflammation?
   Inflammation is the reaction of vascularized tissue to injurious agents, such as microbial infections (like bacteria, viruses, fungi, and parasites), physical injury (like heat, cold, or trauma), chemical injury, necrotic tissue, and immune reaction. Inflammation is a process initiated by an initial inflammatory stimulus that triggers the release of chemical mediators from plasma or connective tissue cells. Such mediators act together or in sequence, amplifying the initial inflammatory response and regulating the vascular and cellular events. The inflammatory response is terminated when the injurious agent is removed and the inflammatory mediators disintegrate, are catabolized, or inhibited.
3. What are the main events involved in the inflammation process?
   The main events involved in inflammation are as follows:
   a. Hemodynamic changes enable inflammatory cells and plasma factors to accumulate to the site of injury
   b. Influx of leukocytes facilitates the destruction and dilution of injurious agents, and creates a barrier that walls off the injurious agents, thereby protecting the host tissues.
   c. The release of mediators ensures that the host tissues are protected from the harmful effects of the inflammatory process.
4. What are the cardinal signs of inflammation?
   The cardinal signs of inflammation are as follows:
   a. Heat (calor): A localized increase in temperature occurs due to the increased blood flow caused by vasodilation.

b. Redness (rubor): The affected area appears red due to increased blood flow.
c. Swelling (tumor): The accumulation of exudates comprising of protein-rich, edema fluid caused by the permeability of blood vessels results in localized swelling of the affected area.
d. Pain (dolor): Pain can result from the pressure exerted on the nerve endings by the swelling and/or by the stimulation elicited by chemical mediators, such as bradykinins and prostaglandins (e.g., PGE2).
e. Loss of function (function laesa): Pain in the inflamed tissues can result in an inability of the tissue to function normally.

5. What is the mode of action of antimicrobial drugs, such as erythromycin, belonging to the group of macrolides?
   Macrolides are inhibitors of protein synthesis that inhibit or kill microorganisms by damaging the cell membrane or by inhibiting the synthesis of cell wall and/or nucleic acids.
6. What is minimum inhibitory concentration and minimum bactericidal concentration?
   Minimum inhibitory concentration is the minimum concentration of an antimicrobial agent at which growth of the target microorganism is inhibited. Minimum bactericidal concentration is the minimum concentration of an antibacterial agent at which the target bacteria is killed.
7. How are bacterial culture and antibiotic sensitivity reports interpreted?
   Bacterial culture reports identify and characterize the bacterial growth seen in culture, generally under a microscope. The antibiotic sensitivity reports are interpreted based on the number of "++," which represents the level of resistance (R) or sensitivity (S) of the target bacteria to the antibacterial agents being tested.
8. What does an elevated WBC count indicate?
   The WBC count is elevated due to the increased production and release of WBCs from the bone marrow caused by the action of the chemical mediators of inflammation, such as TNF. While an increased neutrophil count indicates a bacterial infection, viruses typically cause lymphocytosis. Parasitic infections and allergic inflammatory reactions, on the other hand, stimulate eosinophil production.
9. What are the various outcomes of acute inflammation?
   Acute inflammation may lead to any of the following outcomes:
   a. Resolution: No or minimal tissue destruction, which allows the structure and function of the affected tissue to go back to its normal state.
   b. Abscess formation: Inflammation, resulting from infections involving pyogenic bacteria (pus producing), may result in liquefaction of tissue, resulting in abscess formation.

c. Development of scar tissue: The process of fibrous tissue formation ensures the reorganization and repair of the affected tissue.
d. Chronicity: The inflammatory condition may become chronic, thereby affecting the tissue repeatedly, eventually causing substantial damage.

10. How does acute inflammation progress and develop into chronic inflammation?

    Chronic inflammation is a prolonged inflammatory condition that lasts for months or years, during which active inflammation, injury, and healing proceed simultaneously, causing extensive damage to the host system. Chronic inflammation is one of the complications of acute inflammation, which results from persistent suppuration or presence of indigestible material, such as dead bone or exogenous material like dirt, suture, or other foreign bodies. Chronic inflammation is usually characterized by repeated episodes of acute inflammation. However, certain cases may not have an initial phase of acute inflammation, and the condition in such cases is known as primary chronic inflammation.
    For example, certain infections like tuberculosis, leprosy, and some fungal infections may result in immune reactions to suspected foreign bodies.
    For example, autoimmune diseases.
    For example, granulomatous diseases such as sarcoidosis.

11. What are the characteristic histological features of chronic inflammation?

    Chronic inflammation is characterized histologically by the presence of mononuclear cell infiltrate (lymphocytes, macrophages, and plasma cells) and fibroblasts along with the formation of new blood vessels. Certain chronic inflammatory conditions, like granulomatous inflammation, are characterized by the presence of granulomas consisting of epithelioid cells (activated macrophages), giant cells, and mononuclear cells.

12. What differentiates Streptococcal colonies from *Staphylococcal* colonies in culture?

    Staphylococci grow as single or clustered cell colonies on blood agar and are characterized by golden yellow pigmentation indicative of the presence of staphyloxanthin, a virulence factor.
    Staphyloxanthin enables evasion of the reactive oxygen species and is responsible for the characteristic appearance of an area of ß-hemolysis surrounding the bacterial colonies in culture. The absence of long chains is a distinguishing character between *Streptococci* and *Staphylococci*.

13. How do *Staphylococcal* ß-hemolysin, γ-hemolysin, and Δ-hemolysin mediate their pathological effects?

    ß-hemolysin is a sphingomyelinase responsible for catalyzing the breakdown of sphingomyelin, an important lipid component of the membranous myelin sheaths surrounding the neurons. This bacterial toxin, therefore, elicits neurotoxic effects. γ-hemolysin are synergohymenotropic toxins, which mediate their pathological activity through the synergistic activity of two nonassociated proteins. Δ-hemolysin, on the other hand, is known to mediate its cytotoxic effects through a detergent-like effect on the cell membranes of erythrocytes, leukocytes, macrophages, and platelets.

Learning objectives

1. Determine the appropriate differential diagnosis based on patient history and clinical manifestations.
2. Describe the mechanisms involved in pyrogenic, inflammatory, and pyogenic conditions.
3. Identify the etiological agent responsible for the particular case of posttraumatic foot infection.
4. Describe the characteristics of *S. aureus*, the bacteria responsible for the manifestation of the foot infection.
5. Determine the appropriate course of treatment and the most effective antibiotic regimen based on the clinical manifestations of the disease and culture sensitivity reports to ensure prompt and effective resolution of the infection; and avoid any further debilitating complications.

Discussion

Differential diagnosis

Bacteremia	Bacteremia is a condition caused by a bacterial invasion of the bloodstream. The condition presents with a fever, chills, malaise, an infected wound or site of injury.
Leptospirosis	Leptospirosis is a bacterial infection of the bloodstream triggered from invasion of the bloodstream by the *Leptospira* bacteria in the urine of infected animals through cuts or abrasions in the skin. The disease manifests with a fever, intense headaches, chills, general malaise, myalgia, abdominal pain, nausea, vomiting, diarrhea, photosensitivity, conjunctival suffusion, subconjunctival bleeding, jaundice, meningitis, and renal insufficiency.

Continued

Continued	
Bullous impetigo	Bullous impetigo is an infectious condition presenting with red sores that rupture, ooze pus, and form yellowish-brown crust.
Pediatric osteomyelitis	Pediatric osteomyelitis results from the spread of an infection to the bones, which is frequently seen in children due to an increased blood supply to the growing bones. The condition develops into a fever, irritability, fatigue, swelling, redness, pain, and inability to move the affected limb.
Pediatric serum sickness	Pediatric serum sickness is a condition elicited by a hypersensitivity reaction to certain medications. The condition typically presents with fever, rash, polyarthritis, and polyarthralgias.

Background

S. aureus is a human pathogen of considerable clinical significance due to its implication in several medical conditions. *S. aureus* infections develop a wide spectrum of clinical manifestations. The pathogen is ubiquitously present in the environment and forms an important part of the normal human flora, especially the skin and the mucous membranes. Therefore infections may be community-acquired or nosocomial in nature. While *S. aureus* does not typically cause skin infections in immunocompetent individuals, invasion of the blood stream or internal tissues through cuts or abrasions in the skin may result in serious infections and complications.[39] The emergence of multidrug resistant (MDR) strains, such as MRSA (methicillin-resistant *S. aureus*), has posed numerous challenges in the clinical treatment of *S. aureus* infections.[40]

Etiological agent

Staphylococcus aureus

Classification and morphology

S. aureus are Gram-positive, nonmotile, and nonsporing bacteria, which are characterized by their spherical shape, giving them the suffix "cocci" and clustered arrangement described as being "grape-like" (Fig. 7.9). These bacteria grow as single or clustered cell colonies on blood agar and are identified by their golden yellow pigmentation, indicative of the presence of staphyloxanthin, a virulence factor that enables the evasion of the reactive oxygen species. The presence of staphyloxanthin results in the formation of a characteristic area of ß-hemolysis surrounding the bacterial colonies (Fig. 7.10). While these bacteria may occur singly or in short chains comprising of 3–4 cells, long chains of *S. aureus* are never seen. The absence of long chains is a distinguishing character between *Streptococci* and *Staphylococci*. *S. aureus* are catalase positive, coagulase positive, mannitol fermentation positive, novobiocin sensitive, and facultatively anaerobic organisms.

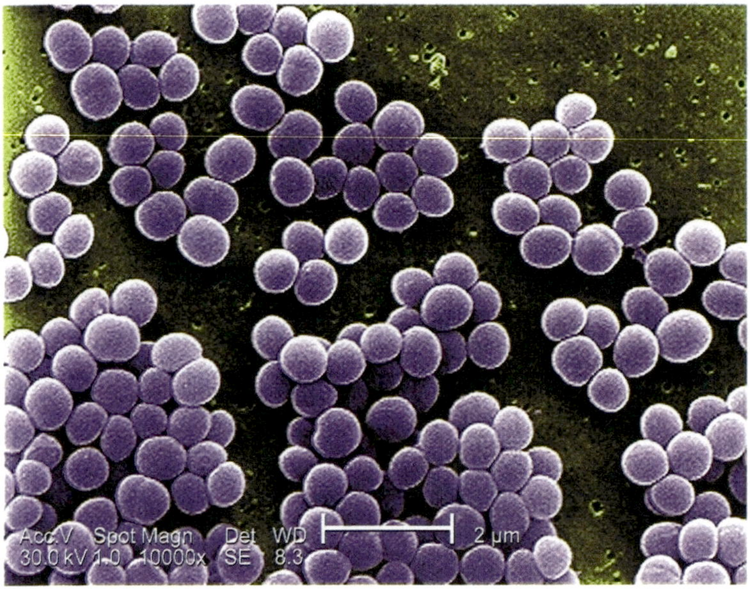

FIGURE 7.9

High magnification image of the *Staphylococcus aureus* bacteria. Source*: Janice Haney Carr, Matthew J. Arduino, DRPH, USCDCP; https://pixnio.com/science/microscopy-images/staphylococcus-aureus/under-a-high-magnification-of-10000x-strain-of-staphylococcus-aureus-bacteria.*

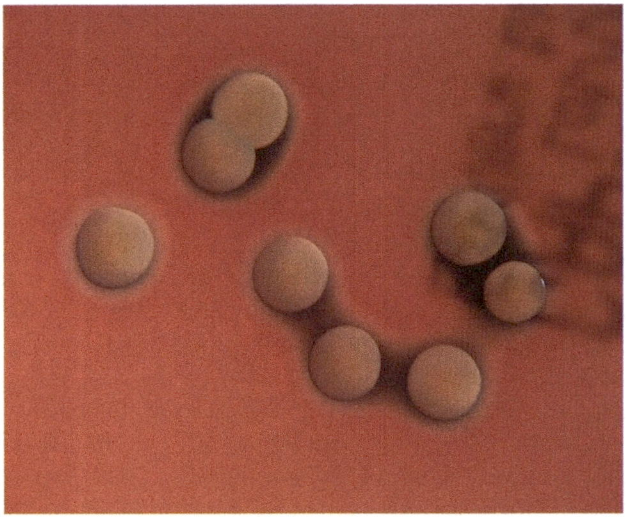

FIGURE 7.10

S. aureus colonies showing golden yellow pigmentation and ß-hemolysis around the colonies. Source*: HansN/Own work/; https://creativecommons.org/licenses/by-sa/3.0/deed.en; https://commons.wikimedia.org/wiki/File:Staphylococcus_aureus_colonies_on_blood_agar.jpg.*

Epidemiology

Humans are the primary reservoirs for normal as well as the MDR *S. aureus* bacteria. These bacterial parasites colonize the skin and mucous membranes of humans at an early stage in life, resulting in a carrier-state, which facilitates the spread of infection in the community as well as in hospital settings.[40] Colonization of the mucous membranes of the respiratory and the genitourinary tract is the most common. Respiratory infections, therefore, increase the rate of transmission of *S. aureus* infections through contaminated droplets. Physiological changes in the genitourinary tract, such as those that occur during menstruation, have been implicated in *S. aureus* infection-associated toxic shock syndrome (TSS) in menstruating females.[41] Most of the global human population is known to harbor the bacteria without any symptoms. While immunocompetent individuals do not develop infections despite colonization, certain individuals are at high-risk of infection, such as immunocompromised individuals, diabetics, regular needle users, and those who need frequent hospitalizations. Transmission of *S. aureus* infections can occur from person-to-person by direct contact, through infected droplets, or through fomites.[39,42] Nosocomial sources of *S. aureus* infections are of special concern, owing to their abilities of multidrug resistance.

Virulence and pathogenicity

S. aureus is known to cause diseases either through infections mediated by the invasion of the bloodstream and the internal tissues or through intoxications mediated by the bacterial toxins. Intact skin acts as an effective anatomical barrier, preventing the cocci from gaining access to the bloodstream and internal tissues. Invasion of the internal mucosal membranes and tissues facilitated by damaged skin is followed by bacterial colonization through adherence of the bacteria to host cells or extracellular matrix, evasion of the host defense mechanisms, and multiplication, all of which eventually result in tissue damage.

The pathogenicity and virulence of *S. aureus* is influenced by numerous cell-associated and extracellular factors, including some exotoxins. The peptidoglycan present in the bacterial cell wall is an important polysaccharide responsible for the rigidity and structural integrity of the bacterial cell. Along with ensuring good structural integrity of the bacterial cell, this peptidoglycan is antigenic in humans and activates the complement cascade, thereby inducing the release of inflammatory cytokines. The adhesion of the cocci to the host cell surface and the evasion of complement-mediated opsonization by the bacteria is facilitated by teichoic acid in the bacterial cell wall and the capsular polysaccharide surrounding it.

The protein A on the cell surface of *S. aureus* strains facilitates chemotaxis in the bacteria and is responsible for immune evasion via antiphagocytic and anticomplementary effects. This cell surface protein also elicits pathological effects by inducing platelet damage and hypersensitivity in the host.

Coagulase type A is an extracellular enzyme that causes clotting of human plasma independently of calcium, or other clotting factors, by interacting with the coagulase reacting factor (CRF) in plasma, binding to prothrombin, and converting fibrinogen to fibrin. Clumping factor, the bound form of coagulase, is another surface protein that causes clotting of human plasma independent of the CRF. The presence of both coagulase and clumping factor is a characteristic feature of *S. aureus* strains, although the test for the clumping factor may sometimes be negative in capsulated strains as it may be enveloped by the capsular polysaccharide.

Lipid hydrolases are instrumental in facilitating the invasion of the skin and subcutaneous tissues. While hyaluronidase breaks down the connective tissue, staphylokinases prevent the formation of the fibrin network, and fatty acid modifying enzymes and proteases help in initiation and spread of infection. Presence of nucleases is an identifying feature of *S. aureus* and plays a significant role in establishing the infection.

Toxins, such as hemolysins and leucocidin, are membrane-active substances and have cytolytic effect. The α-hemolysin is the most important toxin for *S. aureus* bacteria, and is distinguished by its ability to reactivate at 100°C after being inactivated at 70°C. This hemolysin is cytotoxic, especially for leukocytes, macrophages, and lysosomes. Muscle tissues, the renal cortex, and the circulatory system are especially affected by the toxicity of the α-hemolysin. ß-hemolysin is a sphingomyelinase responsible for catalyzing the breakdown of sphingomyelin, an important lipid component of the membranous myelin sheaths surrounding the neurons. This bacterial toxin, therefore, elicits neurotoxic effects. γ-hemolysin and leucocidin are synergohymenotropic toxins, which mediate their pathological activity through the synergistic activity of two nonassociated proteins. Δ-hemolysin, on the other hand, is known to mediate its cytotoxic effects through a detergent-like effect on the cell membranes of erythrocytes, leukocytes, macrophages, and platelets.[43,44]

S. aureus enterotoxins are responsible for the onset of clinical manifestations associated with food poisoning, such as nausea, vomiting, and diarrhea, 2–6 hours after consuming food contaminated with the preformed toxin. The toxin's resistance to high temperatures, even when exposed for prolonged periods of time, make it especially difficult to eradicate it once it has been secreted in carbohydrate- and protein-containing foods, such as meat, fish, and dairy products, contaminated with *Staphylococci*. The food may be infected through unhygienic handling practices, especially if the food handler is a carrier of the pathogen. Although the clinical manifestations indicate an infection of the gastrointestinal mucosa, the toxin actually acts directly on the autonomic nervous system. The toxin is potent in extremely small quantities and is antigenic, eliciting a response from the immune system. Specific antitoxins produced in response to the immune activation can effectively neutralize the toxin. The febrile condition associated with most cases of *Staphylococci* diarrhea is attributable to the pyrogenic nature of the toxin. The toxin also exhibits mitogenic, hypotensive, thrombocytopenic, and cytotoxic effects.

The TSS toxin is responsible for a severe and potentially fatal multisystem disease caused by infections of the skin and mucosa facilitated by trauma and surgical wounds. Although TSS can be a complication of *S. aureus* infections in children and adolescents, it is especially notorious for causing a fatal condition in menstruating women using highly absorbent vaginal tampons. The TSST-1 antibody produced in response to immune activation by the TSST type 1 toxin is protective, and its absence influences the pathogenesis of the condition, as evidenced by the presence of the TSST-1 antibody in the serum of convalescent individuals.[43]

The enterotoxins and TSST-1 secreted by *S. aureus* are potent activators of T-lymphocytes, and hence are responsible for the uncontrolled stimulation of large numbers of T-cells. This results in hyperactivity of the immune response, giving rise to a condition akin to a cytokine storm. The multisystem involvement and wide-spread inflammatory damage inflicted on the systems, characteristic of TSS and *S. aureus* food poisoning, are attributable to this excessively dysregulated activation of the immune system.

The epidermolytic toxin is responsible for *Staphylococcal* scalded skin syndrome (SSSS), an exfoliative skin disease, which causes the outer layer of the epidermis to separate from the underlying tissues. Ritter's disease is the severe form of SSSS observed usually in neonates, while toxic epidermal necrolysis is the manifestation of severe SSSS in older patients. SSSS manifests as pemphigus neonatorum and bullous impetigo in the milder forms.[45,46]

Antibiotic resistance

MRSA strains of *S. aureus* are of extreme clinical significance because of their resistance to treatment with most penicillin-based antibiotics. These MDR strains comprise of a *mec* gene as a part of the larger *Staphylococcal* chromosomal cassette mec (SCCmec) region. The SCCmec type influences the resistance to multiple antibiotics seen in MRSA strains.[47] The penicillin-binding protein 2a (PBP-2a) encoded by the *mec* gene is a modified PBP, which plays a key role in catalyzing the production of the peptidoglycan in the bacterial cell wall. The ability of the modified PBP found exclusively in MRSA strains to catalyze the synthesis of the bacterial cell wall is unaffected by ß-lactam-comprising antibiotics. Therefore MRSA strains thrive even in the presence of certain antibiotics, such as methicillin, nafcillin, oxacillin, and cephalosporins.[47,48] This poses a significant challenge for clinicians to provide effective treatment, and therefore increases the morbidity and mortality levels associated with complications of *S. aureus* infections.

Clinical manifestations

S. aureus is a ubiquitous human pathogen that also forms a significant part of the normal human flora. While an asymptomatic carrier state is common with this bacterium, immunocompetence prevents serious infections. On the other hand, an immunocompromised state or any form of trauma affecting the integrity of the skin, such as injury, wounds, abrasions, surgeries, and regular needle usage, result in an increased vulnerability to bacterial invasion of the bloodstream and the internal tissues, resulting in several conditions, such as bacteremia, infective endocarditis (Fig. 7.11), osteomyelitis (Fig. 7.12), septic arthritis (Fig. 7.13), prosthetic device

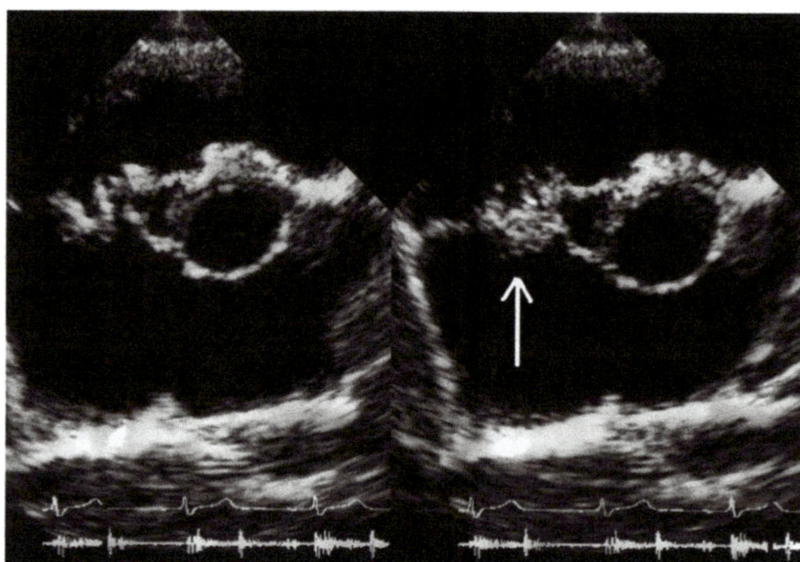

FIGURE 7.11

Echocardiography indicating vegetation on the tricuspid valve. Source: *Daisuke Koya, Kazuyuki Shibuya, Ryuichi Kikkawa and Masakazu Haneda/Koya D, Shibuya K, Kikkawa R, Haneda M. Successful recovery of infective endocarditis-induced rapidly progressive glomerulonephritis by steroid therapy combined with antibiotics: a case report.* BMC Nephrol. *2004;5:18. doi: 10.1186/1471-2369-5-18/; https://creativecommons.org/licenses/by/2.0/deed.en; https://commons.wikimedia.org/wiki/File:Endocarditis_ultrasound.gif.*

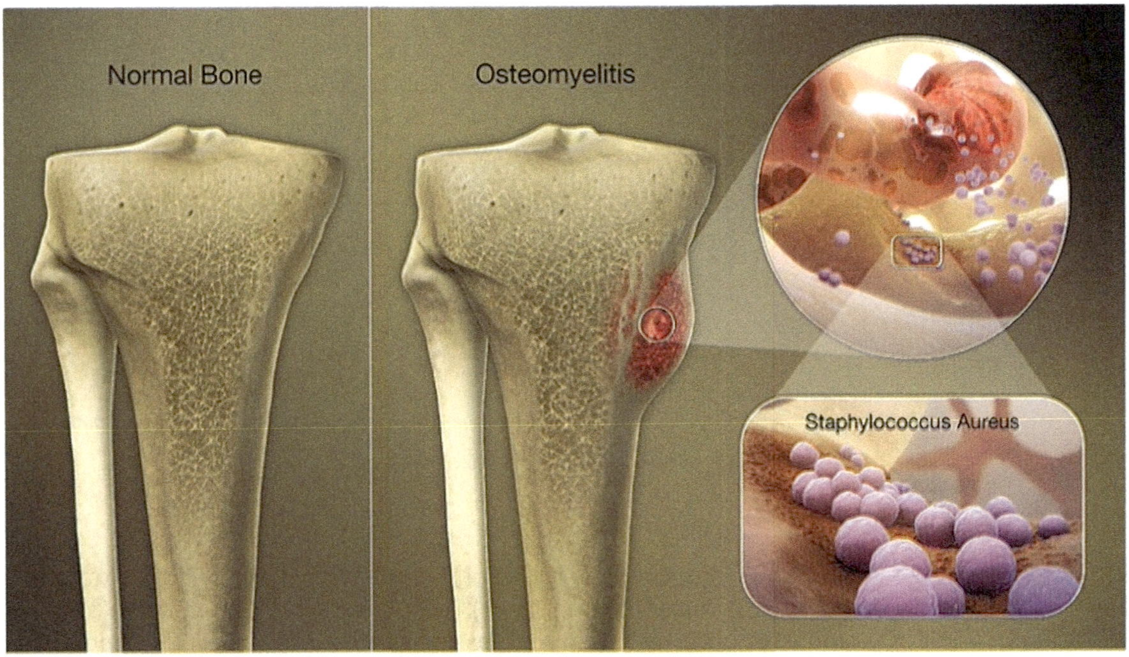

FIGURE 7.12
Osteomyelitis caused by *S. aureus* infection. Source: *https://www.scientificanimations.com/; https://www.scientificanimations.com/wiki-images/; https://creativecommons.org/licenses/by-sa/4.0/deed.en; https://commons.wikimedia.org/wiki/File:3D_Medical_Animation_Staphylococcus_Aureus.jpg.*

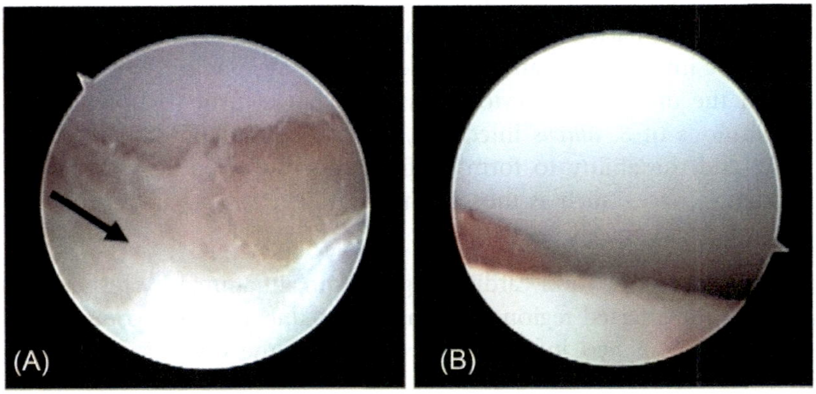

FIGURE 7.13
Septic arthritis of the ankle. (A) Arthroscopic findings at the beginning of surgery showed debris (arrow) in the joint space. (B) Arthroscopic findings after washout showed no irregularity of the articular cartilage. Source: *Tetsuo Hagino, Masanori Wako, Satoshi Ochiai/; https://www.ncbi.nlm.nih.gov/pmc/articles/PMC3192658/; https://creativecommons.org/licenses/by/2.5/deed.en; https://commons.wikimedia.org/wiki/File:SepticArth2011.jpg.*

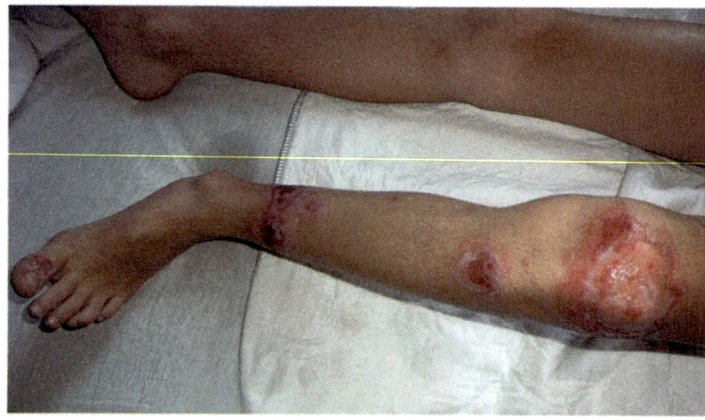

FIGURE 7.14
Bullous impetigo. Source: *Milliejen/Own work/; https://creativecommons.org/licenses/by-sa/3.0/deed.en; https://commons.wikimedia.org/wiki/File:Impetigo_bulleux.jpg.*

infections, impetigo (Fig. 7.14), folliculitis (Fig. 7.15), carbuncles (Fig. 7.16), furuncles (Fig. 7.17), cellulitis (Fig. 7.18), scalded skin syndrome, pulmonary infections, gastroenteritis, meningitis, TSS, and urinary tract infections.[42]

The clinical manifestations of the disease depend on whether the infection is invasive or toxin-mediated, the strain of the etiological agent, the site of infection, and the organ system or tissue involved.[42,49] The varied pathological manifestations are attributable to the numerous mechanisms of immune evasion and the bacteria's ability to adhere to various cell and mucosal surfaces.[49] The immune activation potency of the *Staphylococcal* toxins is responsible for the uncontrolled cytokine storm, characteristic of potentially fatal complications of *S. aureus* infections, such as sepsis and TSS.[50,51] Quorum sensing and the ability to form biofilms allows the bacteria to cling onto inanimate surfaces, such as those of medical devices, and are responsible for prosthetic device infections.[52]

Fever, chills, malaise, and cardinal signs of an inflammatory response, such as pain in the affected region, swelling, and redness, are the common manifestations of *S. aureus* infection, irrespective of the organ system or tissue affected.

Diagnosis

A collation of information, including the patient history, clinical manifestations, and findings of physical examination is essential for a primary clinical diagnosis of an *S. aureus* infection. Culture reports usually confirm the

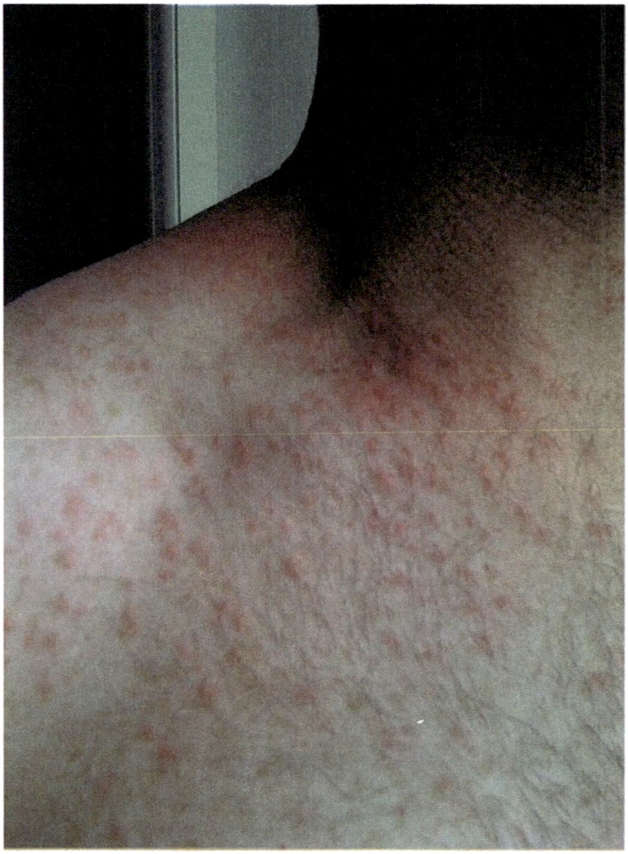

FIGURE 7.15
Folliculitis. Source: *Vivvid/Own work/; https://creativecommons.org/licenses/by-sa/3.0/deed.en; https://commons.wikimedia.org/wiki/File:Folliculitis-1.jpg.*

diagnosis, although the need for real-time reverse transcription–polymerase chain (RT-PCR) for 16S rRNA genes may be indicated in some cases.[53] Culture sensitivity reports are recommended to ensure the prescription of accurate and the most effective antibiotics. As *S. aureus* is ubiquitously present as part of the normal human flora, the differentiation between the nonpathogenic presence of *S. aureus* in a sample from a pathogenic presence associated with the clinical manifestations is important for an accurate diagnosis of the cause.[48]

Treatment and prevention

The parameters of the course of treatment, such as the antibiotic regimen, the mode of therapy, the dosage, and the duration of therapy for *S. aureus*

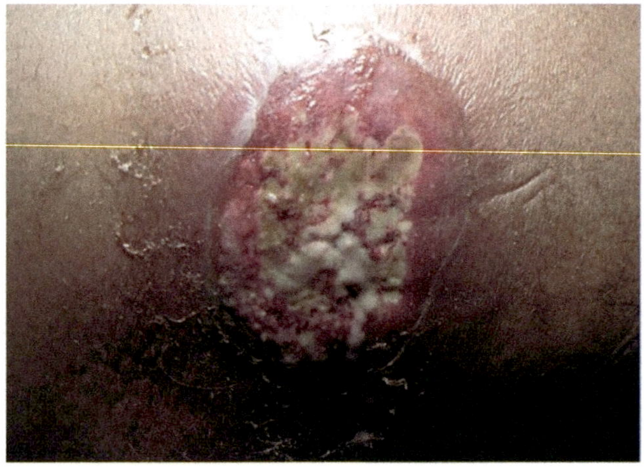

FIGURE 7.16
Carbuncle (cropped image). Source*: Drvgaikwad/Own work/; https://creativecommons.org/licenses/by/3.0/deed.en; https://commons.wikimedia.org/wiki/File:Carbuncle_on_buttok.JPG.*

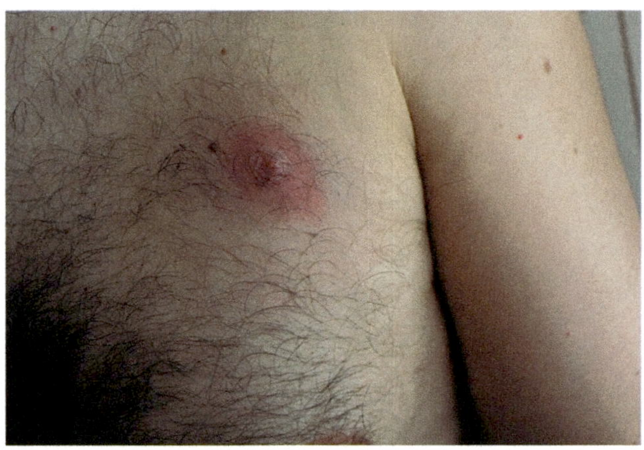

FIGURE 7.17
Furuncle. Source*: El Pantera/Own work/; https://creativecommons.org/licenses/by-sa/3.0/deed.en; https://commons.wikimedia.org/wiki/File:Furuncle-MIN-IMG_2589.jpg.*

infections are influenced by the site of infection, the organ system or tissue affected, and the drug resistance of the infecting strains.[42] Penicillin-based antibiotics are the antibiotics of choice for the treatment of

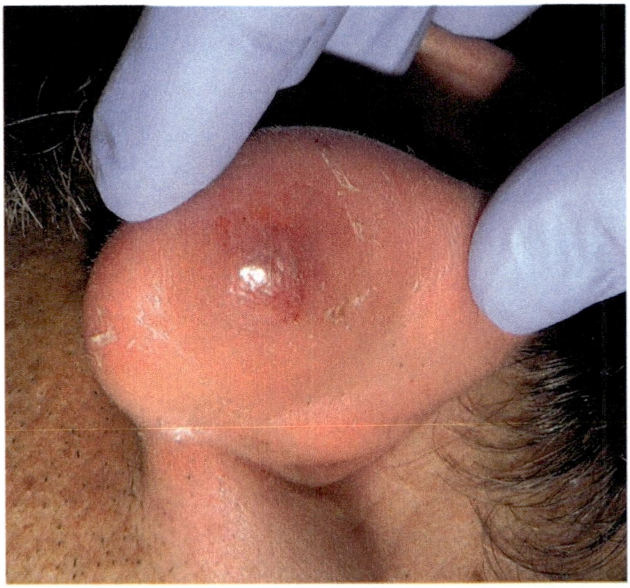

FIGURE 7.18
Cellulitis of the earlobe. Source: *Jmarchn/Own work/; https://creativecommons.org/licenses/by-sa/3.0/deed.en; https://commons.wikimedia.org/wiki/File:EarlobeCellulitis1Back.JPG.*

S. aureus infections that are not caused by MRSA strains, while infections involving MRSA strains are usually treated using vancomycin.[40] However, the emergence of MDR MRSA strains has been increasingly threatening the availability of effective treatment options, thereby resulting in unfavorable prognoses and outcomes. Alternative therapy options, such as fluid-replacement management, may be indicated for toxin-mediated illnesses. The removal of prosthetic devices or medical devices, like pacemakers and in-dwelling catheters, may be required for the effective resolution of infections associated with biofilm formation on the foreign objects.[53]

Due to the ubiquitous presence of *S. aureus* in the environment and as a significant part of the normal human flora, prevention of *S. aureus* infections is challenging. The lack of an effective vaccine has resulted in infection control methods, such as disinfection, sterilization, appropriate asepsis protocols, and hand washing techniques, becoming the mainstay of preventive measures.

Student Notes

Key take-aways

- *S. aureus*, including the MDR strains, are ubiquitous in the environment and forms a significant part of the normal human flora.
- *S. aureus* infections may be community-acquired or nosocomial.
- Colonization of various mucosal surfaces without infection is common, resulting in an asymptomatic carrier state, which facilitates the spread of the pathogen.
- Intact skin acts as an effective anatomical barrier against the invasion of the bloodstream or the internal tissues by the bacteria. However, damaged or traumatized skin may allow entry of the bacteria into the internal tissues and the bloodstream, resulting in clinical manifestations of the infections depending on the organ system or tissue affected.
- An immunocompromised state and underlying medical conditions, such as diabetes, regular needle usage, and frequent hospitalizations, are predisposing factors that increase the risk of *S. aureus* infections.
- *S. aureus* may cause disease by invasion or by toxin-mediated effects.
- *S. aureus* infections are generally localized to the site of infection and are pyogenic in nature, although ineffectively treated cases may progress to systemic manifestations.
- The extreme dysregulation of the immune system caused by some of the *S. aureus* toxins result in a cytokine storm, responsible for the wide-spread inflammatory damage characteristic of *S. aureus* infections.
- Although the clinical manifestations of *S. aureus* infections vary depending on the infecting strain, the site of infection, and the tissue or organ system affected, fever, malaise, chills, and cardinal inflammatory signs, such as swelling, pain, and redness of the affected area are common manifestations.
- TSS and sepsis are potentially fatal complications of *S. aureus* infections.
- A diagnosis of *S. aureus* infections relies on patient history, clinical manifestations, and physical findings. Culture sensitivity reports are important for confirmation of the primary diagnosis and for guiding the course of treatment.
- The emergence of multidrug-resistant strains of *S. aureus*, such as MRSA, poses a huge challenge for the effective clinical treatment of *S. aureus* infections.
- Prevention of *S. aureus* infections is challenging and relies solely on the adherence to appropriate asepsis protocols, effective disinfection, sterilization, and hygienic practices.

Problem 7.4: Septicemia caused by *Acinetobacter baumannii* infection

A 69-year-old woman presented with a fractured hip following a traumatic fall in the shower. The patient had a history of osteoporosis, hypertension, diabetes, and hypercholesterolemia. A trauma CT scan was performed to rule out any other pelvic or abdominal abnormalities and to analyze the nature of the hip fracture. A hip replacement surgery was recommended to the patient, and she was counseled about the outcome of the surgery and the possible risks involved. The patient was admitted, and her vitals, serological parameters, and biochemical parameters were investigated as part of the preanesthetic check-up. Low hemoglobin levels warranted the need for transfusion comprising of 1 unit of packed red blood cells and 1 unit of fresh frozen plasma preceding the surgery. During the surgery, further administration of 1 unit of packed red blood cells was required.

The surgery was uneventful, and the patient was recovering well. As the hemoglobin levels were still low (8.2 g/dL), another unit of packed red blood cells was administered on the third postoperative day. The patient, however, became febrile and experienced severe rigors on the fifth postoperative day.

The following are the reports of the investigations carried out:

1. Vitals

Parameters	Fifth postoperative day	Sixth postoperative day	Eight postoperative day
Temperature (°C)	38	38.4	38.8
Blood pressure (mmHg)	140/90	100/70	75/45
Pulse (/minute)	105	100	90

2. Physical examination (fifth postoperative day)
   Chest: Normal respiratory sounds
   Surgical site: No indication of infection or inflammation
   On the eighth postoperative day, the patient presented with dysuria, fatigue, and increased intensity of rigors.
3. Urine analysis (culture & sensitivity (C&S))
   Fifth postoperative day: Normal
   Eighth postoperative day: 24-hour urinary output—2700 mL; fresh sample collected for C&S: pandrug-resistant *A. baumannii* isolated
4. Inflammatory markers (sixth postoperative day)

Inflammatory markers	Sixth postoperative day	Ninth postoperative day	Reference values
C-reactive protein (mg/L)	45	325	<10
WBCs (/μL)	12,000	21,000	4500–11,000
Erythrocyte sedimentation rate (ESR) (mm/hour)	42	95	0–29

On the ninth postoperative day, the patient developed the onset of septic shock.

5. Blood culture (sixth and eighth postoperative day): pandrug-resistant *A. baumannii* isolated
6. Radiology
   Chest X-ray (eighth postoperative day): Normal

Diagnosis

The final diagnosis concluded septicemia caused by a nosocomial infection of *A. baumannii*.

Treatment

The patient was put on a regimen, including the following antibiotics from the ninth postoperative day:

1. Rifampicin—600 mg once daily
2. Tigecycline—100 mg twice daily
3. Colistin—4.5×10^6 IU twice daily

However, as the patient's condition continued to be critical, the antibiotic regimen was modified on the 11th postoperative day to replace colistin with vancomycin (1000 mg twice daily).

The modified antibiotic regimen resulted in an improvement in the patient's health and the inflammatory markers reduced drastically from the 15th postoperative day. The antibiotic regimen was administered for 14 days. The patient was afebrile by the 19th postoperative day and continued to be afebrile for a week. During this period, three consecutive negative blood culture reports were obtained. The patient was discharged a month after her admission.

Question and answer

1. What does the abbreviation "ESKAPE" stand for?
   ESKAPE is an abbreviation for a group of six human pathogens categorized as "red alert" pathogens, owing to the invasive nature and emerging MDR strains. *Enterococcus faecium, S. aureus, Klebsiella pneumoniae, A. baumannii, Pseudomonas aeruginosa,* and *Enterobacter spp* are the organisms that comprise the "ESKAPE" group.
2. Why is *A. baumannii* also known as "Iraqibacter"?
   The prevalence of *A. baumannii* infections in the injured soldiers of the armed forces deployed to conflict zones, especially Iraq, has led to the bacteria being commonly known as the "Iraqibacter."
3. What is the role played by the pili present on the bacterial surface in the pathogenicity of *A. baumannii*?
   Fimbriae and pili present on the surface of *A. baumannii* are responsible for biofilm formation and adherence to abiotic surfaces, such as those of medical interventional devices. Pili, therefore, facilitate the survival and spread of the bacteria in the hospital environment.
4. How do OMPs such as OmpA contribute to the pathogenicity of *A. baumannii*?
   OmpA plays a role in complement evasion and biofilm formation, but primarily elicits host cell apoptosis. OmpA binds to the epithelium and mitochondria of the host cell, which induces mitochondrial dysfunction and swelling. Cytochrome c, a heme protein, is released from the dysfunctional mitochondria, thereby leading to the formation of apoptosome, which ultimately leads to cell apoptosis.
5. How do phospholipase D and phospholipase C contribute to the virulence of *A. baumannii*?
   Phospholipase D mediates pathogenesis by ensuring resistance to the immune components of the host serum and facilitating epithelial cell evasion, while phospholipase C enhances the cytotoxicity abilities, thereby increasing the damage caused to the epithelial cells.
6. What are OMVs?
   OMVs are spherical vesicles secreted by the bacterial membrane that act as delivery vehicles for several bacterial virulence effectors to host cells, thereby ensuring host – pathogen interaction without close contact between the bacteria and host cells. The OMVs secreted by *A. baumannii* effectively deliver several virulence factors, including porins, such as OmpA, proteases, and phospholipases via lipid rafts to host cells, which induces cytotoxicity. OMVs are also known to elicit a potent host innate immune response through the induction of

proinflammatory cytokines expression and have also been implicated in horizontal transfer of some antibiotic resistance genes.

7. What are the various factors that contribute to antibiotic resistance in *A. baumannii*?

   ß-lactamases, dihydrofolate reductases, and aminoglycoside-modifying enzymes, such as acetyltransferases, adenyltransferases, and phosphotransferases, efflux pumps, like AmvA, porins, and certain other envelope components, for example: lipopolysaccharides (LPSs) and peptidoglycans, modification of antibiotic target sites, and ribosomal protection, contribute to antibiotic resistance in *A. baumannii*.

8. What are the various immune defense mechanisms that attempt to prevent the establishment of *A. baumannii* infections?

   The host immune system responds to *A. baumannii* infections rapidly through interaction of the bacteria with the pattern recognition receptors (PRRs) and induces the production of proinflammatory cytokines and chemokines, such as TNF-α, IL-6, IL-1α, IL-1ß, MIP-2, CCL5, IL-8, IL-10, CXCL-1, MCP-1, IFN-γ, IL-33, and IL-17. Large numbers of macrophages and neutrophils are locally recruited, which attempt to combat the infection at the initial stages. NLRP3 inflammasomes have a critical role in innate defense mechanisms against *A. baumannii* and are responsible for the production of IL-1ß and IL-18.

   Complement-mediated effector mechanisms are also essential in the host defense against *A. baumannii* infections. The interactions of the numerous complement proteins in a regulated cascade results in bacterial elimination through lysis or opsonization of the bacterial cells. The adaptive immune system is responsible for the development of antigen specific T-cell and antibody responses. The specific adaptive immunity is triggered by the effective antigen presentation and the induction of an array of proinflammatory cytokines. While *A. baumannii* infections show high levels of Th1 responses, Th2 responses are relatively lower, indicating a protective role of the Th2 responses.

9. How does *A. baumannii* evade the complement system?

   A. baumannii strains evade the complement system through the acquisition of the alternative complement pathway inhibitor H on the cell surface or through complement system regulators such as OMPs and LPS.

Learning objectives

1. Describe the characteristics of *A. baumannii*.
2. Summarize the pathogenesis of *A. baumannii* infections.
3. Describe the biochemical and pathological changes seen in septicemia.

4. Enumerate the methods involved in the diagnosis of *A. baumannii* infections.
5. Determine the prognosis of *A. baumannii*-associated septicemia.

Discussion

Differential diagnosis

Burkholderia cepacia infection	*B. cepacia* infections present themselves with a fever, cough, signs of respiratory distress such as congestion, shortness of breath, and wheezing.
Nosocomial meningitis	Nosocomial meningitis usually causes a fever, headaches, neck stiffness, and alteration of consciousness.
P. aeruginosa infection of the surgical site	*P. aeruginosa* infection of surgical site is characterized by fruity-smelling discolored pus discharge from the surgical site, redness, rash, pain, and fever with/without chills.
Serratia marcescens infection	*S. marcescens* infection develops as a fever, signs of urinary tract infection, chills, respiratory distress, and indications of septic shock.
Stenotrophomonas maltophilia infection	*S. maltophilia* infections are characterized by a fever, rapid heart rate, low blood pressure, abdominal pain, diarrhea, nausea, vomiting, and an altered mental status.

Background

A. baumannii is a clinically significant pathogen, notorious for its remarkable ability to acquire resistance to antibiotics. The extensive antibiotic resistance spectrum of this organism has resulted in it being designated as the "red alert" human pathogen, which raises an alarm in the medical fraternity.[54] The challenge posed by the emerging antibiotic resistance is compounded by the organism's ability to survive in hospitals, especially in aquatic environments for long periods of time, thereby enhancing its nosocomial spread. The organism usually targets high-risk individuals who are hospitalized, especially in the intensive care units (ICUs). Breaches in intact skin integrity and medical interventional devices, such as catheters, tracheal tubes, nasogastric tubes, ventilators, dialyzers, etc., that compromise the primary defense barriers increase vulnerability to *A. baumannii* infections. Although *A. baumannii* has been primarily implicated in hospital-acquired pneumonia, both nosocomial and community-acquired infections of the CNS, skin, soft tissues, and bone are also being increasingly reported in recent times. *A. baumannii* has been categorized as a common and serious MDR pathogen along with other such pathogens under the acronym, "ESKAPE" standing for *E. faecium, S. aureus, K. pneumoniae, A. baumannii, P. aeruginosa,* and *Enterobacter spp.*[55]

Etiological agent

Acinetobacter baumannii

Classification and morphology

A. baumannii is a Gram-negative, strictly aerobic cocobacilli (Fig. 7.19) belonging to the *Acinetobacter* genus. These are nonfastidious, nonmotile, catalase-positive, oxidase-negative, and nonfermenting organisms. Although the organism is Gram-negative, difficulty involved in de-staining often results in incorrect interpretation of the organism as being Gram-positive. A lack of specific metabolic tests that distinguish *A. baumannii* from other nonfermenting Gram-negative bacteria and ineffective phenotypic tests has warranted the

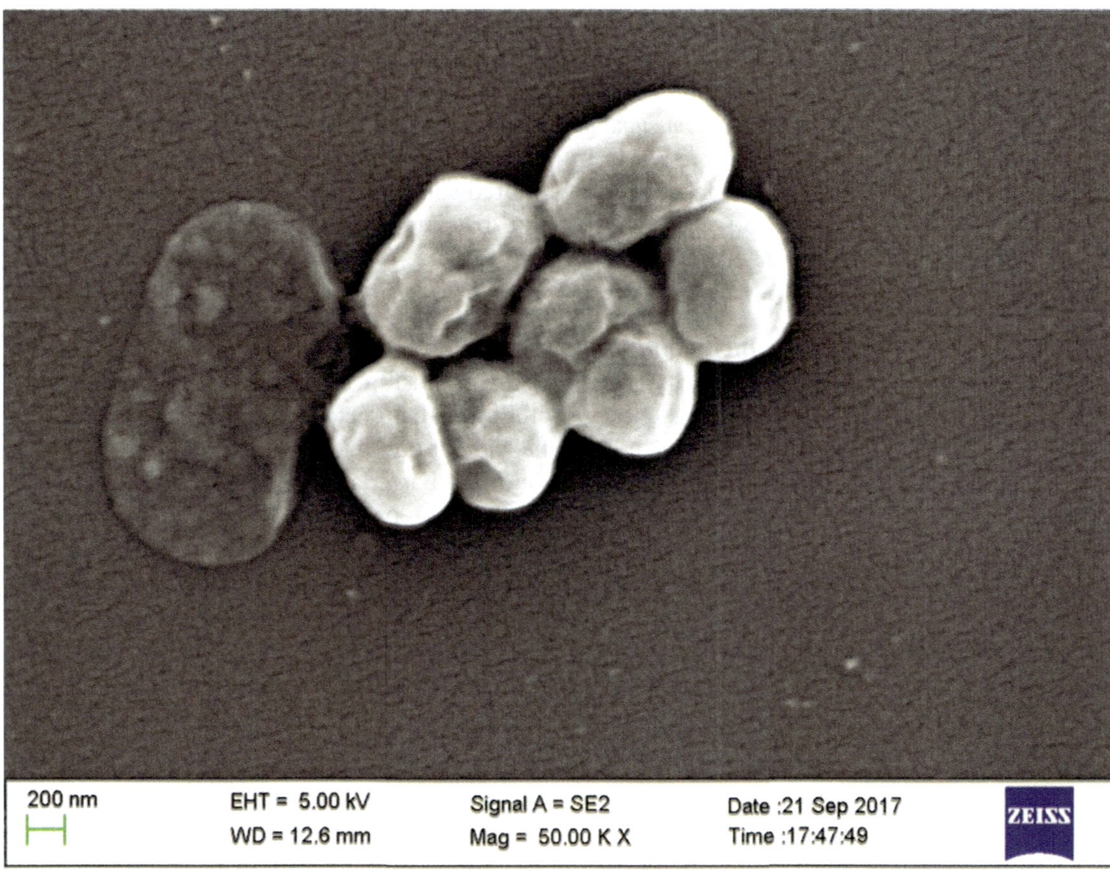

FIGURE 7.19

Scanning electron microscopic image of *Acinetobacter baumannii.* Source: *Vader1941/Visualized on SEM-EDS machine at CeNSE, IISc, Bangalore/; https://creativecommons.org/licenses/by-sa/4.0/deed.en; https://commons.wikimedia.org/wiki/File:A.baumannii_visualized_using_Scanning_electron_microscopy.png.*

need for more advanced molecular diagnostic methods for identifying the various *Acinetobacter* species. Distinction between the various *Acinetobacter* species is important for effective clinical treatment in the face of extensive antibiotic resistance.

Epidemiology

Unlike most species of the *Acinetobacter* genus, *A. baumannii* is not ubiquitous, but has high affinity toward aquatic environments.[56] Although *A. baumannii* has been extensively implicated in skin infections, it generally does not form a part of the normal microflora.[57] The isolation of *A. baumannii* from body lice suggests another potentially important reservoir for the pathogen.[58] *A. baumannii* specifically targets and colonizes moist tissues or areas of the skin prone to the accumulation of sweat, such as the axilla, toe webs, and groin.[59] *A. baumannii* is an opportunistic pathogen and the pathogenicity is often compounded by copathogens, such as *Candida albicans*, *Enterococcus faecalis*, and *K. pneumoniae*, which facilitate the entry of *A. baumannii* into the bloodstream, resulting in potentially fatal septicemia.[60] The prevalence of *A. baumannii* infections in the injured soldiers deployed to conflict zones, particularly Iraq, has led to the bacteria being commonly known as the "Iraqibacter."[61]

Pathogenicity and pathogenesis

The pathogenicity of *A. baumannii* is a function of several virulence factors. The mechanisms elicited by these virulent factors contribute to the pathogenesis of *A. baumannii*. The LPS and lipo-oligosaccharide that present themselves in the outer membrane of the bacteria impart antibiotic resistance and desiccation tolerance.[62,63] Besides, these bacteria also have an external protective capsule that helps them evade the complement-mediated immune responses of the host.[64] Pili on the surface of *A. baumannii* are responsible for biofilm formation and adherence to abiotic surfaces, such as those of medical interventional devices.[63]

These bacteria also secrete various proteins that contribute to antibiotic resistance. OmpA is one such protein belonging to the family of OMPs, also known as porins, which contributes significantly to the pathogenicity of *A. baumannii*. Complement evasion and biofilm formation are two of the most instrumental strategies for the survival of the pathogen both within the host and outside in the inanimate and hostile environments. Although OmpA complements evasion and biofilm formation,[65,66] the protein primarily elicits host cell apoptosis. OmpA binds to the epithelium and mitochondria of the host cell, which induces mitochondrial dysfunction and swelling. Cytochrome c, a heme protein, is released from the dysfunctional mitochondria, thereby leading to the formation of apoptosome, which ultimately leads to cell apoptosis.[67]

The ability of the bacteria to form biofilms is attributed to the availability of nutrients, several Omps, pili, and various macromolecular secretions. Pili and the biofilm-associated protein (BAP) are largely responsible for the initiation of biofilm production and maturation following the adherence of *A. baumannii* to abiotic surfaces, such as medical interventional devices and biotic surfaces, like epithelial cells. Biofilm formation on abiotic surfaces is preceded by the formation of microcolonies facilitated by the attachment of pili to the surfaces. BAP present on the bacterial cell surface, on the other hand, contribute to biofilm development and maturation by mediating the stabilization of the mature biofilm on abiotic or biotic surfaces. The formation of biofilms is also influenced by certain environmental signals, such as metal cations.[63] Along with pili and OmpA, fimbriae are also expressed on the surface of the bacterial cells, which are also known to significantly contribute to the process of bacterial adhesion to the host epithelial cells.

Phospholipase D and C are also some key proteins that contribute to the virulence of *A. baumannii*. While phospholipase D mediates pathogenesis by resisting to the immune components of the host serum and facilitating epithelial cell evasion,[68] phospholipase C enhances the cytotoxicity abilities, thereby increasing the damage caused to the epithelial cells.[69]

OMVs are spherical vesicles secreted by the bacterial membrane. These OMVs act as delivery vehicles for several bacterial virulence effectors to host cells, thereby ensuring host – pathogen interaction without close contact between the bacteria and host cells.[70,71] The OMVs secreted by *A. baumannii* effectively deliver several virulence factors, including porins, such as OmpA, proteases, and phospholipases via lipid rafts, to host cells, which induces cytotoxicity.[72–74] Studies have also indicated that OMVs elicit a potent host innate immune response through the induction of proinflammatory cytokines expression.[71] Therefore the role of OMVs in the cytotoxicity-based pathogenesis of *A. baumannii* is indicated.[71,75] OMVs have also been implicated in the horizontal transfer of some antibiotic resistance genes, and are therefore associated with antibiotic resistance in *A. baumannii*.[76]

Metal acquisition systems are important for the survival of *A. baumannii* in the host system. Acinetobactin, a siderophore, acts as a virulence factor for *A. baumannii* by ensuring the availability of ferric iron required for the bacteria to persist within epithelial cells and induce cytotoxicity.[77–79] The NfuA Fe-S scaffold protein significantly influences the host cell responses to iron chelation and oxidative stress, and has therefore been identified as a virulence factor in *A. baumannii*.[80] Calprotectin is a host metal-chelating protein that limits the availability of zinc and manganese to pathogenic bacteria, thereby inhibiting their growth.[81] ZnuABC is a zinc acquisition system used by *A. baumannii* to overcome zinc limitation and thereby contributes to pathogenesis. ZigA, a metallochaperone has been found to interact tightly with Zn and is important for the bacterial growth and dissemination of *A. baumannii* infection under Zn-starvation conditions.[82] A Mn transporter belonging to the natural resistance-associated macrophage proteins family and a urea carboxylase enzyme (MumC) are important for the growth and virulence of *A. baumannii* in Mn-limited conditions and for urea utilization as a source of nitrogen.[83]

Several protein secretion systems such as the type II secretion system comprising of the T2SS multiprotein complex, the type VI secretion system comprising of the T6SS, and the type V secretion system comprising of the autotransporter Ata have also been identified as important virulence-mediating factors in *A. baumannii.*[84]

Studies have also indicated the PBP 7/8 and the ß-lactamase, PER-1 as *A. baumannii* virulence factors.[85,86] CipA and Tuf are plasminogen-binding proteins that have a complement inhibitory effect, thus contributing to the complement resistance of *A. baumannii.*[87,88] The surface antigen protein, SurA1[89] and the RecA protein[90] responsible for homologous recombination and the SOS response to DNA damage has been found to be important for the fitness and virulence of *A. baumannii.*

Antibiotic resistance

A. baumanni infections pose a formidable clinical challenge owing to the involvement of MDR and pan-drug resistant (PDR) strains. The organism's ability to adapt and evolve rapidly in accordance with the increasing environmental hostilities enables its survival in extreme conditions.

Induction and upregulated expression of all four groups (A, B, C, and D) of ß-lactamases facilitates resistance to penicillins, cephalosporins, cephamycins, and carbapenems.[62] On the other hand, resistance to drugs such as imipenem and tigecycline that are heavily depended upon for clinical treatment of *A. baumannii* infections is a result of efflux pumps.[91,92] Certain efflux pumps have also been implicated in increased risk for biofilm formation caused by low-dose antimicrobial therapy of *A. baumannii* infections.[93] AmvA is an efflux pump that mediates resistance to various detergents, dyes, and disinfectants such as benzalkonium chloride and methyl viologen in addition to antibiotics like erythromycin and acriflavine.[94] Several other efflux pumps have been identified in *A. baumannii* that confer resistance to clinically significant antimicrobial agents such as chloramphenicol, tetracycline, fosfomycin, fluroquinolones, nalidixic acid, netilmicin, tobramycin, gentamicin, kanamycin, chloroxylenol, oxytetracycline, 1,10-phenanthroline, sulfonamide sulfathiazole, and fusidic acid.[62]

Porins influence the permeability of the outer envelope in *A. baumannii*, and play an important role in *A. baumannii* virulence and antibiotic resistance. While porins are implicated in resistance to carbapenems, aztreonam, chloramphenicol, and nalidixic acid,[95–97] certain other envelope components such as LPSs and peptidoglycans influence resistance to colistin in *A. baumannii*.[98]

Acetyltransferases, adenyltransferases, and phosphotransferases are aminoglycoside-modifying enzymes that are highly prevalent in *A. baumannii*, and have been found to confer significant resistance to aminoglycosides.[99,100] *A. baumannii* induces resistance to quinolones through modification of antibiotic target sites, ribosomal protection provides resistance to tetracycline, and dihydrofolate reductases produced by the bacteria mediate resistance to trimethoprim.

Immune responses

A. baumannii is an opportunistic pathogen that does not cause clinically significant disease in immunocompetent individuals; however, immunocompromised individuals and those hospitalized for long periods of time, especially in the ICUs, are at high risk of *A. baumannii* infections. The host immune system responds to *A. baumannii* infections rapidly through interaction of the bacteria with the PRRs, and induces the production of proinflammatory cytokines and chemokines such as TNF-α, IL-6, IL-1α, IL-1ß, MIP-2, CCL5, IL-8, IL-10, CXCL-1, MCP-1, IFN-γ, IL-33, and IL-17. Large numbers of macrophages and neutrophils are locally recruited, which attempt to combat the infection at the initial stages. Strong neutrophil-mediated immune responses to *A. baumannii* infections have been indicated by some studies.[101] However, the bacterial ability to evade the host immune system and high virulence levels of certain *A. baumannii* strains may result in the establishment of a successful infection in spite of the host defenses.[102] Toll-like receptors, TLR2 and TLR4 are the most significantly interacting neutrophil-to-lymphocyte ratios (NLRs) in *A. baumannii* infections. TLR-9 mediated detection of *A. baumannii* has also been suggested to be an important indicator of susceptibility to *A. baumannii* infections.[103] Cytosolic PRRs such as NOD1 and NOD2 play an important role in the recognition of intracellular *A. baumannii*.[104] NLRP3 inflammasomes have a critical role in innate defense mechanisms against *A. baumannii*, and are responsible for the production of IL-1ß and IL-18.[105]

Complement-mediated effector mechanisms play an important role in host defense against *A. baumannii* infections. The interactions of the numerous complement proteins in a regulated cascade results in bacterial elimination through lysis or opsonization of the bacterial cells. C3 deposition in *A. baumannii* strains has indicated the involvement of the alternative complement pathway in complement-mediated host defense mechanisms.[106] However, *A.*

baumannii strains can evade the complement system through the acquisition of the alternative complement pathway inhibitor H on the cell surface,[66] or through complement system regulators such as OMPs and LPS.[102]

While the innate immune responses control the extent of bacterial replication and are responsible for the local inflammatory responses, the adaptive immune system is responsible for the development of antigen-specific T-cell and antibody responses. The specific adaptive immunity is triggered by the effective antigen presentation and the induction of an array of proinflammatory cytokines. Significant levels of specific IgM, IgG, and IgA responses have been noted in *A. baumannii* infections.[107] T-cell responses have been suggested to play a helper role in the induction of antibody responses against *A. baumannii*. While *A. baumannii* infections show high levels of Th1 responses, Th2 responses are relatively lower, thereby indicating a protective role of the Th2 responses.[108]

As *A. baumannii* infections are generally seen in immunocompromised individuals, the possibility of an ineffective induction of adaptive immunity due to a weak innate immune response should be considered.[102]

Clinical manifestations

A. baumannii infections are usually associated with prolonged hospitalization, especially in the ICU requiring the use of medical interventional devices such as nasogastric tubes, ventilators, catheters etc. Antibiotic abuse or prolonged use of antibiotic therapy also predisposes an individual to *A. baumannii* infection.

As the lung is the most commonly *A. baumannii* infected organ in ICU patients on mechanical ventilation, pneumonia is the most common clinical manifestation of *A. baumannii* infections. In addition, *A. baumannii* is also implicated in urinary tract infections, catheter-associated bacteremia, hospital-acquired meningitis, and wound infection.[109] The symptoms of the disease vary depending upon the infected organ system, but the constellation usually comprises of fever, chills, pain in the affected area, dyspnea, arrythmias, pus formation in case of wound infections, alteration of mental status, and convulsions. Although most *A. baumannii* infections are hospital-acquired, community-acquired infections may also occur in high-risk groups of individuals.

Diagnosis

The pathological findings in *A. baumannii* infections cannot be easily differentiated from infections caused by other Gram-negative bacteria such as *S. marcescens*, *Enterobacter aerogenes*, *B. cepacia*, and *P. aeruginosa*. However, this differentiation is clinically significant to ensure the implementation of targeted and effective therapeutic measures. In addition, it is clinically important to prove the direct implication of *A. baumannii* as the etiology of the clinical manifestations in a particular case because although *A. baumannii* colonization is quite common in the hospital environment, such colonization rarely results in infection. Therefore correlating laboratory and microbiological findings with the clinical history and evaluation is very important for ensuring an accurate diagnosis of *A. baumannii* infections.[109]

Elevated inflammatory markers such as leukocytes, erythrocyte sedimentation rate, and C-reactive protein in addition to adequately indicative patient history and clinical presentation of the particular case may be suggestive of *A. baumannii* infections, especially if there have been more cases of proven *A. baumannii* infections in the same hospital set-up. Imaging studies depending on the clinical manifestations are also useful in getting a better perspective of the damage caused by the infection, which may in turn indicate the etiological agent involved in the infection. Culture studies of the urine, blood, and CSF are effective indicators of urinary tract infections, bloodstream infections, and infections of the CNS, respectively. Histopathological examinations, however, are ineffective in differentiating the findings of *A. baumannii* infections from infections caused by other Gram-negative bacteria.

Treatment and management

Aminoglycosides, fourth-generation cephalosporins, tigecycline, carbapenems, and rifampicin are the most commonly used therapeutic antibiotics used to treat *A. baumannii* infections. Treatment with macrolides, third-generation cephalosporins, and penicillin is usually found to make the infection irretractable. As *A. baumannii* infections are often associated with external medical interventional devices such as shunts, catheters, drains, pacemakers, nasogastric tubes, and ventilators, removal of these devices is recommended to ensure effectivity of the therapeutic measures. Surgical debridement of abscesses and necrotic tissues is also indicated in some cases of wound infection.

However, the emergence of MDR and PDR strains of *A. baumannii* has made it increasingly challenging to treat infections. Therefore drug sensitivity reports play a crucial role in ensuring the implementation of the most effective therapeutic methods. It is also important to avoid the use of antibiotics to treat colonization in the absence of infections due to the increased risk of antibiotic resistance. Antibiotic therapies depending on the sensitivity reports usually require to be prescribed for a duration of up to 7 days.

Certain alternative therapeutic and management strategies including the use of bacteriophages, bactericidal gene transfer therapy, cathelicidins, radioimmunotherapy, photodynamic therapy, and nanoparticle technology have been proposed for the treatment and control of MDR and PDR *A. baumannii* infections.[61]

Student Notes

Key take-aways

- *A. baumannii* has been categorized as a "red alert" human pathogen along with other such clinically significant pathogens such as *E. faecium, S. aureus, K. pneumoniae,* and *P. aeruginosa* under the acronym "ESKAPE."
- *A. baumannii* usually causes colonization of biotic and abiotic surfaces in the hospital environment, especially in moist conditions.
- The ability of this organism to form biofilms on abiotic surfaces of medical interventional devices makes it a threat to morbid patients during prolonged hospitalizations, especially in the ICUs.
- Although hospital-acquired *A. baumannii* infections are the most common, community-acquired cases of infection are also seen in immunocompromised individuals.
- While *A. baumannii* colonization is common in hospitals, infection is relatively rare and is generally restricted to immunocompromised and morbid patients.
- *A. baumannii* infections may manifest as urinary tract infections, infections of the CNS, wound infections, respiratory tract infections, or blood stream infections depending upon the affected organ system.
- Diagnosis of *A. baumannii* infections is a clinical challenge because of the pathological similarities with infections caused by other Gram-negative bacteria.
- Ensuring that *A. baumannii* is responsible for the clinical manifestations in a particular case is very important, as asymptomatic *A. baumannii* colonization is common and should not be treated with antibiotics due to the risk of antibiotic resistance.
- The emergence of MDR and PDR strains of *A. baumannii* poses a formidable challenge in the path of effective clinical treatment of these infections.

References

1. Sethia PK, Thukrala R, Sethib NK, Torgovnickc J, Arsurad E, Wasterlainb A. Meningococcal disease: a case report and discussion of clinical presentation and management. *East J Med.* 2012;17:40–45.
2. Stephens DS, Greenwood B, Brandtzaeg P. Epidemic meningitis, meningococcemia, and *Neisseria meningitidis*. *Lancet.*. 2007;369:2196–2210.
3. Rouphael NG, Stephens DS. *Neisseria meningitidis*: biology, microbiology, and epidemiology. *Methods Mol Biol.* 2012;799:1–20. Available from: https://doi.org/10.1007/978-1-61779-346-2_1.
4. McNamara LA, Potts C, Blain AE, et al. Detection of ciprofloxacin-resistant, β-lactamase-producing *Neisseria meningitidis* serogroup Y isolates—United States, 2019–2020. *MMWR Morb Mortal Wkly Rep.* 2020;69:735–739. Available from: http://doi.org/10.15585/mmwr.mm6924a2externalicon.

5. Stephens DS. Biology and pathogenesis of the evolutionarily successful, obligate human bacterium *Neisseria meningitidis*. *Vaccine*. 2009;27(2):B71–B77. Available from: https://doi.org/10.1016/j.vaccine.2009.04.070.
6. Swartley JS, Marfin AA, Edupuganti S, et al. Capsule switching of *Neisseria meningitidis*. *Proc Natl Acad Sci USA*. 1997;94(1):271–276. Available from: https://doi.org/10.1073/pnas.94.1.271.
7. Van Deuren M, van Dijke BJ, Koopman RJ, et al. Rapid diagnosis of acute meningococcal infections by needle aspiration or biopsy of skin lesions. *BMJ*. 1993;306:1229–1232.
8. Coureuil M, Join-Lambert O, Lécuyer H, Bourdoulous S, Marullo S, Nassif X. Pathogenesis of meningococcemia. *Cold Spring Harb Perspect Med*. 2013;3(6):a012393. Available from: https://doi.org/10.1101/cshperspect.a012393. Published June 1, 2013.
9. van de Beek D, de Gans J, Spanjaard L, et al. Clinical features and prognostic factors in adults with bacterial meningitis. *N Engl J Med*. 2004;351:1849.
10. Stein-Zamir C, Abramson N, Zentner G, et al. Invasive meningococcal disease in children in Jerusalem. *Epidemiol Infect*. 2008;136:782–789.
11. London V, Aronowitz P. Purpuric rash of meningococcemia. *J Hosp Med*. 2008;3:169.
12. Pizza M, Rappuoli R. *Neisseria meningitidis*: pathogenesis and immunity. *Curr Opin Microbiol*. 2015;23:68–72. Available from: https://doi.org/10.1016/j.mib.2014.11.006.
13. Visintin C, Mugglestone MA, Fields EJ, et al. National Institute for Health and Clinical Excellence. Management of bacterial meningitis and meningococcal septicemia in children and young people: summary of NICE guidance, Guideline Development Group*BMJ*. 2010;340:c3209. Available from: https://doi.org/10.1136/bmj.c3209.
14. Finne J, Bitter-Suermann D, Goridis C, Finne U. An IgG monoclonal antibody to group B meningococci cross-reacts with developmentally regulated polysialic acid units of glycoproteins in neural and extraneural tissues. *J Immunology*. 1987;138(12):4402–4407. June 15.
15. Holst J, Martin D, Arnold R, et al. Properties and clinical performance of vaccines containing outer membrane vesicles from *Neisseria meningitidis*. *Vaccine*. 2009;27(2):B3–B12. Available from: https://doi.org/10.1016/j.vaccine.2009.04.071. ISSN 0264-410X.
16. Shao ZJ, Li YX. Study progress on serogroup B meningococcal vaccine. *Zhongguo Yi Miao He Mian Yi*. 2009;15:542–546.
17. Aikawa M, Miller LH, Rabbege J. Caveola–vesicle complexes in the plasmalemma of erythrocytes infected by *Plasmodium vivax* and *P. cynomolgi*. Unique structures related to Schüffner's dots. *Am J Pathol*. 1975;79(2):285–300.
18. Udagama PV, Atkinson CT, Peiris JS, David PH, Mendis KN, Aikawa M. Immunoelectron microscopy of Schüffner's dots in *Plasmodium vivax*-infected human erythrocytes. *Am J Pathol*. 1988;131(1):48–52.
19. Gunalan K, Lo E, Hostetler JB, et al. Role of *Plasmodium vivax* Duffy-binding protein 1 in invasion of Duffy-null Africans. *Proc Natl Acad Sci USA*. 2016;113(22):6271–6276. Available from: https://doi.org/10.1073/pnas.1606113113.
20. Price RN, von Seidlein L, Valecha N, Nosten F, Baird JK, White NJ. Global extent of chloroquine-resistant *Plasmodium vivax*: a systematic review and *meta*-analysis. *Lancet Infect Dis*. 2014;14(10):982–991. Available from: https://doi.org/10.1016/S1473-3099(14)70855-2. Epub September 8, 2014.
21. Baird JK, Basri H, Purnomo, et al. Resistance to chloroquine by *Plasmodium vivax* in Irian Jaya, Indonesia. *Am J Trop Med Hyg*. 1991;44(5):547–552. Available from: https://doi.org/10.4269/ajtmh.1991.44.547.

22. Price RN, Auburn S, Marfurt J, Cheng Q. Phenotypic and genotypic characterisation of drug-resistant *Plasmodium vivax. Trends Parasitol.* 2012;28(11):522–529. Available from: https://doi.org/10.1016/j.pt.2012.08.005. Epub October 5, 2012.

23. Carlton JM. Malaria parasite evolution in a test tube. *Science.* 2018;359(6372):159–160. Available from: https://doi.org/10.1126/science.aar4189.

24. Garcia LS. Malaria. *Clin Lab Med.* 2010;30(1):93–129. Available from: https://doi.org/10.1016/j.cll.2009.10.001.

25. Baird JK. Evidence and implications of mortality associated with acute *Plasmodium vivax* malaria. *Clin Microbiol Rev.* 2013;26(1):36–57. Available from: https://doi.org/10.1128/CMR.00074-12.

26. Price RN, Tjitra E, Guerra CA, Yeung S, White NJ, Anstey NM. Vivax malaria: neglected and not benign. *Am J Trop Med Hyg.* 2007;77(6 Suppl):79–87.

27. Kochar DK, Saxena V, Singh N, Kochar SK, Kumar SV, Das A. *Plasmodium vivax* malaria. *Emerg Infect Dis.* 2005;11(1):132–134. Available from: https://doi.org/10.3201/eid1101.040519.

28. Val F, Machado K, Barbosa L, et al. Respiratory complications of *Plasmodium vivax* malaria: systematic review and *meta*-analysis. *Am J Trop Med Hyg.* 2017;97(3):733–743. Available from: https://doi.org/10.4269/ajtmh.17-0131.

29. Boyce MR, O'Meara WP. Use of malaria RDTs in various health contexts across sub-Saharan Africa: a systematic review. *BMC Public Health.* 2017;17(1):470. Available from: https://doi.org/10.1186/s12889-017-4398-1. Published May 18, 2017.

30. Abba K, Kirkham AJ, Olliaro PL, et al. Rapid diagnostic tests for diagnosing uncomplicated non-falciparum or *Plasmodium vivax* malaria in endemic countries. *Cochrane Database Syst Rev.* 2014;2014(12):CD011431. Available from: https://doi.org/10.1002/14651858.CD011431. Published December 18, 2014.

31. Fransisca L, Kusnanto JH, Satoto TB, et al. Comparison of rapid diagnostic test Plasmotec Malaria-3, microscopy, and quantitative real-time PCR for diagnoses of *Plasmodium falciparum* and *Plasmodium vivax* infections in Mimika Regency, Papua, Indonesia. *Malar J.* 2015;14:103. Available from: https://doi.org/10.1186/s12936-015-0615-5. Published March 5, 2015.

32. Tjitra E, Suprianto S, Dyer M, Currie BJ, Anstey NM. Field evaluation of the ICT malaria P.f/P.v immunochromatographic test for detection of *Plasmodium falciparum* and *Plasmodium vivax* in patients with a presumptive clinical diagnosis of malaria in eastern Indonesia. *J Clin Microbiol.* 1999;37(8):2412–2417. Available from: https://doi.org/10.1128/JCM.37.8.2412-2417.1999.

33. Pattanasin S, Proux S, Chompasuk D, et al. Evaluation of a new Plasmodium lactate dehydrogenase assay (OptiMAL-IT) for the detection of malaria. *Trans R Soc Trop Med Hyg.* 2003;97(6):672–674. Available from: https://doi.org/10.1016/s0035-9203(03)80100-1.

34. Dalrymple U, Arambepola R, Gething PW, Cameron E. How long do rapid diagnostic tests remain positive after anti-malarial treatment. *Malar J.* 2018;17(1):228. Available from: https://doi.org/10.1186/s12936-018-2371-9. Published 2018 Jun 8.

35. Recht J, Ashley EA, White NJ. Use of primaquine and glucose-6-phosphate dehydrogenase deficiency testing: divergent policies and practices in malaria endemic countries. *PLoS Negl Trop Dis.* 2018;12(4):e0006230. Available from: https://doi.org/10.1371/journal.pntd.0006230. Published April 19, 2018.

36. Baird JK, Hoffman SL. Primaquine therapy for malaria. *Clin Infect Dis.* 2004;39(9):1336–1345. Available from: https://doi.org/10.1086/424663. Epub October 12, 2004.

37. Hess KM, Goad JA, Arguin PM. Intravenous artesunate for the treatment of severe malaria. *Ann Pharmacother*. 2010;44(7-8):1250–1258. Available from: https://doi.org/10.1345/aph.1M732. Epub June 15, 2010.

38. Goodyer LI, Croft AM, Frances SP, et al. Expert review of the evidence base for arthropod bite avoidance. *J Travel Med*. 2010;17(3):182–192. Available from: https://doi.org/10.1111/j.1708-8305.2010.00402.x.

39. Lowy FD. *Staphylococcus aureus* infections. *N Engl J Med*. 1998;339(8):520–532. Available from: https://doi.org/10.1056/NEJM199808203390806.

40. Boucher HW, Corey GR. Epidemiology of methicillin-resistant *Staphylococcus aureus*. *Clin Infect Dis*. 2008;46(5):S344–S349. Available from: https://doi.org/10.1086/533590.

41. Berger S, Kunerl A, Wasmuth S, Tierno P, Wagner K, Brügger J. Menstrual toxic shock syndrome: case report and systematic review of the literature. *Lancet Infect Dis*. 2019;19(9): e313–e321. Available from: https://doi.org/10.1016/S1473-3099(19)30041-6. Epub May 28, 2019.

42. Tong SY, Davis JS, Eichenberger E, Holland TL, Fowler Jr. VG. *Staphylococcus aureus* infections: epidemiology, pathophysiology, clinical manifestations, and management. *Clin Microbiol Rev*. 2015;28(3):603–661. Available from: https://doi.org/10.1128/CMR.00134-14.

43. Bien J, Sokolova O, Bozko P. Characterization of virulence factors of *Staphylococcus aureus*: novel function of known virulence factors that are implicated in activation of airway epithelial proinflammatory response. *J Pathog*. 2011;2011:601905. Available from: https://doi.org/10.4061/2011/601905. Epub September 14, 2011.

44. Oogai Y, Matsuo M, Hashimoto M, Kato F, Sugai M, Komatsuzawa H. Expression of virulence factors by *Staphylococcus aureus* grown in serum. *Appl Env Microbiol*. 2011;77(22):8097–8105. Available from: https://doi.org/10.1128/AEM.05316-11.

45. Mertz PM, Cardenas TC, Snyder RV, Kinney MA, Davis SC, Plano LR. *Staphylococcus aureus* virulence factors associated with infected skin lesions: influence on the local immune response. *Arch Dermatol*. 2007;143(10):1259–1263. Available from: https://doi.org/10.1001/archderm.143.10.1259.

46. Ananthanarayan R, Paniker CK. 8th ed. *Textbook of Microbiology*. 22. India: Universities Press; 2009:195–202.

47. Centers for Disease Control and Prevention (CDC). Outbreaks of community-associated methicillin-resistant *Staphylococcus aureus* skin infections–Los Angeles County, California, 2002-2003. *MMWR Morb Mortal Wkly Rep*. 2003;52(5):88.

48. Rasigade JP, Vandenesch F. *Staphylococcus aureus*: a pathogen with still unresolved issues. *Infect Genet Evol*. 2014;21:510–514. Available from: https://doi.org/10.1016/j.meegid.2013.08.018. Epub August 29, 2013.

49. DeLeo FR, Diep BA, Otto M. Host defense and pathogenesis in *Staphylococcus aureus* infections. *Infect Dis Clin North Am*. 2009;23(1):17–34. Available from: https://doi.org/10.1016/j.idc.2008.10.003.

50. Salgado-Pabón W, Breshears L, Spaulding AR, et al. Superantigens are critical for *Staphylococcus aureus* infective endocarditis, sepsis, and acute kidney injury. *mBio*. 2013;4(4): e00494. Available from: https://doi.org/10.1128/mBio.00494-13. 13.

51. Musser JM, Schlievert PM, Chow AW, et al. A single clone of *Staphylococcus aureus* causes the majority of cases of toxic shock syndrome. *Proc Natl Acad Sci USA*. 1990;87(1):225–229. Available from: https://doi.org/10.1073/pnas.87.1.225. PMID: 1967495; PMCID: PMC53234.

52. Le KY, Otto M. Quorum-sensing regulation in staphylococci-an overview. *Front Microbiol*. 2015;6:1174. Available from: https://doi.org/10.3389/fmicb.2015.01174. Published October 27, 2015.

53. Taylor TA, Unakal CG. *Staphylococcus aureus* [Updated 2020 August 23, 2020]*StatPearls [Internet]*. Treasure Island (FL): StatPearls Publishing; 2020. Available from: https://www.ncbi.nlm.nih.gov/books/NBK441868/.

54. Cerqueira GM, Peleg AY. Insights into *Acinetobacter baumannii* pathogenicity. *IUBMB Life*. 2011;63(12):1055–1060. Available from: https://doi.org/10.1002/iub.533.

55. Rice LB. Federal funding for the study of antimicrobial resistance in nosocomial pathogens: no ESKAPE. *J Infect Dis*. 2008;197(8):1079–1081. Available from: https://doi.org/10.1086/533452.

56. Fournier PE, Richet H. The epidemiology and control of *Acinetobacter baumannii* in health care facilities. *Clin Infect Dis*. 2006;42(5):692–699. Available from: https://doi.org/10.1086/500202.

57. Seifert H, Dijkshoorn L, Gerner-Smidt P, Pelzer N, Tjernberg I, Vaneechoutte M. Distribution of Acinetobacter species on human skin: comparison of phenotypic and genotypic identification methods. *J Clin Microbiol*. 1997;35(11):2819–2825. Available from: https://doi.org/10.1128/JCM.35.11.2819-2825.1997.

58. La Scola B, Raoult D. *Acinetobacter baumannii* in human body louse. *Emerg Infect Dis*. 2004;10(9):1671–1673. Available from: https://doi.org/10.3201/eid1009.040242.

59. Centers for Disease Control and Prevention (CDC). *Acinetobacter baumannii* infections among patients at military medical facilities treating injured U.S. service members, 2002 – 2004. *MMWR Morb Mortal Wkly Rep*. 2004;53(45):1063–1066.

60. Sebeny PJ, Riddle MS, Petersen K. *Acinetobacter baumannii* skin and soft-tissue infection associated with war trauma. *Clin Infect Dis*. 2008;47(4):444–449. Available from: https://doi.org/10.1086/590568.

61. Howard A, O'Donoghue M, Feeney A, Sleator RD. *Acinetobacter baumannii*: an emerging opportunistic pathogen. *Virulence*. 2012;3(3):243–250. Available from: https://doi.org/10.4161/viru.19700.

62. Lee CR, Lee JH, Park M, et al. Biology of *Acinetobacter baumannii*: pathogenesis, antibiotic resistance mechanisms, and prospective treatment options. *Front Cell Infect Microbiol*. 2017;7:55. Available from: https://doi.org/10.3389/fcimb.2017.00055. Published March 13, 2017.

63. Harding CM, Hennon SW, Feldman MF. Uncovering the mechanisms of *Acinetobacter baumannii* virulence. *Nat Rev Microbiol*. 2018;16(2):91–102. Available from: https://doi.org/10.1038/nrmicro.2017.148.

64. Lees-Miller RG, Iwashkiw JA, Scott NE, et al. A common pathway for O-linked protein-glycosylation and synthesis of capsule in *Acinetobacter baumannii*. *Mol Microbiol*. 2013;89(5):816–830. Available from: https://doi.org/10.1111/mmi.12300.

65. Gaddy JA, Actis LA. Regulation of *Acinetobacter baumannii* biofilm formation. *Future Microbiol*. 2009;4(3):273–278. Available from: https://doi.org/10.2217/fmb.09.5.

66. Kim SW, Choi CH, Moon DC, et al. Serum resistance of *Acinetobacter baumannii* through the binding of factor H to outer membrane proteins. *FEMS Microbiol Lett*. 2009;301(2):224–231. Available from: https://doi.org/10.1111/j.1574-6968.2009.01820.x.

67. Choi CH, Lee EY, Lee YC, et al. Outer membrane protein 38 of *Acinetobacter baumannii* localizes to the mitochondria and induces apoptosis of epithelial cells. *Cell Microbiol*. 2005;7(8):1127–1138. Available from: https://doi.org/10.1111/j.1462-5822.2005.00538.x.

68. Jacobs AC, Hood I, Boyd KL, et al. Inactivation of phospholipase D diminishes *Acinetobacter baumannii* pathogenesis. *Infect Immun*. 2010;78(5):1952–1962. Available from: https://doi.org/10.1128/IAI.00889-09.

69. Camarena L, Bruno V, Euskirchen G, Poggio S, Snyder M. Molecular mechanisms of ethanol-induced pathogenesis revealed by RNA-sequencing. *PLoS Pathog*. 2010;6(4):e1000834. Available from: https://doi.org/10.1371/journal.ppat.1000834. Published April 1, 2010.

70. Ellis TN, Kuehn MJ. Virulence and immunomodulatory roles of bacterial outer membrane vesicles. *Microbiol Mol Biol Rev*. 2010;74(1):81–94. Available from: https://doi.org/10.1128/MMBR.00031-09.

71. Jun SH, Lee JH, Kim BR, et al. *Acinetobacter baumannii* outer membrane vesicles elicit a potent innate immune response via membrane proteins. *PLoS One*. 2013;8(8):e71751. Available from: https://doi.org/10.1371/journal.pone.0071751. Published August 14, 2013.

72. Kwon SO, Gho YS, Lee JC, Kim SI. Proteome analysis of outer membrane vesicles from a clinical *Acinetobacter baumannii* isolate. *FEMS Microbiol Lett*. 2009;297(2):150–156. Available from: https://doi.org/10.1111/j.1574-6968.2009.01669.x.

73. Jin JS, Kwon SO, Moon DC, et al. *Acinetobacter baumannii* secretes cytotoxic outer membrane protein A via outer membrane vesicles. *PLoS One*. 2011;6(2):e17027. Available from: https://doi.org/10.1371/journal.pone.0017027. Published February 28, 2011.

74. Moon DC, Choi CH, Lee JH, et al. *Acinetobacter baumannii* outer membrane protein A modulates the biogenesis of outer membrane vesicles. *J Microbiol*. 2012;50(1):155–160. Available from: https://doi.org/10.1007/s12275-012-1589-4.

75. Li ZT, Zhang RL, Bi XG, et al. Outer membrane vesicles isolated from two clinical *Acinetobacter baumannii* strains exhibit different toxicity and proteome characteristics. *Microb Pathog*. 2015;81:46–52. Available from: https://doi.org/10.1016/j.micpath.2015.03.009.

76. Rumbo C, Fernández-Moreira E, Merino M, et al. Horizontal transfer of the OXA-24 carbapenemase gene via outer membrane vesicles: a new mechanism of dissemination of carbapenem resistance genes in *Acinetobacter baumannii*. *Antimicrob Agents Chemother*. 2011;55(7):3084–3090. Available from: https://doi.org/10.1128/AAC.00929-10.

77. Penwell WF, Arivett BA, Actis LA. The *Acinetobacter baumannii* entA gene located outside the acinetobactin cluster is critical for siderophore production, iron acquisition and virulence. *PLoS One*. 2012;7(5):e36493. Available from: https://doi.org/10.1371/journal.pone.0036493.

78. Ali HM, Salem MZM, El-Shikh MS, Megeed AA, Alogaibi YA, Talea IA. Investigation of the virulence factors and molecular characterization of the clonal relations of multidrug-resistant *Acinetobacter baumannii* isolates. *J AOAC Int*. 2017;100(1):152–158. Available from: https://doi.org/10.5740/jaoacint.16-0139.

79. Gaddy JA, Arivett BA, McConnell MJ, López-Rojas R, Pachón J, Actis LA. Role of acinetobactin-mediated iron acquisition functions in the interaction of *Acinetobacter baumannii* strain ATCC 19606T with human lung epithelial cells, *Galleria mellonella* caterpillars, and mice. *Infect Immun*. 2012;80(3):1015–1024. Available from: https://doi.org/10.1128/IAI.06279-11.

80. Zimbler DL, Park TM, Arivett BA, et al. Stress response and virulence functions of the *Acinetobacter baumannii* NfuA Fe-S scaffold protein. *J Bacteriol*. 2012;194(11):2884–2893. Available from: https://doi.org/10.1128/JB.00213-12.

81. Corbin BD, Seeley EH, Raab A, et al. Metal chelation and inhibition of bacterial growth in tissue abscesses. *Science*. 2008;319(5865):962–965. Available from: https://doi.org/10.1126/science.1152449.

82. Hood MI, Mortensen BL, Moore JL, et al. Identification of an *Acinetobacter baumannii* zinc acquisition system that facilitates resistance to calprotectin-mediated zinc sequestration [published correction appears in PLoS Pathog. 2013;9(1). 10.1371/annotation/2968451e-04b8-4705-bee9-9e40bceffe67]. *PLoS Pathog*. 2012;8(12):e1003068. Available from: https://doi.org/10.1371/journal.ppat.1003068.

83. Juttukonda LJ, Chazin WJ, Skaar EP. *Acinetobacter baumannii* coordinates urea metabolism with metal import to resist host-mediated metal limitation. *mBio*. 2016;7(5):e01475–16. Available from: https://doi.org/10.1128/mBio.01475-16.

84. Weber BS, Harding CM, Feldman MF. Pathogenic Acinetobacter: from the cell surface to infinity and beyond. *J Bacteriol.* 2015;198(6):880–887. Available from: https://doi.org/10.1128/JB.00906-15. Published December 28, 2015.

85. Russo TA, MacDonald U, Beanan JM, et al. Penicillin-binding protein 7/8 contributes to the survival of *Acinetobacter baumannii* in vitro and in vivo. *J Infect Dis.* 2009;199(4):513–521. Available from: https://doi.org/10.1086/596317.

86. Sechi LA, Karadenizli A, Deriu A, et al. PER-1 type beta-lactamase production in *Acinetobacter baumannii* is related to cell adhesion. *Med Sci Monit.* 2004;10(6):BR180–BR184.

87. Koenigs A, Stahl J, Averhoff B, et al. CipA of *Acinetobacter baumannii* is a novel plasminogen binding and complement inhibitory protein. *J Infect Dis.* 2016;213(9):1388–1399. Available from: https://doi.org/10.1093/infdis/jiv601.

88. Koenigs A, Zipfel PF, Kraiczy P. Translation elongation factor Tuf of *Acinetobacter baumannii* is a plasminogen-binding protein [published correction appears in PLoS One. 2015;10(9):e0138398]*PLoS One.* 2015;10(7):e0134418. Available from: https://doi.org/10.1371/journal.pone.0134418.

89. Liu D, Liu ZS, Hu P, et al. Characterization of surface antigen protein 1 (SurA1) from *Acinetobacter baumannii* and its role in virulence and fitness. *Vet Microbiol.* 2016;186:126–138. Available from: https://doi.org/10.1016/j.vetmic.2016.02.018.

90. Aranda J, Bardina C, Beceiro A, et al. *Acinetobacter baumannii* RecA protein in repair of DNA damage, antimicrobial resistance, general stress response, and virulence. *J Bacteriol.* 2011;193(15):3740–3747. Available from: https://doi.org/10.1128/JB.00389-11.

91. Hu WS, Yao SM, Fung CP, Hsieh YP, Liu CP, Lin JF. An OXA-66/OXA-51-like carbapenemase and possibly an efflux pump are associated with resistance to imipenem in *Acinetobacter baumannii. Antimicrob Agents Chemother.* 2007;51(11):3844–3852. Available from: https://doi.org/10.1128/AAC.01512-06.

92. Peleg AY, Adams J, Paterson DL. Tigecycline efflux as a mechanism for nonsusceptibility in *Acinetobacter baumannii. Antimicrob Agents Chemother.* 2007;51(6):2065–2069. Available from: https://doi.org/10.1128/AAC.01198-06.

93. He X, Lu F, Yuan F, et al. Biofilm formation caused by clinical *Acinetobacter baumannii* isolates is associated with overexpression of the AdeFGH efflux pump. *Antimicrob Agents Chemother.* 2015;59(8):4817–4825. Available from: https://doi.org/10.1128/AAC.00877-15.

94. Rajamohan G, Srinivasan VB, Gebreyes WA. Molecular and functional characterization of a novel efflux pump, AmvA, mediating antimicrobial and disinfectant resistance in *Acinetobacter baumannii. J Antimicrob Chemother.* 2010;65(9):1919–1925. Available from: https://doi.org/10.1093/jac/dkq195.

95. Smani Y, Fàbrega A, Roca I, Sánchez-Encinales V, Vila J, Pachón J. Role of OmpA in the multidrug resistance phenotype of *Acinetobacter baumannii. Antimicrob Agents Chemother.* 2014;58(3):1806–1808. Available from: https://doi.org/10.1128/AAC.02101-13.

96. Jeong HW, Cheong HJ, Kim WJ, et al. Loss of the 29-kilodalton outer membrane protein in the presence of OXA-51-like enzymes in *Acinetobacter baumannii* is associated with decreased imipenem susceptibility. *Microb Drug Resist.* 2009;15(3):151–158. Available from: https://doi.org/10.1089/mdr.2009.0828.

97. Fonseca EL, Scheidegger E, Freitas FS, Cipriano R, Vicente AC. Carbapenem-resistant *Acinetobacter baumannii* from Brazil: role of carO alleles expression and blaOXA-23 gene. *BMC Microbiol.* 2013;13:245. Available from: https://doi.org/10.1186/1471-2180-13-245. Published November 6, 2013.

98. Moffatt JH, Harper M, Harrison P, et al. Colistin resistance in *Acinetobacter baumannii* is mediated by complete loss of lipopolysaccharide production. *Antimicrob Agents Chemother.* 2010;54(12):4971–4977. Available from: https://doi.org/10.1128/AAC.00834-10.

99. Zhu J, Wang C, Wu J, Jiang R, Mi Z, Huang Z. A novel aminoglycoside-modifying enzyme gene aac(6′)-Ib in a pandrug-resistant *Acinetobacter baumannii* strain. *J Hosp Infect*. 2009;73 (2):184–185. Available from: https://doi.org/10.1016/j.jhin.2009.05.012.

100. Ploy MC, Giamarellou H, Bourlioux P, Courvalin P, Lambert T. Detection of aac(6′)-I genes in amikacin-resistant Acinetobacter spp. by PCR. *Antimicrob Agents Chemother*. 1994;38 (12):2925–2928. Available from: https://doi.org/10.1128/aac.38.12.2925.

101. Zeng X, Gu H, Peng L, et al. Transcriptome profiling of lung innate immune responses potentially associated with the pathogenesis of *Acinetobacter baumannii* acute lethal pneumonia. *Front Immunol*. 2020;11:708. Available from: https://doi.org/10.3389/fimmu.2020.00708. Published April 22, 2020.

102. Chen W. Host innate immune responses to *Acinetobacter baumannii* infection. *Front Cell Infect Microbiol*. 2020;10:486. Available from: https://doi.org/10.3389/fcimb.2020.00486. Published September 14, 2020.

103. Noto MJ, Boyd KL, Burns WJ, Varga MG, Peek Jr RM, Skaar EP. Toll-like receptor 9 contributes to defense against *Acinetobacter baumannii* infection. *Infect Immun*. 2015;83 (10):4134–4141. Available from: https://doi.org/10.1128/IAI.00410-15.

104. Bist P, Dikshit N, Koh TH, Mortellaro A, Tan TT, Sukumaran B. The Nod1, Nod2, and Rip2 axis contributes to host immune defense against intracellular *Acinetobacter baumannii* infection. *Infect Immun*. 2014;82(3):1112–1122. Available from: https://doi.org/10.1128/IAI.01459-13.

105. Dikshit N, Kale SD, Khameneh HJ, et al. NLRP3 inflammasome pathway has a critical role in the host immunity against clinically relevant *Acinetobacter baumannii* pulmonary infection. *Mucosal Immunol*. 2018;11(1):257–272. Available from: https://doi.org/10.1038/mi.2017.50.

106. King LB, Swiatlo E, Swiatlo A, McDaniel LS. Serum resistance and biofilm formation in clinical isolates of *Acinetobacter baumannii*. *FEMS Immunol Med Microbiol*. 2009;55 (3):414–421. Available from: https://doi.org/10.1111/j.1574-695X.2009.00538.x.

107. Islam AH, Singh KK, Ismail A. Demonstration of an outer membrane protein that is antigenically specific for *Acinetobacter baumannii*. *Diagn Microbiol Infect Dis*. 2011;69(1):38–44. Available from: https://doi.org/10.1016/j.diagmicrobio.2010.09.008.

108. de Breij A, Eveillard M, Dijkshoorn L, van den Broek PJ, Nibbering PH, Joly-Guillou ML. Differences in *Acinetobacter baumannii* strains and host innate immune response determine morbidity and mortality in experimental pneumonia. *PLoS One*. 2012;7(2):e30673. Available from: https://doi.org/10.1371/journal.pone.0030673.

109. Brady MF, Jamal Z, Pervin N. *Acinetobacter* [Updated August 13, 2020]. *StatPearls [Internet]*. Treasure Island (FL): StatPearls Publishing; 2020. Available from: https://www.ncbi.nlm.nih.gov/books/NBK430784/.

CHAPTER 8

Infections in the immunocompromised host and immunization

Problem 23: Sepsis

A 32-year-old man with myelogenous leukemia was admitted for his second course of consolidation chemotherapy. However, the chemotherapy was delayed due to neutropenia. On the fifth day of his hospital stay, the patient presented with a fever, chills, body ache, and vomiting.

The following are the reports of the investigations carried out:

1. Vitals
   a. Temperature: 39.2°C
   b. Blood pressure: 90/60 mm Hg
   c. Pulse: 118/min (thready)
   d. Respiratory rate: 25/min
2. Physical examination
   a. Appearance: Lethargic and pale
   b. Extremities: Cold
   c. Skin: Multiple round-shaped indurated purplish black skin lesions seen on both axillae and extremities
   d. Abdomen: Normal
   e. Chest: Normal
   f. Exit of Hickman catheter: Normal
3. Urine analysis
   a. pH—5
   b. Microscopy
   c. RBC: 2–4/hpf
   d. White blood cells (WBC): ++
   e. Protein: +
   f. Ketone: +
   g. Glucose: Negative
   h. Bilirubin: Negative
   i. Urine culture: Negative

CONTENTS

Problem-Based Learning Approach in Microbiology. DOI: https://doi.org/10.1016/B978-0-323-95092-3.00003-2

4. Complete blood count (CBC)

Test	Values	Reference range
Hemoglobin (g/dL)	10.0	11–13
Hematocrit (%)	29	33–40
WBC (/L)	0.8×10^9	$3.9–11 \times 10^9$
PMNs (%)	22	54–62
Lymphocytes (%)	65	25–33
Blasts (%)	20	1–5
Platelets (/μL)	50,000	150,000–400,000

PMN, Granular leukocyte, granulocyte, and polymorphonuclear leukocyte.

5. Radiology
   a. Chest X-ray: Normal
6. Biochemistry

Test	Values	Reference range
Urea (mmol/L)	10	3–6.3
Sodium (mmol/L)	138	135–145
Potassium (mmol/L)	3.3	3.5–5.3
Chlorine (mmol/L)	102	95–110
Bicarbonate (mmol/L)	16	23–32
Creatinine (μmol/L)	132	60–110
Blood plasma lactate (mmol/L)	3.9	1–2
Anion gap (mmol/L)	20	3–11

7. Arterial blood gases

Test	Values	Reference range
pH	7.29	7.38–7.42
pCO_2 (mmHg)	33	75–100
pO_2 (mmHg)	88	38–42

8. Coagulation studies

Test	Values	Reference range
PT (seconds)	12	11–15
aPTT (seconds)	35	28–38
INR	1.0	0.9–1.2

PT, Prothrombin time; *aPTT,* Activated partial thromboplastin clotting time; *INR,* international normalized ratio.

9. Blood and skin lesion aspirate culture: Gram-negative rod-shaped *Pseudomonas aeruginosa* bacteria

Diagnosis

The final diagnosis revealed *P. aeruginosa* blood infection, which later advanced to opportunistic candidemia and sepsis.

Treatment

The patient was shifted to the intensive care unit (ICU) and was treated as follows:

1. Gentamicin—5 mg/kg/day
   a. Gentamicin trough levels were checked every third day of treatment, and the gentamicin dose was adjusted if the trough level exceeded 2 μg/mL.
2. Piperacillin—4.5 gm/6 h
3. Granulocyte-colony stimulating factor (G-CSF) activated protein C
   a. The intravenous (IV) antibiotics and fluids were administered through the Hickman catheter.
   b. Creatinine levels were checked every third day of treatment.

Follow-up and outcome

The following were the values of some parameters after one day of treatment:

1. Appearance: Reduced lethargy
2. Blood pressure: 90/60 mmHg
3. Pulse: 98/min
4. Temperature: 38.5°C
5. Urine output: 350 mL/8 h
6. Respiratory rate: 18/min

The patient appeared well and had stable vital signs on the fifth day of ICU stay. No new skin lesions were seen.

CBC (on fifth day of ICU stay)

Test	Values	Reference range
Hemoglobin (g/dL)	9.0	11–13
Hematocrit (%)	27	33–40
WBC (/L)	1.9×10^9	$3.9–11 \times 10^9$
PMNs (%)	32	54–62
Lymphocytes (%)	37	25–33

Continued

Continued

Test	Values	Reference range
Blasts (%)	26	1–5
Myelocytes (%)	4	<0.5
Metamyelocytes (%)	4	<1
Platelets (/μL)	40,000	150,000–400,000

On the 13th day, however, the patient presented with a fever and lethargy again.

The vitals were as follows:

1. Blood pressure: 90/60 mmHg
2. Temperature: 39°C
3. Respiratory rate: 23/min
4. Pulse: 118/min

Physical examination of the buccal cavity revealed white cottony patches. Fundoscopic examination of the patches demonstrated the presence of cotton-wooly exudate. Samples were collected for blood and urine cultures, and lesion samples were collected from the buccal cavity. Fungal detection tests using KOH and calcofluor white preparations were prescribed, and the lesion samples were also cultured on selective media.

The antibiotic regimen was changed from gentamicin and piperacillin to Amphotericin B (IV) 0.5 mg/kg/day.

Blood culture: *Candida albicans* (report on 15th day in ICU)

Administration of Amphotericin B was continued; however, the patient's condition progressively deteriorated. Hypoxia, reduced urine output, fall in blood pressure levels, and increased lethargy were noted. The Hickman catheter was removed, and IV fluids and dopamine drip were started. Despite sustained efforts, the patient died on the 19th day in the ICU.

Question and answer

1. What is acute myelogenous leukemia?
   a. Acute myelogenous leukemia is a malignant disease affecting the nonlymphoid cells and is characterized by the failure of maturation and uncontrolled proliferation of these cells.
2. What is chemotherapy and what are its side-effects?
   a. Chemotherapy refers to a therapeutic mode involving cytotoxic drugs. Bone marrow suppression, mucositis, hair loss, and immunosuppression are some of side-effects of chemotherapy.

3. What is neutropenia and how does it affect the normal body functions?
   a. Neutropenia refers to the reduction of the number of PMNs below $1500/mm^3$. Neutrophils are important mediators of the phagocytic attributes of the innate immune system and are important in the process of killing the pathogen. A drastic reduction in the number of neutrophils, therefore, exposes the affected individual to various infections.
4. Which condition explains the worsening symptoms of the patient seen during the ongoing antibiotic therapy?
   a. Sepsis explains the worsening symptoms of the patient seen during the ongoing antibiotic therapy in the best possible way.
5. What is Hickman catheter?
   a. Hickman catheter is a long-term venous access positioned by a vascular surgeon, either having a single lumen or multiple lumena designed to infuse drugs, fluids, and blood, and to withdraw blood samples. The placement of a Hickman catheter avoids multiple venipunctures.
6. What is the diagnostic significance of the symptom of lethargy in this case?
   a. Lethargy could be indicative of anemia caused by bone marrow suppression, which is associated with the medical history in this case. In addition, lethargy could also have been caused by sepsis or septic shock, as indicated by the clinical manifestations, such as low blood pressure, rapid respiratory rate, and pulse seen in this case.
7. What causes the skin lesions?
   a. The skin lesions are septic emboli, known as ecthyma gangrenosum, caused by certain pathogens that infect the skin and are usually seen in immunocompromised individuals.
8. How is the skin lesion helpful in early diagnosis?
   a. The skin lesion samples can be used for culture of the aspiration and Gram stain, which can subsequently facilitate the diagnostic process.
9. What are the possible sources of infection in a hospital environment for immunocompromised patients?
   a. Immunocompromised patients are most often affected by endogenous infections. Malignant conditions and their treatment often cause mucosal damage of the gastrointestinal (GI) tract, allowing normal microbiota to invade various organ systems. The hospital environment, however, makes such patients especially vulnerable because of the high possibility of acquiring nosocomial infections from sources, such as the hands of medical personnel,

medical devices such as catheters, invasive procedures, such as bronchoscopy, dialysis, and exposure to other patients.

10. How can the clinical suspicion of sepsis be confirmed?
    a. The clinical suspicion of sepsis can be confirmed by a combination of patient history, clinical manifestations, blood cultures, and response to therapeutic measures.
11. What is the significance of a platelet count in the diagnosis of sepsis?
    a. The platelet count is reduced due to disseminated intravascular coagulation (DIC) occurring in sepsis and can be an important indication of sepsis.
12. What causes pancytopenia?
    a. Pancytopenia, a reduction of all three types of blood cells, namely the red blood cells (RBCs), white blood cells WBC, and platelets, is caused by a malfunctioning bone marrow, which may be seen in conditions such as autoimmune diseases, malignancies, side-effects of certain therapeutic drugs, and in some bone marrow disorders.
13. What is the diagnostic significance of the presence of ketone bodies in the urine sample and the increased lactate levels in the blood sample?
    a. The presence of ketone bodies is indicative of acidosis, decreased perfusion, and shock, while the increased lactate levels indicate diminished hepatic utilization of lactate that may be a result of sepsis.
14. What are the benefits of using both gentamicin and piperacillin instead of any one of the drugs?
    a. Gentamicin and piperacillin work synergistically, providing more therapeutic benefits than the use of any one drug, and also prevents the appearance of resistant pathogenic strains of the etiological agent.
15. Why is G-CSF added to the treatment regimen?
    a. G-CSF stimulates the proliferation of myeloid progenitor in the bone marrow, which will result in an increase in the numbers of the myeloid cells. The increased number of myeloid cells could improve the outcome in this case by providing better protection against infections.
16. What are the pathologies associated with Gram-negative sepsis responsible for death?
    a. Gram-negative sepsis can result in septic shock, multiple organ failure, and DIC, which are all potentially life-threatening manifestations of the disease.
17. Why were gentamicin and piperacillin removed from the treatment regimen?
    a. Gentamicin and piperacillin are broad spectrum antibiotics and may be responsible for infections by opportunistic pathogens, such

as *C. albicans*. These antibiotics may also demonstrate certain toxicities in some patients. In such cases, it is important to remove gentamicin and piperacillin from the treatment regimen.

18. What are the risk factors resulting in candidemia in this patient?
    a. Neutropenia, broad-spectrum antibiotics, and central venous line are all risk factors resulting in candidemia in this patient.

Learning objectives

1. Pathological implications of neutropenia.
2. Explain opportunistic infections.
3. Define systemic inflammatory response syndrome (SIRS).
4. Define septic shock.
5. Describe multiorgan dysfunction syndrome.
6. Summarize the etiology of sepsis.
7. Summarize the epidemiology of sepsis.
8. Describe the pathophysiology of sepsis.
9. Describe the clinical manifestations of sepsis.
10. Enumerate the methods involved in the diagnosis of sepsis.
11. Determine the prognosis of candidemia.
12. Summarize the components of treatment and management of sepsis.

Discussion

Differential diagnosis

Malignant hyperthermia	Malignant hyperthermia presents itself with a rapid heart rate, rapid and shallow breaths, arrythmia, extremely high-grade fever, excessive sweating, patchy irregular skin color, and severe muscle spasms and rigidity.
Neuroleptic malignant syndrome	Neuroleptic malignant syndrome causes a high-grade fever, muscle stiffness, excessive sweating, altered mental state, rapid heart rate, excessive salivation, and rapid respiratory rate.
Cardiogenic shock	Cardiogenic shock manifests with dyspnea, tachycardia, rapid heartrate, weak thready pulse, hypotension, loss of consciousness, pallor, and sweating.
Obstructive shock	Obstructive shock develops into distended neck veins, muffled heart tones, hypotension, rapid pulse, trouble breathing, and unequal breath sounds.
Hypovolemic shock	Hypovolemic shock presents itself with anxiety, agitation, cool and clammy skin, confusion, decreased urine output, anuria, pallor, rapid breaths, sweating, moist skin, and generalized weakness.

Background

Sepsis is defined as a life-threatening organ dysfunction caused by a dysregulated host response to an infection.[1] The life-threatening pathologies associated with sepsis are responsible for high levels of mortality, despite the advent of new management and treatment strategies. This condition is characterized by a complex disease process elicited by the invasion of the bloodstream by a pathogenic organism. This complex disease process is staged into SIRS, sepsis, severe sepsis, and septic shock (Fig. 8.1). The disease pathology is closely associated with mortality rates in sepsis (Fig. 8.2). SIRS describes the inflammatory process involved in sepsis, and may not always be a result of infection, while septic shock is a result of sepsis-induced hypotension that is irretractable despite adequate fluid resuscitation. Alteration of organ function to an extent that maintenance of homeostasis is not possible without intervention is referred to as multiorgan dysfunction syndrome.

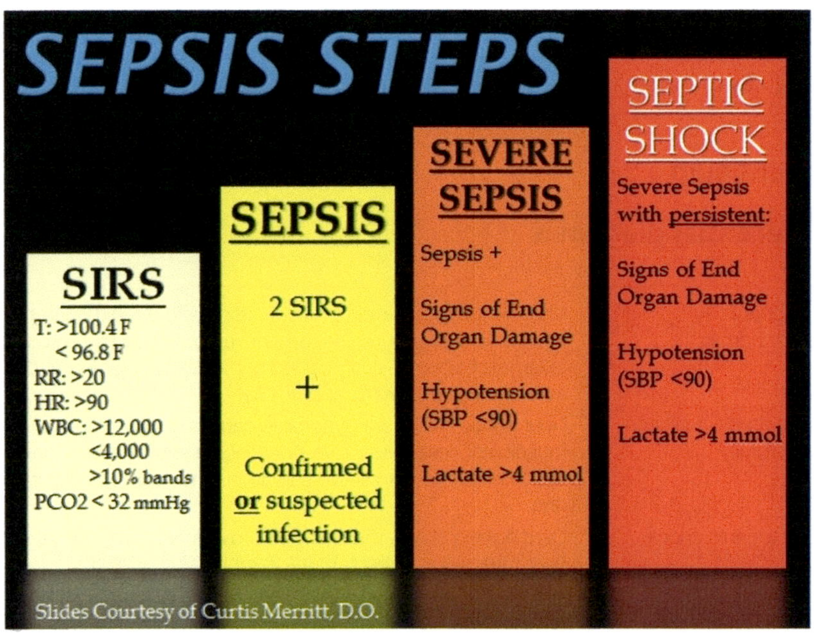

FIGURE 8.1

Clinical parameters for identification of disease stages in sepsis. *Source: From Hadroncastle/Own work/; https://creativecommons.org/licenses/by-sa/4.0/deed.en; https://commons.wikimedia.org/wiki/File:Sepsis_Steps.png.*

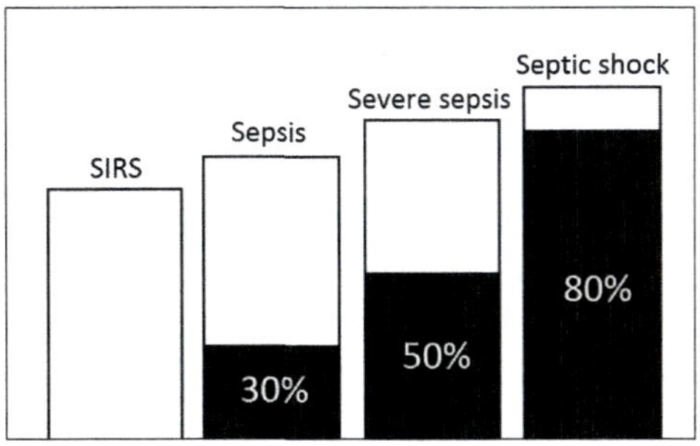

FIGURE 8.2
Graphic representation of mortality rates associated with stages of sepsis. *Source: From Ian Furst/Own work/; https://creativecommons.org/licenses/by-sa/4.0/deed.en; https://commons.wikimedia.org/wiki/File: Sepsis_mortality.png.*

Etiology

Sepsis is a result of various etiologies that threaten the functional integrity of various organ systems. The host responses leading to sepsis result from pro-inflammatory and antiinflammatory responses. The pathological implication of this disease process is a function of pathogen-associated factors and host-associated factors. The competency of the host immune system is essential in the disease process leading to sepsis.

The host immune system recognizes the specific pathogen-associated molecular patterns (PAMPs) and elicits an inflammatory cascade. The cascade begins with the activation of the WBCs followed by the activation of the complement system and coagulation, ultimately resulting in necrotic cell death. The process of respiratory burst mediated by the activated neutrophils results in the generation of toxic metabolites of oxygen, which in combination with the myeloperoxidase enzyme present in neutrophils convert hydrogen peroxidase to hypochlorite. Neutrophil degranulation releases these germicidal metabolites stored in the cytoplasmic granules. These toxic metabolites are proinflammatory, thereby further accelerating the inflammatory pathway and causing a lot of collateral damage. The various components of the immune system function in tandem to keep the proinflammatory mechanisms under control. However, if these proinflammatory mechanisms cannot be kept under control, the inflammatory cascade can be self-sustained, which causes systemic damage and life-threatening multiorgan dysfunction.

An expansion of regulatory T-cells and myeloid suppressor cells facilitates the antiinflammatory mechanisms that regulates the extent of inflammation. The excess oxidants generated by the oxidation processes resulting from inflammation overwhelms the host's antioxidant defenses. The inadequate antioxidant protection is mainly responsible for the collateral damage associated with inflammatory processes.

Host immune responses and pathophysiology

Sepsis research over the years has ensured a better understanding of the complex pathophysiological mechanisms responsible for the clinical manifestations of sepsis. While the manifestations of the disease were previously attributed to an uncontrolled hyperimmune reaction of the host, an appreciable amount of research has revealed a different perspective of the disease. Sepsis, also referred to as distributive, hyperdynamic, or warm shock owing to its systemic hemodynamic manifestations, is now known as a result of a complex and intricate interplay between certain pathogen and host processes.

Inflammatory response

The interaction of the pathogen via PAMPs, such as the antigenic proteins present on the microbial cell membrane and certain endotoxins with the host cell receptors, such as the toll-like receptors, nucleotide-binding oligomerization domain (NOD)-like receptors (NLRs), and retinoic acid inducible gene 1, elicit a strong primary immune response in the host. Effectors of phagocytosis, such as macrophages and cytotoxic executers, like natural killer (NK) cells, are primarily involved in mediating the host defense against the pathogen. Damage-associated molecular patterns constituting released intracellular material from damaged cells are also responsible for similar host responses as those elicited by the interaction with PAMPs. These pathogen–host interactions also activate the intracellular signal transduction pathways, thereby releasing proinflammatory cytokines like tumor necrosis factors alpha (TNF-α), interleukin-1 (IL-1), and IL-6. In addition, NLRs aggregate into large protein complexes known as inflammasomes, which produce apoptotic cytokines, such as IL-1ß, IL-18, and caspases. Proinflammatory cytokines are responsible for the activation of leukocytes and the complement system, upregulation of endothelial adhesion molecules and chemokine expression, tissue factor production, and induction of hepatic acute phase reactants.[2]

The hyperimmune state in sepsis induces a systemic catabolic state to facilitate the fueling of the immune cells by proteins generated from amino acids attained through a rapid and significant breakdown of muscles. The damage caused by the induced catabolic state is further exacerbated by a systemically elevated state of cellular insulin resistance.[2]

Contraindicatory to the hyperimmune state characteristic of sepsis-associated inflammation in the primary stages of the disease, studies have demonstrated prolonged immunosuppression in sepsis cases in the later stages of the disease.[3–5] This results in an inability to resist secondary bacterial, viral, or fungal infections, thereby suggesting that lymphopenia may be a marker for upcoming immunosuppression in sepsis.[6]

Hemodynamic response

While a rapid and aggressive host immune response is vital in neutralizing a pathogenic threat, an exaggerated and uncontrolled immune response can result in extensive collateral damage to the host cells and organs. The systemic hemodynamic manifestations, ranging from mild thrombocytopenia to fulminant DIC, characteristic of sepsis, are caused due to the activation of both the inflammatory and coagulation cascades. This intersection of inflammatory and hemostatic pathways results in dysregulated coagulation. Evidence suggests that the release of certain intracellular material, such as tissue factor from damaged host cells, is responsible for the hypercoagulability seen in sepsis pathology.[7,8] The systemic activation of the coagulation cascade results in the production of thrombin, activation of platelets, and formation of microthrombi (platelet–fibrin clots) (Fig. 8.3), which contribute

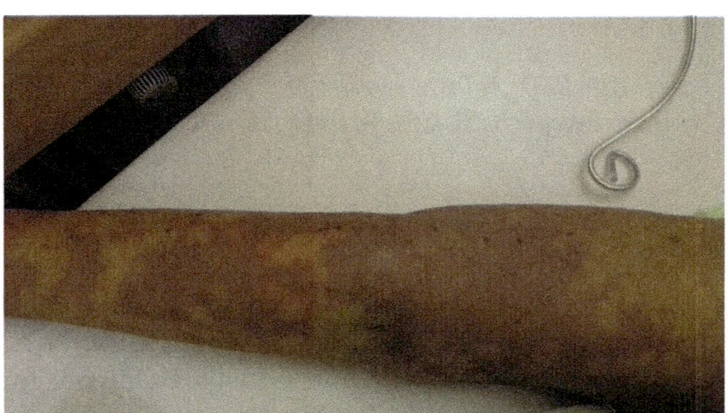

FIGURE 8.3
Patchy skin appearance caused due to microthrombi formation in a patient manifesting symptoms of septic shock. *Source: From Emergency doc/Own work/; https://creativecommons.org/licenses/by-sa/4.0/deed.en; https://commons.wikimedia.org/wiki/File:Sepsis-Mikrothomben1.JPG.*

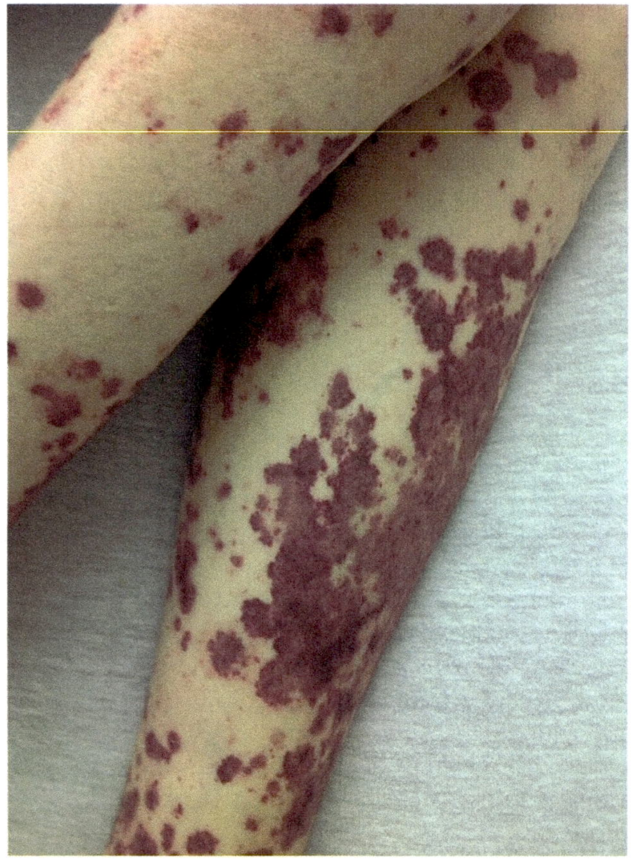

FIGURE 8.4

Petechial rash. *Source: From DrFO. Jr. Tn/Own work/; https://creativecommons.org/licenses/by-sa/3.0/deed.en; https://commons.wikimedia.org/wiki/File:Petechial_rash.JPG.*

to local perfusion defects. DIC is a clinically dreaded complication of sepsis indicated by the appearance of petechiae on the skin (Fig. 8.4), which is caused by the platelets and clotting factors being used up for microthrombi formation. The pathological effects of hypercoagulability are further exacerbated by a suppression of anticoagulability. The severity of systemic inflammation in sepsis drastically reduces the plasma levels of protein C, which in its active form is instrumental in exerting an anticoagulant effect. Therefore the coagulation cascade progresses in an uncontrolled manner, similar to the inflammatory pathways. The increased TNF-α and IL-1ß levels in sepsis also results in a reduction of fibrinolysis and fibrin removal, which contributes to further exacerbation of microvascular thrombosis.[9] Studies have also demonstrated a decreased deformability of RBCs induced by exaggerated inflammatory phase, which worsens the tissue hypoxia. The accumulation of such defects ultimately results in tissue and organ dysfunction.[10,11]

Cellular response

Septic cardiomyopathy is caused by the elevated levels of proinflammatory cytokines, such as TNF-α and IL-1ß. Cardiac dysfunction, characteristic of sepsis, is a result of proinflammatory cytokines compromising mitochondrial function and damaging cardiomyocytes. The resulting cardiovascular dysfunction causes hypoperfusion, which hampers the oxygen delivery to cells, ultimately causing tissue and organ dysfunction.[12] The acute but reversible cardiac damage and the combination of the reduced left ventricular ejection, plus the normal to low left ventricular filling pressure with increased left ventricular compliance, are important features of septic cardiomyopathy.[13] Cardiac pathology in sepsis is associated with systolic and diastolic dysfunction, in addition to decreased stroke volumes and increased end-diastolic and end-systolic volumes.[14,15] Venous return is reduced because of inflammation-mediated dilation of arteries, veins, and the microvasculature comprising of arterioles, venules, and capillaries. Damage to the endothelial cells results in the leakage of intravascular fluid into the interstitial space, which further impairs hemodynamic functioning. Hemodynamic defects and microvascular thrombosis promote hypoperfusion of tissues and organs, resulting in a state of hyperlactatemia due to an excess of lactic acid generated by increased anaerobic glycolysis in cells.[16,17] Cellular damage is worsened by mitochondrial dysfunction and plummeting adenosine triphosphate (ATP) levels mediated by an overwhelming production of reactive oxygen species (ROS) in response to the inflammatory cascade.[2]

Effect on different organ systems

Damage to the alveolar–endothelial barrier results in the accumulation of protein-rich fluid in the interstitial lung spaces and alveoli, compromising the ventilation–perfusion coordination. Acute respiratory distress syndrome may be caused by the hypoxia and decreased lung compliance. Renal damage is mediated by hypoperfusion, hypoxia causing acute tubular necrosis, and defects in microvasculature and tubular functioning. Dysfunction of the GI tract is caused by the increased permeability of the mucosal lining, which enables the translocation of bacteria and luminal enzymes across the bowel wall, increasing predisposition to pathogenic infections and autodigestion of the bowel. Suppressed bilirubin clearance in sepsis causes cholestasis, thus damaging hepatobiliary function. Central nervous system (CNS) dysfunction in sepsis is evident from the common manifestation of altered mental status. Endothelial alterations compromise the blood–brain barrier, facilitating the entry of pathogens, toxins, inflammatory cells, and cytokines. Septic encephalopathy characterized by a spectrum of symptoms ranging from mild confusion to delirium, and coma is a consequence of inflammatory damage, oxidative stress, cerebral edema, and disruption of the neurotransmitter pathway.[2]

Septic shock

A combination of hypovolemia-increased metabolic demands, myocardial depression, and vaso-regulatory perfusion abnormalities results in circulatory insufficiency. Systemic vasodilatation of both arteries and veins, which reduces ventricular preload and afterload due to decreased systemic vascular resistance (SVR), is a characteristic of septic shock. Reduced ventricular preload and afterload causes a compensatory increase in cardiac output and an increased heart rate. Therefore low central venous pressures and low SVR define the hemodynamic pattern of septic shock.[18] The inability to match the increasing oxygen demands combined with hemodynamic instability becomes evident from the increasing lactate levels and decreasing venous oxygen saturation.[11] While the individual pathological changes in sepsis are clinically reversible, the involvement of multiorgan systems usually results in a fatal outcome.

Clinical manifestations and complications

The clinical manifestations of sepsis vary widely depending on the disease stage ranging from SIRS to septic shock. An assessment of risk factors and comorbidities, such as malignancies, cardiac malfunction, renal insufficiency, chronic lung disease, liver pathologies, and old age associated with poor prognosis in sepsis, is the main objective of the emphasis on pertinent patient history in suspected sepsis. Timely determination of the etiology can prevent the loss of crucial time resulting from ineffective empirical treatment and facilitate rapid and customized intervention and management. Although the prognosis and outcome of sepsis are not associated with the etiology of the infection, the site of infection defined by the prevalence of pneumonia, infection of the gastro-intestinal system, or urinary tract infection is a good indication of disease progression.

Detailed physical examination is critical for the utilization of pertinent diagnostic techniques and tools, enabling rapid diagnosis and initiation of effective treatment. The target of the physical examination process would vary depending on the affected organ system indicated by the clinical manifestations of a particular case. A presentation of ARDS would suggest lung pathology, while hypotension would allude to the involvement of the cardiovascular system. Renal involvement would signify signs of renal insufficiency, while an altered mental status could denote septic encephalopathy. Severe anemia, extreme myopathy, and polyneuropathy are generally considered alarming signs of critical illness.

Diagnosis

Although there is no gold standard evaluation method defined for the confirmed diagnosis of sepsis, many scores and criteria, such as the qSOFA score and the SIRS criteria, have been developed to promote an early diagnosis and better prognostic predictions of disease. The triad of systemic clinical manifestations, indications of organ dysfunction or failure, and evidence of disease etiology are considered the foundation for the clinical diagnosis of sepsis. While the SIRS criteria is instrumental in interpreting the systemic clinical manifestations, laboratory parameters, such as CBC, INR, liver function tests, and serum lactate levels, are important markers of organ dysfunction. Microbiological cultures of blood, urine, cerebrospinal fluid, and respiratory, peritoneal and synovial secretions are useful in determining the microbiological etiology of the condition. However, negative microbiological cultures in sepsis cases increase the diagnostic challenge.[10]

Several markers of end-organ dysfunction have been suggested by the sepsis-3 criteria (2001) and the Surviving Sepsis Campaign and CMS Core Measure (SEP-1), which enable the evaluation of sepsis.[1,18,19]

An initial suspicion of sepsis is followed by an array of investigations comprising of CBC with differential, comprehensive metabolic profile, serum lactate, INR/aPTT, chest X-ray, urinalysis, microbiological cultures of various bodily secretions and fluids, procalcitonin, C-reactive protein (CRP), lipase, arterial and venous blood gas, DIC panel comprising of fibrinogen, D-dimer, fibrin degradation products, thyroid stimulating hormone (TSH), CT, and magnetic resonance imaging (MRI) of the organs suspected of being primarily involved in the condition.

Treatment and management

The evolution in the treatment and management strategies employed for sepsis during the recent years has promoted better prognoses and outcomes in septic patients. However, sepsis mortality rates are still considerably high, thereby indicating the scope for further improvement in the presently available treatment and management strategies. The therapies currently available are directed at the management of the initial systemic manifestations of sepsis and organ dysfunction. Initial effective management of sepsis is based on rapid and accurate diagnosis, reports of microbial cultures, and the rapid initiation of appropriate empirical therapeutic measures. Cardio-pulmonary resuscitation essentially comprises of an ongoing process of attenuation of host response and accompanying organ dysfunction.

The complicated nature of sepsis-associated pathologies has enunciated the need for the differentiated 3 and 6-hour care bundles for acute sepsis therapy and a 24-hour sepsis management bundle.[18]

Rapid initiation of broad-spectrum antimicrobial therapy soon after a confirmed diagnosis of severe sepsis or septic shock without waiting for indications of the most appropriate therapy based on microbial culture reports is known to be crucial for a better prognosis and reduced mortality. The acute resuscitation phase should be succeeded by targeted therapy influenced by appropriate culture reports.

Volume resuscitation is also an important component of the acute phase resuscitation and has a critical place in improved disease outcome. Septic shock often indicates the need for vasopressors, such as norepinephrine, epinephrine, vasopressin, and dopamine through a central venous catheter in irretractable cases that do not respond to the intravenous IV fluid bolus to provide additional vascular contraction that promotes the maintenance of optimal blood pressure levels.[19,20]

Regular monitoring and intermittent reevaluation of parameters indicative of volume status and tissue perfusion, such as heart rate, blood pressure, urine output, respiratory rate, arterial and venous blood gases, serum lactate levels, capillary refill, pulse, and skin appearance, are critical for guidance of patient management after the initial resuscitation period.

The magnitude of blood transfusion therapy in a sepsis management regime should be based on the presence or absence of clinical manifestations suggestive of critical coronary disease, myocardial ischemia, or acute hemorrhage.[19,21]

Corticosteroids, such as hydrocortisone as antiinflammatories and vasoconstrictors, have been shown to influence hospitalization periods and dependency on medical equipment in sepsis survivors.[22,23] As ARDS is a commonly encountered clinical manifestation in sepsis, mechanical ventilation and a lung protection strategy are often critical for patient survival.[24]

Administration of monoclonal antibodies against proinflammatory cytokines, such as TNF-α, IL-1 inhibitors, granulocyte colony-stimulating factor, NO inhibitors, antithrombin, N-acetylcysteine, and endotoxin neutralizing

Student Notes

antibodies, has been considered but scratched out due to the failure of demonstrable efficacy of any of the therapies. The development of immunomodulation therapies and strategies continues to be an important area of clinical research as sepsis-associated mortality rate, even with the advent of new therapies and management strategies, remains unacceptably high.

Key take-aways

1. Sepsis is a potentially life-threatening condition that results from dysregulated host immune response, causing end-stage organ dysfunction or failure.
2. Despite the advent of new management strategies and therapies, sepsis continues to have an unacceptably high mortality rate.
3. Risk factors and comorbidities, such as active malignancies, renal insufficiency, cardiac pathologies, liver diseases, and old age, are known to be associated with a poor prognosis and outcome in septic patients.
4. Clinical manifestations, pertinent patient history, and appropriate physical examination form the crucial foundation for an accurate diagnosis of sepsis.
5. Diagnostic parameters for sepsis are not clearly defined and are recommended based on the clinical manifestations.
6. Timely initiation of broad-spectrum antimicrobial therapy and other empirical therapeutic measures followed by targeted therapy based on laboratory investigations is crucial for improved patient survival.
7. Appropriate and timely implementation of acute phase resuscitation measures, regular monitoring of vital physiological parameters, and the intermittent reevaluation of treatment strategies is the mainstay of effective management and treatment of sepsis.
8. The lack of defined diagnostic means and therapeutic measures contributes to the unacceptably high mortality rates associated with sepsis, and drives the pursuit of more effective diagnostic, therapeutic, and management strategies.

Problem 8.2: Progressive multifocal leukoencephalopathy

A 67-year-old man with a 4-year-old history of follicular lymphoma presented with confusion, dysarthria, and aphasia. He had undergone conventional chemotherapy 4 years ago and had been prescribed three courses of rituximab as maintenance therapy, warranted by the persistence of clonal changes in his bone marrow. The presenting symptoms developed after the third course of the prescribed drug.

The following are the reports of the investigations carried out:

1. Vitals
   a. Temperature: 38.5°C
   b. Blood pressure: 90/60 mm Hg
   c. Pulse: 118/min (thready)
   d. Respiratory rate: 25/min
2. Physical examination
   a. Appearance: Lethargic and pale
   b. Extremities: Cold
   c. Skin: Multiple round-shaped indurated purplish black skin lesions on both axillae and extremities
   d. Abdomen: Normal
   e. Chest: Normal
   f. Exit of Hickman catheter: Normal
3. Radiology
   a. MRI—Nonenhancing lesions in the posterior and frontal lobes and in the left centrum semiovale
4. Polymerase chain reaction (PCR) test
   a. Peripheral blood: + ve for JC virus
   b. CSF: Negative
5. Histopathology
   a. Brain biopsy: Changes suggestive of progressive multifocal leukoencephalopathy (PML)
6. CD4 count: 100 cells/mm^3 (500–1200 cells/mm^3)

Diagnosis

The final diagnosis concluded PML.

Treatment

The patient was treated with cidofovir; however, his clinical condition deteriorated. The manifestations of aphasia and confusion worsened, and he soon developed urinary and fecal incontinence with flaccid paralysis of the left arm. He also demonstrated mass effect accompanied by radiologic progression.

The patient finally succumbed to his condition and died 2 months after presentation.

Question and answer

1. What is PML?
   a. PML is a progressive demyelinating disease that preferentially affects the CNS. It is an aggressive, devastating, and a potentially fatal disease with high mortality rates. Although the disease is rare, it is commonly associated with immunocompromised conditions, such as natural and acquired immunodeficiencies, abnormalities of cell-mediated immunity, lymphoproliferative diseases, chronic inflammatory conditions, autoimmune diseases, solid organ and bone marrow transplants, and certain immunosuppressive and immunomodulatory treatment regimens.
2. What is the etiologic agent of PML?
   a. A small, nonenveloped, double-stranded DNA virus belonging to the *Polyomaviridae* family, known as the John Cunningham (JC) virus, is the etiologic agent of PML. The JC virus is ubiquitous in the human population globally, and almost half of the world population demonstrates seropositivity. There are two forms of this virus: the archetypal form that precipitates the primary infection, commonly detected in urine, and the prototypal rearranged form detected in brain tissues of PML patients.
3. How are JC virus infections acquired?
   a. JC virus infections are usually acquired in childhood either through the tonsils or the GI route. Effective host immune responses cause the virus to establish latency and persist in the kidney.
4. What predisposes individuals to reactivation of PML virus?
   a. Cytotoxic T-lymphocytes elicit host immune responses that are instrumental in mediating viral clearance. Several immunocompromised conditions cause CD4 cells to reduce, resulting in inefficient cellular responses, and predisposing individuals to PML. Certain immunomodulatory drugs use the $\alpha4\beta1$ integrin binding mechanism for their function. This mechanism prevents lymphocytes from binding to the vascular cell adhesion molecule 1 (VCAM) on the surfaces of the endothelial cells comprising the CNS. This compromises the host neuro-immunosurveillance capabilities and increases the vulnerability of the CNS to the pathogenicity of the reactivated JC virus.
5. How does the JC virus mediate its neuropathology?
   a. The archetypal JC virus is incapable of effective replication in the glial tissue, and has to develop neurotropism, followed by evasion of the neuro-immunosurveillance and clearance to establish

infection in the brain. Genetic modifications and rearrangements of the viral genome facilitate the transformation of the latent archetypal JC virus to the virulent prototypal JC virus. The genetic transformation influences the viral replication abilities, the transcriptional profiles of several viral genes, tissue tropism, and the pathogenicity of the virus. The transformed virus replicates within cerebral oligodendrocytes and astrocytes, resulting in lysis-mediated damage to oligodendrocytes, and destroying the protective myelin sheaths encasing the neurons. The heavy influx of immune cells, viral proteins, and pathogenic antigens contributes further to the neurologic insult, increasing the extent of demyelination. The rapidly progressive focal neurologic deficits without any increase in intracranial pressure characteristic of PML is a result of the gradual neurological damage.

6. Which are the various forms of clinical manifestations mediated by the reactivation of the JC virus?
   a. PML is the predominant clinical manifestation affected by the reactivation of the JC virus. Granule cell neuronopathy (GCN), demonstrating a pathology of the granule cell neurons of the cerebellum manifested as ataxia, nystagmus, and tremors, John Cunningham virus (JCV) encephalopathy characterized by a lysis-mediated pathology of the cortical gray matter, and isolated JCV meningitis, are some other forms of clinical manifestations mediated by the reactivation of the JC virus.
7. How does the presentation of human immunodeficiency virus (HIV)-associated PML differ from that of immunomodulation-associated PML?
   a. The presentation of HIV-associated PML and immunomodulation-associated PML demonstrates some subtle variations because of the trigger differences. While HIV-associated PML predominantly manifests as pathologies associated with the motor system, immunomodulation-associated PML generally presents with cognitive and language-related deficits.
8. How do CT and MRI scans differentiate between malignancies and infections of the CNS other than PML?
   a. CT and MRI scans detect pathological changes, such as inflammation and mass effect, which are suggestive of malignancies or infections of the CNS other than PML. However, HIV patients undergoing antiretroviral therapy (ART) or antirheumatic therapy may be an exception to this differentiation, as these patients may demonstrate contrast edema and mass effect on MRI scans due to

the involvement of immune reconstitution inflammatory syndrome (IRIS). CT scans showing hypodense confluent lesions without mass effect, contrast enhancement, or risks of herniation, are suggestive of PML, while areas of a decreased signal on T1-weighted images in MRI scans and hyperintense lesions in the subcortical, periventricular, and cerebellum regions on T2-weighted images in MRI with contrast raise are indicative of PML. Multifocal demyelination, indicated by the presence of smaller lesions coalescing into larger foci throughout the white matter with extensive involvement of entire hemispheres, is characteristic of PML. MRI scans are preferred over CT scans for an accurate clinical picture of the extent of the disease pathology involved, and also enable differentiation between brain tumors, abscesses, and inflammatory changes.

9. What are the histopathological changes characteristic of PML pathology?
   a. Histopathological investigations involving brain biopsies demonstrate characteristic cytopathic changes in oligodendrocytes, including enlarged nucleus containing accumulated basophilic material that gets deeply stained and a disruption of normal chromatin pattern. Astrocytes containing lobulated hyperchromatic nuclei may also be seen. The cytopathic changes result in characteristic pathological changes demonstrated in brain biopsies as multiple asymmetric foci of demyelination at various stages of evolution in the cerebral white matter.

Learning objectives

1. Summarize the pathological implications of a low CD4 count.
2. Describe opportunistic infections.
3. Summarize the pathology of follicular lymphoma.
4. Explain the association of PML with HIV infections.
5. Enumerate the risk factors associated with the development of PML.
6. Describe the pathophysiology of PML.
7. Describe the clinical manifestations of PML.
8. Enumerate the methods involved in the diagnosis of PML.
9. Determine the prognosis of PML.
10. Summarize the treatment and management options for PML.

Discussion

Differential diagnosis

HIV encephalitis	HIV encephalitis presents itself with an altered mental state, including focus and concentration issues, forgetfulness, depression, apathy, difficulty following a conversation, cognitive impairment, confusion, lack of coordination, speech-associated problems, tremors, unsteady gait, increasing weakness, psychosis, and inability to perform mundane daily tasks.
Cytomegalovirus (CMV) encephalitis	CMV encephalitis causes delirium, confusion, weakness, urinary retention, and an altered mental status.
IRIS	IRIS may lead to generic symptoms, such as a fever, night sweats, nausea, fatigue, jaundice, lymphadenopathy, and unexplained weight loss. Some cases may also result in abscess development, pleural effusions, and pulmonary infiltrates.
Acute Disseminated Encephalomyelitis (ADEM)	ADEM usually presents itself with a sudden onset of nausea, vomiting, headache, drowsiness, irritability, unsteady gait, inability to walk, vision problems, tingling sensation, and weakness in certain body parts. Seizures may also manifest in certain rare cases.
Multiple sclerosis	Multiple sclerosis presents itself with vision problems, mobility issues, localized unexplained pain, extreme fatigue or lethargy, numbness, tingling sensation, muscle spasms and stiffness, weakness in certain body parts, depression, anxiety, inability to concentrate and focus, difficulty swallowing, bladder and bowel incontinence, sex drive-associated problems, problems associated with speech.

Background

PML is a progressive demyelinating disease. This neurological condition preferentially affects the CNS. Although the disease is rare, it is commonly associated with immunocompromised conditions, such as natural and acquired immunodeficiencies, abnormalities of cell-mediated immunity, lymphoproliferative diseases, chronic inflammatory conditions, autoimmune diseases, solid organ and bone marrow transplants, and certain immunosuppressive and immunomodulatory treatment regimens. PML is an aggressive, devastating, and a potentially fatal disease with high mortality rates. There are no targeted therapies that have proven their effectiveness for PML, and the presently available forms of treatment primarily aim at restoring any defect in the normal functioning of the immune system; elimination of any known factors, such as treatment regimens that may have elicited the condition; and generalized steps of supportive care. The low prevalence of the disease caused

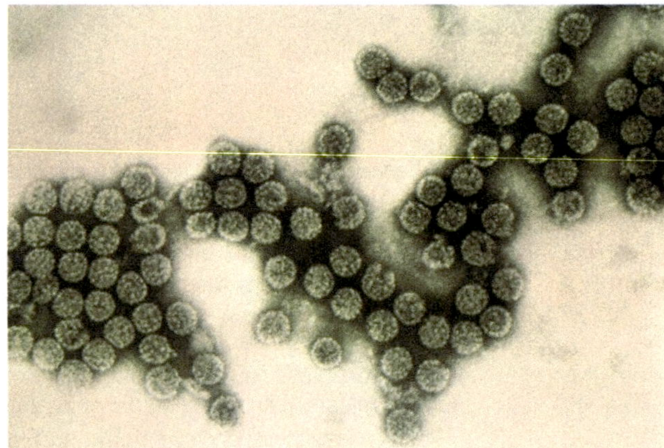

FIGURE 8.5
Microscopic image of polyomaviruses. *Source: From Dr. Erskine Palmer, USCDCP/Own Work/; https://creativecommons.org/licenses/publicdomain/; https://pixnio.com/science/microscopy-images/dna-based-icosahedral-shaped-polyomavirus-virions.*

it to be an orphan disease until the AIDS pandemic, and the advent of immunomodulatory therapies caused the disease incidence to increase drastically.[25,26]

Etiology

PML was discovered in association with chronic leukocytic leukemia and Hodgkin's disease,[27] and it was only after almost two decades of its discovery that the JC virus was demonstrated as the etiologic agent of PML.[28] JC virus is a small, nonenveloped, double-stranded DNA virus belonging to the *polyomaviridae* family (Fig. 8.5). JC virus is known to occur in two forms: the archetypal form that precipitates the primary infection, commonly detected in urine, and the prototypal rearranged form detected in brain tissues of PML patients.[29]

Epidemiology

The JC virus is ubiquitous in the human population globally, and almost half of the world population demonstrates seropositivity. JC virus infections are usually acquired in childhood, either through the tonsils or the GI route. Effective host immune responses cause the virus to establish latency and persist in the kidney.[30] In spite of seropositivity, the virus remains latent and does not manifest pathologically in immunocompetent hosts. Although the ineffective immune responses in immunocompromised individuals provide the ideal environment for the virus to be reactivated and manifest its virulence, the involvement of several complex sequential events in the reactivation of the virus and the development of PML causes it to remain a rare disorder even in the immunocompromised population.

Although the association of PML, with several immunocompromised conditions, has now been recognized, it continues to be an AIDS-defining illness. The advent of the highly active antiretroviral therapy (HAART) has considerably decreased the incidence of PML in HIV-positive individuals. However, HAART has also been known to elicit manifestations, such as focal neurological deficits, altered mental status, and the appearance of noncontrast-enhancing lesions within 6 months of the therapy[30–32] attributed to IRIS associated with PML.[33]

Prolonged use of immunomodulatory therapies involving corticosteroids and certain antirheumatic drugs, such as methotrexate, natalizumab, fingolimod, rituximab, mycophenolate mofetil, belatacept, infliximab, alemtuzumab, fludarabine, leflunomide, and fumaric acid ester, has also been associated with the development of particularly virulent PML.[34,35] Although several autoimmune conditions have demonstrated association with the onset of PML, systemic lupus erythematous shows the highest predisposition to the development of PML.

Pathogenicity, host immune responses, and pathogenesis

The archetypal JC virus is incapable of effective replication in the glial tissue, and has to develop neurotropism, followed by the evasion of the neuro-immunosurveillance and clearance to establish infection in the brain. Genetic modifications and rearrangements of the viral genome facilitate the transformation of the latent archetypal JC virus to the virulent prototypal JC virus.[36] The genetic transformation influences the viral replication abilities, the transcriptional profiles of several viral genes, tissue tropism, and the pathogenicity of the virus.[30]

Host cellular immune responses play a crucial role in the clearance of the reactivated virus, which prevents the virulent manifestations of the disease.[37,38] The immune responses elicited by the cytotoxic T-lymphocytes are indispensable to viral clearance as demonstrated by the elevated levels of these cells in CSF samples of patients who successfully prevented the development of virulent disease.[38] The low levels of CD4 cells in several immunocompromised conditions, resulting in inefficient cellular responses, therefore, predisposes to PML. The viral replication within cerebral oligodendrocytes and astrocytes results in lysis-mediated damage to oligodendrocytes, thereby destroying the protective myelin sheaths encasing the neurons. The heavy influx of immune cells, viral proteins, and pathogenic antigens contributes further to the neurologic insult, increasing the extent of demyelination. The rapidly progressive focal neurologic deficits without any increase in intracranial pressure characteristic of PML is a result of the gradual neurological damage.[26]

The predisposition to PML precipitated by immunomodulatory therapies is attributed to the $\alpha 4\beta 1$ integrin-binding mechanism implemented by certain immunomodulatory drugs. This mechanism prevents lymphocytes from binding to the VCAM on the surfaces of the endothelial cells comprising the CNS. The host neuro-immunosurveillance capabilities are therefore compromised, increasing the vulnerability of the CNS to the pathogenicity of the reactivated JC virus.[39]

Although PML is the predominant clinical manifestation affected by the reactivation of the JC virus, GCN, demonstrating a pathology of the granule cell neurons of the cerebellum manifested as ataxia, nystagmus, and tremors,[40] JCV encephalopathy characterized by a lysis-mediated pathology of the cortical gray matter,[41] and isolated JCV meningitis, are some other forms in which the virulence of the JC virus may be expressed.[42]

Clinical manifestations and complications

The clinical manifestations of PML vary depending on the region of the CNS that has been affected by the disease pathology. The subcortical white matter, periventricular areas, and cerebellar peduncles are the most affected regions, while the disease pathology affecting the optic nerve and the spinal cord is relatively rare. Progressive hemiparesis, visual field deficits, altered mental status, aphasia, and cognitive impairment are the most common clinical presentations of the disease. Pure cerebellar truncal ataxia, appendicular ataxia, and cranial nerve deficits are relatively rare manifestations of PML. Patients in advanced stages of the disease commonly present with severe neurologic deficits, altered vision, quadriparesis, hemianopia, profound dementia, and coma. Cerebral white matter riddled with lesions is characteristic of this severe demyelinating disease (Fig. 8.6), while lesions in the cerebellum, brainstem, and the spinal cord are very rare.

The presentation of PML demonstrates some subtle variations depending on the trigger differences, such as HIV-associated PML and immunomodulation-associated PML. While HIV-associated PML predominantly manifests as pathologies associated with the motor system,[31] immunomodulation-associated PML generally presents itself with cognitive and language-related

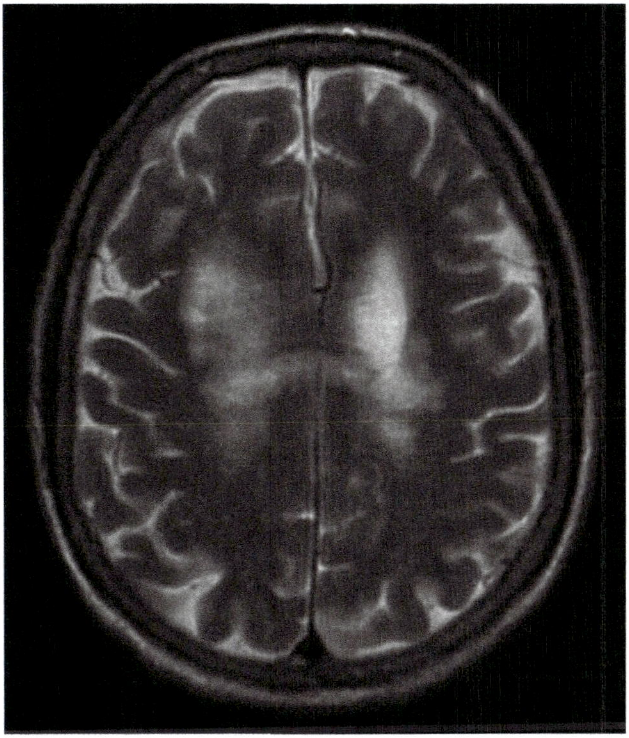

FIGURE 8.6

MRI image of a PML patient. *Source: From http://www.radswiki.net/ https://creativecommons.org/licenses/by-sa/3.0/deed.en; https://commons.wikimedia.org/wiki/File:Progressive_multifocal_leukoencephalopathy_002.jpg.*

deficits.[43] Seizures are frequently associated with patients demonstrating juxtacortical and T1-hyperintense lesions in MRI scans.[41]

Diagnosis

PML is usually considered in the differential diagnosis in high-risk individuals presenting with neurological deficits. CD4 count is a reliable indicator of the degree of immunosuppression, and PML is usually considered in patients presenting with a CD4 count of less than 200, although cases demonstrating the onset of PML pathology even at CD4 counts higher than 200 are known.[44]

The clinical diagnosis of PML requires an amalgamation of information comprising of patient history, clinical manifestations, radiographic evidence, and virological tests. The differentiation of clinical manifestations of PML from those of other demyelinating autoimmune conditions, such as multiple sclerosis, poses a clinical challenge because of the similarities between the general symptoms of PML and an Ms flare.[43,45] Normal CSF cell count and biochemistry profile are common in the early stages of the disease, and can, therefore, be clinically misleading. Although electroencephalogram (EEG) may reveal focal deficits in some cases, early stages of PML usually demonstrate normal EEG reports.

While routine blood counts and tests for specific conditions such as HIV can indicate the possibility of immunosuppression resulting in PML, the clinical diagnosis and evaluation of PML relies on radiological imaging. Contrast-enhanced imaging using CT or MRI scans facilitate the detection of pathological changes, such as inflammation and mass effect, which are suggestive of alternate pathologies like malignancies or infections of the CNS, thereby ruling out the diagnosis of PML.[46] CT scans showing hypodense confluent lesions without mass effect, contrast enhancement, or risks of herniation are suggestive of PML, while areas of a decreased signal on T1-weighted images in MRI scans and hyperintense lesions in the subcortical, periventricular, and cerebellum regions on T2-weighted images in MRI with contrast raise are indicative of PML. CT scans, however, often suggest a lower severity of disease involvement. MRI scans are more reliable for getting a true clinical picture of the extent of the disease pathology involved, and also enable differentiation between brain tumors, abscesses, and inflammatory changes. Multifocal demyelination indicated by the presence of smaller lesions coalescing into larger foci throughout the white matter with extensive involvement of entire hemispheres is characteristic of PML. HIV patients undergoing ART or antirheumatic therapy may develop neurological manifestations and altered mental status suggestive of PML–IRIS. Such patients may demonstrate contrast edema and mass effect on MRI scans.[33,47]

A confirmatory diagnosis of PML relies on a positive PCR test for the JC virus from CSF.[48] PML–IRIS, however, may give a negative PCR test due to the immunocompetence of the host, which counters the viral replication, resulting in an extremely low viral load undetectable on a viral assay. Such patients are therefore diagnosed based on the presence of JCV-specific antibodies. JCV antibody index is used as a prognosis parameter for tracking the progress of the disease. An antibody index exceeding 1.5 has been associated with a higher risk of PML.[49]

Histopathological investigations involving brain biopsies demonstrate characteristic cytopathic changes in oligodendrocytes, including enlarged nucleus containing accumulated basophilic material that gets deeply stained and a disruption of normal chromatin pattern (Fig. 8.7). Astrocytes containing lobulated hyperchromatic nuclei may also be seen.[50] The cytopathic changes result

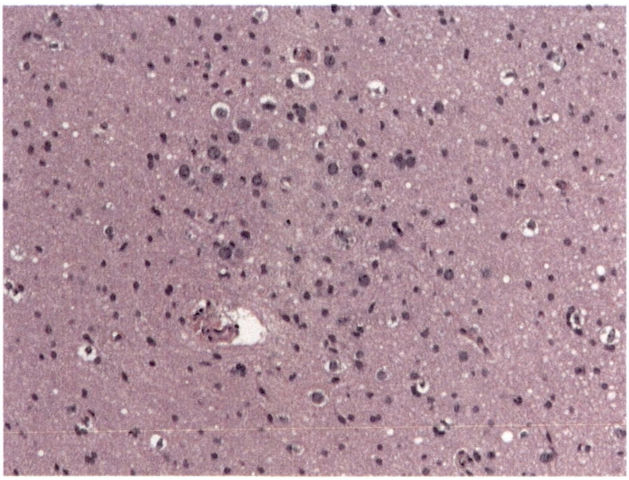

FIGURE 8.7

Brain biopsy showing oligodendrocytes with nuclear enlargement and glassy magenta chromatin. *Source: From Jensflorian/Own work/; https://creativecommons.org/licenses/by-sa/4.0/deed.en; https://commons.wikimedia.org/wiki/File:PML-HE_brain_biopsy.jpg.*

in characteristic pathological changes demonstrated in brain biopsies as multiple asymmetric foci of demyelination at various stages of evolution in the cerebral white matter. Electron microscopic visualization demonstrates the presence of JC virions within reactive astrocytes and enlarged oligodendrocytes.

Treatment and management

PML is a progressive and usually fatal disease. The restoration of an effective immune system as facilitated by ART in AIDS patients and withdrawal of causative drugs and plasma exchange in antirheumatic drug-associated PML cases can slow the disease progression and improve the prognosis.[51,52] The CD4 cell count plays a crucial role in determining the prognosis and survival in HIV cases.[53] Young age, fewer pre-PML debilitating conditions, low viral load, and less extensive brain involvement are also associated with lower mortality.[54] IRIS is often associated with PML due to the worsening clinical manifestations caused during the recovery of the immune system, warranting the need for corticosteroid therapy, in addition to other supportive care and plasma exchange.[33,35] PML–IRIS cases have often shown improvement on cessation of ART;[33] however, the risk of antiretroviral resistance and increase in viral load are contraindications in this treatment method. Glucocorticoids are recommended for PML–IRIS in the event of mass effect and herniation, which are high-risk effects of the damages elicited by immune reconstitution.

Student Notes

Although drugs, such as mefloquine, cytarabine, and cidofovir, have demonstrated anti-JCV properties in vitro, there are no reports of demonstrated efficacy in vivo.[55–57] Dendritic cell vaccines comprising of the JCV antigens have demonstrated the ability to elicit a clinically significant CD8 response. The boosted immune response has prolonged survival in the PML patients involved in the study. However, the extremely low number of subjects involved in the study has warranted the need for studies involving larger cohorts to enable the therapeutic use of these agents.[58] Therefore the mainstays of PML treatment continue to be restoration or bolstering the cellular immune responses in immunocompromised cases, cessation of the inciting drug, and plasma exchange in immunomodulation-associated PML cases, coupled with appropriate supportive patient care.

Key take-aways

1. PML is a progressive neuropathology, affecting the CNS.
2. The disease progression is rapid and associated with poor prognosis and high levels of mortality and morbidity.
3. The JC virus is the etiologic agent of PML.
4. Although the JC virus infects a vast majority of the world's population in childhood, it remains latent in nonneuronal organs such as the kidney.
5. Modulation of the host immune system can cause the JC virus to reactivate and invade the CNS by evading neuro-immunosurveillance.
6. Immunocompromised conditions, such as natural and acquired immunodeficiencies, abnormalities of cell-mediated immunity, lymphoproliferative diseases, chronic inflammatory conditions, autoimmune diseases, solid organ and bone marrow transplants, and certain immunosuppressive and immunomodulatory treatment regimens, increase predisposition to PML.
7. Diagnosis of PML relies on an amalgamation of information obtained from patient history, clinical manifestations, radiological evaluations, specific PCR tests, antibody tests, and microscopic studies.
8. There are no known targeted antiviral therapies for PML. Therefore treatment and management primarily focus on restoration of the immune system, elimination of any known inciting factors, and generalized supportive care.

Problem 8.3: Erythema infectiosum

A 3-year-old female presented with an erythematous rash on her face and trunk, hepatosplenomegaly, and has a history of severe anemia. The patient was severely malnourished at presentation. The patient's birth history was uneventful, with Apgar scores of 4, 5, and 7 at birth, at 5 min, and at 10 min, respectively. Dysmorphic features, such as hypertelorism and small palpebral fissures, were observed soon after birth. The child demonstrated a developmental delay in verbal, gross, and fine motor skills. There were no known food or drug allergies.

The child had been hospitalized with a lower respiratory tract infection (LRTI) at the age of 1.5 years, and was found to be below the 5th centile for height and weight. Investigations for tuberculosis (TB), Epstein–Barr Virus (EBV), Toxoplasmosis, Rubella, CMV, Herpes, Hepatitis B, and HIV had all yielded negative reports, except EBV, which was negative for IgM and positive for IgG. The child had elevated levels of ceruloplasmin, copper, complement C3, and complement C4 at this time. The child had then been rehospitalized at 2.5 years with a presentation of LRTI. Investigations revealed negative TB tests, and elevated levels of complement C3 and C4. There was no history of surgical interventions.

There is no history of maternal alcohol abuse or infections during pregnancy. The patient is the first and only child born to her parents, and had been breastfed until 5 months of age, after which she was put on formula feed and subsequently started on solid food by the age of 10 months. The parents are HIV-negative and are in good health, and there are no known positive TB contacts. There are no other children or pets at home. There is no exposure to smokers and no history of recent travel involved.

The following are the reports of the investigations carried out:

1. General
   a. Weight: 8.4 kg (>5th centile)
   b. Height: 78 cm (>5th centile)
   c. Moderate pallor
   d. Generalized lymphadenopathy with some firm axillary lymph nodes bilaterally
   e. No dehydration
   f. No jaundice
2. Vitals
   a. Afebrile
   b. Blood pressure and heart rate normal
3. Head and neck
   a. Dysmorphic features: hypertelorism and small palpebral fissures
   b. Puffy eyelids
   c. Open anterior and posterior fontanels

   d. Discolored straight hair
   e. Erythematous rash with mild induration observed on parts of the face, forehead, and trunk (indicative of atypical erythema nodosum)
4. Abdomen
   a. Distended
   b. Scratch marks were present
   c. Soft and nontender on palpation
   d. Extensive hepatosplenomegaly
   e. Auscultation revealed bowel sounds
5. Respiratory
   a. No signs of respiratory distress
   b. Chest clear
6. Cardiovascular
   a. Ejection systolic murmur but no significant cardiac lesions
7. Neurological
   a. Delayed milestones with regards to motor skills (fine and gross) and verbal skills
   b. Lack of ability to communicate hindered assessment of higher function
   c. Reflexes: 2/4
   d. Sensation: intact
   e. Tone: normal
   f. Strength: normal
   g. Gait: normal
8. Musculoskeletal
   a. Areas of induration were observed under the skin of both calves (could be attributed to calcification in the calf muscles)
9. Liver function tests

Test	At 1.5 years	At 3 years	Reference values
Total protein (g/L)	66	64	68–80
Albumin (g/L)	39	33	35–55
Total bilirubin (μmol/L)	15.0	17.0	1.71–20.5
Direct bilirubin (μmol/L)	3.0	3.0	–
Alkaline phosphatase (U/L)	297	196	44–147
Gamma GT GGT (U/L)	215	78	9–48
ALT (U/L)	45	47	10–40
AST (U/L)	89	83	8–33

GGT, Gamma-glutamyl transferase; *ALT*, alanine transaminase; *AST*, aspartate aminotransferase.

10. Antibody profiles

Test	At 1.5 years	Reference values
IgG (g/L)	12.90	4.53–9.16
IgA (g/L)	1.30	0.30–1.20
IgM (g/L)	2.30	0.50–2.20

11. Ceruloplasmin: 1.09 (at 1.5 years)
12. ANF: Negative
13. HIV: Negative
14. Urine metabolic screen: Negative
15. Parvovirus PCR (Sample—Bone marrow aspirate): Positive
16. Hematological profile

Test	Values	Reference values
WBC (/L)	4.4×10^9	$5.0–19.5 \times 10^9$
Hemoglobin (g/dL)	6.1	10.7–17.1
Platelet (/L)	45×10^9	$150–400 \times 10^9$
Neutrophils (%)	1.54	20–50
Lymphocytes (%)	0.55	25–33
Monocytes (%)	0.57	2–8
Band (%)	0.56	<10
Myelocytes (%)	0.01	0
Smudge cells (%)	0.01	0
NRBC (%)	1.8	0

NRBC, Nucleated red blood cell.

17. Immunological profile

Test	Values	Reference values
CD3 (%)	0.42	59.7–77.6
CD3 (cells/μL)	88	1578–3707
CD4 (%)	0.16	31.1–47.4
CD4 (cells/μL)	35	870–2144
CD8 (%)	0.257	16.0–26.9
CD8 (cells/μL)	69	472–1107
CD19 (%)	1.4	12.9–29.2
CD19 (cells/μL)	5	434–1274
CD56 (%)	0.53	4.7–16.2
CD56 (cells/μL)	138	155–565
CD10 (%)	1.69	1–3

18. Bone marrow culture
    a. Ziehl–Neelsen stain: Positive for Mycobacteria
    b. Culture: Negative for Mycobacteria
    c. Trephine appearance: Suggestive of secondary marrow fibrosis
19. Liver biopsy
    a. Mild portal tract inflammation demonstrating eosinophils, extramedullary hemopoiesis, and focal lobular inflammation
    b. Mild macrovesicular steatosis
    c. Single nonnecrotizing granuloma
    d. Ziehl–Neelsen stain: Negative
20. Skin biopsy
    a. Deep seated histiocytic infiltrate within the subcutaneous tissue
    b. CD1a and S100: Negative (excluding a Langerhans cell infiltrate)
    c. Chloroacetate esterase stain: Positive in some of the immature myeloid cells
    d. Stains (AFB, acid-fast bacillus and fungi): Negative
21. Chest X-ray: Pneumonitis (both lung fields)

Diagnosis

The patient was diagnosed with severe combined immunodeficiency syndrome (SCID), also known as Omenn syndrome, responsible for the clinical manifestations of erythema infectiosum caused by B19 parvovirus infection.

Treatment

1. Empiric antibiotic therapy
2. Four-drug TB treatment
3. Polyclonal intravenous IV immunoglobulin prophylaxis
4. Transfusion of packed red cells

Question and answer

1. What is SCID?
    a. SCID is a life-threatening condition that predisposes to recurrent infections, and manifests as a failure to thrive. The severe clinical manifestations of the condition are attributed to several deficits of the immune system, which result in severe T- and B-cell dysfunction, also affecting NK cells occasionally. The T- and B-cell

dysfunction usually results in severe infection and death in children by 2 years of age unless medical intervention is involved. Differences in the genetic defects affecting different pathways result in several types of SCID.

2. What is the typical medical history of a child presenting with SCID?
   a. SCID patients often present with failure to thrive and multiple severe and recurrent infections. Such patients typically have resistant forms of pneumonia and mucocutaneous candidiasis, varicella, herpes simplex, respiratory syncytial virus, rotavirus, adenovirus, enterovirus, parainfluenza virus, EBV, and CMV infections.
3. What are the clinical findings on examination of SCID patients?
   a. The clinical findings differ depending on the various infections affecting the patient. Sparse hair, abnormal dentition, osteopetrosis, and common cutaneous findings, such as eczematous dermatitis, recurrent furunculosis, oral thrush, and herpetic dermatitis may be commonly observed. Palpable lymphadenopathy is not presented by SCID patients due to the compromised state of the immune system, and this lack of recognizable peripheral lymph nodes raises a suspicion of SCID in children with multiple and recurrent aggressive infections.
4. What is the pathophysiology of SCID?
   a. SCID results from mutations in genes that are responsible for the differentiation and proliferation of T-cells, B-cells, and NK cells. Deficits in any one of the components also affect the effective functioning of the immune system. Even in conditions that affect only T-cell development and mature B-cells, antibody production is hampered because of the lack of help required from T-cells for antibody production. The NK cells are affected to varying degrees, depending on the type of SCID.
5. How does SCID most commonly occur?
   a. Immunoglobulin diversity is attributed to the somatic recombination of variable (V), diversity (D), and joining (J) gene segments. The recombination activating gene (RAG)-1 and RAG-2 endonuclease proteins initiate this recombination by cleaving DNA at recombination signal sequence sites. The nonlymphoid-specific DNA double-strand break repair machinery subsequently repairs the cleaved DNA. Mutations in RAG-1 or RAG-2 genes cause this process of antigen receptor recombination to be functionally impaired, resulting in various forms of SCID. Certain other proteins involved in the recombination process, including DNA-PKcs, DCLRE1C (also known as Artemis), XLF (also known as Cernunnos), and DNA ligase IV are also prone to acquiring genetic mutations.

6. Does SCID always result in a complete primary immunodeficiency where both B- and T-cells are completely nonfunctional?
   a. SCID typically results in complete deficiency of an effective immune system; however, certain forms of SCID are caused by genetic mutations that permit partial enzymatic activity, precipitating conditions known as "leaky" SCID or Omenn syndrome. The development of a limited repertoire of T- and B-cells is attributed to restrictive V(D)J recombination. The immune responses generated by these T- and B-cells are often directed toward self-antigens precipitating autoimmunity. The T- and B-cell receptors in cases of Omenn syndrome undergo clonal expansion and their levels in blood are often within the normal range or may even be elevated. However, the inability to recognize a wide variety of epitopes results in immunodeficiency.

Learning objectives

1. Define Severe Immunodeficiency Syndrome (SCID).
2. Justify the prefix "leaky" for certain forms of SCID.
3. Explain the association between SCID and multiple recurrent infections.
4. Enumerate the risk factors associated with SCID.
5. Identify the epidemiology of erythema infectiosum.
6. Describe the pathophysiology of erythema infectiosum.
7. Describe the clinical manifestations of erythema infectiosum.
8. Enumerate the methods involved in the diagnosis of erythema infectiosum.
9. Determine the prognosis of erythema infectiosum.
10. Summarize the treatment and management options for erythema infectiosum.

Discussion

Differential diagnosis

HIV	HIV infection usually develops generic symptoms in the initial stages of the disease. The clinical presentation may include a fever, chills, rash, night sweats, muscle aches, sore throat, fatigue, lymphadenopathy, and mouth ulcers. Late stages of HIV infection, also referred to as AIDS, characteristically present with rapid weight loss, recurring fever, night sweats, extreme fatigue, long-lasting diarrhea, extensive lymphadenopathy, recurrent respiratory infections,

Continued

Continued

	sores in the mouth, anus, and genitals, discolored/ pigmented blotches on the skin, or inside the eyelids, nose, and mouth, and an altered mental status.
TB	TB presents itself with a persistent long-lasting cough, loss of appetite, unintentional weight loss, fever, chills, and night sweats.
Hypogammaglobulinemia	Hypogammaglobulinemia can present itself with a wide range of symptoms, depending on the infections involved. Recurrent respiratory tract infections, including bronchitis, pneumonia, and sinus infections, frequent ear infections, skin infections, such as eczema, urinary tract infections, intestinal infections, meningitis, and food allergies within the first 2 years after birth, raise a clinical suspicion for immunodeficiency. Depending on the infections involved, the symptoms may include a persistent cough, sore throat, fever, ear pain, chest congestion, sinus pain, nausea, vomiting, abdominal cramps, refractory diarrhea, myalgia, and joint pain.
Agammaglobulinemia	Agammaglobulinemia is usually characterized by recurrent bacterial infections such as those caused by mycobacteria, *Hemophilus influenzae*, streptococci, staphylococci, pneumococci, and *Pseudomonas* beginning late in the first year of life, although vulnerability to certain viruses, such as the enteroviruses, including the poliomyelitis virus and the echovirus, is also seen.
SCID	SCID generally presents itself with pneumonia, recurrent ear infections, meningitis, chronic skin infections, oral and genital yeast infections, diarrhea, and liver infections.
Measles	Measles often causes a high-grade fever, cough, runny nose, watery eyes, conjunctivitis, tiny white spots referred to as Koplik spots inside the mouth, and flat red spots topped with/ without small raised bumps appearing near the hairline on the face and spreading downward to the neck, trunk, and limbs.
Rubella	Rubella can appear with a low-grade fever, headache, runny nose, cough, mild reddening and swelling of the eye, malaise, and lymphadenopathy.
Roseola	Roseola can cause a sudden high-grade fever, sore throat, runny nose, cough, swollen lymph nodes along the neck, nonitchy rash with flat or raised pink spots or patches that may be surrounded by a white ring on the chest, back, and abdomen, spreading to the neck and arms, while the face and legs may remain unaffected by the rash.
Scarlet fever	Scarlet fever presents itself with an extremely red and sore throat, swollen glands in the neck region, high-grade fever, whitish coating on the tongue/red and bumpy tongue appearance, referred to as "strawberry" tongue, a red skin rash that feels like sandpaper, and bright red skin in the underarms, groin, and elbow fold regions.

Background

Erythema infectiosum, also known as the "slapped cheek syndrome," is one of the six common childhood viral exanthems, hence the disease is also referred to as the "fifth disease."[59] Although erythema infectiosum can affect adults, the occurrence is very rare. The etiological agent responsible for this disease is the human parvovirus B19. Erythema infectiosum is one of the many disorders, including aplastic anemia, polyarthropathy, and hydrops fetalis, attributed to the human parvovirus B19.

Etiology

The human parvovirus B19 is the etiological agent responsible for erythema infectiosum. The virus belongs to the *Parvoviridae* family, and is a nonenveloped, icosahedral, single-stranded linear DNA virus.[60] The virus is strictly restricted to being a human pathogen; therefore zoonotic transmission of the virus does not occur. Although the transmission of the virus is usually through the respiratory route, the tropism of the human parvovirus B19 for the red cell progenitors also facilitates trans-placental transmission and hematogenous spread of the virus.[61]

Epidemiology

Erythema infectiosum occurs worldwide in children, although rare incidences of diseases caused by the virus in adults are also known to occur. The incidence rate of this infection is commonly seen to increase during the spring –summer months of the year.[62]

The etiological agent responsible for erythema infectiosum, human parvovirus B19, is also implicated as the causative agent in several other disorders. The risk of developing a severe disease due to infections caused by this virus in pregnant women and in individuals with underlying hematological disorders, such as sickle-cell anemia, is very high. Human parvovirus B19 infection in pregnant women, especially in the second trimester, can precipitate complications, including miscarriage, intrauterine death, and hydrops fetalis, often resulting in fetal loss.[63] The viral tropism for red cell progenitors results in lytic damage to the reticulocytes, which precipitates a decrease or transient halt in the process of erythropoiesis. Individuals with underlying hematological disorders are prone to developing an aplastic crisis that can lead to severe anemia. The acute anemia often causes such patients to get pallor, tachycardia, tachypnea, and are more prone to fevers, malaise, and lethargy.[64]

Pathogenicity and pathogenesis

The human parvovirus B19 usually infects the cells of the human respiratory tract after being transmitted through direct or indirect contact with contaminated respiratory secretions. The virus also has a tropism for the RBC progenitors, which also makes transmission through contaminated blood products possible. The clinical manifestations of the disease become evident after an incubation period of 5–10 days. The spread of the virus is the highest for about 1 week after the clinical manifestations first begin to appear. The clinical manifestations of the disease differ depending on the immune status of the infected host. Some immunocompetent individuals exposed to the parvovirus B19 may even remain asymptomatic. Others may develop generic flu-like symptoms, such as a fever, cough, malaise, and coryza or more specific symptoms, such as a facial rash or exanthem and arthralgias. The appearance of arthralgias and exanthems marks the end of the infectious period for the host.

Individuals with underlying hematological disorders, such as aplastic anemia, usually demonstrate a very high viral load, and are prone to developing more severe diseases. Trans-placental transmission of the virus in pregnant women results in infection of the fetus. The viral tropism for RBCs compromises the erythropoietic process of the fetus, manifesting high output heart failure, eventually resulting in fetal loss due to hydrops fetalis.[65]

The human parvovirus B19 has an affinity to the cells of the human respiratory tract, where it binds to the cell receptors and enters the cell. The virus then transiently resides in these cells, before migrating to the erythroid progenitor cells, which are the main target cells of the human parvovirus B19. The primary receptor Gb4Cer (globoside), also known as the P antigen, Ku80 autoantigen coreceptors, and a5b1 integrin present on the erythroid progenitor cells are required for attachment of the virus and the endocytic internalization of the virus. In addition to being expressed on the erythrocytes, the globoside receptor is also expressed on platelets and granulocytes, facilitating the dissemination of the virus to the other host tissues. Lungs, heart, liver, kidney, synovium, endothelium, and GI tissues also express the globoside receptor, thus providing reservoirs for the viral DNA to persist long after the active infection has been cleared from the host system.[66] Individuals genetically lacking the P antigen are immune to infection by the human parvovirus B19.[67]

The cytopathic effect of the human parvovirus B19 infection is demonstrated by the appearance of giant pronormoblasts, also referred to as lantern cells[68] in bone marrow aspiration samples. Giant pronormoblasts are progenitor erythroid cells characterized by their large size and prominent eosinophilic nuclear inclusion bodies, cytoplasmic vacuolization, and occasional occurrence of "dog-ear" projections[69] (Fig. 8.8).

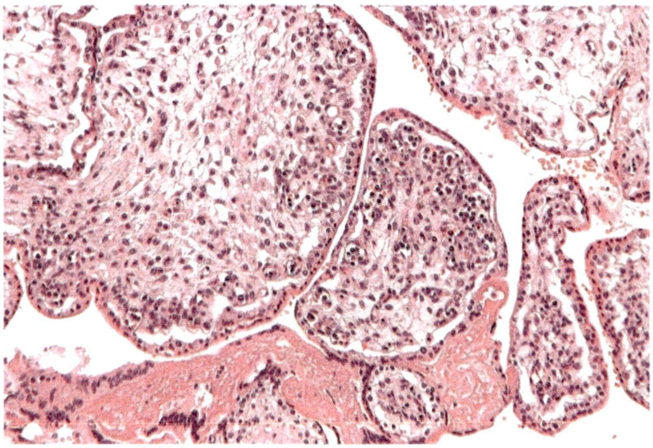

FIGURE 8.8

High magnification micrograph of normal and parvovirus B19 infected red blood cells, showing characteristic nuclear enlargement with glassy (*red*) inclusions. *Source: From Nephron/Own work/; https://creativecommons.org/licenses/by-sa/3.0/deed.en; https://commons.wikimedia.org/wiki/File: Parvovirus_infection_-_high_mag.jpg.*

Host immune response

The immune response to the human parvovirus B19 infection seen in immunocompetent individuals is predominantly the humoral immune response.[70] Effective humoral responses require the activation of CD4 + helper T-cells and B-lymphocytes. The viral infection has also been found to elicit a classic Th1 response via the presentation of the viral capsid proteins to the CD4 T-cells through class II molecules.[71–73] Adaptive immune responses, especially the humoral immune responses generating virus-specific neutralizing antibodies, play an important role in restricting the replication of the virus in the host and eventually resolves the infection.

The prodrome of symptoms associated with the human parvovirus B19 infection usually appears after 5–10 days of exposure to the virus. The infected individuals present with a febrile illness accompanied by coryza, malaise, and myalgia. The shedding of the virus in the respiratory secretions is very high during this period, thereby making the risk of transmission the highest throughout this stage. This phase is also characterized by reticulocytopenia. The production of specific IgM antibodies, approximately 10 days postexposure when the viral load is at peak, marks the beginning of the recovery period. IgM antibodies persist in convalescing serum for about 3 months, but they are also known to be detected for up to several months. About 5 days after the appearance of the IgM antibodies, IgG antibodies appear in the serum. The appearance of the IgG antibodies coincides with the clinical manifestations of rash and arthralgia. The presentation of the rash and arthralgia also marks the end of the viral shedding period, thereby terminating the spread of the virus. The IgG antibodies are detectable in convalescing serum for 2 weeks, after which they persist for life and confer life-long immunity to the host. The IgA antibody is also known to play a significant role in the protection against future human parvovirus B19 infections via the nasopharyngeal route.[74–76]

The clinical course of the human parvovirus B19 infection is dictated by the hematologic and immunologic status of the host. Healthy individuals without any underlying hematological or immunological disorders usually have a self-limiting disease course, often characterized by subclinical erythroid aplasia and the manifestation of a rash and arthralgias during the immunological phase of

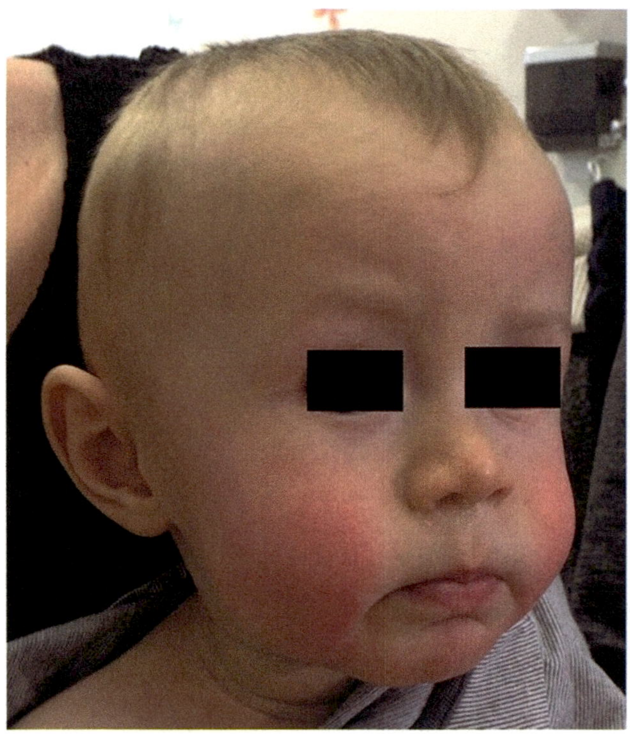

FIGURE 8.9

Erythema infectiosum caused by parvovirus B19 infection, demonstrating the characteristic facial rash, resulting in the reference of "slapped cheek syndrome." *Source: From Gzzz/Own work/; https://creativecommons.org/licenses/by-sa/4.0/deed.en; https://commons.wikimedia.org/wiki/File:Erythema_infectiosum.jpg.*

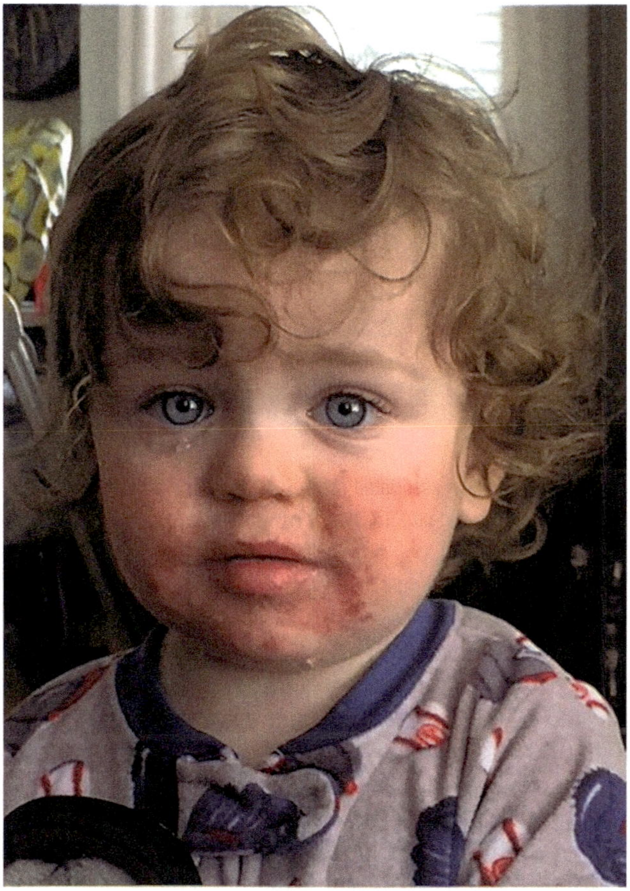

FIGURE 8.10

Pruritic facial rash characteristic of the human parvovirus B19. *Source: From Sandyjameslord/Own work/; https://creativecommons.org/licenses/by-sa/4.0/deed.en; https://commons.wikimedia.org/wiki/File:14_month_old_with_Fifth_Disease.jpg.*

the disease.[77] However, individuals with underlying hematological disorders that cause diminished production or increased destruction of erythrocytes demonstrate severe disease characterized by a dramatic reduction of hemoglobin levels, precipitating an aplastic crisis. Immunocompromised individuals infected by the human parvovirus B19 may even fail to elicit an immune response sufficient for viral clearance, often precipitating a state of chronic anemia.[78]

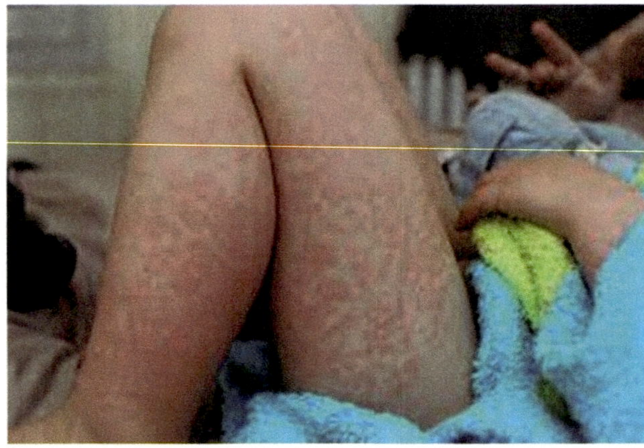

FIGURE 8.11
Late-stage parvovirus B19 infection-associated rash. *Source: From csaveanu/Own work/; https://creativecommons.org/licenses/by-nc-nd/2.0/; https://www.flickr.com/photos/csaveanu/7166778540/.*

Clinical manifestations and complications

Erythema infectiosum classically presents with a biphasic clinical course. The first phase of the disease that manifests immediately after the viral incubation period of 5–10 days, is characterized by nonspecific flu-like symptoms, including a mild fever, coryza, myalgia, nausea, vomiting, diarrhea, malaise, and headache. The classic itchy sensation and erythematous rash with a surrounding region of pallor, especially involving the cheeks, is responsible for the reference of the "slapped-cheek rash" associated with the disease (Figs. 8.9 and 8.10). The appearance of the facial rash and arthralgias marks the immunological phase of the disease, during which the immune responses in an immunocompetent host clear the virus from the system, thus resolving the infection. The facial rash is followed by the development of a nonitchy maculopapular rash appearing on the trunk and limbs (Fig. 8.11). A lacy or reticular appearance of the rash, often seen on the extensor surfaces, marks the resolution of the rash after about a week. The erythema infectiosum rash typically does not affect the palms of the hands and soles of the feet and is often exacerbated by exposure to the sun or heat.[79] Arthralgias typically affects symmetrical joints, such as those of the hands, feet, wrist, knees, and feet. Although joint stiffness is the most common complaint in this phase of the disease, there is no evidence of physical damage to the affected joints. The joint symptoms typically resolve within about 3 weeks of onset. The beginning of arthralgias and appearance of the facial rash marks the end of the contagious phase of the disease.

The lack of an effective immune response in immunocompromised individuals prevents the manifestations of the rash and arthralgias, classically associated with the disease. Such patients are also prone to the development of a chronic form of the infection, resulting in neutropenia, thrombocytopenia, or complete bone marrow suppression due to the insufficient clearance of the virus from the host system by the compromised immune system.[65]

Diagnosis

The self-limiting disease course and mild manifestation of symptoms usually does not necessitate a clinical diagnosis of the human parvovirus B19 infection. However, high-risk individuals with aplastic crisis warrant the need for a laboratory diagnosis of the disease. Laboratory detection usually entails the discovery of specific antibodies for the human parvovirus B19. While IgM antibodies detectable within a week to 10 days of exposure to the virus are indicative of acute infection, IgG antibodies are suggestive of a past infection

Student Notes

as these attain detectable levels about 15 days after exposure to the virus and persist at measurable levels for life.[80] The presence of IgG antibodies specific to the human parvovirus B19 in pregnant women is a reliable determinant of the immunity status that enables a clinical assessment of the risk of potential congenital disabilities in the fetus.[65]

Treatment and management

Erythema infectiosum is typically a self-limiting disease in immunocompetent individuals. Therefore supportive care for symptomatic relief is the mainstay of treatment for erythema infectiosum. Acetaminophen and/or non-steroidal anti-inflammatory drugs (NSAIDs) is recommended for the management of febrile illness, arthralgias, and other flu-like symptoms. Cases demonstrating the development of aplastic crisis warrant the need for hematological testing, followed by RBC transfusions, if indicated. Pregnant women infected with the human parvovirus B19 are especially prone to developing fetus-associated complications, and therefore require medical attention, including frequent ultrasonography scans to prevent conditions like hydrops fetalis.[65]

Key take-aways

1. Erythema infectiosum is one of the six most common childhood exanthems and is caused by the human parvovirus B19.
2. The typical facial rash characteristic of erythema infectiosum causes it to be referred to as "slapped cheek syndrome."
3. Human parvovirus B19, the etiologic agent of erythema infectiosum, is also implicated in several other human diseases, including aplastic anemia, polyarthropathy, and hydrops fetalis.
4. Although parvoviruses are known to cause infections in other animals, the parvovirus B19 is strictly a human pathogen.
5. The transmission of the virus is known to occur through the respiratory route, hematogenous route, and the trans-placental route.
6. The virus has an affinity for the P antigen expressed on the red cell progenitors and migrates to the bone marrow soon after infection through the respiratory route.
7. The pathologies associated with the parvovirus B19 are strongly influenced by the immunological and hematological status of an individual.
8. Erythema infectiosum is a self-limiting disease in immunocompetent individuals with normal hematological profile.
9. Individuals with underlying hematological disorders, such as sickle cell anemia, and pregnant women, are especially prone to developing

severe forms of the disease, for instance, chronic anemia and hydrops fetalis, respectively.

10. The appearance of giant pronormoblasts or RBC progenitors with enlarged nuclei and characteristic inclusions is indicative of parvovirus B19 infection of RBCs.
11. Humoral immune responses play a significant role in the clearance of the virus from the host system and resolution of the infection, although cell-mediated immune responses are also required to facilitate the effectiveness of the humoral immune responses.
12. Specific IgM and IgG antibodies are important immune mediators in the resolution of parvovirus B19 infections and serve as diagnostic markers of the disease stage.
13. While IgM antibodies indicate acute disease, IgG antibodies persist for life and confer immunity against future infections.
14. The early stage of the disease is characterized by flu-like symptoms that sets in immediately after the incubation period of 5–10 days, which is the infectious stage of the disease, while the onset of immune-mediated symptoms, for example, the characteristic rash and arthralgias, marks the end of the transmissible phase of the disease.
15. Treatment and management of erythema infectiosum mainly relies on supportive care for symptomatic relief, although high-risk cases, like individuals with underlying hematological and immunological disorders and pregnant women, need continued and intense medical attention.

Problem 8.4: Chronic granulomatous disease

A 5-year-old male presented with fever, breathing difficulty, and abdominal pain with abdominal guarding. There was also a history of frequent headaches and skin abscesses, which do not heal easily. The patient's perinatal history was uneventful, with breastfeeding up to 6 months of age. All childhood vaccinations had been well-tolerated. There was a history of uncomplicated chicken pox that occurred at 2.5 years of age. The patient, however, measured below the 3rd centile for height and weight, demonstrating failure to thrive. Delayed healing of multiple and frequent skin abscesses had been noted throughout the patient's childhood.

The patient had been hospitalized at 4 years of age for swelling in the right knee and a poorly healed superficial abscess on his left foot, accompanied by fever. Arthrotomy of the affected knee and debridement of the foot had been performed, without any other complications. Sample cultures from the affected knee demonstrated the presence of *Staphylococcus aureus* and granulomas were seen in the histological analysis of the foot lesion. A complete blood count CBC indicated hypochromic microcytic anemia, with a CRP of 200 and erythrocyte sedimentation rate (ESR) of 90, indicative of systemic inflammation. The investigations for serum immunoglobulins were normal. A full investigation of all other organ systems had also been unremarkable. A full course of oral Clindamycin and a 1-week course of nasal mupirocin was the prescribed course of treatment. The parents were counseled with instructions regarding hygiene measures to ensure prevention of further impetigo and boils. The family pet was also investigated for *Staphylococcal* lesions.

The patient is the only child and both his parents are healthy. The family has one pet. Amenities such as running water supply and washroom facilities within the house are available. One of the maternal uncles of the patient was reported to have died in his teenage years due to recurrent infections.

The following are the reports of the investigations carried out:

1. General
   a. Not appearing to be acutely ill
   b. Mild jaundice
   c. Multiple scars from healed lesions/abscesses
2. Vitals
   a. Pyrexial
3. Abdomen
   a. Marked guarding and tenderness of the abdomen
4. Respiratory
   a. Decreased air entry into the right lower lobe
5. Abdominal ultrasound
   a. Loculated fluid in the right hepatic lobe and right lower lung region with communication through the diaphragm
6. Laparotomy and thoracotomy
   a. Purulent material drained
   b. Culture: *Streptococcus milleri*
7. Brain CT scan: Normal
8. Nitroblue tetrazolium (NBT) test (neutrophil function test): 0 (no reduction)

Diagnosis

The patient was diagnosed with chronic granulomatous disease (CGD), as the cause for the vulnerability to recurring infections. The clinical manifestations seen in this presentation resulted from liver abscess caused by *S. milleri*.

Treatment

1. Trimethoprim/Sulphamethoxazole
2. Antifungal prophylaxis
3. Gamma interferon (IFN) treatment

In spite of treatment with prophylactic antimicrobials and IFN therapy, the patient continued to demonstrate signs of being immunodeficient, as indicated by recurring liver abscesses and severe acne over the years.

Question and answer

1. What is CGD?
   a. CGD is an X-linked recessive inherited disorder of the immune system, which prevents phagocytes from killing ingested microbes. Nicotinamide adenine dinucleotide phosphate (NADPH) oxidase enzyme complex present in phagocytes also known as phagocyte oxidase (PHOX) mediates the elimination of the ingested microorganisms. Defects in this enzyme complex results in impaired killing and elimination of microbes. Phagocytosis normally causes neutrophils to increase their oxygen consumption through a process termed as the respiratory or oxidative burst. Phagocytosis acts a physiological stimulus that results in the generation of superoxide anion by the transfer of electrons from reduced NADPH to molecular O_2. This superoxide anion is then converted to bactericidal hydrogen peroxide and other reactive oxidants, causing the oxidative burst. Defects in the enzyme complex results in compromised phagocytosis processes, thereby causing the individual to be susceptible to a broad range of bacterial and fungal infections, which may even be life threatening.
2. What is the significance of the NBT test?
   a. NBT test is a blood test used to screen for CGD. This test measures the ability of the white blood cells WBC to change the colorless NBT to a deep blue by the process of reduction. The white blood cells WBC in a CGD patient are incapable of reducing NBT, thus preventing the indicative color change from occurring. CGD carriers show the presence of a mixture of normal and abnormal neutrophils, thereby equipping the NBT test to detect a carrier state for CGD.
3. Which are the organisms that an individual diagnosed with CGD is especially susceptible to and why?
   a. CGD increases susceptibility to severe and recurrent infections caused by microorganisms that are catalase-positive and are resistant

to nonoxidative killing, such as *S. aureus*, *Burkholderia cepacia*, *Serratia marcescens*, and fungi belonging to the *Aspergillus spp*. The catalase produced by these organisms breaks down endogenously produced hydrogen peroxide, thus preventing the killing of the phagocytosed microbes. Therefore normal functioning of the PHOX system is needed to ensure the death of the infecting microorganisms.

4. Why are patients suffering from CGD not vulnerable to *Streptococcus pneumoniae* infections?
   a. *S. pneumoniae* is a catalase-negative organism. The etiology of CGD is attributed to the inability of the affected system to generate the phagocytosis-associated respiratory burst, which is important for killing the phagocytosed microorganisms. However, the hydrogen peroxide produced by catalase-negative microorganisms can be used by CGD patients to overcome this intrinsic inability to form reactive oxidants, which facilitates the killing of the phagocytosed microorganisms.
5. Which part of the immune system is affected in CGD patients?
   a. CGD is a primary immunodeficiency that affects the ability of the innate immune system to kill micro-organisms by the process of phagocytosis.
6. What causes a B-cell-related cause to be unlikely in this case?
   a. Although the patient had a history of recurrent infections, there were no respiratory infections reported. In addition, the levels of the serum immunoglobulins were found to be normal on investigation. Therefore a B-cell-related cause is unlikely in this case.
7. Why would a T-cell-related deficit be unlikely?
   a. The patient reported to having tolerated all childhood vaccines and also had an adequate immune response to viral infections. This indicates that a T-cell-related deficit is unlikely in this case.
8. What is the association of CGD with granuloma formation?
   a. Microbial infections in CGD patients often result in extensive inflammatory reactions caused by the persistent accumulation of neutrophils resulting from the failure to degrade chemoattractants and antigens. CGD renders the immune system incapable of killing phagocytosed microorganisms leading to a persistent state of cell-mediated immune activation and granuloma formation. Granulomas are typically known to develop in the skin, GI tract, and genitourinary tract (GU), and are associated with related symptoms such as dysphagia, nausea, vomiting, abdominal pain, and obstruction in GI granulomas and dysuria, incontinence, abdominal discomfort, and urinary retention in GU granulomas.
9. What is the treatment for CGD?

a. Treatment and management of CGD revolves around two goals[81]: early diagnosis of the condition to ensure prophylactic use of antibiotics for infection prevention, and[82] awareness and education of patient and caregivers about the condition to ensure timely and appropriate treatment of infections. Bacterial infections are treated using appropriate antibiotics, while fungal infections require the use of antifungal agents. IFN therapy is known to facilitate better immune function and ability to fight infections in CGD patients, thereby reducing the occurrence of infections by almost 70% and also reducing the severity of the infections that may still occur. Therefore IFN therapy is often prescribed for the prevention of infections in CGD patients. The potential of gene therapy for the treatment of CGD is also being explored, as the etiology of CGD is attributed to a mutation in a single gene affecting only the hematopoietic system.

Learning objectives

1. Define CGD.
2. Justify the vulnerability to only certain microbial groups that is typical of CGD.
3. Explain the association between CGD and multiple recurrent infections caused by particular microorganisms.
4. Enumerate the risk factors associated with CGD.
5. Identify the epidemiology of CGD.
6. Describe the role of defective NADPH oxidase in the pathophysiology of CGD.
7. Identify the clinical manifestations of CGD.
8. Enumerate the methods involved in the diagnosis of CGD.
9. Determine the prognosis of CGD.
10. Summarize the treatment and management options for CGD.
11. Justify the use of prophylactic antibacterial and antifungal agents in the management of CGD patients.
12. Discuss the importance of personal hygiene and disease awareness in CGD patients.

Discussion

Differential diagnosis

HIV	HIV infection usually presents with generic symptoms in the initial stages of the disease. The clinical presentation may include fever, chills, rash,

Continued

Continued

	night sweats, muscle aches, sore throat, fatigue, lymphadenopathy, and mouth ulcers. Late stages of HIV infection, also referred to as acquired immune deficiency syndrome (AIDS), characteristically present with rapid weight loss, recurring fever, night sweats, extreme fatigue, long-lasting diarrhea, extensive lymphadenopathy, recurrent respiratory infections, sores in the mouth, anus, and genitals, discolored/pigmented blotches on the skin, or inside the eyelids, nose, and mouth, and an altered mental status.
Malnutrition	Malnutrition usually presents with unintended weight loss, reduced appetite, extreme fatigue, weakness, high frequency of falling sick, prolonged recovery period, long wound healing process, hypothermia, depression, and poor concentration.
Complement deficiency	Complement deficiency affects the ability of the immune system to fight infections, thereby presenting as recurring infections, organ dysfunctions, vasculitis, angioedema, dermatomyositis, glomerulonephritis, manifestations related to the joints, autoimmune disorders, and anaphylactoid purpura.
Cystic fibrosis (CF)	CF presents with wheezing, stuffy nose and sinuses, persistent cough producing thick mucus or phlegm, shortness of breath, especially when exercising, and recurrent lung infections.
Hyperimmunoglobulin E syndrome	Hyperimmunoglobulin E syndrome is characterized by recurring boils, severe rashes, and recurring sinus and lung infections usually presenting in infancy, with high immunoglobulin E levels.
Glucose-6-phosphate dehydrogenase (G6PD) deficiency	G6PD deficiency presents with very broad-spectrum symptoms including rapid heart rate, shortness of breath, dark colored urine, fever, fatigue, dizziness, jaundiced appearance of skin and whites of the eyes, and pallor.
Glutathione synthase (GS) deficiency	GS deficiency presents with symptoms associated with hemolytic anemia such as fatigue, pallor, lightheadedness, irregular heartbeats, and shortness of breath, splenomegaly, dysarthria, ataxia, intellectual disabilities, spasticity, intention tremors, psychomotor retardation, psychosis, seizure disorders, metabolic acidosis, and retinal dystrophies.
Crohn's disease	Crohn's disease is an autoimmune form of inflammatory bowel disease that presents with diarrhea, fever, fatigue, abdominal pain and cramps, blood-tinged stool, mouth sores, reduced appetite

Continued

Continued

and weight loss, and symptoms of anal fistula such as pain or drainage near or around the anus.

Background

CGD is a rare, inherited primary immunodeficiency disorder (PIDD), which affects the ability of certain white blood cells WBC such as neutrophils, monocytes, macrophages, and eosinophils to kill some ingested pathogens. CGD is also known as Bridges–Good syndrome and Quie syndrome.[81] The inability to eliminate certain pathogens is precipitated by a defect in the NADPH oxidase enzyme (NOX) present in the phagocytes, known as phagocyte NADPH oxidase (PHOX). This defect prevents the formation of ROS such as the superoxide radical and the subsequent production of hydrogen peroxide that is required for the elimination of certain pathogens.[82–85] This results in a dysfunctional immune system that increases predisposition to recurrent infections by certain pathogens and a chronic inflammatory state. The systemic distribution of granulomas (Figs. 8.12, 8.13, and 8.14) characteristic of CGD are a result of this chronic and excessive inflammation. In spite of the debilitating defect in the phagocytic cells, the other components of the immune system function normally, thereby increasing the vulnerability of the affected individual only to the pathogens requiring a functional NADPH oxidase enzyme for elimination.

Etiology

The genetic component involved in the etiology of CGD is a result of mutations in any one of five NADPH oxidase genes (NOX 1–5). These mutations are usually inherited in an autosomal recessive or X-linked recessive manner. A defective NOX2 enzyme inherited in the X-linked recessive manner is most commonly implicated in CGD,[86–88] thereby demonstrating a greater prevalence of CGD in males. While the *CYBA*, *NCF1*, *NCF2*, and *NCF4* mutations are inherited in the autosomal recessive manner, the *CYBB* mutation is inherited in an X-linked recessive pattern.

NOX enzymes facilitate the synthesis of ROS in functional immune systems, mediating the formation of superoxide, a particularly potent ROS generated within phagosomes that have engulfed pathogens for elimination. The transfer of electrons from NADPH to the molecular oxygen present within the phagosome results in the formation of an untimely superoxide anion. Subsequently, other ROS such as hydrogen peroxide, hypochlorous acid, hydroxyl radicals, and secondary amines capable of efficiently killing phagocytosed pathogens

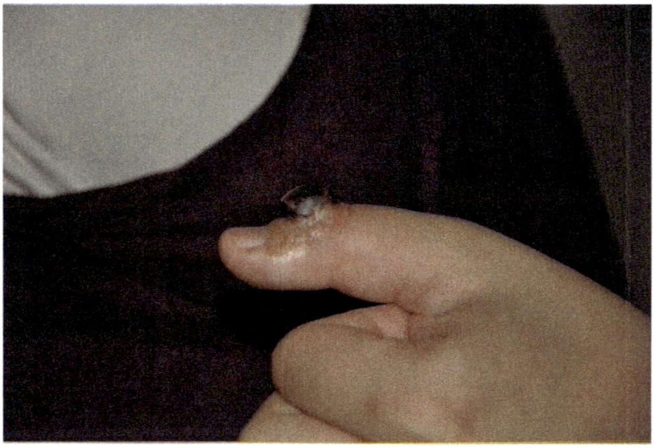

FIGURE 8.12
Pyogenic granuloma on a thumb. *Source: From Makotosan/Own work/; https://creativecommons.org/licenses/by-sa/3.0/deed.en; https://commons.wikimedia.org/wiki/File:Pyogenic_Granuloma_Thumb_2.JPG.*

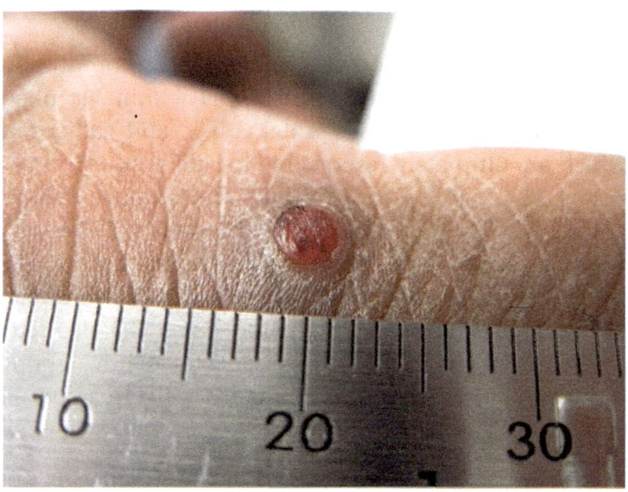

FIGURE 8.13
Pyogenic granuloma on the index finger. *Source: From Kilbad/Own work/; https://creativecommons.org/licenses/by/3.0/deed.en; https://commons.wikimedia.org/wiki/File:Pyogenic_granuloma_on_a_finger-1.jpg.*

are generated. The respiratory burst is a phenomenon resulting from the increased oxygen consumption and rapid generation of ROS.

The engulfment of microorganisms within phagocytic cells results in the formation of an intracellular vesicle formation. Subsequently, ROS such as

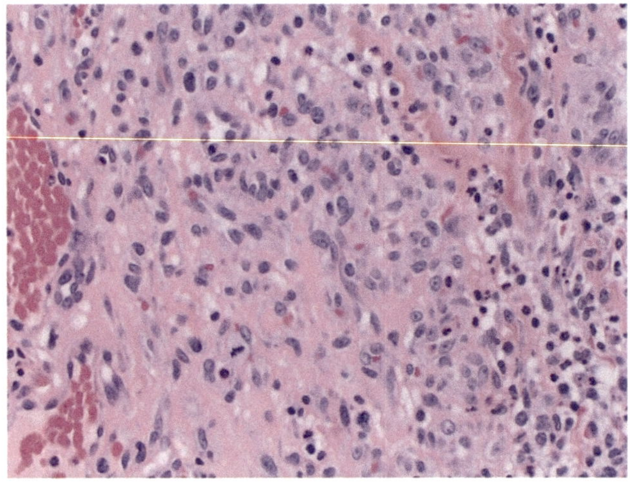

FIGURE 8.14

Histopathology of pyogenic granuloma (high magnification). *Source: From Mikael Häggström/Own work/; https://creativecommons.org/publicdomain/zero/1.0/deed.en; https://commons.wikimedia.org/wiki/File: Histopathology_of_pyogenic_granuloma_-_high_magnification.jpg.*

superoxide and other oxygen-derived metabolites are produced. This results in a potassium influx leading to increased pH within the phagosome. The granule proteases such as elastase, proteinase 3, and cathepsin G are then activated, which are ultimately responsible for the destruction of the ingested microorganisms. The inability to form ROS and produce hydrogen peroxide due to the defect in NOX enzymes in CGD prevents the effective elimination of certain phagocytosed microorganisms. The reduced production of hydrogen peroxide also hampers the enhancement of neutrophil extracellular traps, instrumental in entangling and killing pathogens. CGD also reduces the rate of efferocytosis, the process by which phagocytes remove apoptotic inflammatory cells causing the apoptotic inflammatory cells to accumulate resulting in granulomatous inflammation, characteristic of CGD.

Variations in the mutated gene, the type of mutation, and the position of the mutation within the gene may result in mutations that affect the NOX enzymes only partially, leading to milder forms of the condition.

Epidemiology

CGD is a rare condition affecting 1 in every 200,000–2,50,000 live births, and has an annual incidence of approximately 20 new cases.[89] As the inheritance of X-linked recessive mutation in the NOX2 enzyme is most commonly implicated in CGD, the prevalence of the condition is seen to be higher in males. The X-linked variant of CGD has also been implicated in early disease onset and is known to precipitate a more severe form of the condition than the autosomal recessive form.[90,91] Consanguineous marriages are often associated with the autosomal recessive type of the disease and higher prevalence of the condition. Although CGD may present at any age from infancy to late adulthood, the rate of diagnosis among toddlers is the highest.[92–94] The survival of CGD patients relies on the residual superoxide production independent of the specific mutated gene involved in the etiology.[95]

Pathogenesis and pathophysiology

CGD is a PIDD, characterized by certain inherited genetic mutations in the NOX enzymes that lead to an inability of the phagocytic cells such as the macrophages and neutrophils to kill and eliminate certain phagocytosed microorganisms. The defect in the NOX enzymes caused by the CGD-associated mutations affects the processes involved in the production of ROS and other downstream metabolites such as hydrogen peroxide that are instrumental in killing the phagocytosed microorganisms.

CGD generally increases predisposition to infections caused by catalase-positive organisms rather than catalase-negative organisms. Products of respiratory burst such as hydrogen peroxide are indispensable to the process of elimination of phagocytosed organisms. CGD-associated mutations hamper the normal production of hydrogen peroxide. Microorganisms also produce indigenous hydrogen peroxide as part of their metabolism. The hydrogen peroxide derived from the infecting organisms can be used by CGD patients to eliminate the pathogen, thereby bypassing the intrinsic enzymatic defect. However, some organisms also produce catalase, an enzyme capable of inactivating the indigenously produced hydrogen peroxide. Organisms that produce catalase are known as catalase-positive organisms and those that do not produce catalase are known as catalase-negative organisms. Therefore catalase-negative organisms can be killed by phagocytosis due to the availability of hydrogen peroxide derived from the infecting organisms, while catalase-positive organisms cannot be eliminated due to the catalase-mediated inactivation of hydrogen peroxide.

As CGD results in only the phagocytic function of the immune system being affected, the other aspects of the immune system such as antibody production function effectively. Therefore CGD patients can protect themselves against viral infections and most other bacterial, fungal, and parasitic infections. Infections caused by only a few bacteria and fungi such as *S. aureus*, *S. marcescens*, *Salmonella*, *B. cepacia* complex, *Mycobacterium tuberculosis*, *Chromcobacterium violaceum*, *Francisella philomiragia*, *Granulibacter bethesdensis*, *Burkholderia gladioli*, *Nocardia*, *Aspergillus spp.*, *Rhizo spp.*, *Trichosporon spp.*, *Phellinus tropicalis*, *Paecilomyces variotii*, *Paecilomyces lilacinus*, and *Geosmithia argillacea* are frequently observed in CGD patients.[96–108] The incidence of invasive fungal infections is known to be highest in CGD patients among all other forms of PIDD, and is responsible for a high rate of morbidity and mortality associated with CGD.[96,109,110] Mulch pneumonitis caused by high-level exposure to fungi present in decayed organic matter is observed exclusively in CGD patients, and is associated with high rate of mortality in cases of delayed diagnosis.[111] On the other hand, histoplasmosis, blastomycosis, cryptococcosis, and mucormycosis are relatively rare fungal infections in CGD patients. Due to the predisposition to only selective pathogens, CGD patients do not get infected frequently; there are long periods of time during which they are infection-free and then they may contract a severe infection.

In addition to being infected, a chronic and exaggerated state of systemic inflammation exists even in absence of an infection in CGD patients. The inflammatory manifestations primarily affect the GI tract, the respiratory system, the urogenital tract, and the ocular system.[112] Autoimmune complications are also seen in a small population of CGD patients. The prevalence of complications associated with the inflammatory manifestations have been observed to be higher in X-linked CGD cases than in autosomal recessive CGD cases.

Disorders of the GI tract precipitated by the chronic inflammatory conditions prevalent in CGD patients often manifest within the first decade of life, sometimes even before CGD has been diagnosed and recurrent infections have started manifesting.[113] Symptoms indicative of GI disorders such as abdominal cramps/pain, idiopathic diarrhea, nausea and vomiting, loss of appetite, oral aphthae, and failure to thrive may lead to investigations for inflammatory bowel disease. Cases of early onset of inflammatory bowel disease warrant CGD screening.[112,113] CGD-associated GI manifestations often affect the colon, with perianal diseases such as anal fistulae and perirectal abscesses being the most common presentations.[113,114] GI complications are further compounded by significant and progressive liver pathology manifesting as hepatosplenomegaly, noncirrhotic portal hypertension, nodular regenerative hyperplasia, and splenic sequestration.[115] Mortality associated with liver pathologies in CGD patients is often indicated by the development of thrombocytopenia secondary to splenic sequestration.[116]

CGD-associated inflammatory complications of the GU tract often manifest as bladder granulomata, eosinophilic cystitis, ureteral obstruction, and urinary tract infections.[117,118] Granulomatous lung disease and interstitial pulmonary fibrosis are common manifestations of CGD-associated respiratory tract pathologies.[112,119] The chronic inflammatory state in CGD also affects the ocular system manifesting as chorioretinitis, uveitis, and ocular granulomata.[112] CGD also predisposes patients to a potentially life-threatening complication associated with chronic inflammatory conditions known as macrophage activation syndrome.[120]

Clinical manifestations and complications

Recurrent infections caused by catalase-positive organisms is a hallmark feature of CGD. The infections mostly affect tissues, organs, and organ systems such as the lungs, skin, the GI system, and the urogenital tracts, which are exposed to the outside environment. Infection of the lymph nodes that drain these organ systems and the potential hematogenous spread of the infections poses a threat to other vital organs including the liver, bones, kidney, and brain. Most CGD patients present with hallmark indications such as recurrent infections caused by catalase-positive organisms presenting as cutaneous abscesses, lymphadenitis, cellulitis, impetigo, osteomyelitis, bacteremia, and adenitis, colitis-associated bloody diarrhea (Fig. 8.15), dermatitis, and failure to thrive within the first year of life. The most common symptoms associated with these manifestations are fever, malaise, and weight loss.

Other complications such as portal venopathy, liver abscesses (Fig. 8.16), hepatomegaly, urethral strictures, recurrent urinary tract infections, altered renal function, keratitis, periodontitis, gingivitis, gingival hypertrophy, discoid lupus, photosensitivity, vasculitis, chronic respiratory disease, immune thrombocytopenia, juvenile idiopathic arthritis, and growth retardation may also develop within the first decade of life. X-linked CGD may also be associated with McLeod syndrome resulting in hemolytic anemia, neuroacanthocytosis, elevated creatinine phosphokinases, and late-onset peripheral/CNS manifestations.

Clinical presentation of CGD is highly variable owing to the differences in gene mutations. While some patients are more prone to pyogenic infections, others may develop complications associated with chronic inflammation. The chronic inflammatory state results in the formation of characteristic granulomas, which may cause obstruction in the bladder, bronchus biliary, GI, and uteropelvic tracts. Granulomata may also affect the retina, liver, lungs, and bone. Inflammatory bowel disease that presents with symptoms very similar to those of Crohn's disease such as abdominal cramps, chronic bloody diarrhea, and loss of appetite is a common manifestation of chronic inflammation affecting the GI system.[112,113] Chronic inflammatory conditions such as noninfectious arthritis, gingivitis, chorioretinitis or uveitis, glomerulonephritis, destructive white matter lesions in the brain, and discoid or systemic lupus erythematosus are also seen in CGD patients.[121,122]

A hyperinflammatory state involving autophagy during the microbial killing process and the prevention of excessive T-cell activation by phagocytes has also been implicated in the exaggerated immune activation seen in CGD patients. In addition, the ability of ROS to restrict T-cell responses as well as innate immune responses results in exaggerated lung inflammation, NRF2 activation failure, augmented NF-κB activation, and elevated release of pro-

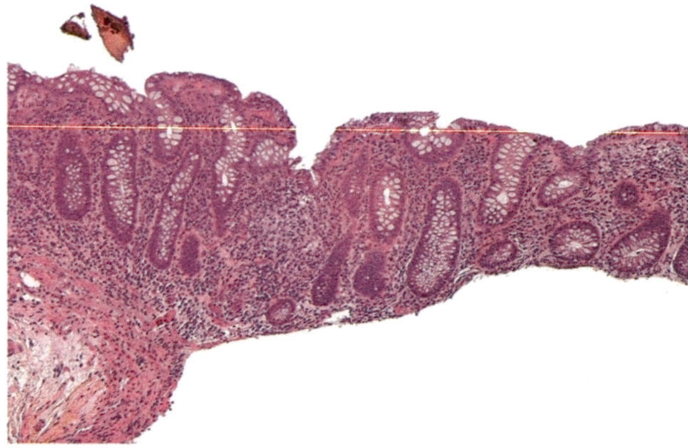

FIGURE 8.15

Micrograph of colitis with granulomatous inflammation from colonic biopsy (HPS, hepatopulmonary syndrome stain). *Source: From Nephron/Own work/; https://creativecommons.org/licenses/by-sa/3.0/deed.en; https://commons.wikimedia.org/wiki/File:Colitis_with_granuloma_low_mag.jpg.*

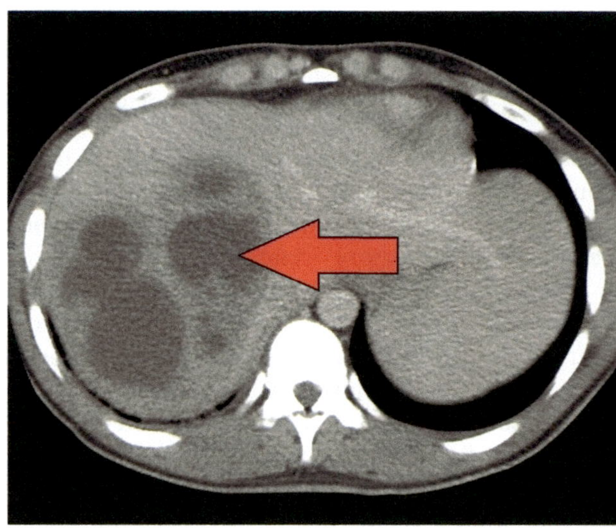

FIGURE 8.16

Large hepatic abscess. *Source: From James Heilman, MD/Own work/; https://creativecommons.org/licenses/by-sa/4.0/deed.en; https://commons.wikimedia.org/wiki/File:LargeHepaticAbscessMark.png.*

inflammatory cytokines (TNF-α, IL-17, and G-CSF). The exaggerated inflammatory response in CGD is attributed to the failure to downregulate NF-κB activation due to the NRF2 activation failure.[123]

Diagnosis

Primary suspicion of CGD is based on clinical manifestations and warrants diagnostic confirmation by cellular assays. The molecular diagnosis of CGD relies on measuring the capacity of phagocytic leukocytes such as macrophages and neutrophils to produce superoxide anions or hydrogen peroxide.[124] Neutrophilic granulocytes are used to detect NADPH oxidase activation in cells that are purified from blood and activated by fluid or particulate stimuli such as phorbol-myristate acetate and the bacterial peptide formyl-methionyl-leucyl-phenylalanine or zymosan (lipid-extracted *Saccharomyces cerevisiae* membranes), living or heat-killed yeasts (*S. cerevisiae, C. albicans*), and bacteria such as *Escherichia coli* or *S. aureus*, respectively. Opsonization of microorganisms by preincubation in human serum results in coating with antibodies and complement components, and ensures adequate oxidase activation. Nonopsonized and opsonized microorganisms bind to different receptors on the neutrophils, and precipitates in a chain of intracellular reactions that ultimately cumulate in NADPH oxidase activation.[123]

NADPH oxidase assays such as the cytochrome c reduction assay and the NBT slide assay measure superoxide levels, while other assays such as the dihydrorhodamine-123 (DHR) assay and the Amplex Red assay measure hydrogen peroxide activity.[124]

Cytochrome c reduction assay enables spectrophotometric measurement, while the NBT assay relies on microscopic observation of dark blue formazan precipitates formed within the cells on superoxide-mediated reduction. NBT assay of X-linked CGD carriers demonstrating a mosaic of cells with and without formazan precipitates is indicative of women who are suspected carriers of X-linked CGD.[123]

The DHR assay demonstrates a fluorescent change that is measurable by flow cytometry (FACS) due to the oxidation of DHR to rhodamine-1,2,3 by H_2O_2 in the presence of peroxidase. As this assay measures NADPH oxidase activity in individual cells, it can be used to diagnose X-linked CGD carriers. However, as this assay relies on peroxidase activity, myeloperoxidase deficiency may lead to a negative DHR report precipitating a false-positive diagnosis of CGD. The Amplex Red assay requires the addition of horse-radish peroxidase to cells so that oxidation can be detected in presence of a peroxidase as fluorescence detectable by a plate reader. However, this assay is known to often overestimate low rates of H_2O_2 generation. The DHR assay is preferred due to its sensitivity and specificity for the reliable detection of all mutations affecting NADPH oxidase in phagocytes.[123]

The molecular diagnosis of CGD needs to be followed up by genetic studies to ensure the implementation of appropriate treatment and management strategies. Although clinical history of individuals can demonstrate the genetic inheritance pattern, the identification of specific genetic mutations requires genetic testing.

Treatment and management

Medical advancements have made CGD a manageable condition, with a high survival rate. Although curative treatments for CGD are not easy to achieve, effective management of the condition can ensure a good quality of life. CGD management relies on the following principles: (1) strict vigilance to ensure early diagnosis of infections, (2) identification of specific infectious agent to ensure appropriate selection of antibiotic or antifungal treatment, and avoid risking further spread of the infection that may be facilitated by empiric treatments, (3) lifelong antibacterial and antifungal prophylaxis, (4) prevention of live bacterial vaccine administration, especially the bacillus Calmette-Guérin (BCG) vaccine, and (5) aggressive management of infectious and inflammatory complications.

Trimethoprim-sulfamethoxazole and itraconazole treatment regimens are the most commonly implemented.[125,126] This treatment combination has been known to successfully reduce the incidence from one severe infection per patient per year to almost one severe infection per patient per 10 years. CGD patients with G6PD deficiency or those who are allergic to sulfamethoxazole can be prescribed dicloxacillin and ciprofloxacin, while patients who are intolerant to itraconazole can be prescribed posaconazole.[127] Prophylaxis using IFN-⊠ has shown varying results,[128,129] and is therefore not considered as the mainstream or preferred management option for CGD.

Severe infections that occur in spite of the recommended prophylactic regimens need to be treated rapidly and aggressively. Initial empiric antibiotic therapy typically proves effective against the most commonly encountered pathogens in CGD patients such as *S. aureus* and Gram-negative bacteria, and refractory infections should be subjected to aggressive diagnosis for identification of specific causative agents within 48 h. Suspicion index for invasive fungal infections should be very high for CGD patients, and empiric treatment for fungal infections should be commenced even before the specific fungal agent has been identified, if pulmonary or pyrogenic symptoms are demonstrated. The challenges associated with low sensitivity of conventional serological tests for *Aspergillus* infections in CGD patients warrants the need for effective first-line therapy using antifungal agents such as voriconazole or combination therapies comprising of different permutations and combinations of voriconazole, liposomal amphotericin B, caspofungin, and posaconazole. Certain refractory fungal infections may warrant surgical interventions.[130]

Inflammatory manifestations of CGD often require treatment using corticosteroids, which is especially effective when used in conjunction with appropriate antimicrobials. Liver abscesses and respiratory infections, such as mulch pneumonitis and *Nocardia* pneumonia, that are refractory to conventional therapy have demonstrated susceptibility to corticosteroid-based therapies.[131–133] Colitis is another long-term and difficult complication of CGD. Although the condition is responsive to treatment with steroids, relapse is common.[96] Although suppression of inflammation using tumor necrosis factor-alpha inhibitors may be effective, this mode of treatment is generally not recommended due to significant complications associated with infections.[134] Salicylic acid derivatives, antimetabolites such as azathioprine, and 6-mercaptopurine are steroid-sparing agents that have shown varying degrees of success in CGD-associated inflammatory conditions. Anakinra, a recombinant human IL-1R antagonist has demonstrated rapid and sustained improvement of CGD colitis.[135]

Early diagnosis of CGD is important for eligibility for hematopoietic cell transplantation (HCT), the only established curative treatment for CGD.[136,137] HCT eligibility depends on several factors including: (1) age, (2) stage of disease when diagnosed, (3) prognosis depending on the specific causative mutation, (4) existing infectious or inflammatory conditions, (5) availability of human leukocyte antigen (HLA)-compatible donor, (6) access to transplantation, and (7) patient preference and consent. As CGD is caused by mutations in single genes, the condition is a good candidate for consideration of gene therapy. Gene therapy may also be a good option for patients who cannot undergo HCT due to the lack of an HLA-compatible donor. However, in spite of the promising results shown by gene therapy in vitro, its success in CGD patients has been limited.[123,130,138]

Key take-aways

1. CGD is a rare, inherited PIDD.
2. CGD affects the ability of certain white blood cells WBC such as neutrophils, monocytes, macrophages, and eosinophils to kill certain phagocytosed pathogens due to a defect in the NOX present in the phagocytes, known as PHOX.
3. The enzymatic defect in CGD prevents the formation of ROS such as the superoxide radical and the subsequent production of hydrogen peroxide that is required for the elimination of certain pathogens.
4. The dysfunctional phagocytosis mechanism of the immune system in CGD increases predisposition to recurrent infections by certain pathogens and a chronic inflammatory state.
5. Systemic distribution of granulomas is characteristic of CGD, and are a result of the chronic and excessive inflammation.

Student Notes

6. Components of the immune system that are not associated with phagocytosis typically demonstrate normal function in CGD patients. This increases the vulnerability of the affected individual exclusively to pathogens requiring a functional NADPH oxidase enzyme for elimination.
7. CGD increases predisposition to catalase-positive organisms, while catalase-negative organisms can be eliminated.
8. Diagnosis of CGD requires both serological testing assays and genetic testing to identify the exact genetic mutation responsible for the condition.
9. Early diagnosis of CGD is important for effective management of the condition and to identify eligibility for HCT, which is the only curative treatment available for CGD.
10. As CGD is caused by mutations in single genes, gene therapy can be considered as a potential therapeutic option; however, its success in CGD cases has been limited.

References

1. Singer M, Deutschman CS, Seymour CW, et al. The Third International Consensus Definitions for Sepsis and Septic Shock (Sepsis-3. *JAMA*. 2016;315(8):801–810. Available from: https://doi.org/10.1001/jama.2016.0287.

2. Gyawali B, Ramakrishna K, Dhamoon AS. Sepsis: The evolution in definition, pathophysiology, and management. *SAGE Open Med*. 2019;7. Available from: https://doi.org/10.1177/2050312119835043. 2050312119835043. Published March 21, 2019.

3. Taylor Jr FB, Chang A, Ruf W, et al. Lethal *E. coli* septic shock is prevented by blocking tissue factor with monoclonal antibody. *Circ Shock*. 1991;33(3):127–134.

4. Eckle I, Seitz R, Egbring R, Kolb G, Havemann K. Protein C degradation in vitro by neutrophil elastase. *Biol Chem Hoppe Seyler*. 1991;372(11):1007–1013. Available from: https://doi.org/10.1515/bchm3.1991.372.2.1007.

5. Biemond BJ, Levi M, Ten Cate H, et al. Plasminogen activator and plasminogen activator inhibitor I release during experimental endotoxaemia in chimpanzees: effect of interventions in the cytokine and coagulation cascades. *Clin Sci (Lond)*. 1995;88(5):587–594. Available from: https://doi.org/10.1042/cs0880587.

6. Hotchkiss RS, Tinsley KW, Swanson PE, et al. Sepsis-induced apoptosis causes progressive profound depletion of B and CD4+ T lymphocytes in humans. *J Immunol*. 2001;166(11):6952–6963. Available from: https://doi.org/10.4049/jimmunol.166.11.6952.

7. Remick DG. Pathophysiology of sepsis [published correction appears in *Am J Pathol*. 2007 Sep;171(3):1078]*Am J Pathol*. 2007;170(5):1435–1444. Available from: https://doi.org/10.2353/ajpath.2007.060872.

8. van der Poll T, Opal SM. Host-pathogen interactions in sepsis. *Lancet Infect Dis*. 2008;8(1):32–43. Available from: https://doi.org/10.1016/S1473-3099(07)70265-7.

9. Heagy W, Nieman K, Hansen C, Cohen M, Danielson D, West MA. Lower levels of whole blood LPS-stimulated cytokine release are associated with poorer clinical outcomes in surgical ICU patients. *Surg Infect (Larchmt)*. 2003;4(2):171–180. Available from: https://doi.org/10.1089/109629603766956960.

10. Angus DC, van der Poll T. Severe sepsis and septic shock. *N Engl J Med*. 2013;369(21):2063. Available from: https://doi.org/10.1056/NEJMc1312359.

11. Rivers EP, Katranji M, Jaehne KA, et al. Early interventions in severe sepsis and septic shock: a review of the evidence one decade later. *Minerva Anestesiol*. 2012;78(6):712–724.

12. Jones AE, Puskarich MA. Sepsis-induced tissue hypoperfusion. *Crit Care Nurs Clin North Am*. 2011;23(1):115–125. Available from: https://doi.org/10.1016/j.ccell.2010.12.007.

13. Vieillard-Baron A. Septic cardiomyopathy. *Ann Intensive Care*. 2011;1(1):6. Available from: https://doi.org/10.1186/2110-5820-1-6. Published April 13, 2011.

14. Poelaert J, Declerck C, Vogelaers D, Colardyn F, Visser CA. Left ventricular systolic and diastolic function in septic shock. *Intensive Care Med*. 1997;23(5):553–560. Available from: https://doi.org/10.1007/s001340050372.

15. Jardin F, Fourme T, Page B, et al. Persistent preload defect in severe sepsis despite fluid loading: a longitudinal echocardiographic study in patients with septic shock. *Chest*. 1999;116(5):1354–1359. Available from: https://doi.org/10.1378/chest.116.5.1354.

16. Okorie ON, Dellinger P. Lactate: biomarker and potential therapeutic target. *Crit Care Clin*. 2011;27(2):299–326. Available from: https://doi.org/10.1016/j.ccc.2010.12.013.

17. Nguyen HB, Rivers EP, Knoblich BP, et al. Early lactate clearance is associated with improved outcome in severe sepsis and septic shock. *Crit Care Med*. 2004;32(8):1637–1642. Available from: https://doi.org/10.1097/01.ccm.0000132904.35713.a7.

18. Bullock B, Benham MD. *Bacterial sepsis* [Updated June 25, 2020]*StatPearls [Internet]*. Treasure Island (FL): StatPearls Publishing; 2020. Available from: https://www.ncbi.nlm.nih.gov/books/NBK537054/.

19. Rhodes A, Evans LE, Alhazzani W, et al. Surviving Sepsis Campaign: International Guidelines for Management of Sepsis and Septic Shock: 2016. *Intensive Care Med*. 2017;43(3):304–377. Available from: https://doi.org/10.1007/s00134-017-4683-6.

20. Asfar P, Meziani F, Hamel JF, et al. High vs low blood-pressure target in patients with septic shock. *N Engl J Med*. 2014;370(17):1583–1593. Available from: https://doi.org/10.1056/NEJMoa1312173.

21. Holst LB, Haase N, Wetterslev J, et al. Lower vs higher hemoglobin threshold for transfusion in septic shock. *N Engl J Med*. 2014;371(15):1381–1391. Available from: https://doi.org/10.1056/NEJMoa1406617.

22. Sprung CL, Annane D, Keh D, et al. Hydrocortisone therapy for patients with septic shock. *N Engl J Med*. 2008;358(2):111–124. Available from: https://doi.org/10.1056/NEJMoa071366.

23. Venkatesh B, Finfer S, Cohen J, et al. Adjunctive glucocorticoid therapy in patients with septic shock. *N Engl J Med*. 2018;378(9):797–808. Available from: https://doi.org/10.1056/NEJMoa1705835.

24. Acute Respiratory Distress Syndrome Network, Brower RG, Matthay MA, et al. Ventilation with lower tidal volumes as compared with traditional tidal volumes for acute lung injury and the acute respiratory distress syndrome. *N Engl J Med*. 2000;342(18):1301–1308. Available from: https://doi.org/10.1056/NEJM200005043421801.

25. Adang L, Berger J. Progressive multifocal leukoencephalopathy. *F1000Res*. 2015;4. Available from: https://doi.org/10.12688/f1000research.7071.1. F1000 Faculty Rev-1424. Published December 10, 2015.

26. Saji AM, Gupta V. *Progressive multifocal leukoencephalopathy* [Updated February 6, 2021] *StatPearls [Internet]*. Treasure Island (FL): StatPearls Publishing; 2021. Available from: https://www.ncbi.nlm.nih.gov/books/NBK560584/.

27. Astrom KE, Mancall EL, Richardson Jr. EP. Progressive multifocal leuko-encephalopathy; a hitherto unrecognized complication of chronic lymphatic leukaemia and Hodgkin's disease. *Brain.* 1958;81(1):93–111. Available from: https://doi.org/10.1093/brain/81.1.93.

28. Padgett BL, Walker DL, ZuRhein GM, Eckroade RJ, Dessel BH. Cultivation of papova-like virus from human brain with progressive multifocal leucoencephalopathy. *Lancet.* 1971;1(7712):1257–1260. Available from: https://doi.org/10.1016/s0140-6736(71)91777-6.

29. Ciappi S, Azzi A, De Santis R, et al. Archetypal and rearranged sequences of human polyomavirus JC transcription control region in peripheral blood leukocytes and in cerebrospinal fluid. *J Gen Virol.* 1999;80(Pt 4):1017–1023. Available from: https://doi.org/10.1099/0022-1317-80-4-1017.

30. White MK, Khalili K. Pathogenesis of progressive multifocal leukoencephalopathy–revisited. *J Infect Dis.* 2011;203(5):578–586. Available from: https://doi.org/10.1093/infdis/jiq097.

31. Berger JR, Pall L, Lanska D, Whiteman M. Progressive multifocal leukoencephalopathy in patients with HIV infection. *J Neurovirol.* 1998;4(1):59–68. Available from: https://doi.org/10.3109/13550289809113482.

32. Bozic C, Subramanyam M, Richman S, Plavina T, Zhang A, Ticho B. Anti-JC virus (JCV) antibody prevalence in the JCV epidemiology in MS (JEMS) trial. *Eur J Neurol.* 2014;21(2):299–304. Available from: https://doi.org/10.1111/ene.12304.

33. Tan K, Roda R, Ostrow L, McArthur J, Nath A. PML-IRIS in patients with HIV infection: clinical manifestations and treatment with steroids. *Neurology.* 2009;72(17):1458–1464. Available from: https://doi.org/10.1212/01.wnl.0000343510.08643.74.

34. Zaheer F, Berger JR. Treatment-related progressive multifocal leukoencephalopathy: current understanding and future steps. *Ther Adv Drug Saf.* 2012;3(5):227–239. Available from: https://doi.org/10.1177/2042098612453849.

35. Williamson EM, Berger JR. Infection risk in patients on multiple sclerosis therapeutics. *CNS Drugs.* 2015;29(3):229–244. Available from: https://doi.org/10.1007/s40263-015-0226-2.

36. Jensen PN, Major EO. A classification scheme for human polyomavirus JCV variants based on the nucleotide sequence of the noncoding regulatory region. *J Neurovirol.* 2001;7(4):280–287. Available from: https://doi.org/10.1080/13550280152537102.

37. Weber F, Goldmann C, Krämer M, et al. Cellular and humoral immune response in progressive multifocal leukoencephalopathy. *Ann Neurol.* 2001;49(5):636–642.

38. Lima MA, Marzocchetti A, Autissier P, et al. Frequency and phenotype of JC virus-specific CD8+ T lymphocytes in the peripheral blood of patients with progressive multifocal leukoencephalopathy [published correction appears in *J Virol.* 2007 Oct;81(19):10835]*J Virol.* 2007;81(7):3361–3368. Available from: https://doi.org/10.1128/JVI.01809-06.

39. O'Connor P, Miller D, Riester K, et al. Relapse rates and enhancing lesions in a phase II trial of natalizumab in multiple sclerosis. *Mult Scler.* 2005;11(5):568–572. Available from: https://doi.org/10.1191/1352458505ms1205oa.

40. Henry C, Jouan F, De Broucker T. JC virus granule cell neuronopathy: a cause of infectious cerebellar degeneration. *J Neurol Sci.* 2015;354(1–2):86–90. Available from: https://doi.org/10.1016/j.jns.2015.05.003.

41. Khoury MN, Alsop DC, Agnihotri SP, et al. Hyperintense cortical signal on magnetic resonance imaging reflects focal leukocortical encephalitis and seizure risk in progressive multifocal leukoencephalopathy. *Ann Neurol.* 2014;75(5):659–669. Available from: https://doi.org/10.1002/ana.24144.

42. Gheuens S, Wüthrich C, Koralnik IJ. Progressive multifocal leukoencephalopathy: why gray and white matter. *Annu Rev Pathol.* 2013;8:189–215. Available from: https://doi.org/10.1146/annurev-pathol-020712-164018.

43. Berger JR. Progressive multifocal leukoencephalopathy. *Handb Clin Neurol.* 2014;123:357–376. Available from: https://doi.org/10.1016/B978-0-444-53488-0.00017-1.

44. Gildenberg PL, Gathe Jr JC, Kim JH. Stereotactic biopsy of cerebral lesions in AIDS. *Clin Infect Dis.* 2000;30(3):491–499. Available from: https://doi.org/10.1086/313685.

45. Sudhakar P, Bachman DM, Mark AS, Berger JR, Kedar S. Progressive multifocal leukoencephalopathy: recent advances and a neuro-ophthalmological review. *J Neuroophthalmol.* 2015;35(3):296–305. Available from: https://doi.org/10.1097/WNO.0000000000000271.

46. Miller RF, Hall-Craggs MA, Costa DC, et al. Magnetic resonance imaging, thallium-201 SPET scanning, and laboratory analyses for discrimination of cerebral lymphoma and toxoplasmosis in AIDS. *Sex Transm Infect.* 1998;74(4):258–264. Available from: https://doi.org/10.1136/sti.74.4.258.

47. Tan IL, McArthur JC, Clifford DB, Major EO, Nath A. Immune reconstitution inflammatory syndrome in natalizumab-associated PML. *Neurology.* 2011;77(11):1061–1067. Available from: https://doi.org/10.1212/WNL.0b013e31822e55e7.

48. Berger JR, Aksamit AJ, Clifford DB, et al. PML diagnostic criteria: consensus statement from the AAN neuroinfectious disease section. *Neurology.* 2013;80(15):1430–1438. Available from: https://doi.org/10.1212/WNL.0b013e31828c2fa1.

49. Plavina T, Subramanyam M, Bloomgren G, et al. Anti-JC virus antibody levels in serum or plasma further define risk of natalizumab-associated progressive multifocal leukoencephalopathy. *Ann Neurol.* 2014;76(6):802–812. Available from: https://doi.org/10.1002/ana.24286.

50. Woodhouse MA, Dayan AD, Burston J, et al. Progressive multifocal leukoencephalopathy: electron microscope study of four cases. *Brain.* 1967;90(4):863–870. Available from: https://doi.org/10.1093/brain/90.4.863.

51. Engsig FN, Hansen AB, Omland LH, et al. Incidence, clinical presentation, and outcome of progressive multifocal leukoencephalopathy in HIV-infected patients during the highly active antiretroviral therapy era: a nationwide cohort study. *J Infect Dis.* 2009;199(1):77–83. Available from: https://doi.org/10.1086/595299.

52. Antinori A, Ammassari A, Giancola ML, et al. Epidemiology and prognosis of AIDS-associated progressive multifocal leukoencephalopathy in the HAART era. *J Neurovirol.* 2001;7(4):323–328. Available from: https://doi.org/10.1080/13550280152537184.

53. Dong-Si T, Gheuens S, Gangadharan A, et al. Predictors of survival and functional outcomes in natalizumab-associated progressive multifocal leukoencephalopathy. *J Neurovirol.* 2015;21(6):637–644. Available from: https://doi.org/10.1007/s13365-015-0316-4.

54. Casado JL, Corral I, García J, et al. Continued declining incidence and improved survival of progressive multifocal leukoencephalopathy in HIV/AIDS patients in the current era. *Eur J Clin Microbiol Infect Dis.* 2014;33(2):179–187. Available from: https://doi.org/10.1007/s10096-013-1941-6.

55. Clifford DB, Nath A, Cinque P, et al. A study of mefloquine treatment for progressive multifocal leukoencephalopathy: results and exploration of predictors of PML outcomes. *J Neurovirol.* 2013;19(4):351–358. Available from: https://doi.org/10.1007/s13365-013-0173-y.

56. De Luca A, Ammassari A, Pezzotti P, et al. Cidofovir in addition to antiretroviral treatment is not effective for AIDS-associated progressive multifocal leukoencephalopathy: a multicohort analysis. *AIDS.* 2008;22(14):1759–1767. Available from: https://doi.org/10.1097/QAD.0b013e32830a5043.

57. Hall CD, Dafni U, Simpson D, et al. Failure of cytarabine in progressive multifocal leukoencephalopathy associated with human immunodeficiency virus infection. *AIDS Clin Trials Group 243 Team N Engl J Med.* 1998;338(19):1345–1351. Available from: https://doi.org/10.1056/NEJM199805073381903.

58. Muftuoglu M, Olson A, Marin D, et al. Allogeneic BK virus-specific T cells for progressive multifocal leukoencephalopathy. *N Engl J Med.* 2018;379(15):1443–1451. Available from: https://doi.org/10.1056/NEJMoa1801540.

59. Mende M, Sockel K. Parvovirus B19 infection. *N Engl J Med.* 2018;379(24):2361. Available from: https://doi.org/10.1056/NEJMicm1807156.

60. Rogo LD, Mokhtari-Azad T, Kabir MH, Rezaei F. Human parvovirus B19: a review. *Acta Virol.* 2014;58(3):199–213. Available from: https://doi.org/10.4149/av_2014_03_199.

61. Macri A, Crane JS. *Parvoviruses* [Updated May 4, 2020]*StatPearls [Internet]*. Treasure Island (FL): StatPearls Publishing; 2021. Available from: https://www.ncbi.nlm.nih.gov/books/NBK482245/.

62. Vafaie J, Schwartz RA. Erythema infectiosum. *J Cutan Med Surg.* 2005;9(4):159–161. Available from: https://doi.org/10.1007/s10227-005-0101-8.

63. Kontomanolis EN, Fasoulakis Z. Hydrops fetalis and THE parvovirus B-19. *Curr Pediatr Rev.* 2018;14(4):239–252. Available from: https://doi.org/10.2174/1573396314666180820154340.

64. Borsato ML, Bruniera P, Cusato MP, Spewien KE, Durigon EL, Toporovski J. Crise aplástica da anemia falciforme condicionada pelo parvovírus B19 [Aplastic crisis in sickle cell anemia induced by parvovírus B19]. *J Pediatr (Rio J).* 2000;76(6):458–460. Available from: https://doi.org/10.2223/jped.94.

65. Kostolansky S, Waymack JR. *Erythema infectiosum* [Updated August 8, 2020]*StatPearls [Internet]*. Treasure Island (FL): StatPearls Publishing; 2021. Available from: https://www.ncbi.nlm.nih.gov/books/NBK513309/.

66. Brown KE, Anderson SM, Young NS. Erythrocyte P antigen: cellular receptor for B19 parvovirus. *Science.* 1993;262(5130):114–117. Available from: https://doi.org/10.1126/science.8211117.

67. Brown KE, Hibbs JR, Gallinella G, et al. Resistance to parvovirus B19 infection due to lack of virus receptor (erythrocyte P antigen. *N Engl J Med.* 1994;330(17):1192–1196. Available from: https://doi.org/10.1056/NEJM199404283301704.

68. Owren PA. Congenital hemolytic jaundice; the pathogenesis of the hemolytic crisis. *Blood.* 1948;3(3):231–248.

69. Koduri PR. Novel cytomorphology of the giant proerythroblasts of parvovirus B19 infection. *Am J Hematol.* 1998;58(2):95–99. Available from: https://doi.org/10.1002/(sici)1096-8652(199806)58:2 < 95::aid-ajh1 > 3.0.co;2-v.

70. Kurtzman GJ, Cohen BJ, Field AM, Oseas R, Blaese RM, Young NS. Immune response to B19 parvovirus and an antibody defect in persistent viral infection. *J Clin Invest.* 1989;84(4):1114–1123. Available from: https://doi.org/10.1172/JCI114274.

71. Corcoran A, Doyle S, Waldron D, Nicholson A, Mahon BP. Impaired gamma interferon responses against parvovirus B19 by recently infected children. *J Virol.* 2000;74(21):9903–9910. Available from: https://doi.org/10.1128/jvi.74.21.9903-9910.2000.

72. Franssila R, Hokynar K, Hedman K. T helper cell-mediated in vitro responses of recently and remotely infected subjects to a candidate recombinant vaccine for human parvovirus b19. *J Infect Dis.* 2001;183(5):805–809. Available from: https://doi.org/10.1086/318819.

73. von Poblotzki A, Gerdes C, Reischl U, Wolf H, Modrow S. Lymphoproliferative responses after infection with human parvovirus B19. *J Virol.* 1996;70(10):7327–7330. Available from: https://doi.org/10.1128/JVI.70.10.7327-7330.1996.

74. Anderson LJ, Tsou C, Parker RA, et al. Detection of antibodies and antigens of human parvovirus B19 by enzyme-linked immunosorbent assay. *J Clin Microbiol.* 1986;24(4):522–526. Available from: https://doi.org/10.1128/JCM.24.4.522-526.1986.

75. Erdman DD, Usher MJ, Tsou C, et al. Human parvovirus B19 specific IgG, IgA, and IgM antibodies and DNA in serum specimens from persons with erythema infectiosum. *J Med Virol.* 1991;35(2):110–115. Available from: https://doi.org/10.1002/jmv.1890350207.

76. Heegaard ED, Brown KE. Human parvovirus B19. *Clin Microbiol Rev.* 2002;15(3):485–505. Available from: https://doi.org/10.1128/cmr.15.3.485-505.2002.

77. Potter CG, Potter AC, Hatton CS, et al. Variation of erythroid and myeloid precursors in the marrow and peripheral blood of volunteer subjects infected with human parvovirus (B19). *J Clin Invest.* 1987;79(5):1486–1492. Available from: https://doi.org/10.1172/JCI112978.

78. Koch WC, Massey G, Russell CE, Adler SP. Manifestations and treatment of human parvovirus B19 infection in immunocompromised patients. *J Pediatr.* 1990;116(3):355–359. Available from: https://doi.org/10.1016/s0022-3476(05)82820-3.

79. Allmon A, Deane K, Martin KL. Common skin rashes in children. *Am Fam Phys.* 2015;92 (3):211–216.

80. Molenaar-de Backer MW, Russcher A, Kroes AC, Koppelman MH, Lanfermeijer M, Zaaijer HL. Detection of parvovirus B19 DNA in blood: viruses or DNA remnants. *J Clin Virol.* 2016;84:19–23. Available from: https://doi.org/10.1016/j.jcv.2016.09.004.

81. Rapini RP, Bolognia JL, Jorizzo JL. *Dermatology: 2-Volume Set.* St. Louis: Mosby; 2007. ISBN 978-1-4160-2999-1.

82. Buchbinder D, Hauck F, Albert MH, et al. Rubella virus-associated cutaneous granulomatous disease: a unique complication in immune-deficient patients, not limited to DNA repair disorders. *J Clin Immunol.* 2019;39(1):81–89. Available from: https://doi.org/10.1007/s10875-018-0581-0.

83. Warris A. Immunopathology of Aspergillus infections in children with chronic granulomatous disease and cystic fibrosis. *Pediatr Infect Dis J.* 2019;38(5):e96–e98. Available from: https://doi.org/10.1097/INF.0000000000002265.

84. Kutukculer N, Aykut A, Karaca NE, et al. Chronic granulamatous disease: two decades of experience from a paediatric immunology unit in a country with high rate of consangineous marriages. *Scand J Immunol.* 2019;89(2):e12737. Available from: https://doi.org/10.1111/sji.12737.

85. Segal BH, Leto TL, Gallin JI, Malech HL, Holland SM. Genetic, biochemical, and clinical features of chronic granulomatous disease. *Medicine (Baltim).* 2000;79(3):170–200. Available from: https://doi.org/10.1097/00005792-200005000-00004.

86. Maydana M, Cabanillas D, Regairaz L, et al. Chronic granulomatous disease: multiple infections as clinical presentation. Pediatric case report. *Arch Argent Pediatr.* 2018;116(6): e744–e748. Available from: https://doi.org/10.5546/aap.2018.e744.

87. Metzger JC, Kurz E, von Spee-Mayer C, et al. Chronic granulomatous disease as a rare differential diagnosis of inflammatory bowel disease. *Z Gastroenterol.* 2018;56(12):1507–1512. Available from: https://doi.org/10.1055/a-0732-5865.

88. Kulkarni M, Hule G, de Boer M, et al. Approach to molecular diagnosis of chronic granulomatous disease (CGD): an experience from a large cohort of 90 Indian patients. *J Clin Immunol.* 2018;38(8):898–916. Available from: https://doi.org/10.1007/s10875-018-0567-y.

89. Pao M, Wiggs EA, Anastacio MM, et al. Cognitive function in patients with chronic granulomatous disease: a preliminary report. *Psychosomatics.* 2004;45(3):230–234. Available from: https://doi.org/10.1176/appi.psy.45.3.230.

90. Thomas DC. How the phagocyte NADPH oxidase regulates innate immunity. *Free Radic Biol Med.* 2018;125:44–52. Available from: https://doi.org/10.1016/j.freeradbiomed.2018.06.011.

91. Winkelstein JA, Marino MC, Johnston Jr RB, et al. Chronic granulomatous disease. Report on a national registry of 368 patients. *Med (Baltim).* 2000;79(3):155–169. Available from: https://doi.org/10.1097/00005792-200005000-00003.

92. Jones LB, McGrogan P, Flood TJ, et al. Special article: chronic granulomatous disease in the United Kingdom and Ireland: a comprehensive national patient-based registry. *Clin Exp Immunol.* 2008;152(2):211–218. Available from: https://doi.org/10.1111/j.1365-2249.2008.03644.x.

93. Martire B, Rondelli R, Soresina A, et al. Clinical features, long-term follow-up and outcome of a large cohort of patients with chronic granulomatous disease: an Italian multicenter study. *Clin Immunol.* 2008;126(2):155–164. Available from: https://doi.org/10.1016/j.clim.2007.09.008.

94. van den Berg JM, van Koppen E, Ahlin A, et al. Chronic granulomatous disease: the European experience. *PLoS One.* 2009;4(4):e5234. Available from: https://doi.org/10.1371/journal.pone.0005234.

95. Kuhns DB, Alvord WG, Heller T, et al. Residual NADPH oxidase and survival in chronic granulomatous disease. *N Engl J Med.* 2010;363(27):2600–2610. Available from: https://doi.org/10.1056/NEJMoa1007097.

96. Marciano BE, Spalding C, Fitzgerald A, et al. Common severe infections in chronic granulomatous disease. *Clin Infect Dis.* 2015;60(8):1176–1183. Available from: https://doi.org/10.1093/cid/ciu1154.

97. Lee PP, Chan KW, Jiang L, et al. Susceptibility to mycobacterial infections in children with X-linked chronic granulomatous disease: a review of 17 patients living in a region endemic for tuberculosis. *Pediatr Infect Dis J.* 2008;27(3):224–230. Available from: https://doi.org/10.1097/INF.0b013e31815b494c.

98. Conti F, Lugo-Reyes SO, Blancas Galicia L, et al. Mycobacterial disease in patients with chronic granulomatous disease: a retrospective analysis of 71 cases. *J Allergy Clin Immunol.* 2016;138(1):241–248, e3. Available from: https://doi.org/10.1016/j.jaci.2015.11.041.

99. Mailman TL, Schmidt MH. *Francisella philomiragia* adenitis and pulmonary nodules in a child with chronic granulomatous disease. *Can J Infect Dis Med Microbiol.* 2005;16(4):245–248. Available from: https://doi.org/10.1155/2005/486417.

100. Greenberg DE, Shoffner AR, Zelazny AM, et al. Recurrent *Granulibacter bethesdensis* infections and chronic granulomatous disease. *Emerg Infect Dis.* 2010;16(9):1341–1348. Available from: https://doi.org/10.3201/eid1609.091800.

101. Ross JP, Holland SM, Gill VJ, DeCarlo ES, Gallin JI. Severe *Burkholderia (Pseudomonas) gladioli* infection in chronic granulomatous disease: report of two successfully treated cases. *Clin Infect Dis.* 1995;21(5):1291–1293. Available from: https://doi.org/10.1093/clinids/21.5.1291.

102. Boyanton Jr BL, Noroski LM, Reddy H, et al. *Burkholderia gladioli* osteomyelitis in association with chronic granulomatous disease: case report and review. *Pediatr Infect Dis J.* 2005;24(9):837–839. Available from: https://doi.org/10.1097/01.inf.0000177285.44374.dc.

103. Dotis J, Pana ZD, Roilides E. Non-Aspergillus fungal infections in chronic granulomatous disease. *Mycoses.* 2013;56(4):449–462. Available from: https://doi.org/10.1111/myc.12049.

104. Haidar G, Zerbe CS, Cheng M, Zelazny AM, Holland SM, Sheridan KR. *Phellinus* species: an emerging cause of refractory fungal infections in patients with X-linked chronic granulomatous disease. *Mycoses.* 2017;60(3):155–160. Available from: https://doi.org/10.1111/myc.12573.

105. Ramesh M, Resnick E, Hui Y, et al. *Phellinus tropicalis* abscesses in a patient with chronic granulomatous disease. *J Clin Immunol.* 2014;34(2):130–133. Available from: https://doi.org/10.1007/s10875-013-9967-1.

106. Wang SM, Shieh CC, Liu CC. Successful treatment of *Paecilomyces variotii* splenic abscesses: a rare complication in a previously unrecognized chronic granulomatous disease child. *Diagn Microbiol Infect Dis.* 2005;53(2):149–152. Available from: https://doi.org/10.1016/j.diagmicrobio.2005.04.015.

107. Silliman CC, Lawellin DW, Lohr JA, Rodgers BM, Donowitz LG. *Paecilomyces lilacinus* infection in a child with chronic granulomatous disease. *J Infect.* 1992;24(2):191–195. Available from: https://doi.org/10.1016/0163-4453(92)92980-w.

108. De Ravin SS, Challipalli M, Anderson V, et al. *Geosmithia argillacea*: an emerging cause of invasive mycosis in human chronic granulomatous disease. *Clin Infect Dis*. 2011;52(6): e136–e143. Available from: https://doi.org/10.1093/cid/ciq250.

109. Falcone EL, Holland SM. Invasive fungal infection in chronic granulomatous disease: insights into pathogenesis and management. *Curr Opin Infect Dis*. 2012;25(6):658–669. Available from: https://doi.org/10.1097/QCO.0b013e328358b0a4.

110. Blumental S, Mouy R, Mahlaoui N, et al. Invasive mold infections in chronic granulomatous disease: a 25-year retrospective survey. *Clin Infect Dis*. 2011;53(12):e159–e169. Available from: https://doi.org/10.1093/cid/cir731.

111. Siddiqui S, Anderson VL, Hilligoss DM, et al. Fulminant mulch pneumonitis: an emergency presentation of chronic granulomatous disease. *Clin Infect Dis*. 2007;45(6):673–681. Available from: https://doi.org/10.1086/520985.

112. Magnani A, Brosselin P, Beauté J, et al. Inflammatory manifestations in a single-center cohort of patients with chronic granulomatous disease. *J Allergy Clin Immunol*. 2014;134 (3):655–662, e8. Available from: https://doi.org/10.1016/j.jaci.2014.04.014.

113. Marciano BE, Rosenzweig SD, Kleiner DE, et al. Gastrointestinal involvement in chronic granulomatous disease. *Pediatrics*. 2004;114(2):462–468. Available from: https://doi.org/10.1542/peds.114.2.462.

114. Marks DJ, Miyagi K, Rahman FZ, Novelli M, Bloom SL, Segal AW. Inflammatory bowel disease in CGD reproduces the clinicopathological features of Crohn's disease. *Am J Gastroenterol*. 2009;104(1):117–124. Available from: https://doi.org/10.1038/ajg.2008.72.

115. Hussain N, Feld JJ, Kleiner DE, et al. Hepatic abnormalities in patients with chronic granulomatous disease. *Hepatology*. 2007;45(3):675–683. Available from: https://doi.org/10.1002/hep.21524.

116. Feld JJ, Hussain N, Wright EC, et al. Hepatic involvement and portal hypertension predict mortality in chronic granulomatous disease. *Gastroenterology*. 2008;134(7):1917–1926. Available from: https://doi.org/10.1053/j.gastro.2008.02.081.

117. Walther MM, Malech H, Berman A, et al. The urological manifestations of chronic granulomatous disease. *J Urol*. 1992;147(5):1314–1318. Available from: https://doi.org/10.1016/s0022-5347(17)37552-3.

118. Barese CN, Podestá M, Litvak E, Villa M, Rivas EM. Recurrent eosinophilic cystitis in a child with chronic granulomatous disease. *J Pediatr Hematol Oncol*. 2004;26(3):209–212. Available from: https://doi.org/10.1097/00043426-200403000-00014.

119. Damen GM, van Krieken JH, Hoppenreijs E, et al. Overlap, common features, and essential differences in pediatric granulomatous inflammatory bowel disease. *J Pediatr Gastroenterol Nutr*. 2010;51(6):690–697. Available from: https://doi.org/10.1097/MPG.0b013e3181dc0d73.

120. Akagi K, Kawai T, Watanabe N, et al. A case of macrophage activation syndrome developing in a patient with chronic granulomatous disease-associated colitis. *J Pediatr Hematol Oncol*. 2014;36(3):e169–e172. Available from: https://doi.org/10.1097/MPH.0b013e31828e5dae.

121. Sillevis Smitt JH, Bos JD, Weening RS, Krieg SR. Discoid lupus erythematosus-like skin changes in patients with autosomal recessive chronic granulomatous disease. *Arch Dermatol*. 1990;126(12):1656–1658. Available from: https://doi.org/10.1001/archderm.126.12.1656b.

122. Manzi S, Urbach AH, McCune AB, et al. Systemic lupus erythematosus in a boy with chronic granulomatous disease: case report and review of the literature. *Arthritis Rheum*. 1991;34(1):101–105. Available from: https://doi.org/10.1002/art.1780340116.

123. Roos D. Chronic granulomatous disease. *Br Med Bull*. 2016;118(1):50–63. Available from: https://doi.org/10.1093/bmb/ldw009.

124. Roos D, de Boer M. Molecular diagnosis of chronic granulomatous disease. *Clin Exp Immunol*. 2014;175(2):139–149. Available from: https://doi.org/10.1111/cei.12202.

125. Margolis DM, Melnick DA, Alling DW, Gallin JI. Trimethoprim-sulfamethoxazole prophylaxis in the management of chronic granulomatous disease. *J Infect Dis*. 1990;162 (3):723–726. Available from: https://doi.org/10.1093/infdis/162.3.723.

126. Gallin JI, Alling DW, Malech HL, et al. Itraconazole to prevent fungal infections in chronic granulomatous disease. *N Engl J Med*. 2003;348(24):2416–2422. Available from: https://doi.org/10.1056/NEJMoa021931.

127. Segal BH, Barnhart LA, Anderson VL, Walsh TJ, Malech HL, Holland SM. Posaconazole as salvage therapy in patients with chronic granulomatous disease and invasive filamentous fungal infection. *Clin Infect Dis*. 2005;40(11):1684–1688. Available from: https://doi.org/10.1086/430068.

128. A controlled trial of interferon gamma to prevent infection in chronic granulomatous disease. The International Chronic Granulomatous Disease Cooperative Study Group. *N Engl J Med*. 1991;324(8):509–516. Available from: https://doi.org/10.1056/NEJM199102213240801.

129. Marciano BE, Wesley R, De Carlo ES, et al. Long-term interferon-gamma therapy for patients with chronic granulomatous disease. *Clin Infect Dis*. 2004;39(5):692–699. Available from: https://doi.org/10.1086/422993.

130. Arnold DE, Heimall JR. A review of chronic granulomatous disease. *Adv Ther*. 2017;34 (12):2543–2557. Available from: https://doi.org/10.1007/s12325-017-0636-2.

131. Leiding JW, Freeman AF, Marciano BE, et al. Corticosteroid therapy for liver abscess in chronic granulomatous disease. *Clin Infect Dis*. 2012;54(5):694–700. Available from: https://doi.org/10.1093/cid/cir896.

132. Freeman AF, Marciano BE, Anderson VL, Uzel G, Costas C, Holland SM. Corticosteroids in the treatment of severe nocardia pneumonia in chronic granulomatous disease. *Pediatr Infect Dis J*. 2011;30(9):806–808. Available from: https://doi.org/10.1097/INF.0b013e318218181d.

133. Yamazaki-Nakashimada MA, Stiehm ER, Pietropaolo-Cienfuegos D, Hernandez-Bautista V, Espinosa-Rosales F. Corticosteroid therapy for refractory infections in chronic granulomatous disease: case reports and review of the literature. *Ann Allergy Asthma Immunol*. 2006;97 (2):257–261. Available from: https://doi.org/10.1016/S1081-1206(10)60023-3.

134. Uzel G, Orange JS, Poliak N, Marciano BE, Heller T, Holland SM. Complications of tumor necrosis factor-α blockade in chronic granulomatous disease-related colitis. *Clin Infect Dis*. 2010;51(12):1429–1434. Available from: https://doi.org/10.1086/657308.

135. de Luca A, Smeekens SP, Casagrande A, et al. IL-1 receptor blockade restores autophagy and reduces inflammation in chronic granulomatous disease in mice and in humans. *Proc Natl Acad Sci USA*. 2014;111(9):3526–3531. Available from: https://doi.org/10.1073/pnas.1322831111.

136. Åhlin A, Fasth A. Chronic granulomatous disease—conventional treatment vs. hematopoietic stem cell transplantation: an update. *Curr Opin Hematol*. 2015;22(1):41–45. Available from: https://doi.org/10.1097/MOH.0000000000000097.

137. Cole T, Pearce MS, Cant AJ, Cale CM, Goldblatt D, Gennery AR. Clinical outcome in children with chronic granulomatous disease managed conservatively or with hematopoietic stem cell transplantation. *J Allergy Clin Immunol*. 2013;132(5):1150–1155. Available from: https://doi.org/10.1016/j.jaci.2013.05.031.

138. Lent-Schochet D, Jialal I. *Chronic granulomatous disease* [Updated February 21, 2021] *StatPearls [Internet]*. Treasure Island (FL): StatPearls Publishing; 2021. Available from: https://www.ncbi.nlm.nih.gov/books/NBK493171/.

Index

Note: Page numbers followed by "*f*" and "*t*" refer to figures and tables, respectively.

F

G

H

N

T

9780323950923